Wireless Communications and Sensing

In this work from leading figures in the field, Lee, Lee, and Lin describe methods of integrated sensing and communication (ISAC) and artificial intelligence-aided radio systems to provide improved efficiency and performance to wireless users using next-generation (6G) communications systems.

The authors provide the tools to master four major considerations of 6G systems: knowledge of communication systems, knowledge of radar systems, understanding of ISAC systems, and machine learning-based enhancements. *Wireless Communications and Sensing: Fundamentals, Signal Processing, and Machine Learning Solutions* begins by providing preliminary information regarding communication and radar systems, including descriptions of system architecture and design concepts (e.g., wireless propagation, waveforms, MIMO signal processing, and multi-user scenarios) for entry-level readers. Throughout, the authors describe novel artificial intelligence-enabled approaches to aid the design of communication and radar systems. Furthermore, this book also describes the architecture and design concepts of ISAC systems and offers an in-depth discussion of the applications of machine learning to communication, radar, and ISAC systems. Readers will be able to use their comprehensive understanding of this field to enhance ISAC system performance or aid the deployment of ISAC systems in real-world scenarios to fulfill 6G visions.

This book is especially valuable for graduate and higher-level students, scholars, and engineers who want to step into the research areas of integrated sensing and communications and artificial intelligence-aided radio systems. Readers should have at least a bachelor's degree in engineering and have had some exposure to signals and systems, communication theory, and radar principles.

Wireless Communications and Sensing

Fundamentals, Signal Processing, and Machine Learning Solutions

Ta-Sung Lee, Ming-Chun Lee, and
Chia-Hung Lin

CRC Press
Taylor & Francis Group
Boca Raton London New York

CRC Press is an imprint of the
Taylor & Francis Group, an **informa** business

First edition published 2026
by CRC Press
2385 NW Executive Center Drive, Suite 320, Boca Raton FL 33431

and by CRC Press
4 Park Square, Milton Park, Abingdon, Oxon, OX14 4RN

CRC Press is an imprint of Taylor & Francis Group, LLC

ISBN: 978-1-032-52756-7 (hbk)
ISBN: 978-1-032-52757-4 (pbk)
ISBN: 978-1-003-40823-9 (ebk)

DOI: 10.1201/9781003408239

Typeset in CMR10 font
by KnowledgeWorks Global Ltd.

Contents

Preface and Acknowledgments

Over the past few decades, wireless communications and sensing technologies have experienced unprecedented advancements, driven by the increasing demand for high-speed, intelligent, and highly integrated systems. With the ongoing evolution toward 6G and beyond, there is a growing need to develop more efficient and adaptive solutions that can seamlessly integrate communication and sensing capabilities. This integration is expected to play a crucial role in enabling next-generation applications such as autonomous driving, intelligent transportation systems, smart cities, industrial automation, and digital healthcare.

This book was conceived with the goal of providing a comprehensive and structured introduction to the fundamentals of wireless communications and sensing, along with cutting-edge signal processing and machine learning techniques that enhance system performance. It aims to serve as a bridge between theoretical foundations and practical implementations, offering readers both the fundamental principles and advanced research trends in the field.

Scope and Structure of the Book

To facilitate a clear understanding of the subject, the book is organized into four major parts, each focusing on a specific aspect of wireless communication and sensing:

1. *Wireless Communication Basics* – The first part lays the foundation by covering essential concepts such as wireless channel modeling, signal propagation, large- and small-scale fading effects, multiple-input multiple-output (MIMO) systems, and orthogonal frequency division multiplexing (OFDM) principles. It also delves into interference management and multiple access techniques, which are critical for ensuring reliable and high-capacity wireless networks.
2. *Radar Sensing Basics* – The second part introduces the fundamental principles of radar sensing, including waveform design, range and Doppler estimation, target detection, and MIMO radar processing. Given the increasing importance of radar-based sensing in applications like automotive safety and environmental monitoring, this section provides a solid understanding of how radar systems function and how they can be optimized for various scenarios.

3. *Integrated Sensing and Communication (ISAC) Systems* – Recognizing the growing interest in jointly designing communication and sensing systems, the third part of the book focuses on ISAC architectures, design principles, and performance trade-offs. Topics include radar-assisted communications, communication-aided sensing, waveform design strategies, and resource allocation techniques. These discussions provide insights into how wireless networks can be leveraged not only for data transmission but also for high-precision environmental awareness.
4. *Learning-Based Enhancements* – The final part explores the role of deep learning in advancing wireless communications and sensing. It introduces key machine learning techniques, discusses their applications in channel estimation, beam selection, interference mitigation, and spectrum sensing, and highlights how learning-based approaches can enhance ISAC systems. With the rapid growth of artificial intelligence (AI)-driven solutions, this section provides a timely perspective on how emerging technologies can reshape the future of wireless networks.

Intended Audience and Applications

This book is intended for graduate students, researchers, and industry professionals who seek an in-depth understanding of modern wireless communications and sensing. The material is designed to be accessible to those with a background in electrical engineering, signal processing, or computer science, and it can serve as a valuable reference for both academic and practical applications.

- For students, this book provides a structured learning path that starts with fundamental concepts and progresses to state-of-the-art techniques, making it suitable for courses on wireless communications, radar sensing, and machine learning applications in signal processing.
- For researchers, it offers a comprehensive review of current challenges and emerging solutions, making it a useful resource for exploring new research directions in ISAC and learning-based wireless technologies.
- For industry professionals, particularly those working in telecommunications, automotive sensing, and AI-driven wireless systems, this book provides practical insights, which can help in designing next-generation communication and sensing solutions.

Contributions and Acknowledgments

The development of this book has been a collaborative effort, and we are grateful to the many individuals and organizations that have supported this work. We extend our sincere appreciation to our colleagues and research collaborators for their valuable feedback and insights, which have helped refine the technical content of the book. We also acknowledge the support of our institutions and funding agencies, which have played a crucial role in enabling our research and academic endeavors.

Finally, we are deeply appreciative of our families and friends for their unwavering encouragement and patience throughout the writing process. Their support has been instrumental in bringing this book to completion.

We hope that this book serves as a useful and inspiring resource for readers interested in the intersection of wireless communications, sensing, and machine learning. As these fields continue to evolve, we look forward to seeing new innovations and contributions from the next generation of engineers and researchers.

About the Authors

Ta-Sung Lee is senior vice president, chair professor of the Department of Electronics and Electronical Engineering, and director of the IoT Intelligent Systems Research Center at National Yang Ming Chiao Tung University, Taiwan. He is a fellow of the IEEE, and he earned his PhD from Purdue University in 1989.

Ming-Chun Lee is an associate professor at the Institute of Communications Engineering at National Yang Ming Chiao Tung University, Taiwan. He earned his PhD from the University of Southern California in electrical engineering in 2020.

Chia-Hung Lin is a research assistant at North Carolina State University and is completing his PhD in electrical and computer engineering at North Carolina State University, USA.

1

Introduction

1.1 6G visions

In recent years, information and communication technology (ICT) has been developing at an unprecedented rate in human history. Many highly anticipated applications such as fully automated driving [1–3], full sensory digital sensing and reality [4–6], eHealth [7,8], and smart manufacturing/construction [9–13] are actively being developed. Although some of these applications may still be in their infancy, it is certain that they will transform our daily lives fundamentally and lead human history into a new chapter [14–16]. Upon analyzing these applications, one common factor becomes apparent: A powerful and reliable networking and communication system must act as a backbone to support these applications and comply with any service requirements they may have. As a result, even though we have just entered a new era of 5G communications and are currently enjoying the benefits of 5G commercial uses, experts from academia and industry are already thinking about what 6G will be and investigating the roadmap toward 6G. They are also exploring emerging trends and requirements, as well as various enabling techniques and architectures to make the aforementioned applications come true [17–20].

The Technologies for Network 2030 group within the International Telecommunication Union-Telecommunication (ITU-T) has outlined a vision for 6G communications, with the goal of providing a highly digitalized, intelligent, and globally data-driven platform enabled by near-instant and unlimited full wireless connectivity [18,21]. This platform should integrate different ICT functionalities such as sensing (including following applications, such as positioning, navigation, and imaging), communication, computing, caching, and control to support all full-vertical applications. In Figure 1.1, we provide a capability comparison between the 5G communication system and the 6G communication system to highlight the contributions of the 6G communication system and illustrate the main quality of service (QoS) requirements of different 6G usage scenarios. The basic goal of the 6G communication system is to provide data rate improvements, including 5x average spectrum efficiency, 10x user-experienced data rate, and at least 50x peak data rate. Furthermore, the 6G communication system also promises to offer seamless and low-latency communications, enabling more novel applications. Depending on the

DOI: 10.1201/9781003408239-1

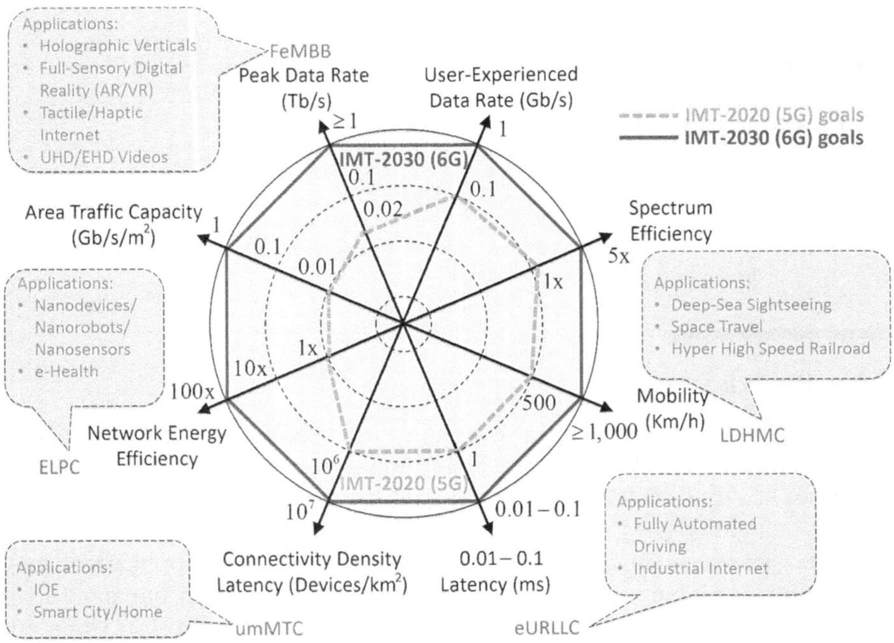

FIGURE 1.1
The promised QoS requirements and typical scenarios of 5G and 6G communication systems.

requirements of different applications, 6G defines five typical usage scenarios: further enhanced mobile broadband (FeMBB), ultra-massive machine-type communications (umMTC), extremely reliable and low-latency communications (ERLLC), long-distance and high-mobility communications (LDHMC), and extremely low-power communications (ELPC). In light of this direction, the 6G communication system aims to provide appropriate QoS promises to support different types of applications. Some applications may only have a single QoS requirement to be satisfied, while others may need several QoS guarantees to function appropriately. For example, with the enhanced area traffic capacity and network energy efficiency, fine-grained Internet of Everything [22] can be realized, monitoring a large area by connecting to a number of high-resolution sensors simultaneously to allow precise operations. In addition, with the improved connectivity to offer low-latency communication to moving targets, fully automated driving [1] is starting to be realized to provide real-time centralized driving operations.

Although 6G presents an exciting blueprint for the future intelligent information society, meeting its harsh QoS requirements is crucial to fulfill its promise of providing a reliable and stable platform to support upcoming applications. Generally, there are two straightforward directions to improve communication systems' capabilities: (i) *increasing system bandwidth* and/or (ii) *improving spectrum efficiency*. On the one hand, technologies like Terahertz (THz) communications [23, 24] and laser communications [25–27] can expand the resulting spectrum resources for the 6G communication system. On the other hand, technologies like very-large-scale antenna arrays, for example, spatial modulation multiple-input multiple-output (SM-MIMO) [28, 29], large intelligent surfaces (LIS) [30,31], and orbital angular momentum (OAM) multiplexing [32, 33] aim to provide improved spectrum efficiency for superior capabilities of 6G communication systems. In addition to these two directions, novel ICT technologies like advanced sensing [34, 35], blockchain-based spectrum sharing [36, 37], and quantum computing [38, 39] will also be employed to enhance the security and efficiency of the 6G communication system. For interested readers, we provide a brief discussion of each technology below.

Increasing system bandwidth: Expanding system bandwidth is essential to enhance communication system capability. The millimeter-wave band (mmWave) from 24.25 GHz to 52.6 GHz was first employed in the 5G communication system, resulting in 5G QoS improvements. In the 6G communication system, researchers are exploring the use of higher-frequency bands ranging from 0.1 to 10 THz (i.e., THz band from 10 GHz to 1000 GHz) to achieve multi-Tb/s data transmission. The THz band provides the potential for hundreds of giga-hertz bands to users, making it a crucial aspect of 6G development. Due to the short wavelength of THz bands, a single base station can utilize more than 10,000 antenna elements for wireless communication. Highly directional transmission using super-narrow beams can concentrate transmit power in a narrow direction to overcome the high path-loss property of THz communications. This results in improved interference control and transmission security. Consequently, THz communication [23, 24] has become a major research topic in the development of the 6G communication system. In addition to THz communications, the 6G communication system aims to provide seamless connectivity to multiple devices simultaneously (i.e., umMTC). To achieve this, the system aims to integrate terrestrial communications with space/air/underwater communications to form a four-tier integrated communication platform. Laser communication is more suitable than electromagnetic wave signals for free-space and underwater propagation environments, and it will be employed to support wireless communications in those environments. Interested readers can refer to [23–27] for more details on the current development of these technologies.

Improving spectrum efficiency: In addition to increasing radio resources, the 6G communication system aims to improve spectrum efficiency with a fixed amount of resources. One way to achieve this is by utilizing SM-MIMO with over 10,000 antenna elements, which is well-suited for THz communications. Interested readers can refer to [28,29] for more information on this topic. Two new design concepts, LIS and OAM, are also being developed to improve spectrum efficiency in the 6G communication system. The LIS concept utilizes massive passive reflecting elements with controllable phase or amplitude on a spatially continuous aperture, optimizing performance based on surrounding radio environments. Several studies have demonstrated superior performance compared to conventional communication systems by incorporating the LIS design concept. For more information, readers can refer to refs. [30,31]. In the OAM multiplexing technology, multiple data streams are transmitted simultaneously on the same radiofrequency resources by using a set of orthogonal electromagnetic waves, exploiting the angular momentum of the wave as a new degree of freedom. Interested readers can refer to refs. [32,33] for more information on this topic.

Novel ICT technologies: After selecting appropriate technologies based on the above two perspectives, 6G communication systems can be further enhanced by incorporating novel ICT technologies. For example, in terms of blockchain-based spectrum sharing, the conventional centralized spectrum sharing approach becomes impractical when billions of machine-type devices are connected to 6G communication systems. As an alternative, original spectrum-sensing mechanisms can be employed to enable the sharing of underutilized spectrum resources with numerous devices, which not only increases system bandwidth from the device's perspective but also improves spectrum efficiency from the system's perspective. By combining blockchain technology, spectrum sharing can be implemented on a distributed platform that is more resilient to cyberattacks and malicious devices than traditional centralized spectrum-sharing mechanisms. In addition, a financial compensation mechanism can be introduced to incentivize more devices to participate in the spectrum-sharing mechanism. Interested readers can find more details on this topic in refs. [36, 37]. The other example is in terms of computing resources; quantum computing can be used to provide real-time computing results to optimize 6G communication systems' planning and operations to serve users. Unlike traditional computing, quantum computing employs qubits to represent information, enabling parallel computing to be performed easily. Quantum computing can aid communication system decisions by providing incredible accelerations compared to traditional methods. Interested readers can find more details on this topic in refs. [38, 39]. By leveraging the collaboration of these technologies, 6G communication systems can meet the harsh QoS requirements of future applications.

Besides other ICT technologies, such as the aforementioned blockchain and quantum computing technologies, we put our main focus on the intersection of wireless communication and advanced sensing in this book because of the importance and practicality of this specific emerging research direction. From the perspective of its importance, traffic injuries are a major concern worldwide [40, 41], causing the loss of about 2.4 lives every minute, as reported by the World Health Organization. To mitigate this situation, a range of existing and forthcoming applications, such as advanced driver assistance systems (ADAS) [42,43], intelligent transportation systems (ITS) [44,45], and even fully automated driving [1], have been developed to improve road user safety and efficiency. By supporting information acquisition and exchange of

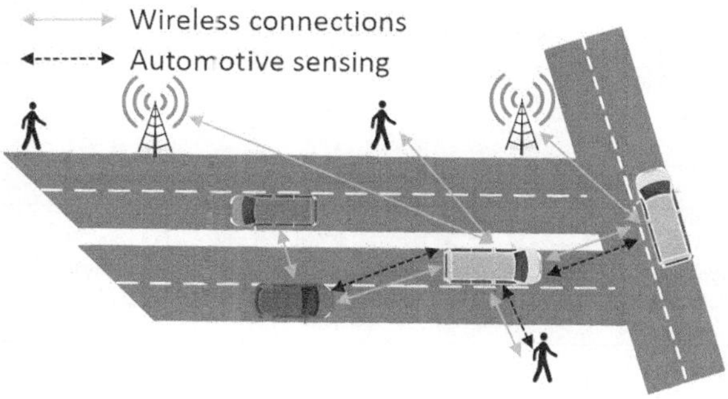

FIGURE 1.2
The considered scenario with wireless connections and automotive sensing.

these applications, a deeply integrated system with wireless communication and advanced sensing capability can facilitate their deployment and operation in vehicular scenarios, thereby contributing to the enhancement of road safety and efficiency. On the other hand, from the perspective of its practicality, vehicular applications such as ADAS and ITS are more mature than other applications and are already available to the public, with some vehicles featuring level-3 automated driving assistance functionalities (i.e., conditionally automated driving assistant). In addition, the United States Department of Transportation has deployed several ITS applications, including speed management, human service transportation coordination, and automated speed enforcement systems, to assist road users in the country. As a result, research outcomes related to such systems in vehicular scenarios can be implemented immediately, benefiting society by significantly improving the safety and efficiency of road users. Therefore, this book aims to provide good material to let readers catch up with this emerging and promising research topic.

Focusing on real-world vehicular scenarios, Figure 1.2 illustrates the considered scenario where multiple vehicles are operating on roads. To support various applications, such as ADAS and ITS, automotive sensing and wireless connections are extensively utilized to obtain information from the environment and surrounding objects, including pedestrians, roadside units, and other vehicles. While other key technologies can still be incorporated, sensing and communications are undoubtedly crucial components of such systems. In particular, they must work together seamlessly to provide ADAS functionality and protect road users. For example, an industrial report [46, 47] suggests that to enable basic ADAS functions, such as rear-end collision, intersection, and pedestrian warnings, the system must achieve 100 ms end-to-end latency and 150 cm positioning accuracy, providing sufficient response time to drivers. More advanced ADAS warnings, such as lane change assistance (requiring 100 ms end-to-end latency and 30 cm positioning accuracy) and vehicle platooning (requiring 20 ms end-to-end latency and 30 cm positioning accuracy), have even more stringent requirements to function effectively. In Figure 1.3, we can

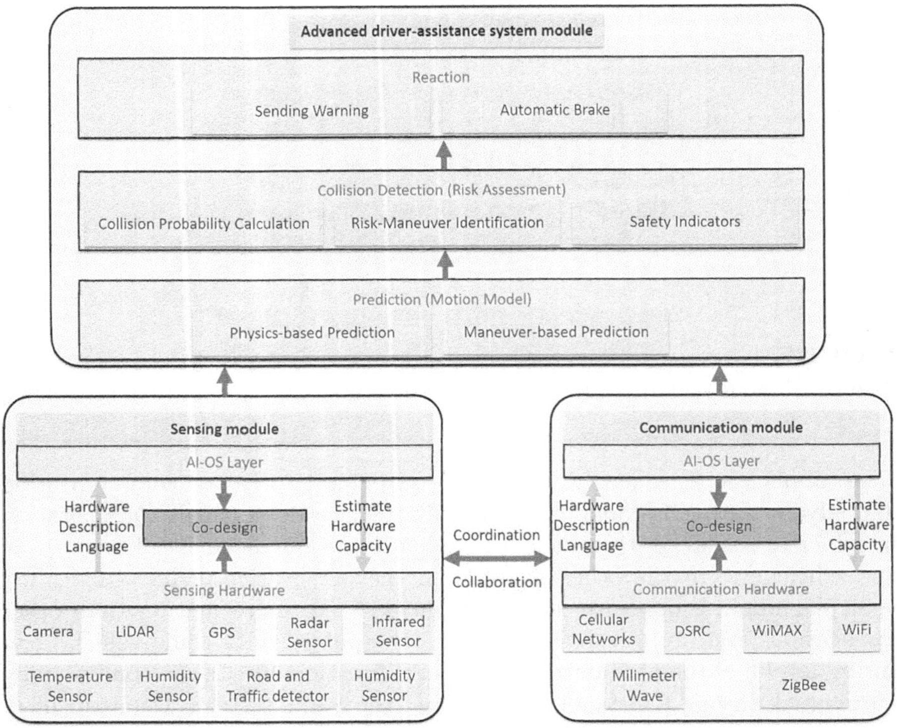

FIGURE 1.3
The collaboration between communication and sensing modules to realize the above ADAS functionalities in 6G communication systems.

see how the collaboration between communication and sensing modules can enable the ADAS functionalities for 6G communication systems. The sensing module employs various sensors as sensing hardware to obtain environmental information. This information is processed by different algorithms placed in the artificial intelligence-operation system (AI-OS) layer, which provides comments to sensing hardware and results to the ADAS module. Similarly, the same workflow can also be observed in the communication module. Note that communication hardware and sensing hardware can work together to process information to enable raw data-level coordination/collaboration. After that, the ADAS module can further utilize the processed information from both modules through decision-level coordination/collaboration to predict risks for drivers and trigger warning or assistance systems when necessary. Those different levels of coordination or collaboration can further improve the ADAS system performance but will also require advanced algorithms to enable raw data-level or decision-level fusion and joint decision-making. On the one hand, communications can aid sensing by allowing vehicles to share sensing results and safety-related data, creating a wider field of view and improving the positioning accuracy of targets by aggregating environmental information from several sensors. On the other hand, sensing can aid communication by using environmental information as input to improve communication operations. For example, vehicle density can be considered when allocating resources to roadside units [48]. In the aforementioned process, in literature [49–53], one can notice that machine learning (ML)-based solutions play an important role in such systems to further aid the signal processing in communication and sensing systems (i.e., AI-OS layer in communication and sensing module). Moreover, in the ADAS module, ML-based solutions are also the mainstream methods for heterogeneous data fusion from both layers and enabling advanced decision-making such as ADAS predictions and reactions. Thus, we will discuss the current status of wireless communications and automotive sensing in Sections 1.2 and 1.3, respectively. Furthermore, we will discuss the motivation and current status of the coordination/collaboration between current sensing and communication modules, also known as integrated sensing and communication (ISAC) systems in Section 1.4. Finally, Section 1.5 will briefly introduce novel ML solutions to enhance the performance and efficiency of such systems to conclude this chapter.

1.2 Overview of Wireless Communications

Figure 1.2 illustrates different wireless connections utilized in vehicular scenarios to enable various functionalities. Vehicular-to-vehicular (V2V) connections allow multiple vehicles to connect, enabling the sharing of sensing information collected by each vehicle. Similarly, vehicular-to-pedestrian (V2P) connections can be employed to broadcast vehicle movements in advance to surrounding pedestrians, serving as a warning system. Lastly, vehicular-to-infrastructure (V2I) connections can provide traffic light warnings or route recommendations to each vehicle, improving overall traffic efficiency. Collectively, these connections are known as vehicular-to-everything (V2X) connections which can further enhance ADAS capabilities as discussed earlier. Currently, two primary communication standards are utilized to provide V2X connectivity to road users in vehicular scenarios: cellular-V2X (C-V2X) and dedicated short-range communications (DSRC) [54,55]. DSRC can serve as a short-range and uncoordinated communication tool for V2V and V2I connections to establish vehicular ad hoc networks (VANETs). In contrast, C-V2X can provide connectivity with superior throughput and lower latency owing to the aid of the surrounding infrastructure. Moreover, considering the fundamental differences and inherent benefits of both standards, current literature is exploring the possibility of developing a dual-interface-enabled V2X communication system that leverages the advantages of both standards in appropriate scenarios [56]. Regardless of which standard is adopted, in the United States, the Federal

Communications Commission (FCC) has allocated the frequency range 5850–5925 MHz for ITS applications. This band is divided into 7 channels, with 10 MHz allocated to each channel (5855–5925 MHz), while the remaining 5 MHz is reserved for future usage. Specifically, channels 172 (5855–5865 MHz) and 184 (5915–5925 MHz) are assigned for safety applications; channels 174 (5865–5875 MHz), 176 (5875–5885 MHz), 180 (5895–5905 MHz), and 182 (5905–5915 MHz) are for non-safety applications; and channel 178 (5885–5895 MHz) is designated for control purposes. Given the limited frequency resources mentioned above, providing wireless connectivity to numerous road users that satisfies the stringent requirements of ITS applications can be really challenging. However, over the course of several decades, wireless communication has developed a mature framework with dedicated operations that meet the connectivity needs of road users. To introduce the framework that enables real vehicular communications, this book first explains the wireless propagation nature of radiofrequency bands in Chapter 2. Then, Chapter 3 discusses the fundamental radiofrequency waveform, which serves as the wireless communication specification starting from the 4th-generation communications, to provide readers with an understanding of how a real communication system operates. In Chapter 4, the signal processing operations of single-user scenarios are explored. Given the limited frequency resource, a spatial resource offered by multiple antennas in the transmitters and receivers (i.e., multiple-input-multiple-output (MIMO)) architecture is crucial for enabling high-throughput transmissions in the considered vehicular scenarios. Consequently, this chapter provides an introduction to typical operations in such systems. Moreover, extending from single-user scenarios, interference management and multiple access are both important considerations when serving several users simultaneously. Chapter 5 concludes with a discussion of how to divide available resources to create diversity in frequency, spatial, and code domains to resolve several wireless connections for multiple-user scenarios.

1.3 Overview of Automotive Sensing

As also illustrated in Figure 1.2, vehicles require sensing capabilities to perceive environmental information for ADAS functionalities. To meet this requirement, modern vehicles are often equipped with three types of sensors, namely radar [57], Lidar [58], and vision sensors [59, 60]. While these three sensors can provide similar functionalities to ADAS, we argue that radar sensors are indispensable for automotive sensing due to their relatively low-cost and robustness. Specifically, radar sensors are essential because they can provide range and radial velocity information of surrounding objects to ADAS systems, while Lidar and vision systems require special designs to offer velocity information. Moreover, radar is the only sensor in this list that can function normally without any special designs in challenging light conditions, such as in bad weather, dark or bright. As ADAS or even fully automated driving systems should remain functional in any scenario, solely Lidar or vision sensors cannot be employed alone for this purpose, and radar sensing is still an integral part when it comes to automotive sensing. Therefore, we will primarily focus on the current situation of radar systems while discussing automotive sensing topics in this book. In the United States, the FCC has approved the use of the 76–77 GHz and 77–81 GHz mmWave spectrum for automotive long-range radar and short-range radar usages, respectively. With this ample frequency resource, we will introduce how modern radar systems utilize those frequency bands to obtain environmental information by employing different waveforms in Chapter 6. Similarly, multiple antennas can benefit radar systems by providing improved estimation results compared to single antenna cases or even offering further environmental information such as direction of arrival. We will explain these mechanisms in detail in Chapter 7. Finally, we aim to devote a chapter to discussing interference mitigation schemes when multiple radars employ the same resources in Chapter 8 to conclude the discussion of modern radar systems.

1.4 Overview of Integrated Sensing and Communications

In Figure 1.2, one can also see that modern vehicular scenarios involve multiple wireless connections and radar sensing at the same time to provide the desired functionalities. While both applications require frequency resources, the current approach of assigning dedicated bands to each of them may not be sufficient as the number of connected vehicles increases and the demand for ADAS and fully autonomous driving applications grows. To address this challenge, researchers are exploring the potential of ISAC systems [61–63], which allow the ample frequency resources from automotive sensing systems to be shared with communication systems. Moreover, different levels of integration can also be realized in such a framework for joint benefits, such as reduced mutual interference and consequently improved performance. In light of this direction, ISAC systems can be categorized into co-existence and coordinated architectures depending on the level of interaction between the communication and sensing systems. Co-existence ISAC systems provide a platform for communication and sensing systems with different hardware components to be co-designed and realize collision avoidance. On the other hand, coordinated ISAC systems are capable of simultaneously providing benefits to communication and sensing purposes, allowing for more complex co-designs. In Chapter 9, we will delve into the architecture of ISAC systems, exploring both co-existence and coordinated models. We will also discuss the design concepts of the ISAC transceiver in Chapter 10.

1.5 Overview of Learning-Based Enhancements

Finally, over the past decade, we also noticed that the development of learning-based enhancements has provided new momentum to human technologies, offering promising alternatives to traditional optimization-based algorithms. ML technologies have the potential to provide more efficient solutions to better solve current bottlenecks and enable novel applications that were previously thought to be unsolvable by optimization-based algorithms. Computer vision [64, 65] and natural language processing [66, 67] are two ML-dominated research areas demonstrating ML algorithms' value. ML algorithms provide new state-of-the-art solutions to classic computer vision applications [68] and enable new applications in computer vision society, such as highly convincing video editing and video generating in medical, business, and manufacturing domains [69]. Reinforcement learning solutions also accelerate the development of natural language processing research [70, 71], providing killer applications, such as ChatGPT, to enable highly human-like chatboxes and text generations. Researchers are also investigating the possibilities of utilizing ML solutions to aid communication/sensing system designs, and have already made exciting progress [72–75]. In 3rd Generation Partnership Project (3GPP) Release 18, three practical usages of ML-aided communication and sensing system designs are listed for researchers to further study, including channel state information feedback, beam management, and fine-grained target positioning [76]. Moreover, we can also see that increasingly more communication and sensing operations have involved ML algorithms in the design loop for enhanced performance. Generally speaking, ML algorithms are particularly valuable in cases with model deficits and algorithm deficits [77], where traditional optimization-based algorithms heavily rely on precise mathematical models to be developed. In cases with imprecise mathematical models or even without mathematical models (i.e., model deficits), the development of optimization-based algorithms is extremely challenging and the achieved performance will be limited significantly. As an alternative, given ample training samples as a precondition, ML algorithms can learn the underlying system model automatically by observing the input/output relationship of training samples to tackle the cases without mathematical models. In addition, even with precise mathematical models for the considered problems, computationally demanding optimization-based algorithms are still widely employed in the designs of communication operations (i.e., algorithm deficits). With appropriate training, ML algorithms provide a good approximation in building a function to directly output the desired output based on a given input. Thus, complicated calculations can be ignored and similar performance can be obtained to facilitate real-time designs of communication system operations. To introduce readers to recent developments in ML algorithms, Chapter 11 provides an overview of mainstream ML solutions and typical usages. The book also includes chapters that discuss ML-based signal processing and interference mitigation topics in communication systems (Chapters 12 and 13) and radar systems (Chapters 14 and 15) by further considering recent learning-based enhancements and their applications in both communication and automotive sensing systems. Finally, Chapter 16 discusses the intersection between the main topic of this book, ML-enabled design and ISAC systems to conclude this book. This book aims to encourage more researchers to contribute their efforts in the direction of ML-enabled design toward efficient wireless communications, advanced sensing, and ISAC operations in 6G systems.

Part I

Wireless Communication Basics

2

Wireless Channels and Signal Models

2.1 Wireless Propagation Overview

When we discuss wireless transmissions, we are referring to the utilization of electromagnetic (EM) waves to deliver signals, or information, from the transmitter to the receiver. Unlike wired transmissions, where the signal follows a stable and fixed path, the behavior of signals in wireless communication scenarios is more complex and variable. For instance, the signal strength significantly varies based on the distance traveled by the signal. Additionally, the interaction between the input signals from the transmitter and the surrounding environment causes reflection, diffraction, and scattering, which also affect the behavior of the received signals at the receiver's end. These interactions are influenced by the carrier wavelength of the transmitted signals and the size and material composition of encountered obstacles, as shown in Figure 2.1. Considering these factors, signals transmitted through different paths can result in constructive or destructive superpositions, ultimately forming the final received signals. Furthermore, in real-world wireless transmissions, environmental changes such as the movement of people and vehicles lead to variations in transmission behaviors, including the presence of obstacles and changes in the distance between the transmitter and receiver. Consequently, the wireless channel becomes a highly time-variant system. Mathematically speaking, a wireless communication system can be characterized as a linear system. In this framework, the output signals at the receiver side can be described as the convolution of the input signals from the transmitter side with the wireless channel response in the time domain. In order to successfully decode the desired information at the receiver side, it is imperative to have a comprehensive understanding of the underlying wireless channel response and this understanding actually plays a pivotal role in the design of modern transceivers.

To do so, we will approach the problem from a statistical perspective in this chapter, bridging the gap between real-world physics and the corresponding channel behavior in modern transceiver designs. This chapter starts by introducing various types of channel behaviors, such as large-scale/small-scale

DOI: 10.1201/9781003408239-2

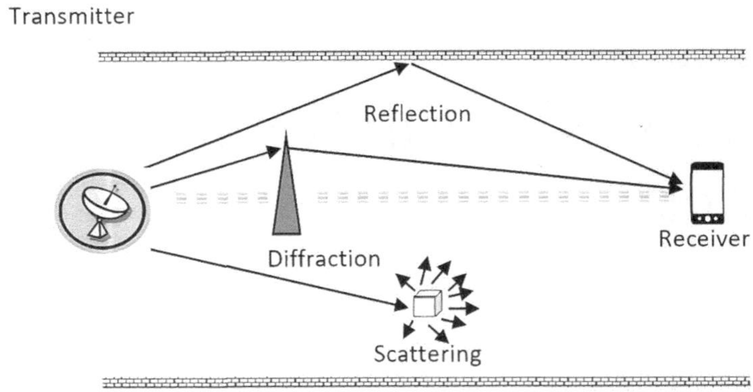

FIGURE 2.1
Illustration of a wireless communication transmission environment.

fading, frequency-selective/non-selective channel models, and fast/slow fading behaviors, beginning with single antenna systems. Furthermore, we extend the scope to multiple-input-multiple-output (MIMO) systems by deploying multiple antennas on both the transmitter and receiver sides. We introduce the MIMO channel model, which takes into account the extended spatial degree of freedom. Finally, this chapter provides an explanation of the typical transmission environment settings defined by international standard groups, using the 3GPP proposed spatial channel model (SCM) and SCM-extension (SCME) as examples.

2.2 Large-Scale Fading

This section focuses on the phenomenon of large-scale fading in wireless transmission. Specifically, when EM waves are transmitted, there are two primary factors that contribute to significant power attenuation. First, as EM waves traverse a medium that is not transparent to them, power losses occur and these losses are directly related to the distance traveled by the EM waves, known as path loss. Second, when obstacles are present between the transmitter and receiver, the EM waves experience substantial power losses due to the shadowing effect caused by passing through these obstacles. In the following explanations, we will delve into the following two types of large-scale fading phenomena.

Path loss: In terms of path loss attenuation, let's begin by considering the simplest that is free-space transmission. In this scenario, the relationship between the transmit power and the receive power can be described using the

TABLE 2.1
The path loss exponent of common environments.

Environment	n
Free-space	2
Urban	2.7–3.5
Indoor (with Line of Sight)	1.6–1.8
Indoor (without Line of Sight)	4–6

free-space propagation model:

$$P_{\mathrm{R}}(d) = \frac{P_{\mathrm{T}}G_{\mathrm{T}}G_{\mathrm{R}}\lambda^2}{(4\pi)^2 d^2 L} = P_{\mathrm{R}}(\frac{d_0}{d})^2, \quad d > d_0 \tag{2.1}$$

where P_{R} and P_{T} represent received and transmitted power and G_{R} and G_{T} stand for receive and transmit antenna gain, respectively. λ is the carrier frequency, d is the distance between transmit antenna and receive antenna, L is the system loss factor, and d_0 is a reference distance.[1] Eq. (2.1) describes the relationship between received and transmitted power and the equation of path loss can be inferred consequently. As aforementioned, path loss refers to the power losses which EM waves consume to travel in the medium. Using dB unit, the path loss in free-space scenario can be expressed as:

$$PL(d)[\mathrm{dB}] = 10\log_{10}\frac{P_{\mathrm{T}}}{P_{\mathrm{R}}} = -10\log_{10}\left[\frac{G_{\mathrm{T}}G_{\mathrm{R}}\lambda^2}{(4\pi)^2 d^2}\right]. \tag{2.2}$$

After introducing the free-space path loss model, it is important to note that in realistic environments, which are considerably more complex compared to the free-space case, we often formulate the path loss model as a random variable. Consequently, the average path loss can be expressed as follows:

$$\overline{PL}(d)[\mathrm{dB}] = \overline{PL}(d_0) + 10n\log_{10}(\frac{d}{d_0}). \tag{2.3}$$

In Eq. (2.3), it is evident that the path loss is proportional to the distance raised to the power of n, denoted as $\overline{PL}(d) \propto (\frac{d}{d_0})^n$. Here, n is a predetermined parameter known as the path loss exponent which depends on the specific environment being considered. The values of n for common environments are listed in Table 2.1 as a reference.

[1]Note that if without the constraint $d > d_0$, the received power in a place, where receiver is very close to the transmitter, will be infinite. However, in real scenarios, following the law of conservation of energy, the received power should never exceed the transmitted power. Hence, the constraint should be added to ensure the law of conservation of energy.

Furthermore, let's explore two more complex path loss models for wireless communications that take into account additional physical parameters. The first model is Hata model [78], which is expressed as follows:

$$PL(d)[\text{dB}] = 69.55 + 26.16\log_{10}(f_c) - 13.82\log_{10}h_{\text{te}} - a(h_{\text{re}}) \\ + (44.9 - 6.55\log_{10}h_{\text{te}})\log_{10}(d), \tag{2.4}$$

where f_c is the carrier frequency, h_{te} and h_{re} are the height of transmit antenna and receive antenna in meters, respectively (Note that $30 < h_{\text{te}} < 200$ and $1 < h_{\text{re}} < 10$ should be held to match the setting of measurement data), and $a(h_{\text{re}})$ is a predefined parameter based on the height of the receive antenna. Another commonly used path loss model is the Walfisch–Bertoni model [79], which is expressed as follows:

$$PL[\text{dB}] = L_0 + L_{\text{rts}} + L_{\text{ms}}, \tag{2.5}$$

where L_0, L_{rts}, and L_{ms} represent free-space path loss, the excess path loss caused by the curvature of the Earth, and the excess path loss caused by building geometry. Interested readers can find detailed parameter settings of these two models in their original papers [78, 79].

Shadowing effect: Besides the path loss caused by the travel distance of EM waves, obstacles along the path of EM waves can lead to significant attenuation and affect the strength of the EM waves. This phenomenon is commonly referred to as the shadowing effect. In the large-scale fading model, we often represent the shadowing fading effect as a random variable following a log-normal distribution. It is worth noting that when expressing the path loss formula in dB units, the shadowing effect is transformed into a normal distribution, as the logarithmic operation is canceled out. Therefore, the total large-scale fading model, which considers both the path loss and shadowing effects, can be expressed as follows:

$$PL[\text{dB}] = \overline{PL}(d) + X_{\text{SF}}. \tag{2.6}$$

In the expression, the variable X_{SF} follows a normal distribution $\mathcal{N}(0, (\sigma_{\text{SF}})^2)$, representing the shadowing effect. A commonly used range for the standard deviation σ_{SF} is between 4 and 10 which helps to incorporate an appropriate and realistic level of shadowing effect. In Figure 2.2, the solid line represents the path loss effect, which consistently decreases the received signal power with distance. On the other hand, the dotted line represents the total large-scale effect, taking into account the fluctuation caused by the random presence of obstacles between the transmitter and receiver. By incorporating the shadowing effect, the total large-scale fading exhibits additional variations around the path loss trend. Finally, it is worth mentioning that there exist advanced large-scale propagation models that incorporate additional terms to capture specific behaviors in particular frequency bands and environments. These models build upon the fundamental principles and real measurement data, expanding on

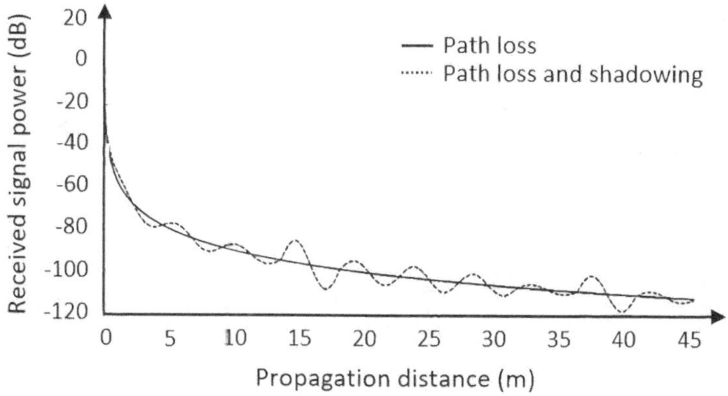

FIGURE 2.2
Illustration of the impact of environmental fading and distance on signal strength.

the concepts introduced in the previously discussed models. However, it is important to note that despite the existence of these advanced models, the three models discussed earlier (free-space model, Hata model, and Walfisch–Bertoni model) remain the most commonly used ones. These advanced models are developed based on similar logical foundations and utilize real measurement data to refine and enhance the accuracy of path loss predictions.

2.3 Small-Scale Fading

Besides the large-scale fading caused by distance and encountered obstacles during the travel of EM waves, the multi-path between transmitter and receiver and the relative movement of transmitter and receiver also cause different signal behaviors. We will introduce how the current wireless communication system considers those effects in this section.

Multi-path effect and delay spread: Imagine speaking loudly in a spacious hall, where the sound waves you produce result in an echo that persists for a period of time. This phenomenon can be explained by the behavior of acoustic waves which traverse various angles, undergo different reflections, and ultimately reach the listener's location with varying arrival times and strengths. Similar effects can be observed in the transmission of EM waves. A visual representation of a sample channel effect is presented in Figure 2.3. In our analysis, we will initially focus on a specific moment, denoted as t_0, assuming that the magnitude of the initial impulse in this instance is given by $h(t_0, \tau_1) = 0.2$. This implies that after a time delay of τ_1 seconds, the first echo

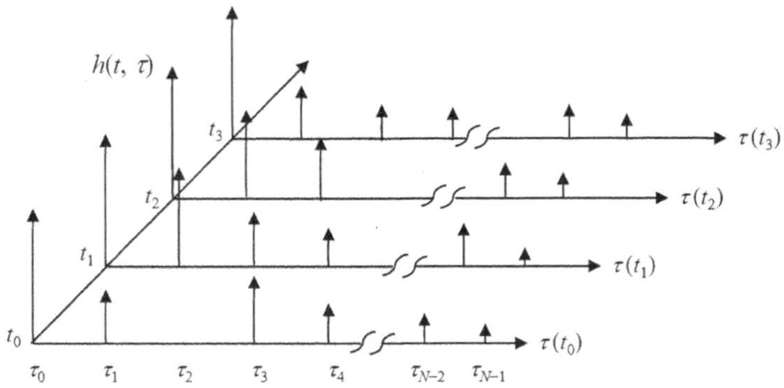

FIGURE 2.3
Illustration of a time-varying channel impulse response.

will reach the listener's position with an intensity of 0.2, corresponding to an attenuation of 80% during transmission. Subsequently, additional echoes will arrive at the listener's location with varying delays and magnitudes. Furthermore, in more complex scenarios where the environment undergoes continuous changes, such as people entering and leaving the hall, the resulting channel becomes time-variant, as indicated by the t axis in Figure 2.3. Therefore, a comprehensive channel model should incorporate both the multi-path effect and the time-variant nature of the channel. To do so, the impulse response of a wireless channel can be expressed as:

$$h(t,\tau)=\sum_{i=1}^{N}\alpha_i(t)\exp[-j(2\pi f_c\tau_i(t)+\phi_i(t))]\delta(\tau-\tau_i(t)), \tag{2.7}$$

where N is the number of paths. $\alpha_i(t)$, $\tau_i(t)$, and $\phi_i(t)$ represent the magnitude, delay time, and phase of the ith path at timeslot t, respectively. Based on Eq. (2.7), the relationship between input signal $u(t)$ and output signal $x(t)$ can be obtained as:

$$x(t)=\int h(t,\tau)u(t-\tau)d\tau=\sum_{i=1}^{N}\alpha_i(t)\exp[-j(2\pi f_c\tau_i(t)+\phi_i(t))]u(t-\tau_i(t)). \tag{2.8}$$

In the case where the considered wireless channel is time-invariant, Eq. (2.7) can be simplified as:[2]

$$h(\tau)=\sum_{i=1}^{N}\alpha_i\exp[-j\theta_i]\delta(\tau-\tau_i), \tag{2.9}$$

where θ_i is the phase of the ith path in such case.

[2]Note that the resulting impulse response is no longer a function of t.

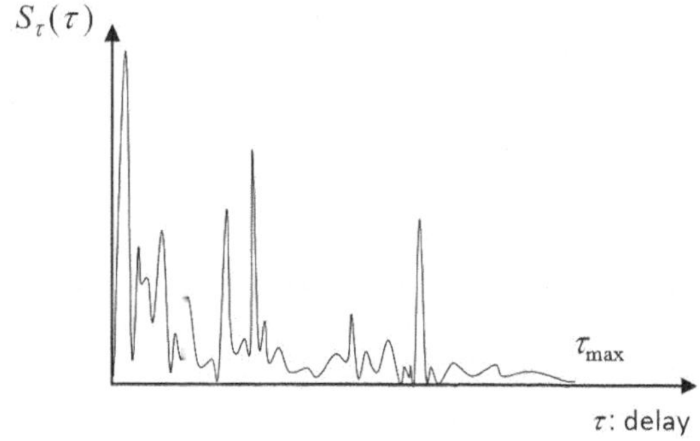

FIGURE 2.4
Illustration of a power delay profile.

In wireless channels characterized by multi-path effects, the power delay profile plays a crucial role in determining the relationship between power and the relative delay of each path. This profile is often visualized as shown in Figure 2.4. In order to simplify our analysis, we initially assume that the wireless channels are time-invariant. For a specific carrier frequency, different power delay profiles result in varying degrees of constructive or destructive interference. In essence, the frequency response of a particular frequency is influenced by the power delay profiles of the underlying channel. Since it is challenging to express the power delay profiles of all paths using simple mathematical formulations, similar to the shadowing effect, statistical methods are commonly employed to describe these profiles. Specifically, the delay and delay spread can be utilized to represent the relationship between power and relative delay within a given channel. Given a power delay profile $S_\tau(\tau)$, we can express the maximum excess delay as follows:

$$T_{\mathrm{m}} = \tau_{\max}, \tag{2.10}$$

and average excess delay as:

$$\overline{\tau} = \frac{\int_0^{\tau_{\max}} \tau S_\tau(\tau) d\tau}{\int_0^{\tau_{\max}} S_\tau(\tau) d\tau} \tag{2.11}$$

In Eq. (2.10), the maximum excess delay corresponds to the delay of the longest path within the channel, while the average excess delay, as given by Eq. (2.11), represents the average delay across all paths. In addition to the excess delay measures, the delay spread serves as a descriptor for the distribution of delays among the paths within the considered channel. A smaller delay spread indicates that the differences in delay between the paths are also smaller. Consequently, the arrival times of signals from all paths are relatively similar. Conversely, a larger delay spread indicates significant differences in the arrival times of signals from the various paths. The delay spread is commonly quantified using the root mean square (RMS) approach, defined as follows:

$$\tau_{\mathrm{RMS}} = \sqrt{\overline{\tau^2} - (\overline{\tau})^2};\ \overline{\tau^2} = \frac{\int_0^{\tau_{\max}} \tau S_\tau(\tau)^2 d\tau}{\int_0^{\tau_{\max}} S_\tau(\tau) d\tau}. \tag{2.12}$$

The coherence bandwidth of a channel can be defined based on the concept of excess delay and delay spread. The coherence bandwidth represents the maximum bandwidth within which a transmitted signal experiences similar channel characteristics. To illustrate this concept, let's consider a hypothetical scenario where the coherence bandwidth of a channel at a carrier frequency of 1.9 GHz is determined to be 500 kHz. In this case, any signal with a total bandwidth smaller than 500 kHz, transmitted within the 1.9 GHz frequency band, would be expected to undergo the same channel behavior, as depicted in Figure 2.5(a). Conversely, if a signal with a total bandwidth larger than 500 kHz is transmitted through the 1.9 GHz band, different parts of the signal will encounter distinct channel behaviors, as illustrated in Figure 2.5(b). In general, the coherence bandwidth is inversely proportional to the delay of a channel. A larger delay leads to a greater phase difference across different frequencies, resulting in a reduced bandwidth with similar channel behavior. Based on this principle, three commonly used definitions for coherence bandwidth are as follows:

$$B_{\mathrm{c}} \approx \frac{1}{\tau_{\max}}, \tag{2.13}$$

$$B_{\mathrm{c}} \approx \frac{1}{50\tau_{\mathrm{RMS}}}, \tag{2.14}$$

and

$$B_{\mathrm{c}} \approx \frac{1}{5\tau_{\mathrm{RMS}}}, \tag{2.15}$$

Specifically, Eq. (2.13) defines it as the reciprocal of the maximum excess delay. Eq. (2.14) and Eq. (2.15) define it as the bandwidth where the correlation between different frequency components exceeds 0.9 and 0.5, respectively.[3]

Time-varying channels and Doppler spread: Wireless channels typically exhibit variability due to changes in the relative positions of the transmitter, receiver, and the surrounding environment, including obstacles. These variations are a primary cause of time-varying channels. However, considering the movements of all three entities throughout the entire transmission duration is exceedingly complex. In practical scenarios, the movements of the transmitter and receiver have a more significant impact than the environmental changes. Therefore, most time-varying channel models assume a fixed environment and focus solely on the channel effects caused by the movements of the transmitter and receiver, specifically the channel effects resulting from the Doppler effect. The Doppler effect is utilized to describe the changes in signal carrier frequency induced by the movements of the transmitter and receiver,

[3]Please refer to ref. [80] for more information.

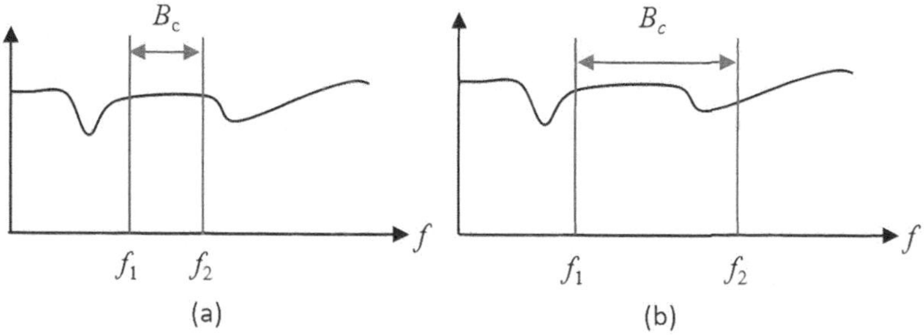

FIGURE 2.5
(a) Illustration of the case when the signal bandwidth is smaller than coherent bandwidth. (b) Illustration of the case when the signal bandwidth is larger than coherent bandwidth.

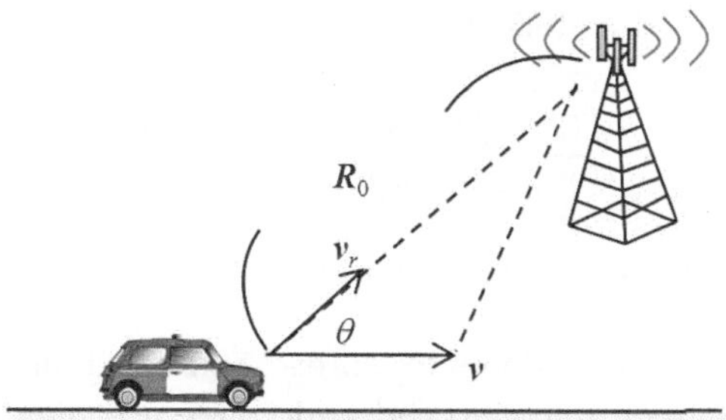

FIGURE 2.6
Illustration of the Doppler shift.

as depicted in Figure 2.6. In this figure, R_0 represents the distance between the transmitter and receiver, v denotes the velocity of the transmitter, and θ represents the angle between the wireless transmission and the direction of the transmitter's movement. It is important to note that we are considering a single-path example in this scenario. Within this model, the change in signal carrier frequency f_{D} is determined by the transmitter velocity v, the angle θ, and the wavelength λ of the signal carrier, as expressed in the following equation:

$$f_{\mathrm{D}} = \frac{v}{\lambda}\cos(\theta). \tag{2.16}$$

Similar to the concept of delay spread, the Doppler spread is utilized to characterize the channel effects caused by the Doppler effect. Specifically, the Doppler spread represents the frequency range or bandwidth that is influenced by the Doppler effect. For instance, if a single-tone signal with a bandwidth of zero passes through a channel with a Doppler spread of 200 kHz, the resulting received signal will have a bandwidth of 200 kHz. The Doppler spread is commonly quantified using the RMS approach, which provides a measure of

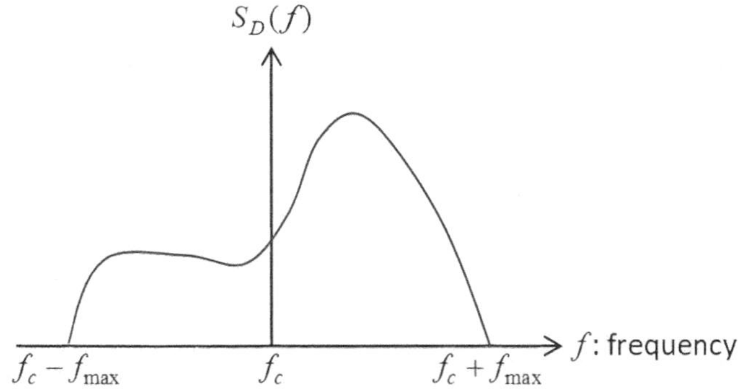

FIGURE 2.7
Illustration of a Doppler power spectrum.

the spread. It is defined as follows:

$$f_{\text{RMS}} = \sqrt{\frac{\int (f-\overline{f})^2 S_{\text{D}}(f)df}{\int S_{\text{D}}(f)df}}; \quad \overline{f} = \frac{\int f S_{\text{D}}(f)df}{\int S_{\text{D}}(f)df}, \tag{2.17}$$

where S_{D} is the Doppler power spectrum of the channel, as shown in Figure 2.7.

The relationship between Doppler spread and coherence time is similar to the relationship between delay spread and coherence bandwidth. In a time-varying channel, coherence time refers to the duration during which the channel remains statistically unchanged. Consequently, the coherence time is inversely proportional to the Doppler shift of the channel. A larger Doppler shift, indicating a higher velocity of the transmitter, leads to rapid changes in channel characteristics. As a result, the duration during which the channel remains unchanged becomes shorter. Similar to coherence bandwidth, there are three commonly used definitions for coherence time. These definitions are as follows:

$$T_{\text{c}} \approx \frac{1}{f_{\max}}, \tag{2.18}$$

$$T_{\text{c}} \approx \frac{9}{16\pi f_{\max}}, \tag{2.19}$$

and

$$T_{\text{c}} \approx \sqrt{\frac{9}{16\pi f_{\max}^2}} = \frac{0.423}{f_{\max}}. \tag{2.20}$$

Specifically, Eq. (2.18) sets it as the reciprocal of the maximum Doppler shift $f_{\max}$. Eq. (2.19) defines it as the duration with correlation greater than 0.5. Eq. (2.20) utilizes the geometric mean to calculate the coherence time.[4] One commonly used channel model for simulating time-varying multi-path wireless

[4] Please refer to ref. [80] for more information.

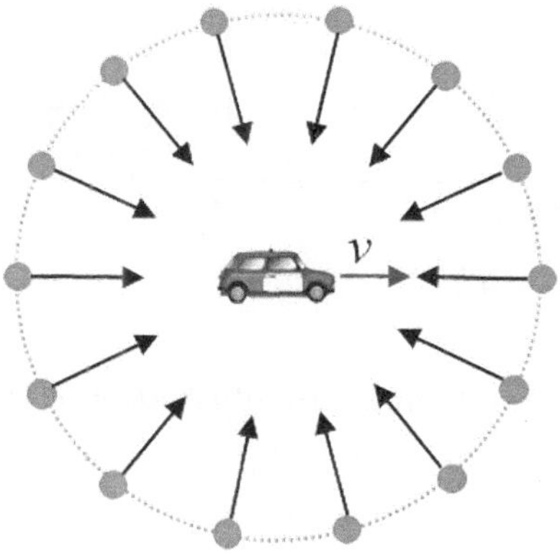

FIGURE 2.8
Illustration of the Jakes model environment.

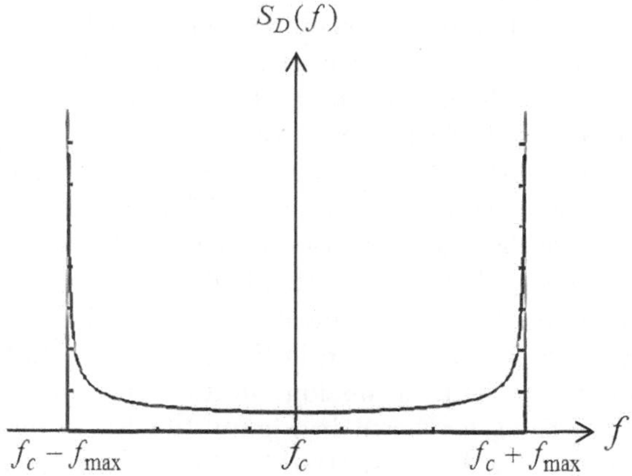

FIGURE 2.9
Illustration of a Doppler power spectrum under Jakes model.

connections is the Jakes model. The Jakes model is often employed to capture the effects of transmitter or receiver movements, as depicted in Figure 2.8. This model focuses on small-scale fading and does not consider large-scale path loss or shadowing effects. In the Jakes model, it is assumed that the receiver is moving with a velocity of v, and a signal with a carrier frequency of f is arriving from all angles uniformly. Consequently, the signal strength from each path is equal and follows an independent and identically distributed Rayleigh distribution. As illustrated in Figure 2.9, when a large number of signals arrive, the power spectrum exhibits a relationship with frequency. Notably, the majority of the signal power is concentrated at the largest frequency offset, denoted as $f_{\max}$.

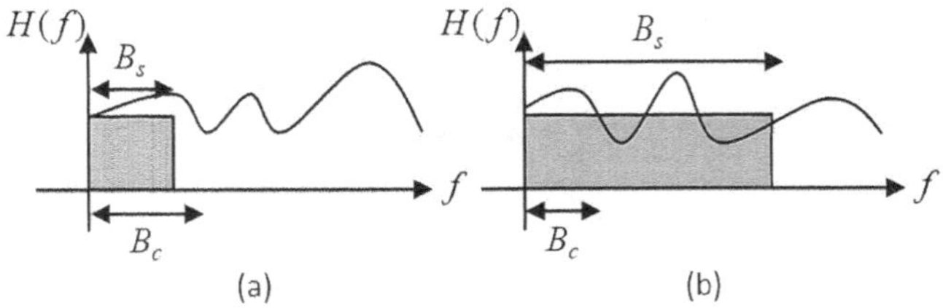

FIGURE 2.10
(a) Illustration of the flat fading. (b) Illustration of the frequency-selective fading.

Summary of small-scale fading: Wireless channels can be classified into two categories based on frequency-varying effects: flat fading and frequency-selective fading. In flat fading, the signal experiences similar channel behavior within its bandwidth, even though different parts of the signal may have different carrier frequencies. This occurs when the signal bandwidth is smaller than the coherence bandwidth of the wireless channel, as shown in Figure 2.10(a). In frequency-selective fading, different parts of the signal encounter varying channel behaviors due to the signal bandwidth exceeding the coherence bandwidth, as shown in Figure 2.10(b). From a time-varying perspective, wireless channels can be categorized as slow fading and fast fading. Slow fading refers to the situation where the signal experiences similar channel behavior, even if different parts of the signal are transmitted in different time slots. This occurs when the transmission duration of a single signal is smaller than the coherence time of the wireless channel, as shown in Figure 2.11(a). On the other hand, fast fading occurs when the transmission duration of a single signal is longer than the coherence time, as shown in Figure 2.11(b), resulting in different channel behaviors for signals transmitted in different time slots.

The choice of which type of fading to consider in a specific scenario depends not only on the channel behaviors discussed above but also on the characteristics of the transmitted signal. For example, a channel with a coherence bandwidth of 50 kHz would be considered a flat fading channel for LTE-A subcarriers (with a bandwidth of 15 kHz per subcarrier). However, the same channel would be considered frequency-selective for signals utilizing Universal Mobile Telecommunications Systems (UMTS). Figure 2.12 provides a summary of the four types of small-scale fading discussed in this section.

2.4 Wireless Signal Models with Multiple Antennas

In the previous section, we examined the channel modeling of wireless connections between the transmitter and the receiver. The focus was primarily on elucidating the influence of various environmental parameters on the

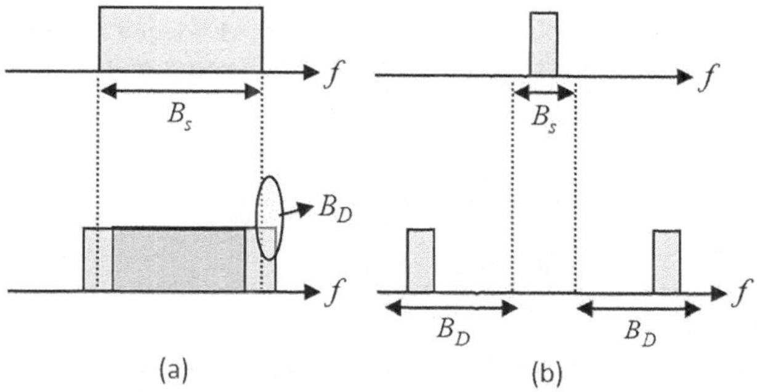

FIGURE 2.11
(a) Illustration of the slow fading in frequency domain. (b) Illustration of the fast fading in frequency domain.

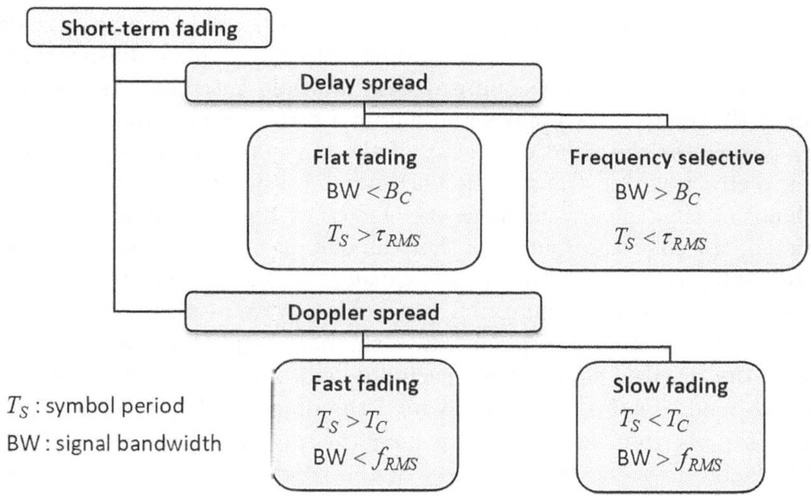

FIGURE 2.12
Illustration of the four types of different small-scale fading in this section.

resultant wireless channel. Nevertheless, it is crucial to acknowledge that the wireless connections are also significantly affected by the implementation of diverse antenna configurations at the transmitter and receiver, which was not addressed in the previous section. In reality, contemporary wireless communication systems, including commercial mobile communication systems, have already embraced the utilization of multi-antenna structures to fully exploit the advantages offered by spatial diversity. Consequently, recent advancements in channel modeling have incorporated these multi-antenna structures at both the transmitter and receiver ends, resulting in the development of more intricate, multidimensional, and time-varying channel models. The purpose of this section is to present some frequently employed signal models for multi-antenna

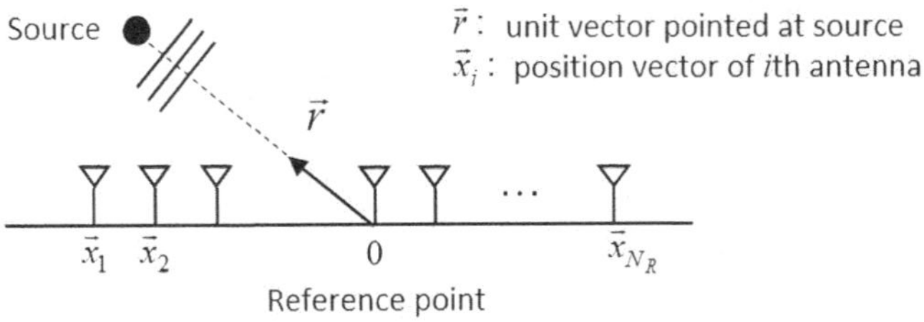

FIGURE 2.13
Illustration of the scenario with a single information source and multiple antennas in the receiver.

systems, thereby enabling readers to gain a comprehensive understanding of current wireless systems.

Antenna array and beamforming technology: We initially consider a simplified scenario in which a point transmitting source establishes wireless connections with a receiver equipped with multiple antennas. As depicted in Figure 2.13, the receiver employs an antenna array comprising N_{R} elements. Here, $\vec{x}_i$ denotes the position vector of the ith antenna, while $\vec{r}$ represents a unit vector pointing from a reference point toward the point source. It is worth noting that the arrangement of elements within the antenna array is not restricted; the sole requirement is that the distance between the point source and the antenna array should be significantly large, adhering to the far-field assumption. Under this assumption, the incident wave resulting from the point source can be treated as a plane wave, thereby simplifying the relationship between the received signals from each antenna. Another assumption is that a narrow-band signal is transmitted from the point source, ensuring that the received signals from each antenna possess the same baseband representation, further facilitating the simplification of the received signals of the antenna array. Based on the aforementioned model, the equivalent baseband signals received from each element in the antenna array can be represented as a vector, as indicated below:

$$\mathbf{x}(t) = \begin{bmatrix} x_1(t) \\ \vdots \\ x_{N_{\mathrm{R}}}(t) \end{bmatrix} = \begin{bmatrix} e^{j\frac{2\pi}{\lambda}\vec{x}_1 \cdot \vec{r}} \\ \vdots \\ e^{j\frac{2\pi}{\lambda}\vec{x}_{N_{\mathrm{R}}} \cdot \vec{r}} \end{bmatrix} u(t) + \begin{bmatrix} n_1(t) \\ \vdots \\ n_{N_{\mathrm{R}}}(t) \end{bmatrix} = \mathbf{a}(\vec{r})u(t) + \mathbf{n}(t), \tag{2.21}$$

where λ is the carrier wavelength, $u(t)$ is the baseband signal, and $\mathbf{a}(\vec{r})$ is the array steering vector which is also known as spatial signature vector (SSV). The array steering vector characterizes the spatial response of the antenna array. Furthermore, by examining the definition of the array steering vector, we can discern how the far-field assumption and narrow-band assumption contribute to obtaining an elegant result in Eq. (2.21). First, the plane wave assumption, derived from the far-field assumption, enables us to represent the

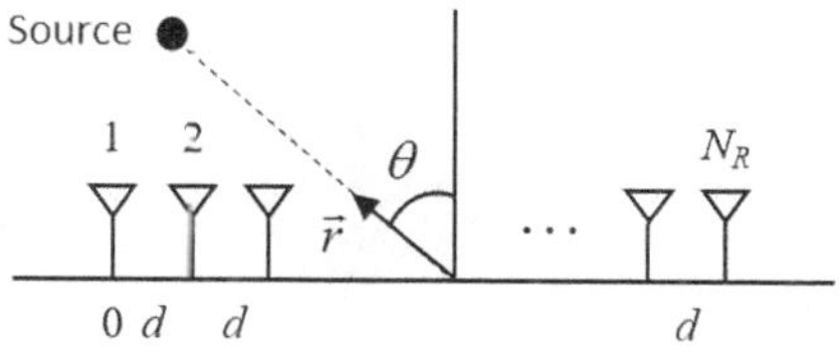

FIGURE 2.14
Illustration of the uniform linear array (ULA).

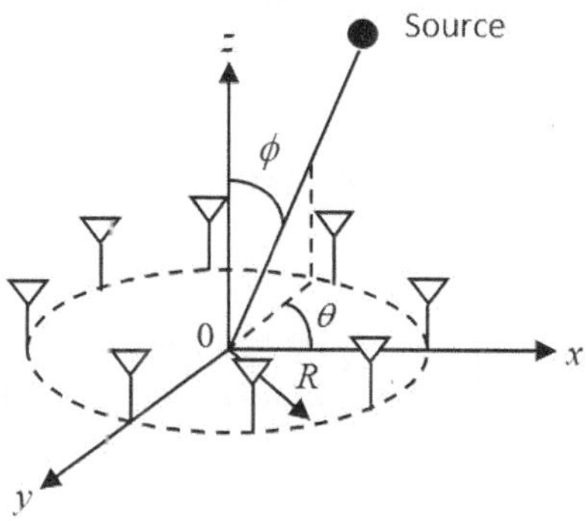

FIGURE 2.15
Illustration of the uniform circular array (UCA).

path of the wave from the point source to the ith element of the antenna array as $\vec{x}_i \cdot \vec{r}$. Second, the narrow-band assumption permits the extraction of the common baseband signal from the array steering vector, thereby enabling each element within the array steering vector to represent the resulting phase response of each antenna, i.e., the spatial response.

Two commonly employed permutation schemes for forming antenna arrays are the uniform linear array (ULA) and the uniform circular array (UCA). When considering the number of antennas, N_{R}, the ULA structure arranges all antennas in a linear configuration with equal spacing, as illustrated in Figure 2.14. Conversely, the UCA structure distributes all antennas evenly along a circular ring, as depicted in Figure 2.15. The array steering vector for the ULA structure can be mathematically represented as follows:

$$\mathbf{a}(\theta) = \begin{bmatrix} 1 \\ e^{-j\frac{2\pi}{\lambda} d \sin\theta} \\ \vdots \\ e^{-j\frac{2\pi}{\lambda}(N_{\mathrm{R}}-1) d \sin\theta} \end{bmatrix}, \tag{2.22}$$

where θ is the angle between $\vec{r}$ and the normal line of this array and d is the antenna spacing. The array steering vector of the UCA structure can be

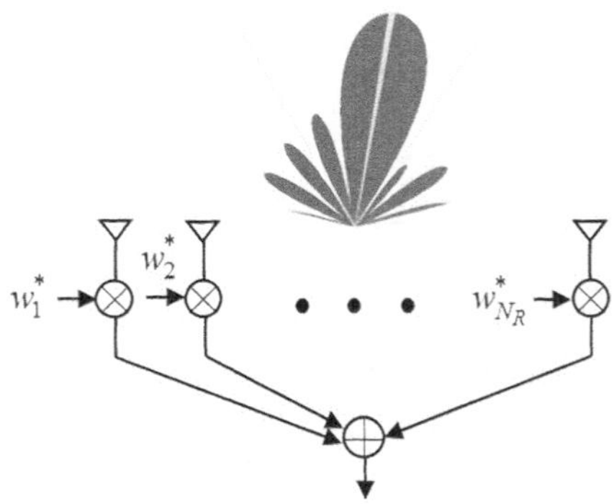

FIGURE 2.16
Illustration of a beamformer and the corresponding spatial response.

expressed as:

$$\mathbf{a}(\theta) = \begin{bmatrix} e^{j\frac{2\pi}{\lambda}R\sin\phi\cos\theta} \\ e^{j\frac{2\pi}{\lambda}R\sin\phi\cos(\theta-\frac{2\pi}{N_\mathrm{R}})} \\ \vdots \\ e^{j\frac{2\pi}{\lambda}R\sin\phi\cos(\theta-\frac{2(N_\mathrm{R}-1)\pi}{N_\mathrm{R}})} \end{bmatrix}, \tag{2.23}$$

where R is the radius of the circular ring, θ is the azimuth angle measured from the X-axis counter-clockwise, and ϕ is the elevation angle measured from the Z-axis.

One of the most commonly employed applications of antenna arrays is beamforming which can be understood as a spatial filtering technique. By applying specific weights to the received signal at each antenna and summing the processed signals, a spatial filtering structure known as a beamformer is obtained, as depicted in Figure 2.16. It is evident that different weights yield different spatial responses, and the goal of beamforming design is to determine the optimal weights for a given scenario. It is worth noting that the principle of beamforming bears resemblance to that of finite impulse response (FIR) filters in digital signal processing. However, instead of frequency responses, beamforming generates spatial responses by associating spatial angles with frequencies. The beamforming principle can be employed in both the transmitter and receiver, referred to as transmit beamforming and receive beamforming, respectively. Mathematically, receive beamforming can be formulated as follows:

$$y(t) = \sum_{i=1}^{N_\mathrm{R}} w_i^* x_i(t) = \mathbf{w}^H \mathbf{x}(t), \tag{2.24}$$

where $\mathbf{w} = [w_1, w_2, ..., w_{N_\mathrm{R}}]^T$ is the weighting vector. Imagine that a convex lens can be utilized to focus a straight beam of light. Similarly, the beamformer, through appropriate weights, can concentrate on receiving signals from

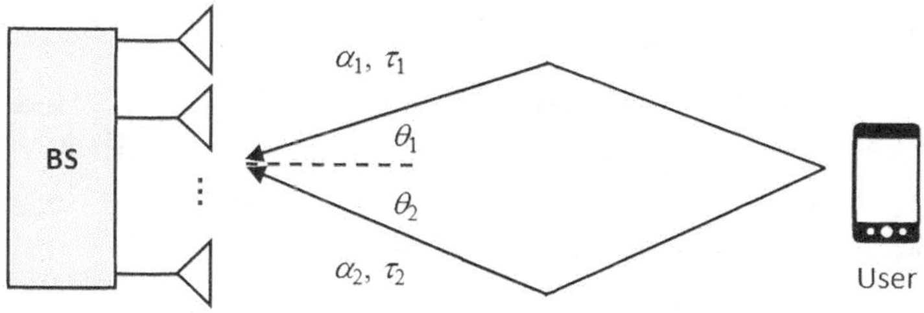

FIGURE 2.17
Illustration of a SU-SIMO scenario.

a specific direction, thereby enhancing the received signal power and improving transmission quality. Analogously, in the case of transmit beamforming, assuming an antenna array with N_{T} antennas is employed, it can be expressed as follows:

$$\mathbf{x}(t) = \mathbf{w}^H u(t); \quad \begin{bmatrix} x_1(t) \\ \vdots \\ x_{N_{\mathrm{T}}}(t) \end{bmatrix} = \begin{bmatrix} w_1^* \\ \vdots \\ w_{N_{\mathrm{T}}}^* \end{bmatrix} u(t). \tag{2.25}$$

The concept of transmit beamforming can be likened to a searchlight, where signals from different antennas are concentrated toward a specific direction to enhance signal power. Both transmit and receive beamforming offer two key advantages by boosting power in a particular direction. First, the transmission quality can be improved, leading to superior quality of service (QoS). Second, it enables the mitigation of mutual interference between the desired signal and other signals. These two benefits serve as strong motivations for system designers to widely adopt beamforming technology in modern wireless communications.

SIMO, MISO, and MIMO signal models: In this section, we will introduce several commonly used multi-antenna signal models, building upon the aforementioned basic signal model. We begin by considering the single-user single-input-multi-output (SU-SIMO) model, which pertains to the uplink scenario where a user is connecting to a base station, as illustrated in Figure 2.17. The receive signal in this model can be expressed as follows:

$$\mathbf{x}(t) = \sum_{j=1}^{J} \alpha_j \mathbf{a}(\theta_j) u(t - \tau_j) + \mathbf{n}(t), \tag{2.26}$$

where J is the number of paths. α_j, θ_j, and τ_j represent the fading gain, angle of arrival (AoA), and delay of the jth path, respectively. $\mathbf{a}(\theta_j)$ is the array steering vector and $u(t)$ is the transmitted baseband signal. Additionally, the single-user multiple-input-single-output (SU-MISO) model addresses the downlink scenario, wherein the base station transmits signals to a user, as depicted in Figure 2.18. The receive signal in this model can be expressed as

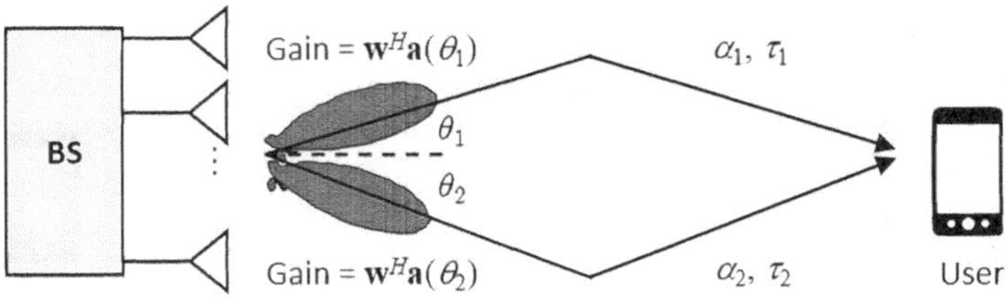

FIGURE 2.18
Illustration of a SU-MISO scenario.

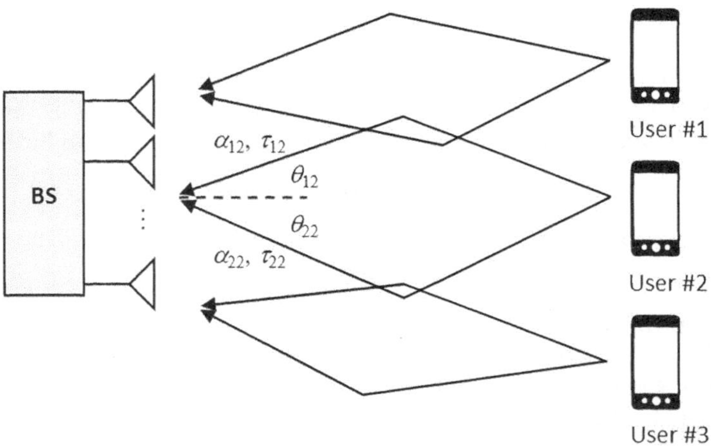

FIGURE 2.19
Illustration of an MU-SIMO scenario.

follows:

$$\mathbf{x}(t) = \sum_{j=1}^{J} \alpha_j \mathbf{w}^H \mathbf{a}(\theta_j) u(t - \tau_j) + \mathbf{n}(t), \tag{2.27}$$

where $\mathbf{w}$ is the transmit beamforming weight and θ_j is the angle of departure (AoD). Furthermore, the multi-user single-input-multiple-output (MU-SIMO) model encompasses a more intricate scenario. It involves multiple users, each equipped with a single antenna, transmitting simultaneously to a base station equipped with multiple antennas during the uplink phase, as illustrated in Figure 2.19. The receive signal can be expressed as:

$$\mathbf{x}(t) = \sum_{q=1}^{Q} \sum_{j=1}^{J_q} \alpha_{jq} \mathbf{a}(\theta_{jq}) u_q(t - \tau_{jq}) + \mathbf{n}(t), \tag{2.28}$$

where Q is the number of transmitters and $u_q(t)$ is the transmitted signal from qth transmitter. Finally, the multi-user multiple-input-single-output

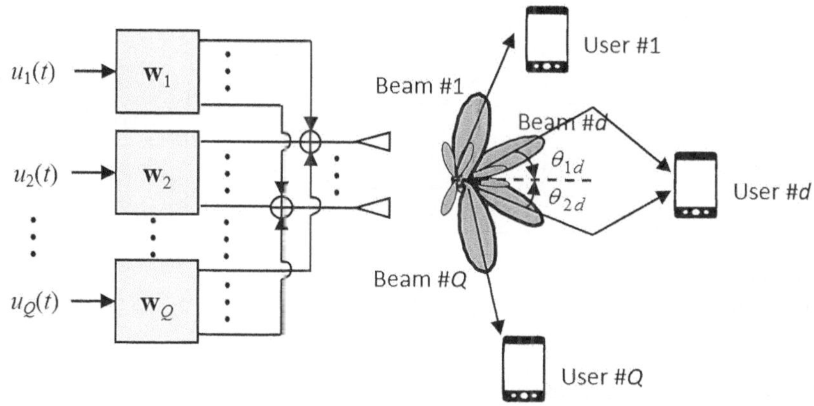

FIGURE 2.20
Illustration of an MU-MISO scenario.

(MU-MISO) model describes the converse scenario to the MU-SIMO model. In this case, a base station equipped with multiple antennas serves several users, each with a single antenna, during the downlink phase, as depicted in Figure 2.20. The receive signal at the dth user can be expressed as:

$$x_d(t) = \sum_{q=1}^{Q}\sum_{j=1}^{J_d} \alpha_{jd}\mathbf{w}_q^H \mathbf{a}(\theta_{jq})u_q(t-\tau_{jd}) + n_d(t), \tag{2.29}$$

where $\mathbf{w}_q$ is the beamforming weight that the base station utilized to conduct transmission to the qth receiver. In this scenario, designing beamforming weights is a better strategy to let each user receive the desired signal and minimize mutual interference between users simultaneously.

In contrast to the previous scenarios, when multiple users are replaced by a single device with multiple antennas, it constitutes a MIMO system. The MIMO system describes a scenario where a transmitter with N_T antennas communicates with a receiver equipped with N_R antennas. To streamline the analysis, we assume that the transmitter and receiver employ a single device to accomplish wireless transmission, known as point-to-point transmission, as illustrated in Figure 2.21. The receive signal can be expressed as:

$$\mathbf{y}(t) = \int \mathbf{H}(t,\tau)\mathbf{x}(t-\tau)d\tau + \mathbf{n}(t), \tag{2.30}$$

where $\mathbf{H}(t,\tau) = [h_{ij}(t,\tau)] \in \mathbb{C}^{N_\mathrm{R}\times N_\mathrm{T}}$ is a MIMO channel matrix and the ith element of the receive signal vector can be expressed as:

$$y_i(t) = \sum_{j=1}^{N_\mathrm{T}} \int h_{ij}(t,\tau)x_j(t-\tau)d\tau + n_i(t). \tag{2.31}$$

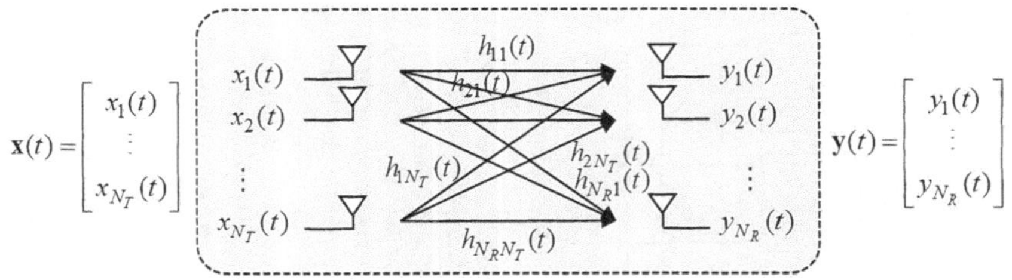

FIGURE 2.21
Illustration of the MIMO signal model.

In Eq. (2.31), h_{ij} represents the element at the ith row and jth column of the matrix $\mathbf{H}(t,\tau)$. x_j denotes the jth element of the vector $\mathbf{x}$, and n_i represents the ith element of the vector $\mathbf{n}$. In the previously mentioned MU-SIMO signal model, a group of users with single antennas can also be considered as a transmitter with multiple antennas. Consequently, this can be expressed in the form of an equivalent MIMO signal model as follows:

$$\begin{aligned}\mathbf{x}(t) &= \sum_{q=1}^{Q}\sum_{j=1}^{J_q}\alpha_{jq}\mathbf{a}(\theta_{jq})u_q(t-\tau_{jq})+\mathbf{n}(t)\\ &= \sum_{q=1}^{Q}\sum_{j=1}^{J_q}\alpha_{jq}\mathbf{a}(\theta_{jq})\delta(t-\tau_{jq})*u_q(t)+\mathbf{n}(t)\\ &= [\mathbf{h}_1(t)\cdots\mathbf{h}_Q(t)]*\begin{bmatrix}u_1(t)\\ \vdots\\ u_Q(t)\end{bmatrix}+\mathbf{n}(t)=\mathbf{H}(t)*\mathbf{u}(t)+\mathbf{n}(t),\end{aligned} \tag{2.32}$$

where each row of channel matrix $\mathbf{H}(t)$ corresponds to the equivalent channel vector of each transmitted signal, expressed as:

$$\mathbf{h}_q(t)=\sum_{j=1}^{J_q}\alpha_{jq}\mathbf{a}(\theta_{jq})\delta(t-\tau_{jq}). \tag{2.33}$$

When applying the assumptions of slow fading and flat fading, the channel matrix $\mathbf{H}(t)$ is no longer a function of t and τ, but instead becomes a constant matrix denoted as $\mathbf{H}(t)=[h_{ij}]$. By making this simplification, as depicted in Figure 2.22, the MIMO signal model can be further simplified as follows:

$$\mathbf{y}(t)=\mathbf{H}\mathbf{x}(t)+\mathbf{n}(t), \tag{2.34}$$

where the ith element of receive signal $y_i(t)$ can be expressed as:

$$y_i(t)=\sum_{j=1}^{N_\mathrm{T}}h_{ij}x_j(t)+n_i(t). \tag{2.35}$$

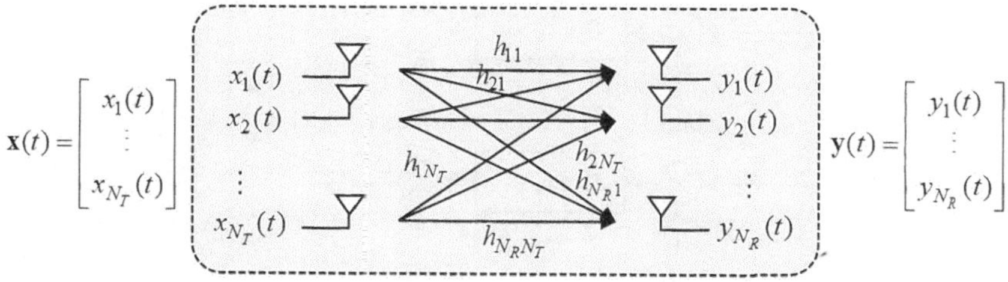

FIGURE 2.22
Illustration of the MIMO signal model under slow fading and flat fading assumptions.

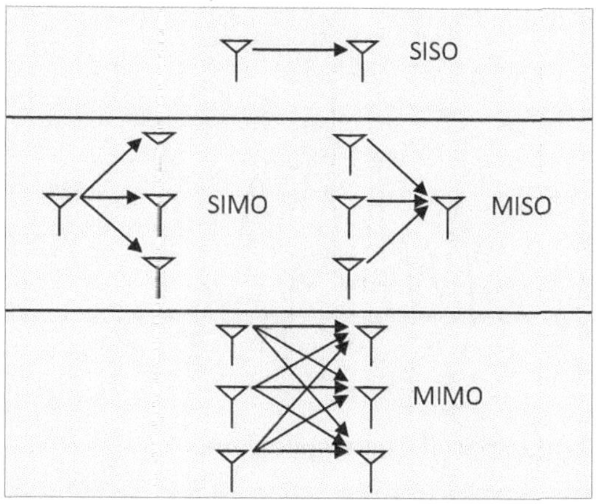

FIGURE 2.23
Evolution from single-antenna systems to multi-antenna systems.

This simplified model can be applied to consider several widely used wireless communication models, including multiple-input-multiple-output orthogonal frequency division multiplexing (MIMO-OFDM). Please note that the MIMO-OFDM system will be introduced in Chapter 4. To summarize this section, Figure 2.23 illustrates the evolution from single-antenna systems to multi-antenna systems.

Angle spread: In this section, we apply the concept of multi-antennas to the previously introduced multi-path channel and discuss the resulting effects. Figure 2.24 depicts a channel model that considers both multi-path and multi-antenna effects simultaneously. On the receiver side, an antenna array with multiple antennas is utilized to receive signals from different directions. Each path in the channel has a distinct incident angle, and the angle spread is used to describe the range of incident angles. The concept of angle spread is analogous to delay spread and Doppler spread, which describe the spread of delays and frequencies, respectively, in a multi-path channel. For instance,

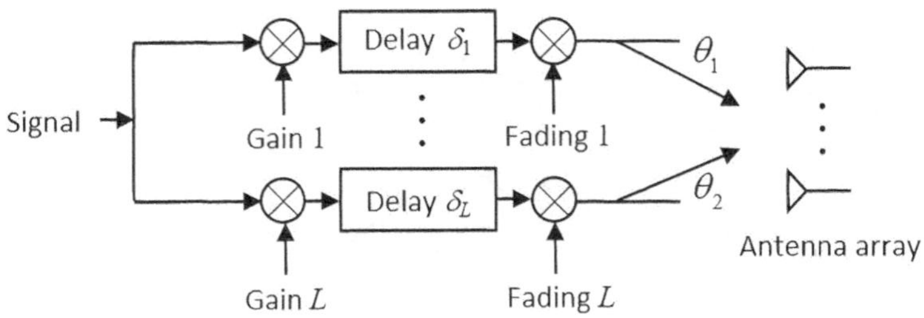

FIGURE 2.24
Illustration of the channel model considering multi-path effect and multiple antennas effect.

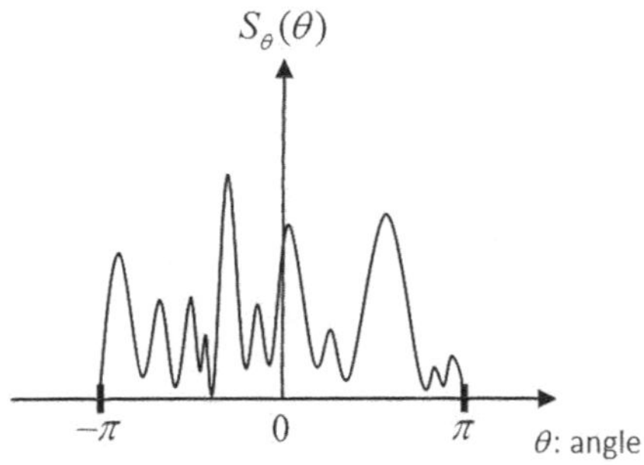

FIGURE 2.25
Illustration of a signal spatial angle spectrum.

consider a channel with three paths. The angles at which these paths leave the transmitter antenna are $-10°$, $0°$, and $30°$, respectively, while the angles at which they enter the receiver antenna are $3°$, $4°$, and $6°$, respectively. Consequently, the angle spread observed by the transmitter antenna and the receiver antenna would be $3°$ and $40°$, respectively. In practical scenarios, a realistic signal will exhibit multiple paths during transmission. Therefore, we employ the signal spatial angle spectrum $S_\theta(\theta)$ to analyze the angle spread of signals, as illustrated in Figure 2.25. To quantify the angle spread, the RMS is a commonly used measure, defined as follows:

$$\theta_{\text{RMS}} = \sqrt{\frac{\int_{-\pi}^{\pi}(\theta-\overline{\theta})^2 S_\theta(\theta)d\theta}{\int_{-\pi}^{\pi} S_\theta(\theta)d\theta}}; \quad \overline{\theta} = \frac{\int_{-\pi}^{\pi}\theta S_\theta(\theta)d\theta}{\int_{-\pi}^{\pi} S_\theta(\theta)d\theta}. \tag{2.36}$$

Following a similar concept to the coherence time and coherence bandwidth, we can define the coherence distance D_C based on the angle spread. In particular, the coherence distance D_C is expected to be inversely proportional

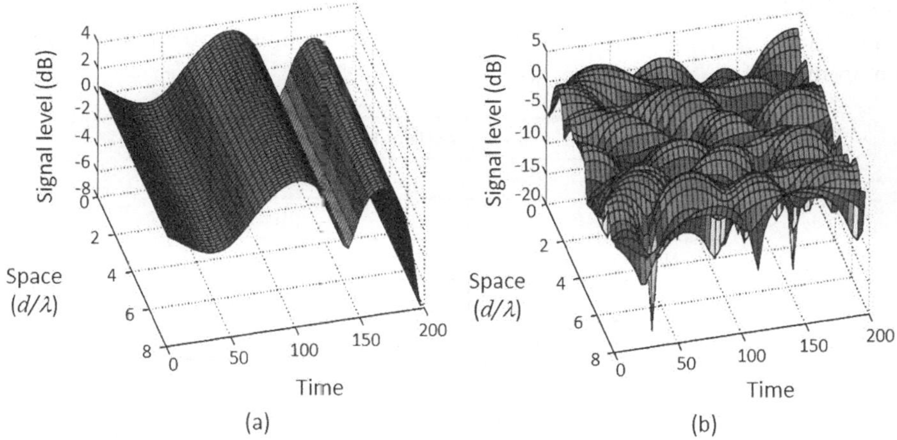

FIGURE 2.26
(a) Illustration of a time-variant but spatial-invariant channel. (b) Illustration of a time-variant and spatial-variant channel.

to the angle spread. Mathematically, this relationship can be expressed as:

$$D_{\mathrm{C}} \propto \frac{1}{\theta_{\mathrm{RMS}}}. \tag{2.37}$$

A larger angle spread implies a larger phase difference caused by multi-paths when observed at different spatial locations. Consequently, the distance at which the same channel is observed will be shorter. Up to this point, we have discussed how the wireless channel varies in both time domain and spatial domain. Figures 2.26(a) and 2.27(a) illustrate different channel behaviors. Specifically, Figure 2.26(a) depicts a time-varying but spatially invariant channel, while Figure 2.26(b) illustrates a time-varying and spatially varying channel. To provide further insight, Table 2.2 presents the typical values of delay spread, Doppler spread, and angle spread in common environments.

In order to analyze the reasons for delay spread, Doppler spread, and angle spread in real environments, let's consider a scenario where a user is transmitting signals to a base station located at a distant location, as depicted in Figure 2.27. From the perspective of the base station, the user and its nearby environment can be considered as a single point due to the long distance between them. Therefore, the resulting delay spread and angle spread are not influenced by the user's nearby environment. However, if there are moving objects in the vicinity of the user, such as vehicles, Doppler spread will be caused due to the relative motion between the user and the objects. As the transmitted signal travels away from the user's nearby environment, it may encounter distant scattering objects such as high buildings or mountains. At this stage, since the distances to different scattering objects are usually

TABLE 2.2
The values of delay spread, angle spread, and Doppler spread in common environments.

Environment	Delay spread	Angle spread	Doppler spread
Flat Rural	0.5 μs	1°	190 Hz
Urban	5 μs	20°	120 Hz
Hilly	20 μs	30°	190 Hz
Mall	0.3 μs	120°	10 Hz
Indoors	0.1 μs	360°	5 Hz

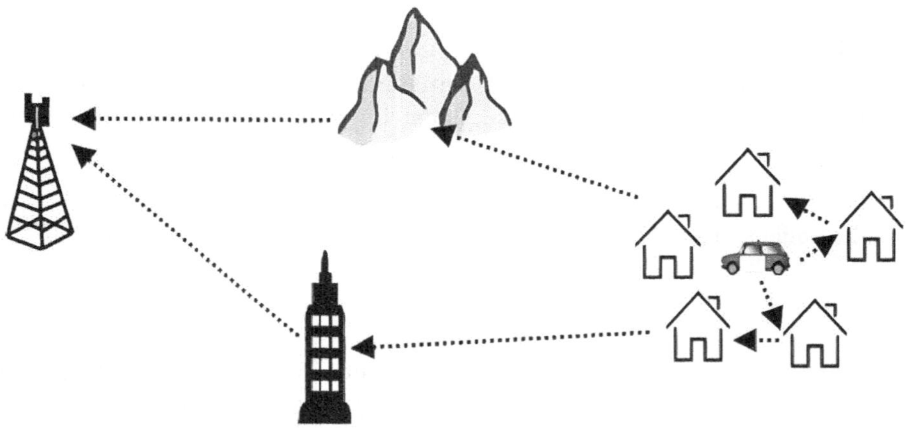

FIGURE 2.27
Illustration of a scenario with different channel spreads.

significant, the arrival times and incident angles of the signal from these objects can vary significantly, leading to delay spread and angle spread. However, as these scattering objects are typically stationary, Doppler spread is not caused during this stage. Before reaching the base station, the transmitted signals will interact with the nearby environment surrounding the base station. Since the travel distances of different multi-paths caused by the nearby environment are similar, the arrival times of these paths are also similar, resulting in no delay spread in this stage. Additionally, considering that base stations are often situated on higher ground, there are usually no moving objects in close proximity, and therefore no Doppler spread is caused during this stage. However, since the arrival angles of the different multi-paths are distinct, angle spread at the receiver side is generated during this stage.

2.5 Wireless Channel for Communication Systems

After completing the introduction on multi-antenna multi-path time-varying channels, our objective in this section is to present the channel models put forth by the 3GPP. Specifically, we will focus on the spatial channel model (SCM). The SCM is a ray-based channel model designed to simulate the behavior of signals in complex environments with multiple antennas and paths. It takes into account various factors, including angle spread, delay spread, and shadowing effect, in order to generate representative paths for accurate signal simulation.

Application scenarios and large-scale fading models: The SCM supports a signal bandwidth of up to 5 MHz, operating at a carrier frequency of 1900 MHz. It enables the simulation of three distinct scenarios: urban microcell, urban macrocell, and suburban macrocell. Table 2.3 presents important parameters for comparison across these scenarios. In the macrocell scenario, where the base station covers a large area, the likelihood of receiving line-of-sight (LoS) signals is considerably low. Consequently, LoS signals are not considered in this scenario. Conversely, in the microcell scenario, where the base station covers a smaller area, LoS signals can potentially be received. Thus, we define two additional scenarios: microcell LoS and microcell non-line-of-sight (NLoS), based on whether LoS signals are received or not. The path loss models utilized in the three scenarios differ, and their details are presented in Table 2.4. Within this table, the constant C assumes values of 0 dB and 3 dB, respectively. Moreover, h_{BS} and h_{MS} represent the heights of the base station and the mobile station (user), respectively, in meters. The parameter d corresponds to the distance between the base station and the user, also measured in meters, while f_c denotes the carrier frequency in MHz. The specific parameters for the macrocell and microcell scenarios are listed in Table 2.5. By substituting these values into the equations presented in Table 2.4 and simplifying, we obtain the simplified path loss models, as displayed in Table 2.6. Furthermore, Figure 2.28 provides a graphical representation of the relationship between path loss and distance in the different scenarios.

Small-scale fading models: In terms of small-scale fading models, the multi-antennas multi-paths model SCM employed is depicted in Figure 2.29. Within the figure, the base station (BS) serves as the transmitter, while the mobile station (MS) functions as the receiver. The SCM incorporates the contributions of multiple paths to accurately represent small-scale fading in realistic scenarios. Additionally, the AoD, denoted as $\theta_{n,m,\text{AoD}}$, for the m-th subpath of the n-th path originating from the BS, as well as the AoA, denoted as $\theta_{n,m,\text{AoA}}$, for the m-th subpath of the n-th path arriving at the user can be expressed as:

$$\begin{aligned} \theta_{n,m,\text{AoD}} &= \theta_{\text{BS}} + \delta_{n,\text{AoD}} + \Delta_{n,m,\text{AoD}}, \\ \theta_{n,m,\text{AoA}} &= \theta_{\text{MS}} + \delta_{n,\text{AoA}} + \Delta_{n,m,\text{AoA}}, \end{aligned} \tag{2.38}$$

TABLE 2.3
The main parameter settings of the three simulation scenarios in the SCM.

	Distance of BS to BS	BS antenna position	LoS component
Suburban macrocell	3 km	High (above local clutter)	No
Urban macrocell	3 km	High (above rooftop)	No
Urban microcell	1 km	Rooftop level	Yes / No

TABLE 2.4
The path loss models of the three simulation scenarios in the SCM.

Macrocell
$PL[\text{dB}] = [44.9-6.55\log(h_{BS})]\log(d/1000) + 45.5 + 0.7h_{MS} + (35.46-1.1h_{MS})\log(f_c) - 13.82\log(h_{BS}) + C$
Microcell (NLoS)
$PL_{MI\text{-}NLoS}[\text{dB}] = -55.9 + 38\log(d) + [24.5+(1.5f_c/925)]\log(f_c)$
Microcell (LoS)
$PL_{MI\text{-}LoS}[\text{dB}] = -35.4 + 26\log(d) + 20\log(f_c)$

TABLE 2.5
The parameter settings of the three macrocell and microcell scenarios in the SCM.

	h_{BS}	h_{MS}	d	f_c
Macrocell	32 m	1.5 m	$\geq$ 35 m	1900 MHz
Microcell	12.5 m	1.5 m	$\geq$ 20 m	1900 MHz

TABLE 2.6
The simplified path loss models of different scenarios in the SCM.

		Path loss
Macrocell	Suburban	$PL_{SM} = 31.5 + 35\log(d)$
	Urban	$PL_{UM} = 34.5 + 35\log(d)$
Microcell	NLoS	$PL_{MI\text{-}NLoS} = 34.53 + 38\log(d)$
	LoS	$PL_{MI\text{-}LoS} = 30.18 + 26\log(d)$

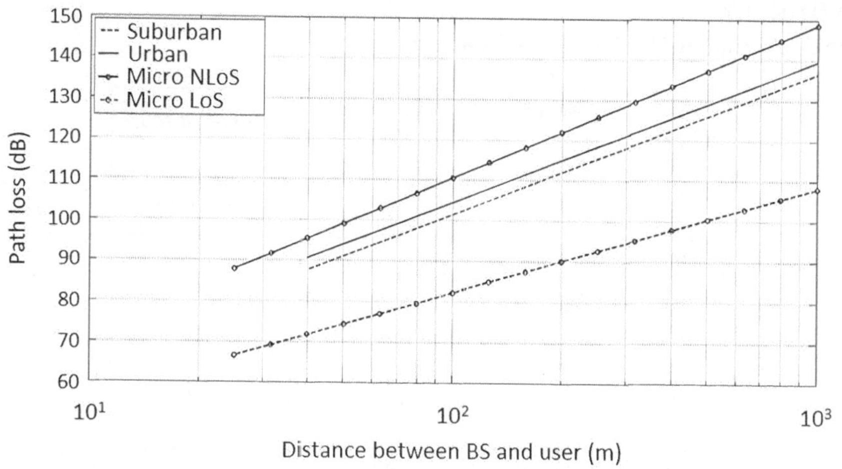

FIGURE 2.28
Illustration of the relationship between distance and path loss in different scenarios of SCM.

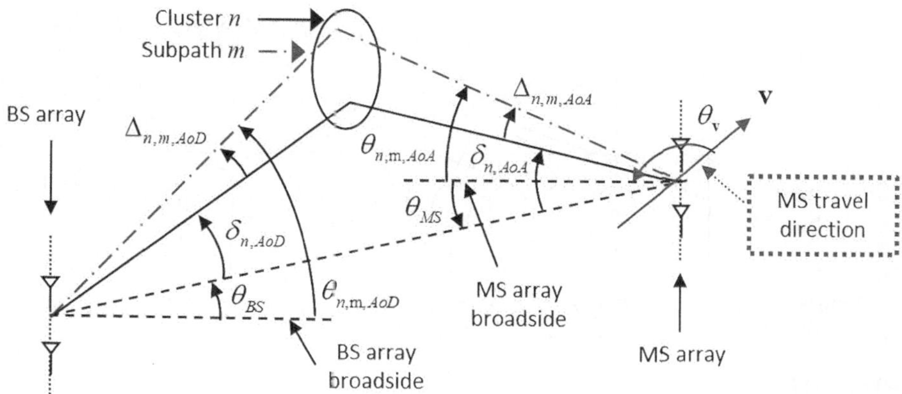

FIGURE 2.29
Illustration of the multi-antennas multi-path small-scale fading models in SCM.

where θ_{BS} and θ_{MS} are the AoD and AoA of LoS path, $\delta_{n,\mathrm{AoD}}$ and $\delta_{n,\mathrm{AoA}}$ are the AoD difference and AoA difference between nth path and LoS path, which will be introduced later. $\Delta_{n,m,\mathrm{AoD}}$ and $\Delta_{n,m,\mathrm{AoA}}$ are the AoD offset and AoA offset between m subpath of n path and n path, as listed in Table 2.7. Note that the subpaths between the base station and the user will be aligned randomly to present the signal phase close to the uniform distribution, as shown in Figure 2.30.

TABLE 2.7
The AoA and AoD offsets of subpath in the SCM.

Subpath number (m)	Offset at BS AS = 2° Macrocell $\Delta_{n,m,AoD}$ (degree)	Offset at BS AS = 5° Microcell $\Delta_{n,m,AoD}$ (degree)	Offset at MS AS = 35° $\Delta_{n,m,AoD}$ (degree)
1, 2	0.0894	0.2236	1.5649
3, 4	0.2826	0.7064	4.9447
5, 6	0.4984	1.2461	8.7224
7, 8	0.7431	1.8578	13.0045
9, 10	1.0257	2.5642	17.9492
11, 12	1.3594	3.3986	23.7899
13, 14	1.7688	4.4220	30.9538
15, 16	2.2961	5.7403	40.1824
17, 18	3.0389	7.5974	53.1816
19, 20	4.3101	10.7753	75.4274

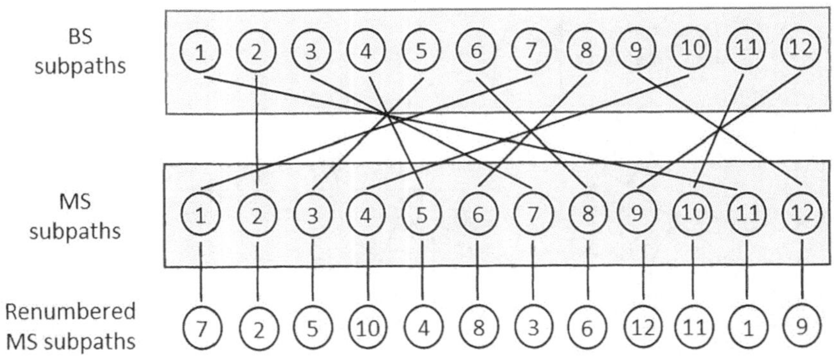

FIGURE 2.30
Illustration of the random subpath association between the base station and mobile station in SCM.

Figure 2.29 presents the small-scale fading models utilized in the SCM. In the subsequent discussion, we will introduce the parameter settings involved in these models. Empirically derived from real measurements, the angle spread, delay spread, and shadowing effect are considered as correlated random variables following a log-normal distribution. Specifically, the angle spread and delay spread exhibit a positive correlation with a coefficient of $\rho_{\text{DA}} = 0.5$. Conversely, negative correlations can be observed between the delay spread and shadowing effect, as well as between the angle spread and shadowing effect, with correlation coefficients of $\rho_{\text{AF}} = \rho_{\text{DF}} = -0.6$, as illustrated in

TABLE 2.8
Illustration of the random subpath association between the base station and mobile station in SCM.

Correlation matrix	
Intra BS	Inter BS
$\begin{bmatrix} 1 & \rho_{DA} & \rho_{DF} \\ \rho_{DA} & 1 & \rho_{AF} \\ \rho_{DF} & \rho_{AF} & 1 \end{bmatrix}$	$\begin{bmatrix} 0 & 0 & 0 \\ 0 & 0 & 0 \\ 0 & 0 & \zeta \end{bmatrix}$
$\rho_{DA} = +0.5$ $\rho_{AF} = r_{DF} = -0.6$	$\zeta = +0.5$

Table 2.8. Based on these parameter settings, the SCM incorporates three independent Gaussian random variables and the specified correlations to generate values for the angle spread, delay spread, and shadowing effect, as described in Eqs. (2.39) and (2.40):

$$\begin{bmatrix} \alpha_j \\ \beta_j \\ \gamma_j \end{bmatrix} = \begin{bmatrix} 1 & \rho_{\text{DA}} & \rho_{\text{DF}} \\ \rho_{\text{DA}} & 1 & \rho_{\text{AF}} \\ \rho_{\text{DF}} & \rho_{\text{AF}} & 1 \end{bmatrix}^{1/2} \begin{bmatrix} \omega_{j1} \\ \omega_{j2} \\ \omega_{j3} \end{bmatrix} + \begin{bmatrix} 0 & 0 & 0 \\ 0 & 0 & 0 \\ 0 & 0 & \xi \end{bmatrix}^{1/2} \begin{bmatrix} \xi_1 \\ \xi_2 \\ \xi_3 \end{bmatrix}; \tag{2.39}$$

$$\begin{aligned} \sigma_{\text{DS},j} &= 10^{\varepsilon_{\text{DS}}\alpha_j + \mu_{\text{DS}}} \\ \sigma_{\text{AS},j} &= 10^{\varepsilon_{\text{AS}}\beta_j + \mu_{\text{AS}}} \\ \sigma_{\text{SF},j} &= 10^{0.1(\sigma_{\text{SH}}\gamma_j)} \end{aligned} \tag{2.40}$$

By utilizing these generated results, additional parameters such as delay, AoD, and AoA can be computed to account for the combined effect of multiple paths in the channel model.

In Eq. (2.39), ω_{j1}, ω_{j2} and ω_{j3} denote three independent Gaussian random variables, while ξ_1, ξ_2, and ξ_3 represent another set of three independent Gaussian random variables. In Eq. (2.40), the subscript j corresponds to the j-th base station. The variables σ_{DS}, σ_{AS}, and σ_{SF} denote the delay spread, angle spread, and shadowing effect, respectively. Furthermore, μ_{DS} and μ_{AS} represent the mean values of σ_{DS} and σ_{AS}, respectively, after applying a common logarithm operation. For example, μ_{DS} can be calculated as $\mu_{\text{DS}} = \mathbb{E}\{\log_{10}(\sigma_{\text{DS}})\}$. Additionally, ε_{DS}, ε_{AS}, and σ_{SH} denote the variances of σ_{DS}, σ_{AS}, and σ_{SF}, respectively, after applying a common logarithm operation. For instance, $(\varepsilon_{\text{DS}})^2$ can be computed as $(\varepsilon_{\text{DS}})^2 = \mathbb{E}\{[\log_{10}(\sigma_{\text{DS},j}) - (\mu_{\text{DS}})]^2\}$.

First, let us discuss the generation of parameters for each path, including delay, AoD, and AoA. Once the values of σ_{DS}, σ_{AS}, and σ_{SF} have been determined, the delay, power, AoD, and AoA of each path can be generated and subsequently combined to simulate the multi-path effects. The generation of the delay for a specific path can be expressed as follows:

$$\tau_n^{'} = -\gamma_{\text{DS}}\ln(z_n), \quad n = 1, 2, \cdots, N, \tag{2.41}$$

where N is the number of paths, $z_n \sim \mathcal{U}(0,1)$ is a standard uniformly distributed random variable, and γ_{DS} is a constant ratio factor related to the considered scenario. To obtain the delay values for all paths, we compute Eq. (2.41) and arrange the results in ascending order. Let's denote the computed delays as $\tau_{(1)}^{'}, \tau_{(2)}^{'}, \ldots, \tau_{(n)}^{'}$, where the subscript (1) represents the path with the smallest delay, and subscript (n) represents the path with the largest delay. Next, we normalize these delay values by subtracting $\tau_{(1)}^{'}$, resulting in $\tau_n = \tau_{(n)}^{'} - \tau_{(1)}^{'}$ to obtain a set of normalized delays, denoted as $0 = \tau_1 < \tau_2 < \ldots < \tau_n$. Now, let's discuss the generation of power for each path. The power generation for a particular path can be expressed as:

$$P_n^{'} = e^{\frac{(1-\gamma_{\text{DS}})\tau_n}{\gamma_{\text{DS}}\sigma_{\text{DS}}}} \cdot 10^{-0.1\xi_n}, \quad n = 1, 2, \cdots, N, \tag{2.42}$$

where $\xi_n \sim \mathcal{N}(\mu, \sigma^2)$ is i.i.d.[5] Gaussian random variable and σ is given as 3 dB. After obtaining $P_n^{'}$ from Eq. (2.42), we normalize the results to ensure the power sum of all paths will become 1, that is,

$$P_n = \frac{P_n^{'}}{\sum_{i=1}^{N} P_n^{'}}. \tag{2.43}$$

Then, we can compute the AoD and AoA according to Eqs. (2.44) and (2.45).

$$\delta_n^{'} \sim \mathcal{N}[0, (\gamma_{\text{AS}} \cdot \sigma_{\text{As}})^2], \quad n = 1, 2, \cdots, N. \tag{2.44}$$

$$\begin{aligned} &\delta_{n,\text{AoA}} \sim \mathcal{N}(0, \sigma_{n,\text{AoA}}^2); \\ &\sigma_{n,\text{AoA}} = 104.12^\circ \cdot (1 - e^{-0.2175|10\log_{10}(P_n)|}) \end{aligned} \tag{2.45}$$

In Eq. (2.44), γ_{AS} is a constant ratio factor. After generating $\delta_{(n)}^{'}$ randomly, we also compute the absolute value of the results and arranged them in ascending order to obtain $|\delta_{(1)}^{'}| < |\delta_{(2)}^{'}| < \cdots < |\delta_{(N)}^{'}|$, where the AoD of nth path can be obtained as $\delta_{n,\text{AoD}} = \delta_{(n)}^{'}$.

After discussing the parameter generation in the macrocell scenario, it is important to note that the parameter generation in the urban microcell scenario exhibits a slight difference. Specifically, in Eq. (2.46), the method employed by the urban microcell scenario to generate the delay for each path differs from Eq. (2.41). The generation of delay for each path in the urban microcell scenario can be expressed as follows:

$$\tau_n^{'} \sim N(0, 1.2\mu s), \quad n = 1, 2, \cdots, N. \tag{2.46}$$

[5]The abbreviation of independent, identically distributed.

Similarly, we proceed to normalize the results by subtracting the minimum value and arranging them in ascending order to obtain $0 = \tau_1 < \tau_2 < \ldots < \tau_n$. However, if the LoS scenario is considered, the delay of the LoS path will be set as 0, indicating that it has the shortest delay among all paths. Next, let's discuss the generation of power for each path. The power generation for a particular path can be expressed as:

$$P_n^{'} = 10^{-(\tau_n + 0.1 z_n)}, \quad n = 1, 2, \cdots, N, \tag{2.47}$$

where $z_n \sim \mathcal{N}(\mu, \sigma^2)$ is the i.i.d. Gaussian random variable and σ is set as 3 dB. If LoS scenario is considered, the power generation of each path can be expressed as:

$$P_n = \frac{P_n^{'}}{(K+1)\sum_{i=1}^{N} P_i^{'}}. \tag{2.48}$$

In Eq. (2.48), since the power of LoS path $P_{\text{LoS}} = \frac{K}{K+1}$ will be much greater than other paths, Ricean K factor is introduced to simulate the ratio between LoS path and total power. Specifically, the probability of having LoS path and factor K can be computed using Eqs. (2.49) and (2.50).

$$Pr_{\text{LoS}} = (\frac{300 - (d_{\text{MS-BS}})}{300}), \quad d_{\text{MS-BS}} < 300\ (m). \tag{2.49}$$

$$K[\text{dB}] = 13.0 - 0.03 \times (d_{\text{MS-BS}}). \tag{2.50}$$

In Eqs. (2.49) and (2.50), $d_{\text{MS-BS}}$ represents the distance between the base station and the user. Finally, AoD and AoA can be computed as:

$$\delta_{n,\text{AoD}} \sim \mathcal{U}(-40^\circ, 40^\circ), \quad n = 1, 2, \cdots, N. \tag{2.51}$$

$$\begin{aligned} &\delta_{n,\text{AoA}} \sim \mathcal{N}(0, \sigma_{n,\text{AoA}}^2), \quad n = 1, 2, \cdots, N; \\ &\sigma_{n,\text{AoA}} = 104.12^\circ \cdot (1 - e^{-0.265|10\log_{10}(P_n)|}). \end{aligned} \tag{2.52}$$

The generation method described earlier pertains to the macrocell scenario. However, if the LoS scenario is considered, the AoD and AoA of the LoS path follow the LoS direction between the base station and the user. The detailed parameter settings of SCM are provided in Table 2.9. In addition to the scenarios, the antenna properties of both the transmitter and receiver sides also play a crucial role in determining the relationship between the transmitted signal and the received signal. These properties are also defined in the SCM. Specifically, SCM defines the antenna pattern and antenna gain of the base station side as shown in Eq. (2.53) and Eq. (2.54), respectively.

$$A(\theta)[\text{dB}] = -\min[12(\frac{\theta}{\theta_{\text{3-dB}}})^2, A_m], -180^\circ \leq \theta < 180^\circ. \tag{2.53}$$

$$G(\theta) = 10^{0.1A(\theta)}. \tag{2.54}$$

TABLE 2.9
The channel scenarios and corresponding parameters in the SCM.

Channel scenario	Suburban macro	Urban macro	Urban micro
Number of paths (N)	6	6	6
Number of sub-paths (M) per path	20	20	20
Mean RMS AS at BS	$E(\sigma_{AS}) = 5^\circ$	$E(\sigma_{AS}) = 8^\circ, 15^\circ$	NLoS: $E(\sigma_{AS}) = 19^\circ$
AS at BS	$\mu_{AS} = 0.69$ $\varepsilon_{AS} = 0.13$	8° $\mu_{AS} = 0.810$ $\varepsilon_{AS} = 0.34$ 15° $\mu_{AS} = 1.18$ $\varepsilon_{AS} = 0.210$	N/A
$r_{AS} = \sigma_{AoD} / \sigma_{AS}$	1.2	1.3	N/A
Per-path AS at BS	2°	2°	5° (LoS and NLoS)
BS per-path AoD distribution standard deviation	$N(0, \sigma^2_{AoD})$ $\sigma_{AoD} = r_{AS} \cdot \sigma_{AS}$	$N(0, \sigma^2_{AoD})$ $\sigma_{AoD} = r_{AS} \cdot \sigma_{AS}$	$U(-40^\circ, 40^\circ)$
Mean RMS AS at MS	$E(\sigma_{AS,MS}) = 68^\circ$	$E(\sigma_{AS,MS}) = 68^\circ$	$E(\sigma_{AS,MS}) = 68^\circ$
Per-path AS at MS	35°	35°	35°
MS per-path AoA Distribution	$N(0, \sigma^2_{AoA})$	$N(0, \sigma^2_{AoA})$	$N(0, \sigma^2_{AoA})$
Delay spread	$\mu_{DS} = -6.80$ $\varepsilon_{DS} = 0.288$	$\mu_{DS} = -6.18$ $\varepsilon_{DS} = 0.18$	N/A
Mean total RMS delay spread	$E(\sigma_{DS}) = 0.17\ \mu s$	$E(\sigma_{DS}) = 0.65\ \mu s$	$E(\sigma_{DS}) = 0.251\ \mu s$
$r_{DS} = \sigma_{delays} / \sigma_{DS}$	1.4	1.7	N/A
Lognormal shadowing standard deviation	8 dB	8 dB	NLoS: 10 dB LoS: 4 dB
Pathloss model (dB) (d in meters)	$31.5 + 35\log_{10}(d)$	$34.5 + 35\log_{10}(d)$	NLoS: $34.53 + 38\log_{10}(d)$ LoS: $30.18 + 26\log_{10}(d)$

In Eqs. (2.53) and (2.54), θ represents the main radiation angle of the antenna, $\theta_{3\text{-dB}}$ is the 3-dB beamwidth, and A_m is the maximum attenuation. Considering an antenna pattern supporting three sectors, $\theta_{3\text{-dB}}$ will be set as 70° and A_m will be set as 20 dB, as shown in Figure 2.31(a). Considering an antenna pattern supporting six sectors, $\theta_{3\text{-dB}}$ will be set as 35° and A_m will be set as 23 dB, as shown in Figure 2.31(b). The antenna of the user is usually assumed to be omnidirectional. After completing the aforementioned parameter computations, the obtained results will be utilized to generate multi-antenna MIMO channel coefficients. In the context of a MIMO channel, where the transmitter is equipped with N_{T} antennas and the receiver has N_{R} antennas, as illustrated in Figure 2.32, the channel coefficient corresponding to the n th

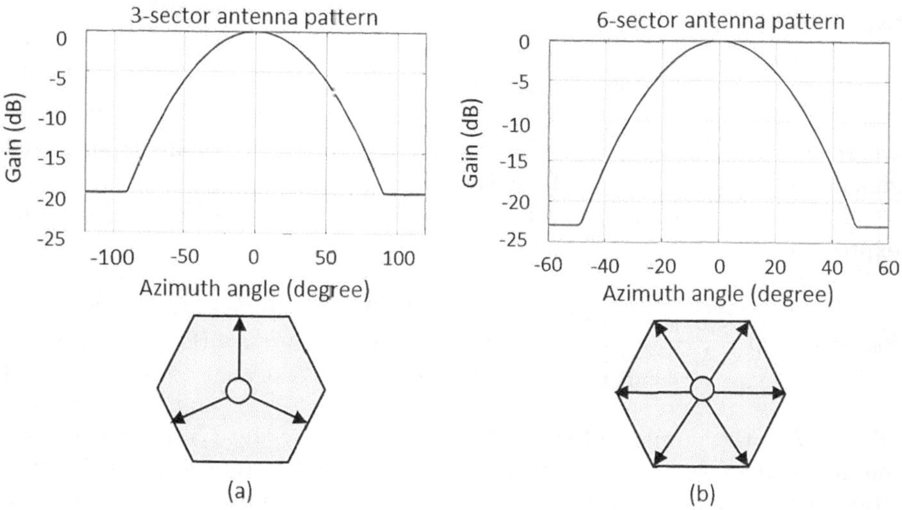

FIGURE 2.31
(a) Antenna pattern of a three-sector antenna. (b) Antenna pattern of a six-sector antenna.

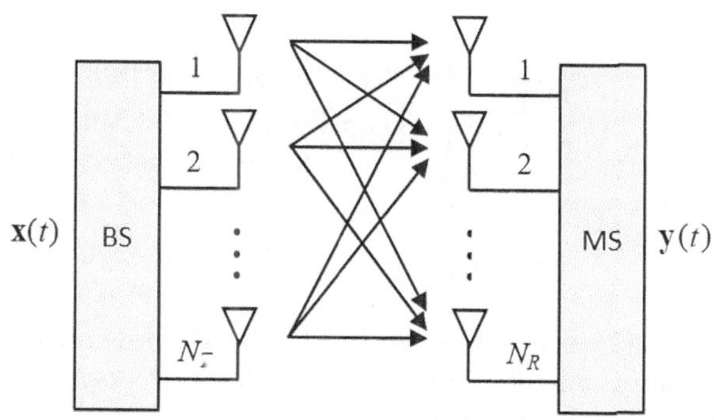

FIGURE 2.32
Illustration of a MIMO system with N_{T} transmitting antennas and N_{R} receiving antennas.

path can be represented by a $N_{\mathrm{T}} \times N_{\mathrm{R}}$ matrix, as shown in Eq. (2.55).

$$\mathbf{H}_n(t) = \begin{bmatrix} h_{1,1,n}(t) & h_{1,2,n}(t) & \cdots & h_{1,N_{\mathrm{T}},n}(t) \\ h_{2,1,n}(t) & h_{2,2,n}(t) & \cdots & h_{2,N_{\mathrm{T}},n}(t) \\ \vdots & \vdots & \ddots & \vdots \\ h_{N_{\mathrm{R}},1,n}(t) & h_{N_{\mathrm{R}},2,n}(t) & \cdots & h_{N_{\mathrm{R}},N_{\mathrm{T}},n}(t) \end{bmatrix}. \tag{2.55}$$

The relationship between the transmitted signal and the received signal is expressed as:

$$\mathbf{y}(t) = \sum_{n=1}^{N} \mathbf{H}_n(t)\mathbf{x}(t-\tau_n) + \mathbf{n}(t), \tag{2.56}$$

where $\mathbf{y}(t) \in \mathbb{C}^{N_{\mathrm{R}}\times 1}$ is the receive signal, $\mathbf{x}(t) \in \mathbb{C}^{N_{\mathrm{T}}\times 1}$ is the transmitted signal, and $\mathbf{n}(t) \in \mathbb{C}^{N_{\mathrm{R}}\times 1}$ is the noise.

If the NLoS scenario is considered, the (u, s) element of Eq. (2.55) can be expressed as:[6]

$$h_{u,s,n}(t) = \sqrt{\frac{P_n\sigma_{\mathrm{SF}}}{M}} \sum_{m=1}^{M} \begin{bmatrix} \sqrt{G_{\mathrm{BS}}(\theta_{n.m.\mathrm{AoD}})}\exp(j[kd_s\sin(\theta_{n.m.\mathrm{AoD}}) + \Phi_{n,m}])\times \\ \sqrt{G_{\mathrm{MS}}(\theta_{n.m.\mathrm{AoS}})}\exp(j[kd_u\sin(\theta_{n.m.\mathrm{AoA}})])\times \\ \exp(jk||v||\cos(\theta_{n.m.\mathrm{AoA}} - \theta_v)t) \end{bmatrix}, \tag{2.57}$$

where M is the number of subpaths of a path, G_{BS} and G_{MS} stand for the antenna gain of the base station and the user, d_s and d_u represent the distance from a reference point to the sth antenna of the base station and the distance from a reference point to the uth antenna of the user, $\Phi_{n,m}$ is the phase of mth subpath, $||v||$ and θ_v are the magnitude and direction of user velocity, and $k = 2\pi/\lambda$ is the number of waves. On the other hand, in the urban microcell LoS scenario is considered, the (u, s) element of Eq. (2.55) can be expressed as:

$$h^{\mathrm{LoS}}_{u,s,1}(t) = h_{u,s,1}(t) + \sqrt{\frac{K\sigma_{\mathrm{SF}}}{K+1}} \times \begin{bmatrix} \sqrt{G_{\mathrm{BS}}(\theta_{\mathrm{BS}})}\exp(j[kd_s\sin(\theta_{\mathrm{BS}})])\times \\ \sqrt{G_{\mathrm{MS}}(\theta_{\mathrm{MS}})}\exp(j[kd_u\sin(\theta_{\mathrm{MS}}) + \Phi_{\mathrm{LoS}}])\times \\ \exp(jk||v||\cos(\theta_{\mathrm{MS}} - \theta_v)t) \end{bmatrix}. \tag{2.58}$$

and

$$h^{\mathrm{LoS}}_{u,s,n}(t) = h_{u,s,n}(t), \quad n \neq 1. \tag{2.59}$$

In Eq. (2.58), $h_{u,s,n}(t)$ is as defined in Eq. (2.57). As previously mentioned, K is the Ricean K factor, θ_{BS}, θ_{MS}, and Φ_{LoS} are the AoD, AoA, and phase of LoS path, respectively. The biggest difference between Eq. (2.57) and Eq. (2.58) is that the LoS path is considered in $n = 1$ case with a power ratio $K/(K+1)$ of the total power, so that the sum power of remaining NLoS paths is $1/(K+1)$ of the total power. The whole parameter generation procedure is summarized in Figure 2.33.

[6]Please refer to ref. [81] for more details.

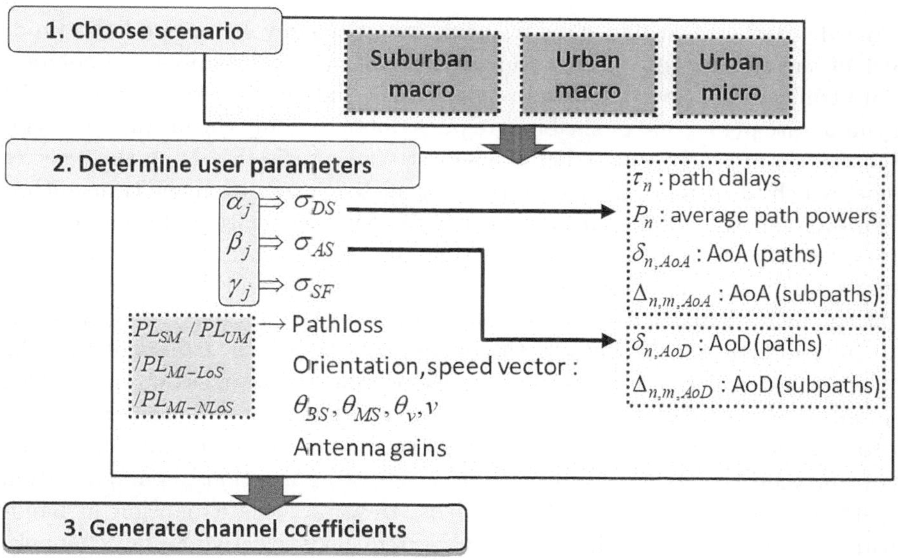

FIGURE 2.33
Illustration of the model parameter generation process in SCM.

3

OFDM Principles

3.1 OFDM Basics

Considerations regarding data transmission speed revolve around the product of the symbol rate per second and the bit rate per symbol. To enhance data transmission speed, two approaches can be adopted: either increasing the symbol rate per second or the bit rate per symbol. Raising the bit rate per symbol necessitates achieving a higher signal-to-noise ratio (SNR) on the receiver side, which in turn requires a higher transmitted power from the transmitter. However, such an increase may be limited by cost constraints and regulatory provisions. Conversely, augmenting the symbol rate per second entails employing a communication system with a wider bandwidth, thereby necessitating the mitigation of more pronounced frequency-selective fading channels to fully exploit the benefits of the expanded bandwidth. Consequently, the appropriate design of transceivers to ensure robust transmission quality even in the presence of severe frequency-selective fading channels has emerged as a prominent research area. To address the growing need for increased data transmission speeds to support a wide range of applications, communication systems must employ larger bandwidths. However, this expansion in bandwidth introduces the challenge of severe frequency-selective fading channels, resulting in inter-symbol interference (ISI). Traditional single-carrier systems require time-domain equalizers with longer durations to combat ISI, placing additional demands on the transceiver. Alternatively, the adoption of a multi-carrier system can alleviate the ISI effect by reducing the bandwidth assigned to each carrier, mitigating the need for demanding time-domain equalizers. In multi-carrier systems, as shown in Figure 3.1, the bandwidth allocated to each subcarrier is reduced, thereby reducing the impact of frequency-selective fading. In fact, the fading experienced by each subcarrier can be approximated as flat-fading, significantly reducing the complexity of the equalizer. In summary, as communication systems gravitate toward larger bandwidths, multi-carrier systems, particularly orthogonal frequency division multiplexing (OFDM) systems, are preferred over single-carrier systems. OFDM systems offer higher frequency efficiency and simpler transceiver architectures compared to other implementations. In this section, our objective is to introduce the concept and design considerations of OFDM systems, including

DOI: 10.1201/9781003408239-3

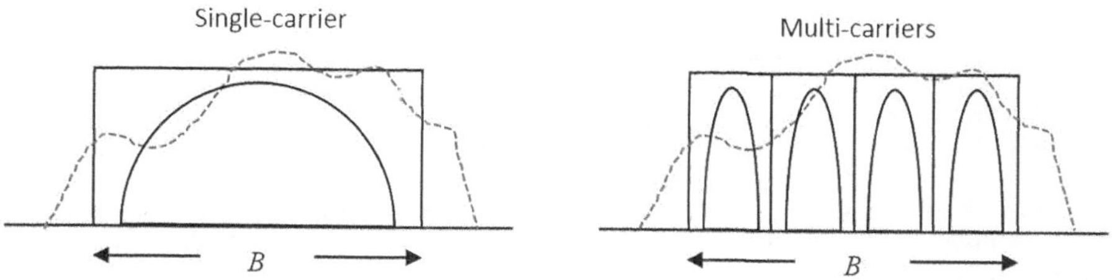

FIGURE 3.1
Illustration of a single-carrier signal spectrum and a multi-carrier signal spectrum.

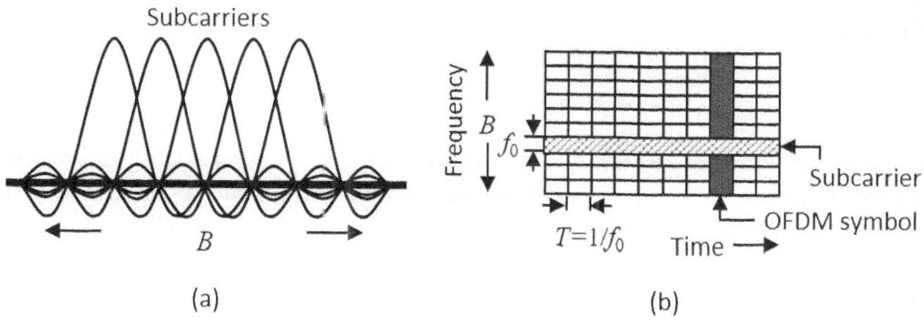

FIGURE 3.2
(a) Illustration of an OFDM signal spectrum. (b) Illustration of an OFDM signal allocation in the frequency and time domains.

peak-to-average power ratio (PAPR), signal detection, and channel estimation. We will also present mathematical models of OFDM systems for comprehensive understanding. Finally, we will expand our discussion to multi-user systems by introducing orthogonal frequency division multiplexing access (OFDMA) and single-carrier frequency division multiple access (SC-FDMA).

OFDM transceivers: As a multi-carrier transmission technology, OFDM systems allocate transmitted symbols to a set of orthogonal subcarriers. In particular, for two subcarriers with central frequencies f_1 and f_2, the orthogonality condition can be defined as follows:

$$\int_0^T e^{j2\pi f_1 t} \cdot e^{-j2\pi f_2 t} dt = 0. \tag{3.1}$$

In Eq. (3.1), we can further deduce that the orthogonality condition is satisfied when $f_1 - f_2 = n/T$, where $n \in \mathbb{R}$. This implies that the minimum frequency spacing between two different subcarriers is $f_0 = 1/T$. Therefore, when utilizing a set of orthogonal subcarriers for data transmission, the allocation of resources in the frequency and time domains is structured as depicted in Figure 3.2. In Figure 3.2(a), we illustrate the spectrum of an OFDM signal using

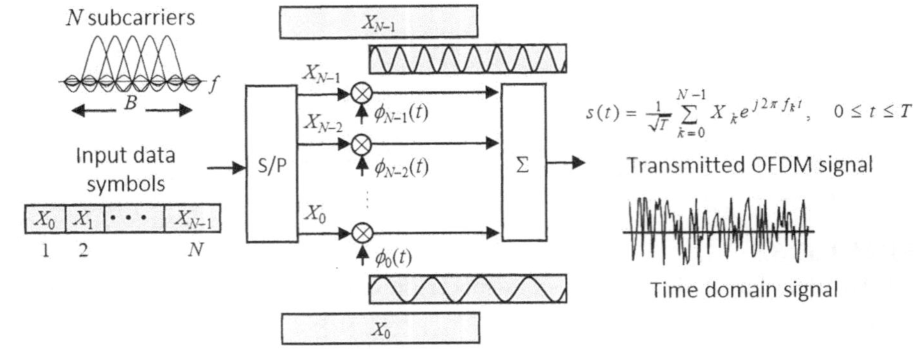

FIGURE 3.3
Illustration of the OFDM transmitter structure.

a rectangular window function. It is worth noting that the rectangular window function, which will be discussed later in this section, is the most fundamental window function used in OFDM systems. In such a system, the spectrum of each subcarrier takes the shape of a sinc function. Notably, the peak value of each sinc function occurs at the null values of the other sinc functions, thereby ensuring no mutual interference between any two subcarriers. Consequently, there is no requirement for an additional guard band to prevent interference. As a result, OFDM systems can achieve a higher spectrum utilization rate, which is a significant advantage of this system. The time-frequency resource allocation of a typical OFMA system is also illustrated in Figure 3.2(b) for reference, which will be further discussed in later sections.

The structure of an OFDM transmitter is depicted in Figure 3.3. The process of generating OFDM signals can be divided into the following steps: (1) Combination of N data symbols: The N data symbols, denoted as $X_0, X_1, \cdots, X_{N-1}$, are combined into a block. This block of symbols is then subjected to a serial-to-parallel conversion (S/P) operation to parallelize the symbols. (2) Orthogonal subcarrier modulation: The parallelized symbols are modulated using an orthogonal set of subcarriers, denoted as $\phi_0(t), \phi_1(t), \cdots, \phi_{N-1}(t)$. Each subcarrier is responsible for carrying one symbol, and the modulation process generates individual subcarrier signals. (3) Superposition of subcarrier signals: The individual subcarrier signals are superimposed to generate the final OFDM signal, denoted as $s(t)$. Based on the aforementioned steps, the transmitted OFDM signals can be expressed as follows:

$$s(t) = \begin{cases} \sum_{k=0}^{N-1} X_k \phi_k(t), & 0 \leq t \leq T, \\ 0, & \text{otherwise}, \end{cases} \tag{3.2}$$

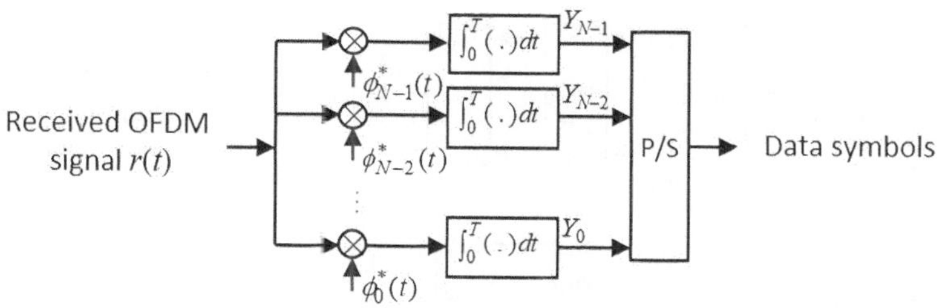

FIGURE 3.4
Illustration of the OFDM receiver structure.

where T is the OFDM signal length and $\phi_k(t)$ is the subcarrier function as:

$$\phi_k(t) = \frac{1}{\sqrt{T}} e^{j2\pi \frac{k}{T} t}. \tag{3.3}$$

In contrast, the structure of an OFDM receiver is illustrated in Figure 3.4. The steps involved in receiving OFDM signals can be divided as follows: (1) Matched filtering: The received signals are fed into the corresponding matched filters of the N subcarriers. Each subcarrier signal is processed by its respective matched filter. (2) Combination of matched filter outputs: The outputs of the N matched filters are combined to form a block of received symbols, denoted as $Y_0, Y_1, \cdots, Y_{N-1}$. This block of symbols is then subjected to a parallel-to-serial conversion (P/S) operation to convert them into a sequentialized format. If we assume that there is no channel distortion or noise present at this stage, the received signal can be considered as equal to the transmitted signal, i.e., $r(t) = s(t)$. In this case, the operation of the matched filter can be expressed as follows:

$$Y_i = \int_0^T s(t)\phi_i^*(t)dt = \frac{1}{T}\sum_{k=0}^{N-1} X_k \int_0^T e^{j2\pi \frac{k-i}{T} t} = X_i. \tag{3.4}$$

In Eq. (3.4), it is evident that only the symbol corresponding to $k = i$ can be successfully received. This is a consequence of the orthogonality property inherent in the subcarrier sets used in OFDM systems. The orthogonality ensures that symbols transmitted by different subcarriers do not interfere with each other during transmission. As a result, the OFDM system can employ corresponding matched filters to receive symbols from all subcarriers without any mutual interference. This property greatly simplifies the implementation of OFDM systems and contributes to their ease of use.

After introducing the concept of OFDM transceivers for analog signal transmission, we now extend the application of OFDM systems to digital signal transmission. Digital signal transmission is the predominant mode of data communication in contemporary wireless and mobile communications.

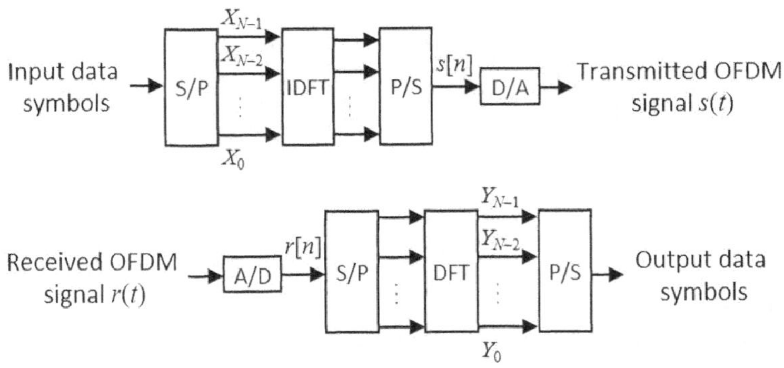

FIGURE 3.5
Illustration of the transceiver structure for digital OFDM systems.

The OFDM systems designed for digital signal transmission can be mathematically expressed as follows:

$$s[n] = s(t)|_{t=\frac{nT}{N}} \begin{cases} \frac{1}{N} \sum_{k=0}^{N-1} X_k e^{j2\pi \frac{k}{N} n}, & 0 \leq n \leq N-1, \\ 0, & \text{otherwise}, \end{cases} = \text{IDFT}\{X_k\}. \quad (3.5)$$

In Eq. (3.5), it is evident that $s[n]$ represents the inverse discrete Fourier transform (IDFT) results of X_k. Likewise, if we disregard the influence of the channel and noise at this stage, the received signal can be represented as $r[n] = s[n]$. Consequently, the operation of the matched filter, as described in Eq. (3.4), can be adapted to the digital domain as follows:

$$Y_i = \text{DFT}\{r[n]\} = \sum_{n=0}^{N-1} s[n]e^{-j2\pi \frac{i}{N} n} = \sum_{k=0}^{N-1} X_k \delta[k-i] = X_i. \quad (3.6)$$

In the digital version, the matched filters in the analog domain are substituted with DFT operations in the digital OFDM systems. This replacement maintains the property that symbols transmitted by different subcarriers do not interfere with each other during transmission, enabling successful reception in OFDM systems. The transceiver structure for digital OFDM systems is depicted in Figure 3.5. The generation of OFDM signals involves the following steps: (1) Combining N data symbols into a block $X_0, X_1, \cdots, X_{N-1}$ and performing S/P to parallelize the symbols. (2) Utilizing IDFT operations to generate N time-domain signals $s[0], s[1], \cdots, s[N-1]$, followed by P/S to generate digital signals. (3) Converting the digital signals to analog signals using a digital-to-analog converter (DAC) to obtain the analog OFDM signals for actual data transmission. The reception of OFDM signals involves the following steps: (1) Converting the received signals to digital form using

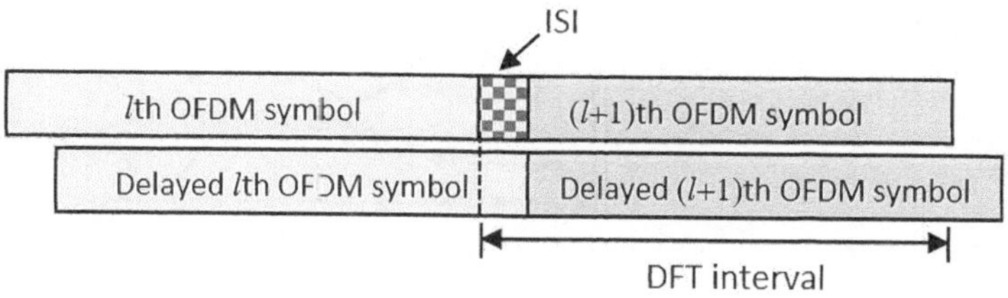

FIGURE 3.6
Illustration of the ISI in OFDM systems.

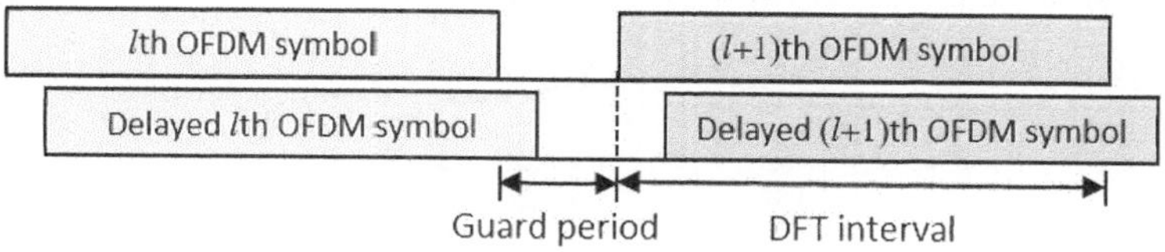

FIGURE 3.7
Illustration of guard periods in OFDM systems.

an analog-to-digital converter (ADC) to obtain digital received signals. (2) Sequentializing the digital received signals using P/S, and applying DFT operations to obtain the parallelized transmitted symbols. (3) Performing P/S on the parallelized transmitted symbols to output the data symbol sequences. In the literature, the terms "OFDM signals" and "OFDM symbols" are used interchangeably. Therefore, the processes described above can be referred to as OFDM modulation and demodulation processes. In other words, OFDM transmitters modulate N data symbols into an OFDM symbol, while OFDM receivers demodulate the OFDM symbol to retrieve the N data symbols. In this book, "OFDM signals" refer to signals without a specific format, whereas "OFDM symbols" refer to a data block containing N data symbols (e.g., $s(t)$ in Eq. (3.2) or $s[n]$ in Eq. (3.5)).

In the previous discussions of OFDM systems, we focused on the ideal scenario without considering the effects of the underlying channel. However, in practical implementations, the presence of multipath channels can introduce interference between adjacent OFDM symbols, leading to ISI, as shown in Figure 3.6. To mitigate the ISI effect, a guard period, also known as an idle period, is introduced between adjacent OFDM symbols, as depicted in Figure 3.7. As long as the duration of the guard period is longer than the channel duration, OFDM symbols will not interfere with each other. While the guard period effectively addresses the ISI effect, it also disrupts the orthogonality of OFDM subcarriers, as shown in Figure 3.8. Consider a channel with two paths: if the OFDM receiver uses the arrival time of the first path as the reference for performing DFT demodulation, the subcarrier corresponding to the second path will encounter the idle period during the DFT duration,

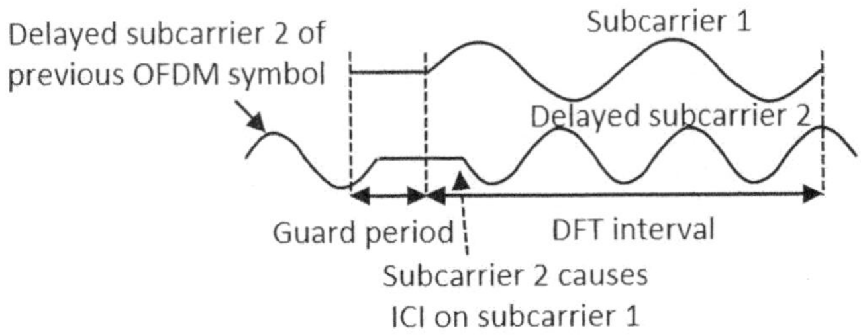

FIGURE 3.8
Illustration of guard periods and ICI in OFDM systems.

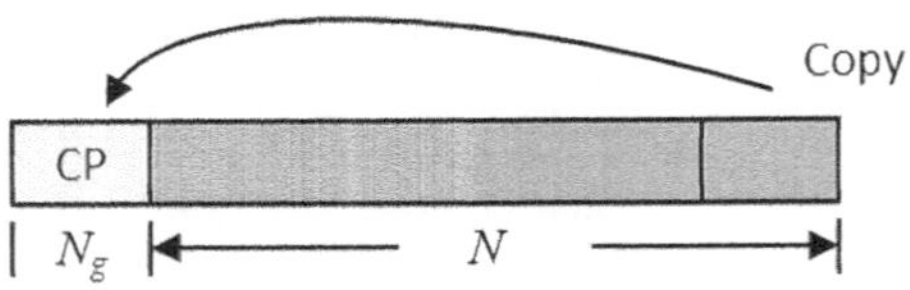

FIGURE 3.9
Illustration of CP in OFDM systems.

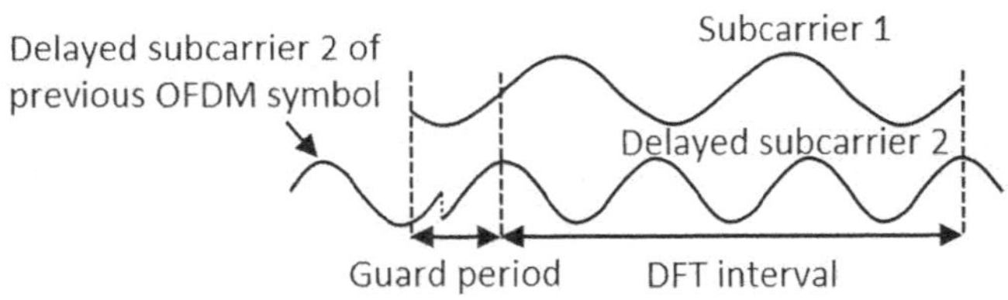

FIGURE 3.10
The impact of CP to subcarriers in OFDM systems.

resulting in an incomplete subcarrier. Consequently, the orthogonality between different subcarriers from different paths is compromised, leading to intercarrier interference (ICI) and affecting the accurate reception of symbols from different subcarriers.

To maintain the orthogonality among subcarriers and ensure the integrity of subcarrier waveforms across different paths during the DFT duration, OFDM transceivers employ a technique known as cyclic prefix (CP). This involves appending a sequence of N_{g} symbols, copied from the tail of the signals, to the beginning of the transmitted signals, as depicted in Figure 3.9. The CP acts as a guard interval that eliminates ISI and helps restore the orthogonality of subcarriers. Specifically, by incorporating the CP, the subcarriers traversing different paths can maintain their completeness, thereby preserving the orthogonality among subcarriers and preventing interference within the DFT duration, as shown in Figure 3.10. It is important to note that the duration of the CP in OFDM systems should exceed the channel duration to

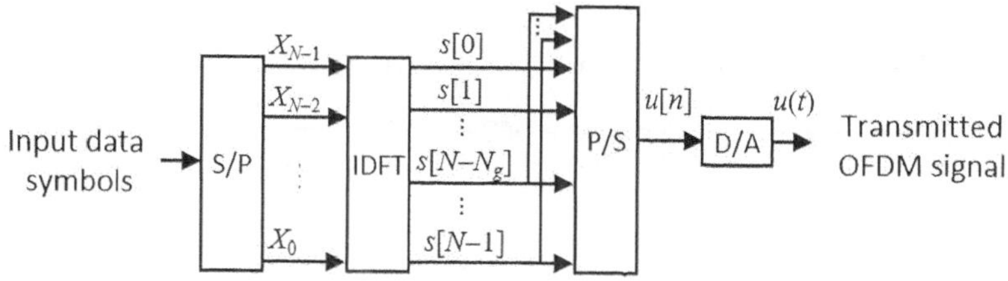

FIGURE 3.11
Illustration of the digital OFDM transmitter structure with CP mechanism.

avoid the presence of ISI due to an insufficient CP length. By properly designing the duration of the CP, the receiver can remove the interference caused by the channel and achieve accurate symbol detection from each subcarrier.

The transmitter architecture for digital OFDM systems incorporating the CP functionality is illustrated in Figure 3.11. In this configuration, the last N_g symbols of the IDFT time-domain signals $s[0], s[1], \cdots, s[N-1]$ are redundantly placed at the beginning of the transmitted signals. Subsequently, the signals undergo DAC to generate analog signals. The resulting waveform, consisting of the original OFDM symbols preceded by the CP, is referred to as a complete OFDM symbol, as depicted in Figure 3.12. The signal model for this configuration can be represented as follows:

$$u[n] = \begin{cases} \dfrac{1}{N} \displaystyle\sum_{k=0}^{N-1} X_k e^{j2\pi \frac{k}{N} n - N_g}, & 0 \leq n \leq N + N_g - 1, \\ 0, & \text{otherwise}, \end{cases} \tag{3.7}$$

The portion of the transmitted signal containing the original OFDM symbols is referred to as the useful part. The receiver architecture for digital OFDM systems incorporating the CP functionality is depicted in Figure 3.13. Once the received signals are converted into digital form through ADC, the CP is removed before inputting the signals into the DFT operations to extract the OFDM symbols in each subcarrier. It is important to note that precise timing synchronization is crucial in this step to accurately determine the starting position of the CP. Inaccurate timing synchronization can result in performance degradation as it may lead to the removal of some portions of the useful parts along with the CP.

The CP mechanism in OFDM systems serves multiple purposes, including the mitigation of ISI and ICI, while also simplifying the complexity involved in the equalization process of OFDM transceivers. When disregarding the influence of noise, the received signal, denoted as $v[n]$, is produced after the transmission of complete OFDM symbols through the channel. This can be mathematically expressed as follows:

$$v[n] = u[n] * h[n], 0 \leq n \leq N + N_g + L_h - 2, \tag{3.8}$$

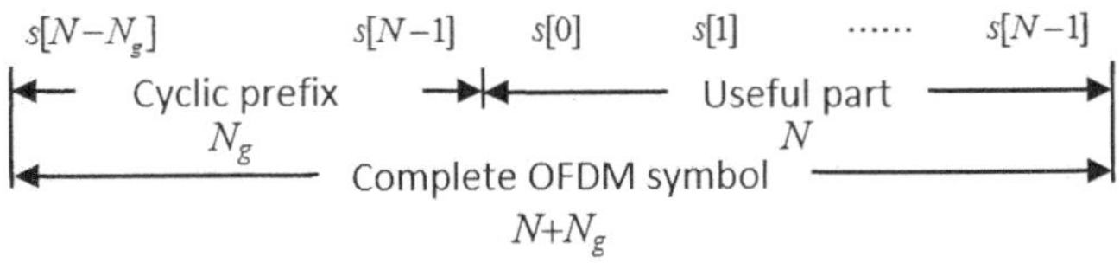

FIGURE 3.12
Illustration of the complete OFDM symbols.

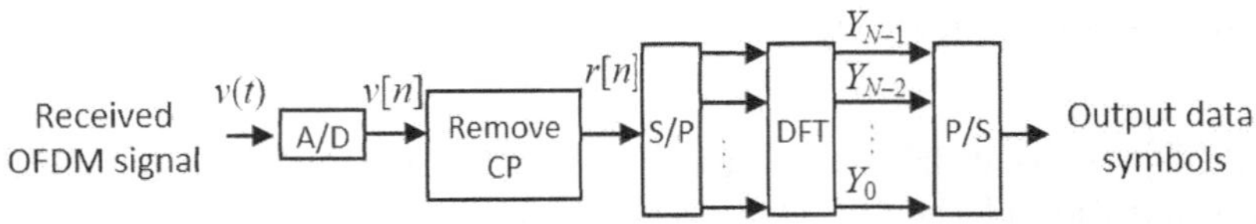

FIGURE 3.13
Illustration of the digital OFDM receiver structure with CP mechanism.

where $h[n]$ is the channel impulse response and L_h is the channel length. When the length of the CP exceeds the length of the channel, upon removal of the CP, the linear convolution between the transmitted signal $u[n]$ and the channel impulse response $h[n]$ as depicted in Eq. (3.8) undergoes a transformation. Specifically, this linear convolution is converted into the circular convolution operation between the original OFDM symbol $s[n]$ and $h[n]$. Mathematically, this can be expressed as follows:

$$r[n] = s[n] \odot_N h[n], 0 \leq n \leq N-1. \tag{3.9}$$

By applying DFT operations on $r[n]$ in Eq. (3.9), we can exploit the property of circular convolution to derive the following result:

$$Y_k = \text{DFT}\{r[n]\} = X_k H_k, 0 \leq k \leq N-1, \tag{3.10}$$

where

$$H_k = \text{DFT}\{h[n]\}\text{and } X_k = \text{DFT}\{s[n]\};, 0 \leq k \leq N-1. \tag{3.11}$$

From Eq. (3.10), it is worth noting that to recover the transmitted data symbol, the transceiver only needs to eliminate the channel's effect on each subcarrier.

$$\hat{X}_k = Y_k / H_k. \tag{3.12}$$

The aforementioned discussions highlight an important observation: in cases where the CP length surpasses the channel length, OFDM systems can employ a simple one-tap equalizer to mitigate the channel's effect. This serves as another significant function of the CP in OFDM systems. By leveraging the orthogonality between subcarriers and the CP mechanism, OFDM systems effectively mitigate the impact of ISI in frequency-selective fading environments.

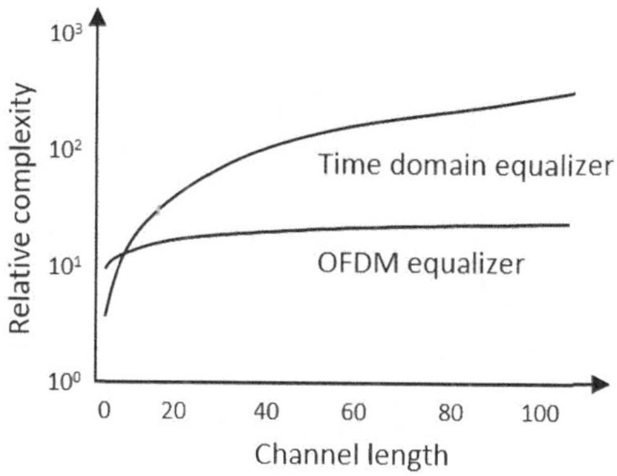

FIGURE 3.14
Illustration of the equalization complexity of OFDM systems and single-carrier systems.

Moreover, OFDM systems significantly reduce the complexity of equalization by simplifying signal models. This stands in stark contrast to single-carrier systems. Consequently, OFDM systems enhance the performance and flexibility of communication systems with wider bandwidth, which is a key factor contributing to their status as the predominant technology for high-speed mobile communications. Figure 3.14 illustrates a comparison of equalization complexity between single-carrier and OFDM systems. Notably, the equalization complexity of OFDM systems increases linearly with the channel length, whereas that of single-carrier systems exhibits an exponential growth with the channel length.

In the previous derivations, we made the assumption that OFDM systems employ rectangular window functions, resulting in the subcarrier spectrum being a superposition of sinc functions. However, sinc functions exhibit slow sidelobe decay, which can potentially cause interference to users operating in adjacent bandwidths when transmitting OFDM signals. This interference is illustrated in Figure 3.15. To address this limitation, conventional OFDM systems utilize a carefully designed window function, which is multiplied with the transmitted signals. This window function aims to reduce the sidelobe effect and minimize interference with other users. Consequently, the OFDM symbols can be expressed as follows when the window function is applied:

$$u[n] = \frac{1}{N} g_T[n] \sum_{k=0}^{N-1} X_k e^{j2\pi \frac{k}{N}(n-N_{\text{g}})}, 0 \leq n \leq N + N_{\text{g}} + N_{\text{w}} - 1, \tag{3.13}$$

where $g_T[n]$ is the window function with length $N + N_{\text{g}} + N_{\text{w}} - 1$, and N_{w} is the edge length of the smooth extension part on both sides. The raised cosine function is the commonly used window function of OFDM systems,

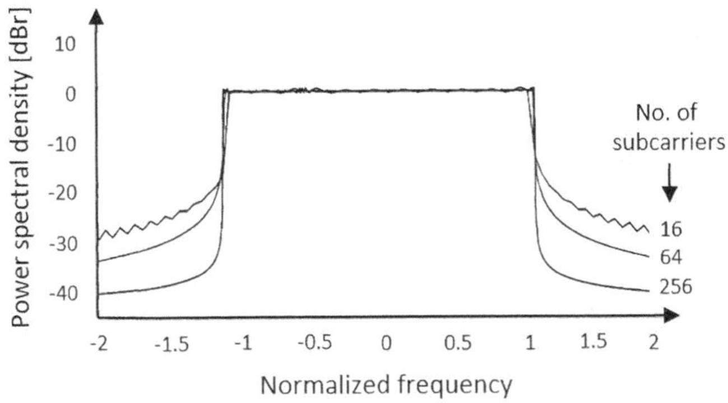

FIGURE 3.15
Illustration of the signal power spectrum in OFDM systems.

being expressed as:

$$g_T[n] = \begin{cases} \frac{1}{2} + \frac{1}{2}\cos[\pi + \frac{n\pi}{\beta(N+N_{\mathrm{g}})}], & 0 \leq n \leq N_{\mathrm{w}} - 1, \\ 1, & N_{\mathrm{w}} \leq n \leq N + N_{\mathrm{g}} - 1, \\ \frac{1}{2} + \frac{1}{2}\cos[\frac{[n-(N+N_{\mathrm{g}})]\pi}{\beta(N+N_{\mathrm{g}})}], & N + N_{\mathrm{g}} \leq n \leq N + N_{\mathrm{g}} + N_{\mathrm{w}} - 1, \end{cases} \tag{3.14}$$

In the given expression, the parameter $\beta = N_{\mathrm{w}}/(N + N_{\mathrm{g}})$ represents the roll-off factor, which can be utilized to control the rate at which the sidelobes decrease. A higher roll-off factor results in a more rapid decrease in sidelobe levels. However, it also leads to a larger OFDM symbol size and more pronounced distortion at the symbol edges. Conversely, a lower roll-off factor causes slower sidelobe decay, which can result in increased interference to other users. Therefore, OFDM system designers must strike a suitable balance between these two considerations during implementation. Figure 3.16 illustrates a comparison of the OFDM spectrum using different window functions with varying roll-off factors, showcasing the impact of these factors on sidelobe characteristics.

The complete architecture of an OFDM transceiver is depicted in Figure 3.17. By utilizing orthogonal subcarriers for data transmission, the adverse effects of frequency-selective fading on each subcarrier are significantly mitigated. As a result, OFDM systems can accommodate larger delay spreads and employ longer OFDM symbol lengths. Additionally, the use of well-established DFT and IDFT operations facilitates efficient digital implementation of OFDM systems, enabling faster execution times. The introduction of the CP mechanism in OFDM systems effectively eliminates ISI between adjacent OFDM symbols while maintaining the orthogonality between subcarriers. This, in turn, simplifies the equalization process by enabling the use

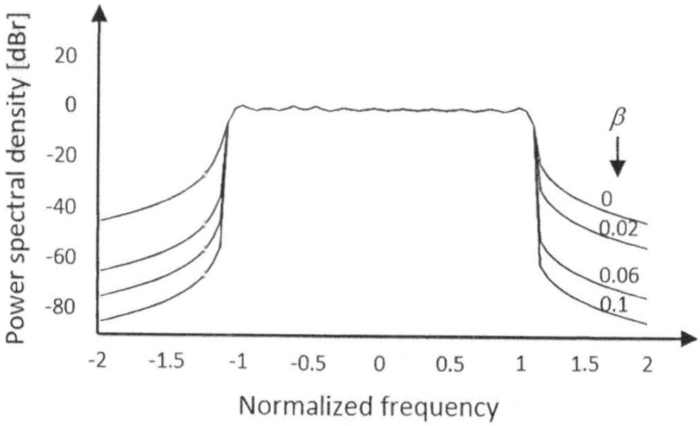

FIGURE 3.16
Illustration of the roll-off factor effect to signal power spectrum in OFDM systems.

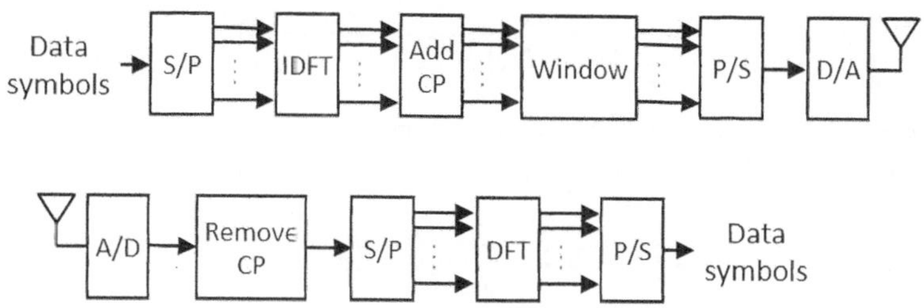

FIGURE 3.17
Illustration of the comprehensive transceiver structure in OFDM systems.

of one-tap equalizers, leading to a substantial reduction in the complexity of OFDM receivers. In conclusion, OFDM systems offer significant flexibility in the implementation of mobile communication systems. Given that different subcarriers experience distinct channel conditions, dynamic transmission control techniques can be employed to optimize system performance based on the prevailing channel conditions. For example, dynamic and adaptive adjustment of frequency and/or power resources can enhance system throughput. It is worth noting that the SNR definition of OFDM systems differs from that of conventional communication systems due to the presence of the CP and guard bands. Specifically, the SNR definition in conventional communication systems can be expressed as follows:

$$\mathrm{SNR} = \frac{P_{\mathrm{signal}}}{P_{\mathrm{noise}}} = \frac{E_{\mathrm{s}} R_{\mathrm{s}}}{N_0 W}, \tag{3.15}$$

where E_{s} is the data symbol power, R_{s} is the data symbol transmission rate, N_0 is the noise power spectrum density, and W is the system bandwidth.

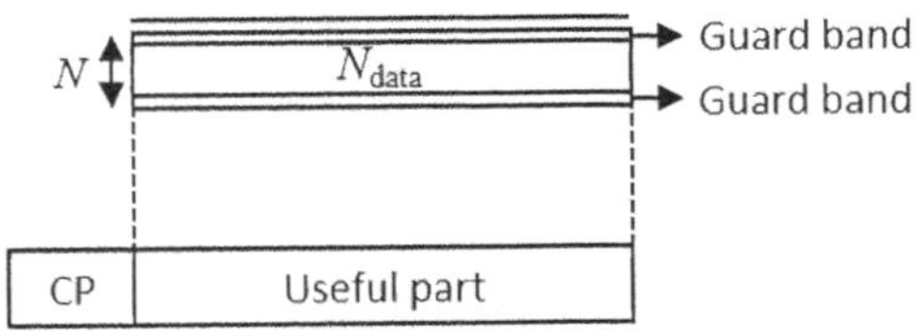

FIGURE 3.18
Illustration of the useful part for data transmission in OFDM systems.

However, in OFDM systems, a certain number of subcarriers are intentionally reserved and not utilized for signal transmission, forming the guard band. This is done to mitigate interference. As shown in Figure 3.18, assuming the number of effective subcarriers used for data transmission is denoted as N_{data}, the length of the DFT is denoted as N, and the length of the CP is denoted as N_{g}, the data symbol transmission rate can be expressed as follows:

$$R_{\text{s}} = \frac{N_{\text{data}}}{(N+N_{\text{g}})\frac{T}{N}} = \frac{N_{\text{data}}}{(N+N_{\text{g}})}W, \tag{3.16}$$

where $T = N/W$ is the OFDM symbol period.

By taking Eq. (3.15) into Eq. (3.16), we can obtain

$$\text{SNR} = \frac{E_{\text{s}}}{N_0}\frac{N_{\text{data}}}{N+N_{\text{g}}}. \tag{3.17}$$

In Eq. (3.17), it is evident that the effective SNR of OFDM systems is slightly reduced due to the energy allocated to the CP and guard band mechanisms. This reduction in effective SNR is attributed to the fact that OFDM systems allocate a portion of the overall transmitted energy for these purposes, resulting in a slightly diminished SNR available for data transmission.

3.2 The PAPR Problems in OFDM Systems and Solutions

In the previously discussed OFDM systems, we established that an OFDM symbol comprises N data symbols and is transmitted by the superposition of N subcarriers. Due to the random nature of the data symbols, the superposition of these N subcarriers will also exhibit randomness, resulting in random constructive or destructive interference. As a consequence, the strength of the OFDM signals will fluctuate, giving rise to what is known as the PAPR issue. Mathematically, the PAPR is defined as the ratio between the maximum instantaneous power and the average power, and it can be expressed as follows:

$$\text{PAPR}\{s[n]\} = \frac{|s[n]|^2}{\mathbb{E}\{|s[n]|^2\}}. \tag{3.18}$$

A larger PAPR corresponds to greater variations in signal strength. Consequently, to ensure the proper functioning of the power amplifier, its operational range needs to be reduced to avoid distortions. This reduction in operational range leads to decreased efficiency of the power amplifier, as it operates at a lower average power level. Moreover, the increased PAPR also poses challenges for the ADC and DAC. To maintain similar levels of quantization errors, the ADC and DAC must utilize a larger number of bits, which results in increased costs associated with employing these components. In summary, a larger PAPR in OFDM systems can have implications such as reduced power amplifier efficiency, narrower operational range, and increased costs for ADC/DAC due to the need for higher-resolution quantization. Managing the PAPR is an important consideration in OFDM system design to mitigate these challenges. Considering an OFDM symbol comprising of N subcarriers, the instantaneous power of the OFDM symbol can reach up to N times the average power of the OFDM signal. To illustrate this, let's consider an example where $N = 4$ and the OFDM symbol is generated by 4 binary phase shift keying (BPSK) symbols. Table 3.1 presents all possible PAPR values for this scenario. It can be observed that the maximum PAPR is 6 dB, which corresponds to the logarithm of $N = 4$. In practical OFDM systems, where larger values of N are employed (often reaching hundreds or even thousands), severe PAPR issues need to be addressed. Here, we introduce three solutions to mitigate PAPR in current OFDM systems: PAPR lowering solutions with signal distortion, PAPR lowering solutions without signal distortion, and the linear amplification with nonlinear components (LINC) solution.

TABLE 3.1
The PAPR values with different BPSK symbol combinations in OFDM systems.

Data block X	PAPR of IDFT{X} (dB)	Data block X	PAPR of IDFT{X} (dB)
[1,1,1,1]	6.0	[−1,1,1,1]	0
[1,1,1,−1]	0	[−1,1,1,−1]	3.0
[1,1,−1,1]	0	[−1,1,−1,1]	6.0
[1,1,−1,−1]	3.0	[−1,1,−1,−1]	0
[1,−1,1,1]	0	[−1,−1,1,1]	3.0
[1,−1,1,−1]	6.0	[−1,−1,1,−1]	0
[1,−1,−1,1]	3.0	[−1,−1,−1,1]	0
[1,−1,−1,−1]	0	[−1,−1,−1,−1]	6.0

PAPR lowering solution with signal distortion: The PAPR lowering solution with signal distortion aims to mitigate PAPR by removing or reducing signal components above a certain threshold, which may introduce signal distortions. The key advantage of this approach is its simplicity and ease of implementation. One common method used in practice is the clipping technique, where portions of the signal exceeding a specified threshold, denoted as A, are replaced with the threshold value A. Mathematically, this can be expressed as follows:

$$s[n] = \begin{cases} s[n], & \|s[n]\| \leq A, \\ Ae^{j\phi\{s[n]\}}, & \|s[n]\| > A. \end{cases} \tag{3.19}$$

The clipping method, although effective in reducing PAPR, introduces nonlinear distortions to the signal. An improved approach is the peak windowing method, which aims to reduce signal strength above a given threshold while mitigating distortions. Instead of directly removing the signal portions exceeding the threshold, the peak windowing method applies a predefined window function to these portions, effectively reducing their signal strength to the threshold value. The window function introduces a smoothing effect, helping to mitigate distortions caused by the signal processing. However, it is important to note that this method may still introduce out-of-band interference. Figure 3.19 provides an illustration of the results obtained using the peak windowing method. A comparison with the clipping method reveals that the peak windowing method effectively reduces the out-of-band interference of OFDM symbols. However, it is also evident that the mitigated interference is still higher compared to distortionless signals. In summary, the peak windowing method offers an improved solution compared to the clipping method by reducing distortions and out-of-band interference. However, it is important to consider the trade-off between PAPR reduction and the level of interference introduced in OFDM systems.

Another solution for reducing PAPR with minimal signal distortion is the peak cancellation method, illustrated in Figure 3.20. In this method, the signal $s[n]$ obtained after applying the IDFT undergoes a peak detection operation to identify signal portions that exceed a specified threshold A in terms of signal strength. The detected peaks provide information about their amplitudes and locations, which are then utilized to generate a cancellation signal with corresponding amplitude and phase values. By subtracting this cancellation signal from the original signal, the peak values can be reduced, thereby achieving the desired goal. The operations can be expressed as:

$$z[n] = s[n] - \sum_i a_i e^{j\phi_i} f[n - n_i], \tag{3.20}$$

where $f(.)$ is the pre-designed impulse function, a_i is the amplitude of the peak signal portions above the signal strength A, and ϕ_i is the phase of the peak signal portions above the signal strength A. The peak cancellation method,

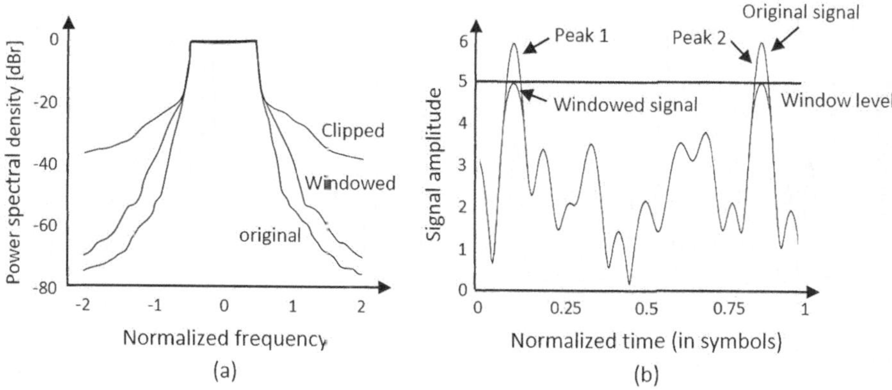

FIGURE 3.19
Illustration of the results obtained using the peak windowing method in OFDM systems. (a) OFDM symbol power spectrum. (b) OFDM time-domain signal.

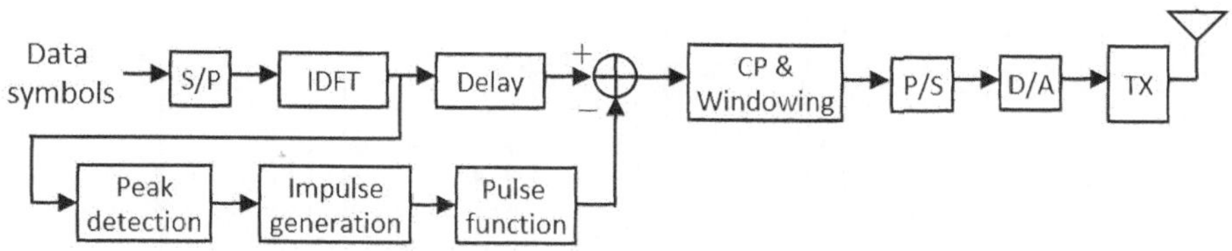

FIGURE 3.20
Illustration of the peak cancellation method structure when implemented.

as demonstrated in Figure 3.21, can effectively mitigate out-of-band interference associated with PAPR problems through careful design of the impulse function. By appropriately designing the impulse function, this method ensures that any undesired out-of-band interference is minimized or eliminated. The figure visually illustrates how the peak cancellation method achieves this objective.

PAPR lowering solution without signal distortion: The objective of PAPR reduction solutions without signal distortion is to modify the original signal according to certain rules, resulting in a form with reduced PAPR. However, these modifying rules need to be known by the receiver, requiring additional information to be transmitted during implementation. This section introduces three PAPR reduction techniques without signal distortion: partial transmit sequence (PTS), selected mapping (SLM), and tone reservation (TR). The architecture of PTS[1] is shown in Figure 3.22. In PTS, the original data symbol blocks are divided into non-overlapping blocks $\boldsymbol{X}_1, \boldsymbol{X}_2, \ldots, \boldsymbol{X}_M$,

[1]Please refer to ref. [82] for more details.

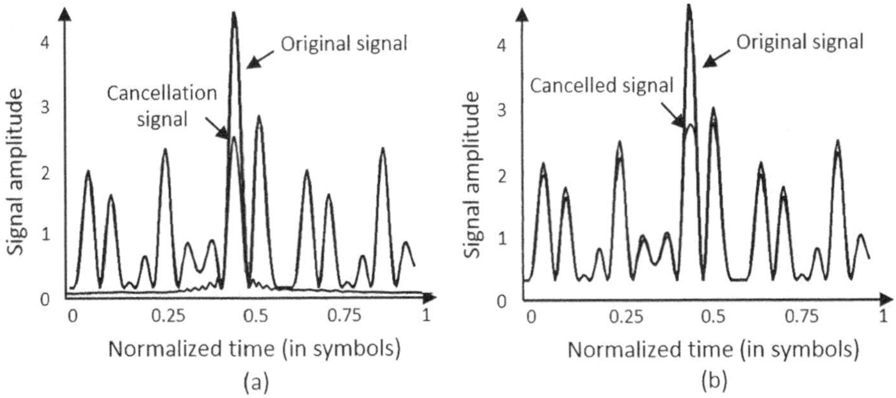

FIGURE 3.21
Illustration of the results obtained using the peak cancellation method in OFDM systems. (a) OFDM time-domain signal before cancellation. (b) OFDM time-domain signal after cancellation.

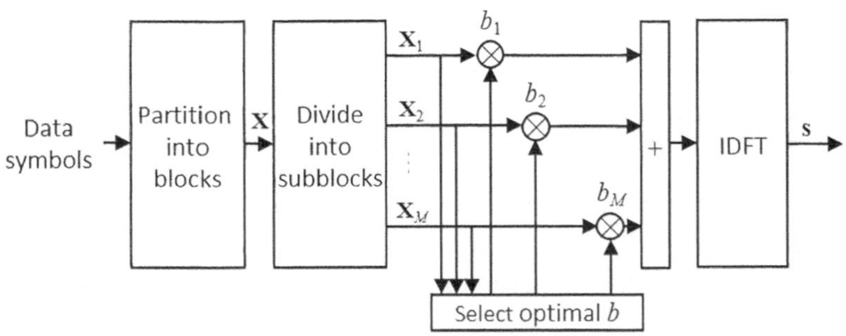

FIGURE 3.22
Illustration of the partial transmit sequence structure when implemented.

which are then multiplied by different coefficients $b_1, b_2, \ldots, b_M$ and fed into the IDFT to generate OFDM symbols. By optimizing the coefficients, PAPR can be reduced through the corresponding transformations. A larger value of M generally leads to lower PAPR but also increases complexity. The receiver also requires the coefficients sequence to successfully restore the original signals, and this is achieved in current OFDM systems by using a codebook stored in advance at both the transmitter and receiver. The transmitter selects a sequence from the codebook and informs the receiver of the sequence number, significantly reducing the overhead of transmitting the entire coefficient sequence. The architecture of SLM[2] is shown in Figure 3.23. SLM multiplies the original data symbol blocks $\boldsymbol{X}$ by different phase

[2]Please refer to ref. [82] for more details.

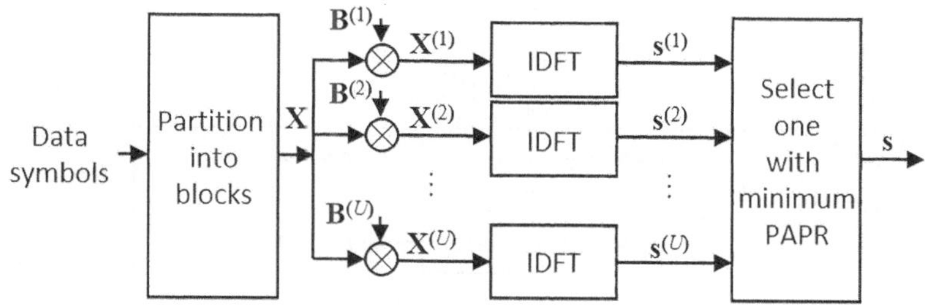

FIGURE 3.23
Illustration of the selected mapping structure when implemented.

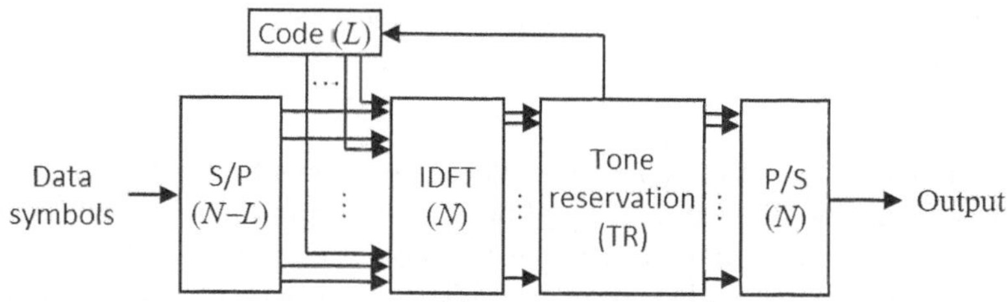

FIGURE 3.24
Illustration of the tone reservation structure when implemented.

sequences $\boldsymbol{B}^{(1)}, \boldsymbol{B}^{(2)}, \ldots, \boldsymbol{B}^{(M)}$ to generate different candidate symbol blocks $\boldsymbol{X}^{(1)}, \boldsymbol{X}^{(2)}, \ldots, \boldsymbol{X}^{(M)}$. These candidates are then fed into the IDFT, and the one with the lowest PAPR is selected as the transformed OFDM symbol $\boldsymbol{s}$. Similar to PTS, SLM requires the transmission of phase sequence information to the receiver in advance. Finally, the architecture of TR[3] is shown in Figure 3.24. TR preserves a portion of subcarriers to reduce the degree of freedom for PAPR reduction. In TR-based OFDM systems, only $N-L$ subcarriers out of the total N subcarriers are used for data transmission, while L subcarriers are reserved to transmit non-data symbols, thus lowering PAPR. However, this design compromises spectrum efficiency since some subcarriers are dedicated to non-data transmission. Consequently, adjusting the value of L to achieve a good trade-off between PAPR and spectrum efficiency becomes an important design consideration in such systems.

Linear amplification with nonlinear components: The LINC solution[4] utilizes a special transformation to convert high-PAPR original signals into equivalent signals with lower PAPR. This technique employs power amplifiers to amplify these equivalent signals, as depicted in Figure 3.25, and combines the amplified results to form the final OFDM signals. LINC is capable of decomposing the original varying-envelope OFDM signals into two signal

[3]Please refer to ref. [82] for more details.
[4]Please refer to ref. [83] for more details.

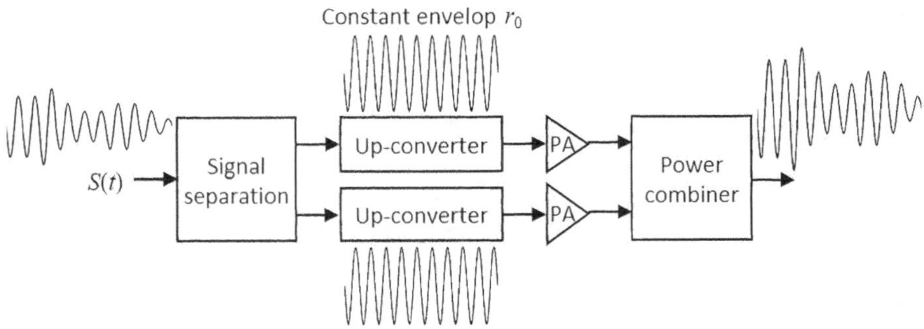

FIGURE 3.25
Illustration of the linear amplification with nonlinear components structure when implemented.

components with constant envelope. These components are then separately passed through identical power amplifiers, and a combiner is used to reconstruct the amplified signals. Since the two signal components have constant envelope characteristics, the power amplifiers can operate within a fixed linear range. An additional advantage of the LINC solution is the potential to achieve higher amplification gains by employing nonlinear amplifiers without introducing signal distortion. In Figure 3.25, the transmit signal $s(t)$ can be decomposed as follows:

$$s(t) = A(t) \cdot e^{j\phi(t)} = \underbrace{\frac{1}{2} r_0 e^{j(\phi(t)+\theta(t))}}_{s_1(t)} + \underbrace{\frac{1}{2} r_0 e^{j(\phi(t)-\theta(t))}}_{s_2(t)}, \tag{3.21}$$

In the LINC solution, the original signals are represented by their envelope $A(t)$, phase $\phi(t)$, signal component 1 $s_1(t)$, and signal component 2 $s_2(t)$. The phase relation is given by $\theta(t) = \cos^{-1}(A(t)/r_0)$, where r_0 is a constant satisfying $r_0 \geq A_{\max}$ and $A_{\max}$ is the maximum value of $|A(t)|$. Eq. (3.21) represents a general expression that can be applied to decompose any signals. Following amplification by the power amplifiers and combining of the amplified components, the final transmitted signals can be expressed as:

$$L \cdot [G \cdot s_1(t) + G \cdot s_2(t)] = L \cdot G \cdot 2 \cdot s(t), \tag{3.22}$$

In the LINC solution, the amplifier gain is denoted by G, and L represents the signal loss after the combiner processing. For instance, in the case of a common Wilkinson combiner, the value of L is set as $1/\sqrt{2}$ in Eq. (3.22). Consequently, the final transmit signals can be expressed as $\sqrt{2}Gs(t)$. The combiner efficiency can be defined as:

$$\eta_c(t) = \frac{|\sqrt{2}Gs(t)|^2}{|Gs_1(t)|^2 + |Gs_2(t)|^2} = \cos^2\theta(t) = (\frac{A(t)}{r_0})^2. \tag{3.23}$$

TABLE 3.2
The comparison of different PAPR lowering solutions in OFDM systems.

	Power increase	Data rate loss	Required processing
Signal-distortion method	No	No	Tx: amplitude clipping, filtering Rx: none
PTS	No	Yes	Tx: M IDFTs, side information delivery Rx: inverse PTS
SLM	No	Yes	Tx: U IDFTs, side information delivery Rx: inverse SLM
TR	Yes	Yes	Tx: non-data subcarriers processing Rx: ignore non-data subcarriers
LINC	No	No	Tx: additional hardware Rx: none

In Eq. (3.23), it can be observed that the amplifier gain does not affect the PAPR, while the combiner efficiency varies with the magnitude of the envelope. Consequently, a higher PAPR results in a less stable combiner efficiency. This issue becomes a key research focus when developing LINC solutions.

The selection of PAPR lowering solutions: When selecting a PAPR reduction solution for OFDM systems, it is important to consider practical aspects beyond just PAPR performance. Factors such as the impact on transmit power, transmit speed, error rate, and implementation complexity should also be taken into account. For example, in the case of the TR solution, a portion of subcarriers is reserved for transmitting non-data symbols, which results in a reduction in transmit speed. Additionally, to maintain a similar transmit speed and error rate, OFDM systems employing TR may need to increase the transmit power. When using PTS or SLM solutions, extra bandwidth is required to transmit additional information to the receiver, leading to a decrease in transmit speed. If the received extra information is not error-free, it may result in the incorrect restoration of the original signals at the receiver and an increase in the error rate. Moreover, PTS and SLM entail higher computational complexity compared to other solutions, making them less suitable for implementation on the user side. On the other hand, LINC solutions do not impose such negative impacts on OFDM systems, but they do require additional hardware for execution. In summary, PAPR reduction solutions with signal distortion and LINC solutions can be directly employed in existing OFDM systems without modifications to the receiver, whereas other solutions necessitate adjustments to system specifications. Table 3.2 provides an overview of the impact on OFDM systems and the required resources when employing different PAPR reduction solutions.

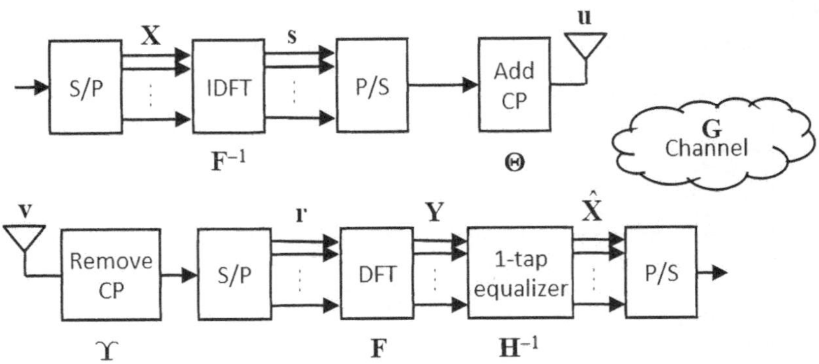

FIGURE 3.26
Illustration of the OFDM transceivers and the matrix signal models.

3.3 OFDM Transceiver Designs

In this section, we will introduce the OFDM signal model to provide a mathematical perspective on OFDM systems. We will analyze the properties and performance of these systems. Additionally, we will discuss methods for improving OFDM systems and delve into the issues of ISI and ICI in OFDM systems.

OFDM systems signal model: Figure 3.26 illustrates the OFDM system architecture that we will discuss in this section. The distinction between Figures 3.26 and 3.17 lies in the representation of underlying signals and operations in a matrix format. In this representation, the transmitted and received OFDM signals are denoted as vectors, while the operations within the OFDM systems are expressed as matrices. Additionally, uppercase letters are used to indicate signals in the frequency domain, while lowercase letters represent signals in the time domain. Throughout this section, we assume the utilization of a rectangular window function and disregard the influence of noise. Moreover, we do not differentiate between sequential and parallel OFDM signals in this discussion. In the transmitter depicted in Figure 3.26, the data symbol vector is represented as $\mathbf{X}$, with dimensions of $N \times 1$. This vector can be expressed as:

$$\mathbf{X} = [X_0, X_1, \cdots, X_{N-1}]^T, \tag{3.24}$$

where $\mathbf{X}_n$ represents the data symbol placed in the n subcarrier. After IDFT operation, we obtain "useful" OFDM symbol vector $\mathbf{s}$, being expressed as:

$$\mathbf{s} = \mathbf{F}^{-1} \cdot \mathbf{X} = [s[0], s[1], \cdots, s[N-1]]^T, \tag{3.25}$$

In Eq. (3.25), $\mathbf{F}$ is a $N \times N$ DFT matrix. Hence, the matrix $\mathbf{F}^{-1}$ represents the IDFT operation. The mathematical expression of the DFT matrix is given

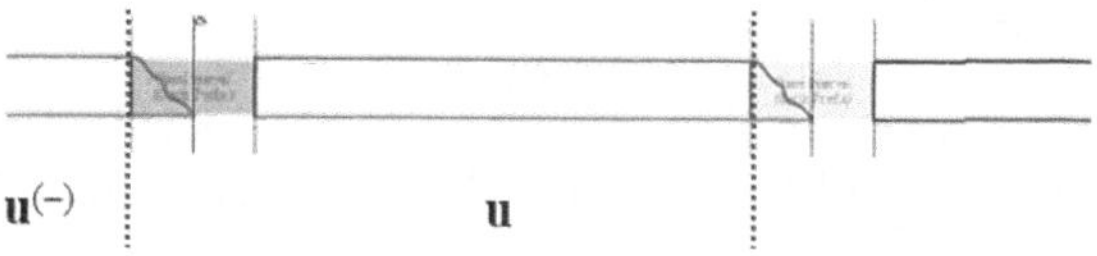

FIGURE 3.27
Illustration of the OFDM received signals and ISI effect.

by:

$$\mathbf{F} = \begin{bmatrix} 1 & 1 & \cdots & 1 \\ 1 & W^{1\cdot 1} & \cdots & W^{(N-1)\cdot 1} \\ \vdots & \vdots & \ddots & \vdots \\ 1 & W^{1\cdot(N-1)} & \cdots & W^{(N-1)\cdot(N-1)} \end{bmatrix}, W = e^{-j\frac{2\pi}{N}}. \tag{3.26}$$

By adding a CP with a length of N_g to the vector $\mathbf{s}$, the resulting "completed" OFDM symbol vector $\mathbf{u}$, with dimensions of $(N + N_g) \times 1$, can be expressed as:

$$\mathbf{u} = \mathbf{\Theta} \cdot \mathbf{s} = [s[N - N_g], \cdots, s[N-1], s[0], \cdots, s[N-1]]^T, \tag{3.27}$$

where

$$\mathbf{\Theta} = \begin{bmatrix} \mathbf{O}_{N_g\times(N-N_g)} & \mathbf{I}_{N_g} \\ \mathbf{I}_N & \end{bmatrix}_{(N+N_g)\times N}. \tag{3.28}$$

To implement the CP operations, the most recent N_g symbols are redundantly placed at the beginning of the signal to form the vector $\mathbf{u}$. This vector represents the final transmitted signal from the transmitter. At the receiver, the received signal is denoted as $\mathbf{v}$ and has dimensions of $(N + N_g) \times 1$. Please note that the actual length of the received signal may be longer than $N + N_g$ due to the effects of the channel. However, for the purpose of our discussions, we will disregard this portion since we are employing CP. The received signal $\mathbf{v}$ can be expressed as:

$$\mathbf{v} = \mathbf{G} \cdot \mathbf{u} + \mathbf{G}^{(-)} \cdot \mathbf{u}^{(-)} = [\mathbf{v}[0], \mathbf{v}[1], \cdots, \mathbf{v}[N + N_g - 1]]^{\mathrm{T}} \tag{3.29}$$

where $\mathbf{G}$ is a channel matrix with the dimension of $(N + N_g) \times (N + N_g)$, representing the channel impact on the transmitted vector. $\mathbf{G}^{(-)}$ corresponds to the channel matrix associated with the previously transmitted signal vector $\mathbf{u}^{(-)}$. Eq. (3.29) takes into account the influence of the previous OFDM signal on the current transmitted signal, resulting in the presence of the ISI effect, as illustrated in Figure 3.27. The region between the two dotted lines represents $\mathbf{u}$, and it can be observed that $\mathbf{u}^{(-)}$ extends into this region due to channel delay, thereby generating ISI effect. In the received signal $\mathbf{v}$, each element $v[n]$ can be described as the convolution result of the transmitted signal and channel vector, being expressed as:

$$v[n] = \sum_{i=0}^{L_h-1} h[i]u[n-i] + h[i]u^{(-)}[n+N+N_{\mathrm{g}}-i], \tag{3.30}$$

where L_h is the channel length. By expressing Eq. (3.30) in a matrix form, we can obtain a channel matrix $\mathbf{G}$ as:

$$\boldsymbol{G} = \begin{bmatrix} h[0] & 0 & \cdots & 0 & 0 & 0 & 0 \\ h[1] & h[0] & \ddots & \vdots & 0 & 0 & 0 \\ \vdots & h[1] & \ddots & 0 & \vdots & 0 & 0 \\ h[L_h-1] & \vdots & \ddots & h[0] & 0 & \vdots & 0 \\ 0 & h[L_h-1] & \vdots & h[1] & h[0] & 0 & \vdots \\ \vdots & \vdots & \ddots & \vdots & h[1] & \ddots & 0 \\ h[0] & 0 & \cdots & h[L_h-1] & \cdots & h[1] & h[0] \end{bmatrix}. \tag{3.31}$$

From Eq. (3.31), it can be observed that $\mathbf{G}$ is a Toeplitz matrix with dimensions $(N+N_{\mathrm{g}}) \times (N+N_{\mathrm{g}})$, meaning that each column can be obtained by shifting down one slot, reflecting the effect of channel delay. Following the same concept, we can also obtain $\mathbf{G}^{(-)}$, which is also a Toeplitz matrix with dimensions $(N+N_{\mathrm{g}}) \times (N+N_{\mathrm{g}})$, given by:

$$\mathbf{G} = \begin{bmatrix} 0 & \cdots & h[L_h-1] & \cdots & h[1] \\ \vdots & \vdots & \vdots & \ddots & \vdots \\ 0 & \cdots & 0 & \vdots & h[L_h-1] \\ \vdots & \vdots & \vdots & \ddots & \vdots \\ 0 & \cdots & 0 & \cdots & 0 \end{bmatrix}. \tag{3.32}$$

One can also observe that $\mathbf{G}^{(-)}$ consists of the lower right part of $\mathbf{G}$ and is padded with zeros. This reflects the fact that ISI is caused by the end portion of the previous OFDM symbol. At the receiver, by removing the CP portion of $\mathbf{v}$, we can obtain a "useful" received signal vector $\mathbf{r}$ with dimensions $N \times 1$, which can be expressed as:

$$\begin{aligned} \mathbf{r} &= \boldsymbol{\Upsilon} \cdot \boldsymbol{v} = \boldsymbol{\Upsilon} \cdot \mathbf{G} \cdot \mathbf{u} + \boldsymbol{\Upsilon} \cdot \mathbf{G}^{(-)} \cdot \mathbf{u}^{(-)} \\ &= \boldsymbol{\Upsilon} \cdot \mathbf{G} \cdot \mathbf{u} = [r[0], r[1], \cdots, r[N-1]]^T, \end{aligned} \tag{3.33}$$

where

$$\boldsymbol{\Upsilon} = [\mathbf{O}_{N \times N_g} \ \ \mathbf{I}_{N \times N}] \tag{3.34}$$

is applied to remove the CP operation. By examining the structure of $\mathbf{G}^{(-)}$, we can determine that the ISI effect caused by the previous OFDM symbol is

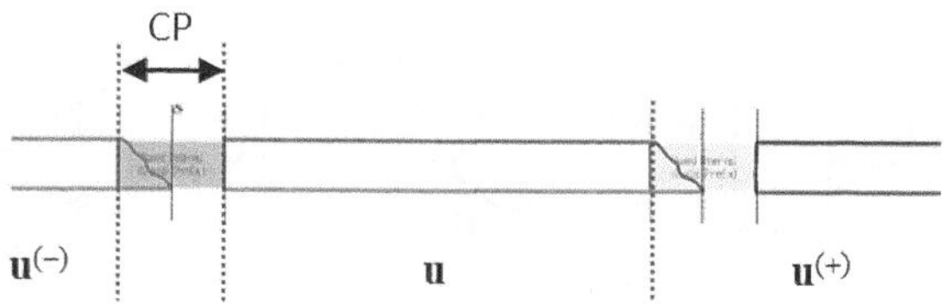

FIGURE 3.28
Illustration of the OFDM received signals and ISI removal process.

limited to the first $L_h - 1$ elements of the current OFDM symbol. Therefore, if $N_g > L_h - 1$ is satisfied, the ISI effect can be fully addressed after removing the cyclic prefix CP, as depicted in Figure 3.28. (Since the impact of the previous signal does not extend beyond the CP region, indicated by the gray region, the ISI effect can be eliminated after CP removal.) By applying the DFT operation to the received signal $\mathbf{r}$, we can obtain a frequency domain signal block vector $\mathbf{Y}$ with dimensions of $N \times 1$, which can be expressed as:

$$\begin{aligned}\mathbf{Y} &= \mathbf{F}\cdot\mathbf{r} = \mathbf{F}\cdot\boldsymbol{\Upsilon}\cdot\mathbf{G}\cdot\mathbf{u} = \mathbf{F}\cdot\boldsymbol{\Upsilon}\cdot\mathbf{G}\cdot\boldsymbol{\Theta}\cdot\mathbf{F}^{-1}\cdot\mathbf{X}\\ &= [Y_0, Y_1, \cdots, Y_{N-1}]^T,\end{aligned} \tag{3.35}$$

In Eq. (3.36), the effect of adding CP, removing CP, and channel can be expressed as an equivalent channel as:

$$\mathbf{H}_e = \boldsymbol{\Upsilon}\cdot\mathbf{G}\cdot\boldsymbol{\Theta}$$
$$= \begin{bmatrix} h[0] & 0 & 0 & 0 & h[L_h-1] & \cdots & h[1] \\ h[1] & h[0] & 0 & \vdots & 0 & 0 & h[2] \\ \vdots & h[1] & \ddots & 0 & \vdots & 0 & \vdots \\ h[L_h-1] & \vdots & \ddots & h[0] & 0 & \vdots & h[L_h-1] \\ 0 & h[L_h-1] & \vdots & h[1] & h[0] & 0 & \vdots \\ \vdots & \vdots & \ddots & \vdots & \ddots & \ddots & 0 \\ 0 & 0 & \cdots & h[L_h-1] & h[L_h-2] & \cdots & h[0] \end{bmatrix}\cdot \tag{3.36}$$

$\mathbf{H}_e$ is a $N \times N$ circulant matrix, which can be diagonalized by a DFT matrix as:

$$\mathbf{G} = \begin{bmatrix} H_0 & 0 & \cdots & 0 \\ 0 & H_1 & \vdots & 0 \\ \vdots & \vdots & \ddots & \vdots \\ 0 & 0 & \vdots & H_{N-1} \end{bmatrix} \tag{3.37}$$

and the diagonal element $\{H_k\}$ is the eigenvalues of $\mathbf{H}_e$. By substituting Eqs. (3.36) and (3.37) back intc Eq. (3.35), we can derive a simplified model for

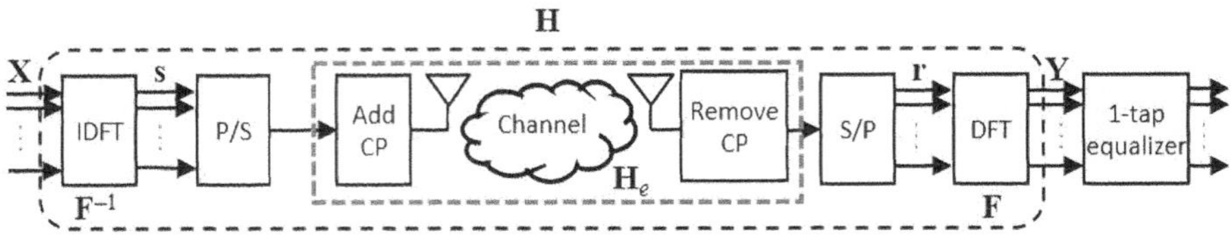

FIGURE 3.29
Illustration of an OFDM transceiver and the corresponding equivalent matrix signal model.

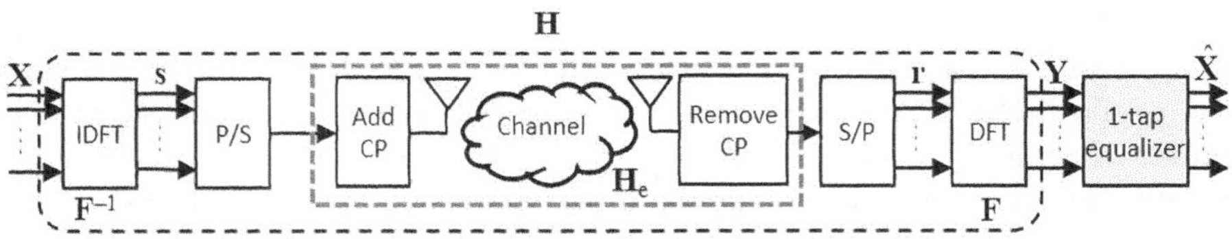

FIGURE 3.30
Illustration of an OFDM transceiver and the equalizer architecture.

the OFDM transceiver as:

$$\mathbf{Y} = \mathbf{H} \cdot \mathbf{X}, \tag{3.38}$$

where $\mathbf{H}$ is a diagonal matrix, indicating that the data symbols on different subcarriers do not interfere with each other. This result aligns with Eq. (3.10). Further analysis of Eq. (3.37) reveals that the diagonal elements of $\mathbf{H}$ correspond to the frequency response of the channel, representing the channel's behavior at different subcarriers.

$$H_k = \text{DFT}\{h[n]\}, \quad k = 0, \cdots, N-1. \tag{3.39}$$

The architecture of the previously discussed OFDM systems is illustrated in Figure 3.29.

OFDM systems equalization: In this section, we will discuss OFDM equalizers based on the aforementioned OFDM signal model. The objective of OFDM equalizers is to recover the transmitted data symbol vector $\mathbf{X}$ from the received signal vector $\mathbf{Y}$, as depicted in Figure 3.30. In this context, $\mathbf{D}$ represents the matrix representation of the equalizer. When taking into account the presence of noise, the received signal vector $\mathbf{Y}$ can be expressed as:

$$\mathbf{Y} = \mathbf{H} \cdot \mathbf{X} + \mathbf{N}, \tag{3.40}$$

where $\mathbf{N}$ is the noise vector, following independent and identically distributed (i.i.d.) complex Gaussian distribution with zero mean and variance σ_n^2, and has dimensions of $N \times 1$. After passing through the equalizer $\mathbf{E}$, the estimated data symbol vector can be obtained as:

$$\hat{\mathbf{X}} = \mathbf{E} \cdot \mathbf{X}. \tag{3.41}$$

When it comes to equalizer designs, let's begin by introducing the zero-forcing (ZF) equalizer, which aims to completely eliminate the channel effect. The mathematical expression of the ZF equalizer is the inverse matrix of the channel matrix $\mathbf{H}$ and can be represented as:

$$\mathbf{E}_{\text{ZF}} = \mathbf{H}^{-1}. \tag{3.42}$$

By applying the ZF equalizer, the data symbol estimate vector can be expressed as:

$$\begin{aligned} \hat{\mathbf{X}} &= \mathbf{H}^{-1} \cdot \mathbf{Y} = \mathbf{X} + \mathbf{H}^{-1} \cdot \mathbf{N} \\ \hat{X}_k &= H_k^{-1} Y_k = X_k + H_k^{-1} N_k, \quad k = 0, \cdots, N-1. \end{aligned} \tag{3.43}$$

In Eq. (3.43), it can be observed that the ZF equalizer can be simplified to a first-order equalizer when the equivalent channel matrix $\mathbf{H}$ is diagonal. After equalization, the SNR of the data symbol on the k-th subcarrier is given by:

$$\text{SNR}_{\text{ZF},k} = |H_k|^2 \frac{\sigma_X^2}{\sigma_n^2}, \tag{3.44}$$

where $\sigma_X^2 = \mathbb{E}\{|X_k|^2\}$ represents the power of the data symbol on the k-th subcarrier. Unlike the ZF equalizer, the minimum mean square error (MMSE) equalizer is designed to minimize the square error between the data symbol vector and the data symbol estimate vector. Mathematically, it can be expressed as:[5]

$$\mathbf{E}_{\text{MS}} = \arg \min_{\mathbf{E}} \mathbb{E}\{||\mathbf{X} - \mathbf{E}\mathbf{Y}||^2\} = (\mathbf{H}^H \mathbf{H} + \frac{\sigma_n^2}{\sigma_X^2} \mathbf{I}_N)^{-1} \mathbf{H}^H, \tag{3.45}$$

The key distinction between the ZF equalizer and the MMSE equalizer lies in their considerations of the channel and noise effects. The MMSE equalizer takes into account both the channel and noise effects simultaneously, and its mathematical expression is given as:

$$\begin{aligned} \hat{\mathbf{X}} &= (\mathbf{X}^H \mathbf{X} + \frac{\sigma_n^2}{\sigma_X^2} \mathbf{I}_N)^{-1} \mathbf{H}^H \mathbf{Y} \\ \hat{X}_k &= \frac{\sigma_X^2 |H_k|^2}{\sigma_X^2 |H_k|^2 + \sigma_n^2} X_k + \frac{\sigma_X^2 |H_k|^2}{\sigma_X^2 H_k^* + \sigma_n^2} N_k, \quad k = 0, \cdots, N-1. \end{aligned} \tag{3.46}$$

After the equalization process, the SNR of the data symbol on the k subcarrier can be expressed as:

$$\text{SNR}_{\text{MS},k} = |H_k|^2 \frac{\sigma_X^2}{\sigma_n^2}. \tag{3.47}$$

[5]Here, min represents the operation of minimizing the objective function, while arg min indicates selecting the input of the function that results in the minimum value of the objective function.

By comparing Eqs. (3.44) and (3.47), it becomes apparent that the SNR after ZF equalization and MMSE equalization are equal. This finding contradicts the common understanding that the MMSE equalizer outperforms the ZF equalizer in terms of performance. The reason behind this phenomenon is the absence of ISI during OFDM transmission. Consequently, the ZF equalizer does not introduce any noise enhancement during the ISI cancellation process. With this understanding, it is possible to simplify the equalization process by adopting a ZF equalizer instead of an MMSE equalizer in OFDM systems.

OFDM systems enhancement: According to the signal model of the OFDM transceiver presented earlier, we understand that each subcarrier forms an independent transmission channel without mutual interference. Exploiting this property, OFDM systems can enhance system performance by appropriately allocating transmit power to different subcarriers, maximizing the overall system capacity. This can be formulated as the following optimization problem:[6]

$$\begin{aligned} &\max_{p_0,p_1,\cdots,p_{N-1}} \sum_{k=0}^{N-1} \log_2(1+\frac{p_k|H_k|^2}{\sigma_n^2}) \\ &\text{subject to} \sum_{k=0}^{N-1} p_k \leq P,\ p_k \geq 0,\ k=0,\cdots,N-1, \end{aligned} \tag{3.48}$$

where p_k is the transmit power of kth subcarrier and P is the maximum transmit power constraint. Using the Lagrange multiplier method, Eq. (3.48) can be solved and the solution can be expressed as:[7]

$$p_k = \frac{1}{\lambda} - \frac{\sigma_n^2}{|H_k|^2}; \sum_{k=0}^{N-1}(\frac{1}{\lambda} - \frac{\sigma_n^2}{|H_k|^2})^+ = P. \tag{3.49}$$

The solution presented in Eq. (3.49) follows the principle of water-filling. In this approach, the transmit power allocation to each subcarrier is determined based on the inverse proportionality between $\sigma_n^2/|H_k|^2$ and the channel gain. Specifically, subcarriers with better channel conditions are allocated higher transmit power, while subcarriers with poorer channel conditions receive lower transmit power or may even be abandoned. The analogy of a container with an uneven bottom surface helps illustrate the concept. Imagine that the water level represents the available transmit power, and the bottom surface of the container corresponds to the channel gains. The height between the water surface and the bottom surface indicates the power allocation for each subcarrier. With a higher bottom surface (weaker channel gain), the allocated power will be smaller. If the bottom surface is higher than the water surface, no transmit power will be allocated to that subcarrier, as shown in Figure 3.31.

[6] Please refer to ref. [84] for more details; The term max is an abbreviation for "maximize," indicating the objective function seeks to achieve the maximum value.

[7] $(x)^+ = x$ if $x > 0$, $(x)^+ = 0$ if $x < 0$.

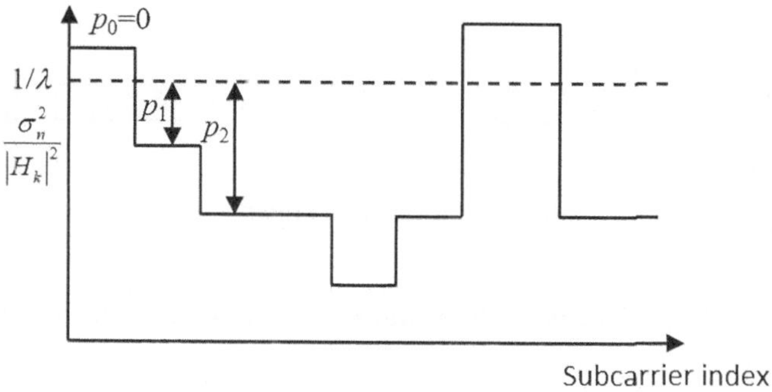

FIGURE 3.31
Illustration of the water-filling power allocation in OFDM systems.

This water-filling solution ensures that transmit power is efficiently allocated across subcarriers based on their channel conditions, optimizing system capacity while adhering to the power constraint. The analogy provides a visual representation of this power allocation process.

OFDM systems ISI and ICI issues: In the previous discussion, we assumed the perfect case of the OFDM transceiver without considering the presence of ISI and ICI effects. However, in practical OFDM systems, these effects can arise due to various imperfections, such as insufficient CP length or carrier frequency offset (CFO). Insufficient CP length results in both ISI and ICI effects, while CFO primarily causes ICI effect. These effects will be the main focus of this section. When the CP length is insufficient, we have the condition $N_g < L_h - 1$. The received signal in Eq. (3.35) can be revised using Eqs. (3.29) and (3.33) as follows:

$$\mathbf{Y} = \mathbf{F} \cdot \boldsymbol{\Upsilon} \cdot \mathbf{G} \cdot \mathbf{u} + \mathbf{F} \cdot \boldsymbol{\Upsilon} \cdot \mathbf{G}^{(-)} \cdot \mathbf{u}^{(-)} + \mathbf{N}. \tag{3.50}$$

By substituting $\mathbf{u}$ and $\mathbf{u}^{(-)}$ into Eq. (3.50), we can derive

$$\begin{aligned} \mathbf{Y} &= \mathbf{F} \cdot \boldsymbol{\Upsilon} \cdot \mathbf{G} \cdot \boldsymbol{\Theta} \cdot \mathbf{F}^{-1} \cdot \mathbf{X} + \mathbf{F} \cdot \boldsymbol{\Upsilon} \cdot \mathbf{G}^{(-)} \cdot \boldsymbol{\Theta} \cdot \mathbf{F}^{-1} \cdot \mathbf{X}^{(-)} + \mathbf{N} \\ &= \mathbf{F} \cdot \mathbf{H}_{\mathrm{e,cp}} \cdot \mathbf{F}^{-1} \cdot \mathbf{X} + \mathbf{F} \cdot \mathbf{H}_{\mathrm{e,cp}}^{(-)} \cdot \mathbf{F}^{-1} \cdot \mathbf{X}^{(-)} + \mathbf{N}. \end{aligned} \tag{3.51}$$

In Eq. (3.51), due to the insufficient CP length, the matrix **He,cp** is not circular and the product $\mathbf{F} \cdot$ **He,cp** $\cdot\, \mathbf{F}^{-1}$ is no longer a diagonal matrix. This results in the appearance of the ICI effect. Additionally, the insufficient CP length prevents OFDM receivers from fully eliminating the interference caused by the previous OFDM symbol, leading to the presence of the ISI effect. By rearranging Eq. (3.51), we obtain

$$\begin{aligned} \mathbf{Y} &= \mathbf{F} \cdot (\mathbf{H}_{\mathrm{e}} - \mathbf{H}_{\mathrm{ICI}}) \cdot \mathbf{F}^{-1} \cdot \mathbf{X} + \mathbf{F} \cdot \mathbf{H}_{\mathrm{ISI}} \cdot \mathbf{F}^{-1} \cdot \mathbf{X}^{(-)} + \mathbf{N} \\ &= \mathbf{F} \cdot \mathbf{H}_{\mathrm{e}} \cdot \mathbf{F}^{-1} \cdot \mathbf{X} - \mathbf{F} \cdot \mathbf{H}_{\mathrm{ICI}} \cdot \mathbf{F}^{-1} \cdot \mathbf{X} + \mathbf{F} \cdot \mathbf{H}_{\mathrm{ISI}} \cdot \mathbf{F}^{-1} \cdot \mathbf{X}^{(-)} + \mathbf{N} \\ &= \mathbf{H} \cdot \mathbf{X} - \mathbf{F} \cdot \mathbf{H}_{\mathrm{ICI}} \cdot \mathbf{F}^{-1} \cdot \mathbf{X} + \mathbf{F} \cdot \mathbf{H}_{\mathrm{ISI}} \cdot \mathbf{F}^{-1} \cdot \mathbf{X}^{(-)} + \mathbf{N}. \end{aligned} \tag{3.52}$$

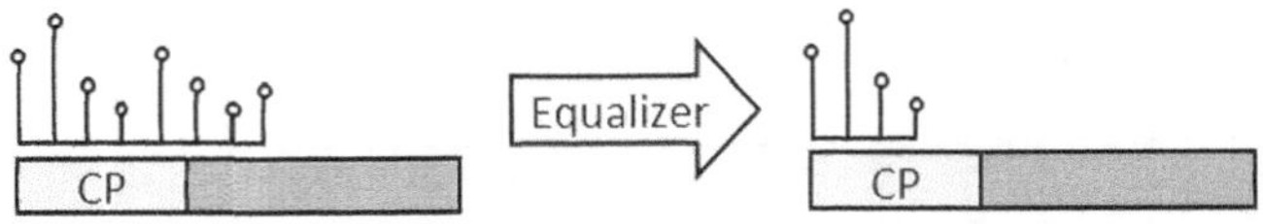

FIGURE 3.32
Illustration of the time-domain equalizer designs.

In Eq. (3.52), $\mathbf{H}_{\text{ICI}}$ represents the channel matrix for ICI effect, which can be expressed as:

$$\mathbf{H}_{\text{ICI}} = \begin{bmatrix} 0 & \cdots & h[L_h-1] & h[L_h-2] & \cdots & h[N_g+1] & 0 & \cdots & 0 \\ 0 & \cdots & 0 & h[L_h-1] & \cdots & h[N_g+2] & 0 & \cdots & 0 \\ \vdots & \cdots & \vdots & \vdots & \ddots & \cdots & \cdots & \cdots & \cdots \\ 0 & \cdots & 0 & 0 & \cdots & h[L_h-1] & 0 & \cdots & 0 \\ \vdots & \cdots & \vdots & \vdots & \vdots & \vdots & \vdots & \vdots & \vdots \\ 0 & \cdots & 0 & 0 & \cdots & 0 & 0 & \cdots & 0 \end{bmatrix}, \tag{3.53}$$

In Eq. (3.53), $\mathbf{H}_{\text{ISI}}$ represents the channel matrix for the ISI effect, which can be expressed as:

$$\mathbf{H}_{\text{ISI}} = \begin{bmatrix} 0 & \cdots & 0 & h[L_h-1] & h[L_h-2] & \cdots & h[N_g+1] \\ 0 & \cdots & 0 & 0 & h[L_h-1] & \cdots & h[N_g+2] \\ \vdots & \cdots & \vdots & \vdots & \vdots & \ddots & \vdots \\ 0 & \cdots & 0 & 0 & 0 & 0 & h[L_h-1] \\ \vdots & \cdots & \vdots & \vdots & \vdots & \vdots & \vdots \\ 0 & \cdots & 0 & 0 & 0 & \cdots & 0 \end{bmatrix}. \tag{3.54}$$

By substituting Eqs. (3.53) and (3.54) into Eq. (3.55), we can obtain the components in different subcarrier as:

$$Y_k = H_k X_k - \sum_{k'=0}^{N-1} [\mathbf{F}\mathbf{H}_{\text{ICI}}\mathbf{F}^{-1}]_{k+1,k'+1} X_{k'} + \sum_{k'=0}^{N-1} [\mathbf{F}\mathbf{H}_{\text{ISI}}\mathbf{F}^{-1}]_{k+1,k'+1} X_{k'}^{(-)} + N_k, \tag{3.55}$$

where $[\cdot]_{i,j}$ stands for the i row j column value. In order to mitigate the interference caused by insufficient CP length, a solution is to design a special time-domain equalizer that reduces the effective channel length at the receiver, as depicted in Figure 3.32. It's important to note that this time-domain equalizer is specifically designed to address the insufficient CP length problem, whereas OFDM systems typically employ frequency-domain equalizers to remove channel effects. The architecture of the time-domain equalizer is illustrated in Figure 3.33(a). Prior to the DFT operation at the receiver, the received signal undergoes filtering by a finite impulse response (FIR) filter represented by $w[n]$. This filter is responsible for reducing the effective

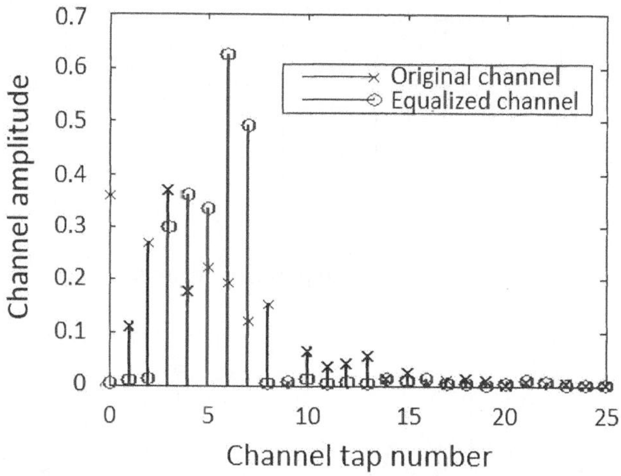

FIGURE 3.33
(a) Illustration of the architecture of the time-domain equalizer. (b) Illustration of the equivalent architecture of the time-domain equalizer.

channel length. Consequently, the effective channel $b[n]$ has a length shorter than the CP length, as depicted in Figure 3.33(b). It should be noted that the time-domain equalizer introduces an additional signal delay denoted as d. In other words, the time-domain equalizer reduces the effective channel length at the expense of increased signal delay.

In order to address the insufficient CP length issue, an alternative approach is to directly mitigate the ISI and ICI effects in the frequency domain. One commonly used method is to employ a linear equalizer, such as a ZF equalizer or a MMSE equalizer, to eliminate the ICI effect. Additionally, a decision feedback equalizer can be utilized to remove the ISI effect. By further analyzing Eq. (3.51), we can derive the following mathematical model:

$$\begin{aligned}\mathbf{Y} &= \mathbf{F}\cdot\mathbf{H}_{\mathrm{e,cp}}\cdot\mathbf{F}^{-1}\cdot\mathbf{X}+\mathbf{F}\cdot\mathbf{H}^{(-)}_{\mathrm{e,cp}}\cdot\mathbf{F}^{-1}\cdot\mathbf{X}^{(-)}+\mathbf{N}\\ &= \mathbf{Q}_{\mathrm{e,cp}}\cdot\mathbf{X}+\mathbf{Q}^{(-)}_{\mathrm{e,cp}}\cdot\mathbf{X}^{(-)}+\mathbf{N}.\end{aligned} \tag{3.56}$$

In the equalizer design, we initially employ a decision feedback equalizer to mitigate the ISI effect. The mathematical expression for the DFE can be given as follows:

$$\mathbf{Y}^{\text{ISI-free}} = \mathbf{Y}-\mathbf{Q}^{(-)}_{\mathrm{e,cp}}\cdot\mathbf{X}^{(-)} = \mathbf{Q}_{\mathrm{e,cp}}\cdot\mathbf{X}+\mathbf{N}. \tag{3.57}$$

After mitigating the ISI effect using the decision feedback equalizer, we can further employ a linear equalizer to eliminate the ICI effect and recover the transmitted data symbol. The mathematical expression for the data symbol estimate using the MMSE equalizer can be given as follows:

$$\hat{\mathbf{X}} = \mathbf{E}_{\mathrm{MS}}\mathbf{Y}^{\text{ISI-free}} = (\mathbf{Q}^{H}_{\mathrm{e,cp}}\mathbf{Q}_{\mathrm{e,cp}}+\frac{\sigma_n^2}{\sigma_X^2}\mathbf{I}_N)^{-1}\mathbf{Q}^{H}_{\mathrm{e,cp}}\mathbf{Y}^{\text{ISI-free}}. \tag{3.58}$$

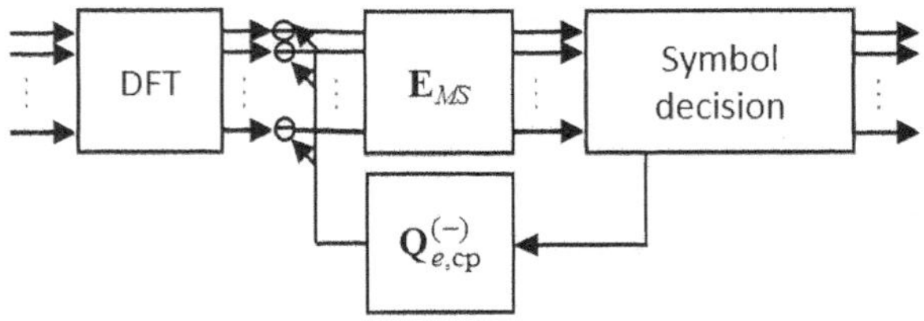

FIGURE 3.34
Illustration of the frequency-domain equalizer designs.

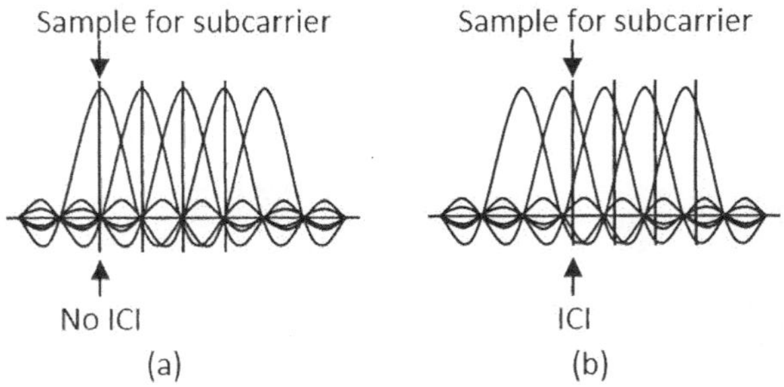

FIGURE 3.35
Illustration of the CFO effect and consequent ICI effect. (a) Without ICI effect. (b) With ICI effect.

Finally, we can utilize the decision result to restore the transmitted data symbol, as shown in Figure 3.34.

The ICI effect in OFDM systems can also be caused by a CFO, which arises due to a frequency mismatch between the oscillators of the transmitter and receiver. In the absence of CFO, the DFT frequency sample points in the receiver align precisely with the peak frequencies of the subcarriers, resulting in no ICI effect, as illustrated in Figure 3.35(a). However, when CFO is present, the shift in the DFT frequency sample points at the receiver leads to mutual interference among the subcarriers, causing the ICI effect, as depicted in Figure 3.35(b). The CFO can be quantified as the product of the frequency spacing between subcarriers, denoted as Δf, and a coefficient α. Thus, the CFO can be expressed mathematically as:

$$f_{\text{offset}} = \alpha\Delta f = k_0\Delta f + \epsilon\Delta f, \ |\epsilon| \leq 0.5. \tag{3.59}$$

In Eq. (3.59), the CFO coefficient α can be separated into two parts: an integer component k_0 and a decimal component ϵ. When $k_0 \neq 0$, the receiver encounters the issue of incorrect data symbol ordering, as depicted in Figure 3.36. For instance, when $k_0 = 1$, the original data symbol order of $0-1-2-3$ becomes $1-2-3-4$, where symbol 4 is unknown, leading to incorrect demodulation results that cannot be utilized. In the case where $k_0 = 0$ but $\epsilon \neq 0$, which is the more common scenario in OFDM systems, the receiver experiences the

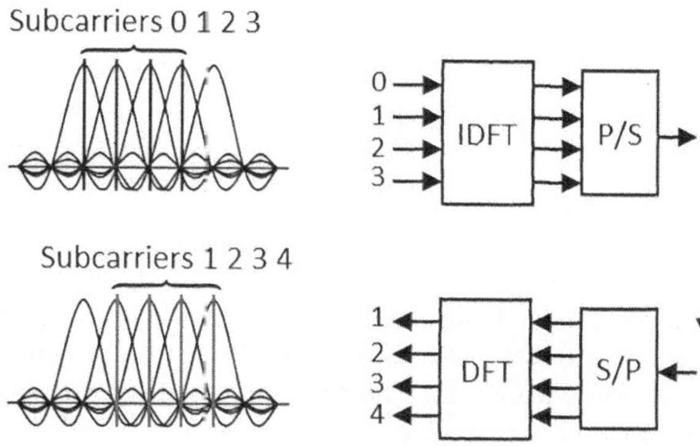

FIGURE 3.36
Illustration of the CFO effect caused by integer CFO.

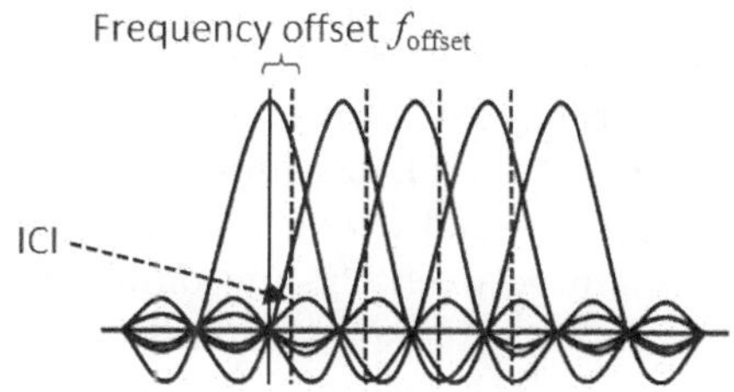

FIGURE 3.37
Illustration of the CFO effect caused by decimal CFO.

ICI effect and a degradation in symbol reception gain, as illustrated in Figure 3.37. In this figure, the dotted lines represent the sampling points with lower DFT reception gain compared to the solid line (which represents the absence of CFO effect). Additionally, the reception phase is also influenced and subject to interference from other subcarriers. Generally, OFDM systems rarely encounter integer CFO, and mechanisms are in place to detect and correct it promptly. As a result, existing literature focuses more on the effects of decimal CFO cases. Considering the CFO effect, the received signal can be expressed as:

$$v[n] = e^{j2\pi f_{\text{offset}} \frac{nT}{N}} (u[n] * h[n]) + n[n]. \tag{3.60}$$

Using similar derivation in this section,[8] we can obtain the OFDM received signal block vector in frequency-domain as:

$$\mathbf{Y} = \mathbf{F} \cdot \mathbf{\Phi}_f \cdot \mathbf{\Upsilon} \cdot \mathbf{G} \cdot \mathbf{\Theta} \cdot \mathbf{F}^{-1} \cdot \mathbf{X} + \mathbf{N} = \mathbf{F} \cdot \mathbf{H}_{\text{e, cfo}} \cdot \mathbf{F}^{-1} \cdot \mathbf{X} + \mathbf{N}, \tag{3.61}$$

[8] Assuming CP length is sufficient.

where

$$\mathbf{\Phi}_f = \begin{bmatrix} 1 & & & 0 \\ & e^{j2\pi f_{\text{offset}} \frac{T}{N}} & & \\ & & \ddots & \\ 0 & & & e^{j2\pi f_{\text{offset}} \frac{(N-1)T}{N}} \end{bmatrix}. \tag{3.62}$$

In Eq. (3.61), we introduce $\mathbf{\Phi}_f$ to describe the CFO effect. The presence of $\mathbf{\Phi} f$ causes $\mathbf{H}_{\text{e, cfo}}$ to deviate from a circular matrix, leading to the occurrence of the ICI effect. To mitigate the impact of the CFO, a common approach is to estimate and compensate for the CFO effect prior to performing DFT operations. This can be expressed as:

$$\mathbf{Y} = \mathbf{F} \cdot \underbrace{\hat{\mathbf{\Phi}}_f^{-1}}_{\text{CFO compensation}} \cdot \mathbf{\Phi}_f \cdot \mathbf{\Upsilon} \cdot \mathbf{G} \cdot \mathbf{\Theta} \cdot \mathbf{F}^{-1} \cdot \mathbf{X} + \mathbf{N}. \tag{3.63}$$

There are two classic methods for estimating the CFO effect in order to facilitate subsequent compensation. Data-driven approaches utilize training symbols or pilot symbols to estimate CFO, providing precise estimation results but compromising with the extra signal overheads. Non-data-driven approaches only consider the correlation between CP and data symbol to conduct CFO estimation.[9]

OFDM systems channel estimation: Existing communication systems commonly employ coherent detection for channel estimation, which requires the receiver to have knowledge of the channel information in order to correctly detect the transmitted symbols. In this section, we will provide a brief introduction to the existing methods for channel estimation in OFDM systems. The general approach is to use training symbols or pilot symbols to estimate the channels in either the time-domain or frequency-domain. The choice of using training symbols or pilot symbols depends on the precision requirements of the communication system, as they incur different overheads. Training symbols are transmitted on all subcarriers in specific timeslots prior to transmitting data symbols. These training symbols provide the receiver with the necessary channel information to detect the subsequent data symbols. On the other hand, pilot symbols are training signals that are placed on specific subcarriers and share the bandwidth with data symbols. The placement of training symbols and pilot symbols is illustrated in Figure 3.38. While training symbols generally result in more accurate channel estimation compared to pilot symbols, they also occupy more subcarriers and therefore, incur larger overhead. As a result, OFDM systems need to strike a balance between using training symbols and pilot symbols. Typically, during the initial stage or asynchronous transmission, OFDM systems tend to utilize training symbols to achieve accurate transmission settings. However, during the stable transmission stage, pilot symbols are employed to reduce system overhead.

[9]Please refer to refs. [85, 86] for more details.

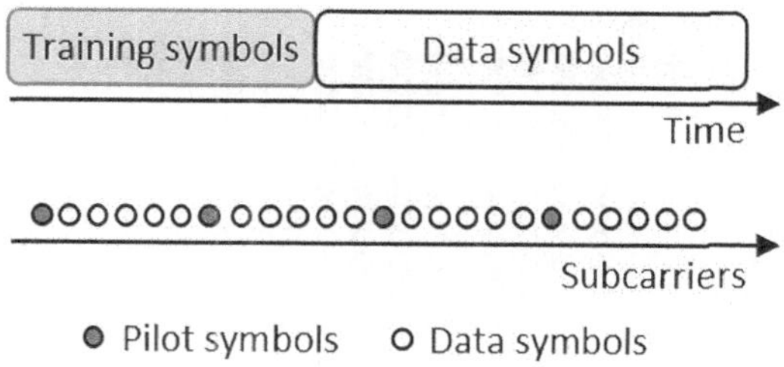

FIGURE 3.38
Illustration of the training symbols and data symbols.

In Eq. (3.40), the received signal block vector on different subcarriers can be expressed as:

$$Y_k = H_k X_k + N_k, \; k = 0, \cdots, N-1. \tag{3.64}$$

In the period of training symbols transmission, since the transmitted symbols X_k are known at the receiver, a straightforward channel estimation result can be obtained as follows:

$$\hat{H}_k = Y_k X_k^{-1}, \; k = 0, \cdots, N-1. \tag{3.65}$$

The channel estimation based on training symbols can also be carried out in the time-domain. Initially, the time-domain channel response $h[n]$ is estimated, which enables the subsequent calculation of the corresponding frequency-domain channel estimates. By combining Eqs. (3.27), (3.31), and (3.33), the received signals can be expressed as:

$$\mathbf{r} = \boldsymbol{\Upsilon} \cdot \mathbf{G} \cdot \boldsymbol{\Theta} \cdot \mathbf{s} + \mathbf{n} = \mathbf{H}_\mathrm{e}\mathbf{s} + \mathbf{n} = \mathbf{S}\mathbf{h} + \mathbf{n}, \tag{3.66}$$

where $\mathbf{S}$, $\mathbf{h}$, and $\mathbf{n}$ is defined as:

$$\begin{aligned} \mathbf{S} &= \begin{bmatrix} s[0] & s[N-1] & \cdots & s[N-L_h+1] \\ s[1] & s[0] & \cdots & s[N-L_h+2] \\ \vdots & \vdots & \vdots & \vdots \\ s[N-1] & s[N-2] & \cdots & s[N-L_h] \end{bmatrix}. \\ \mathbf{h} &= [h[0], h[1], \cdots, h[L_h-1]]^T \\ \mathbf{n} &= [n[0], n[1], \cdots, n[L_h-1]]^T \end{aligned} \tag{3.67}$$

The main concept behind Eq. (3.66) is to interchange the roles of the channel matrix and symbol vector, transforming them into a symbol matrix and a channel vector, respectively. This transformation allows the straightforward channel estimation result to be expressed as the least squares solution, given by:

$$\hat{\mathbf{h}} = (\mathbf{S}^H\mathbf{S})^{-1}\mathbf{S}^H\mathbf{r} = \mathbf{h} + (\mathbf{S}^H\mathbf{S})^{-1}\mathbf{S}^H\mathbf{n}. \tag{3.68}$$

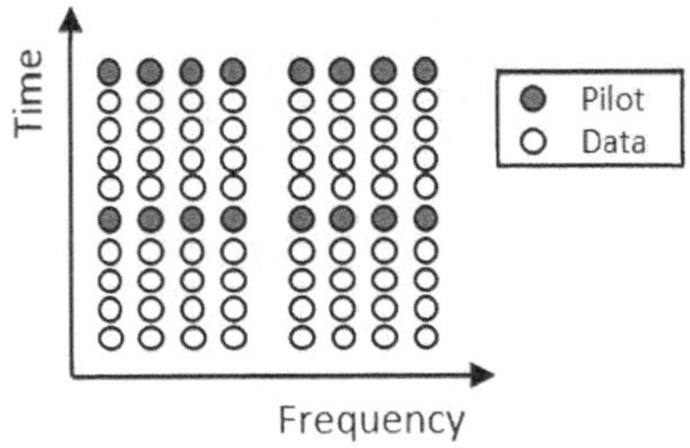

FIGURE 3.39
Illustration of the block-type pilot symbols.

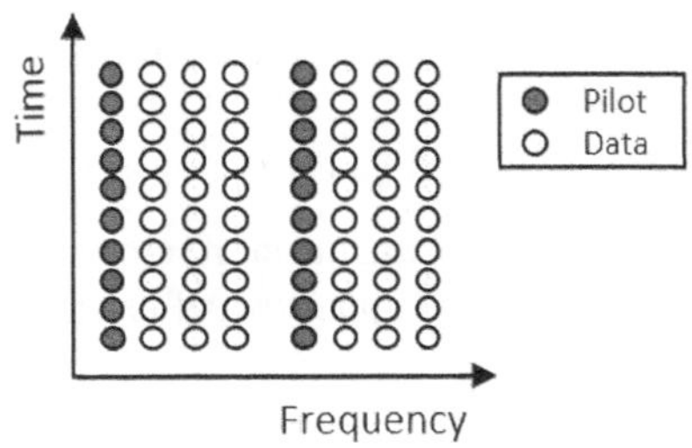

FIGURE 3.40
Illustration of the comb-type pilot symbols.

Finally, frequency-domain channel response estimates can be obtained by performing N-points DFT operations on $\hat{\mathbf{h}}$, resulting in the expression

$$\hat{H}_k = \text{DFT}\{\hat{h}[n]\},\ k = 0, \cdots, N-1. \tag{3.69}$$

In cases where the subcarrier spacing is smaller than the channel's coherent bandwidth[10], the fading characteristics of adjacent subcarriers are likely to be similar. This similarity allows OFDM systems to perform channel estimation with reduced overhead. Channel estimation based on pilot symbols involves placing known pilot symbols at specific locations to estimate their channel response. Interpolation or other methods can then be employed to infer the channel response of non-pilot symbols. There are several common methods for placing pilot symbols, including block type, comb type, mixed type, and scatter type. The block-type pilot symbols, illustrated in Figure 3.39, involve placing pilot symbols on all subcarriers with the same time-interval spacing. This enables accurate estimation of channel variations in the time domain. However, it does not handle frequency-domain channel variations well, making block-type pilot symbols more suitable for slow fading channels. On the other hand, comb-type pilot symbols, depicted in Figure 3.40, are the opposite of block-type pilot symbols. They place pilot symbols on all time-slots with

[10]The idea of coherent bandwidth can be found in Chapter 2.

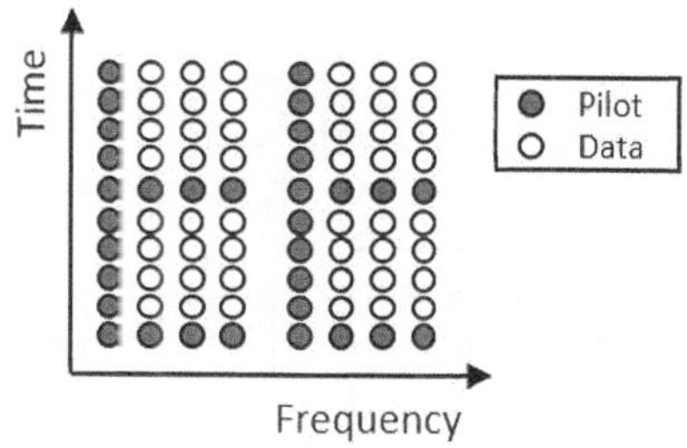

FIGURE 3.41
Illustration of the mixed-type pilot symbols.

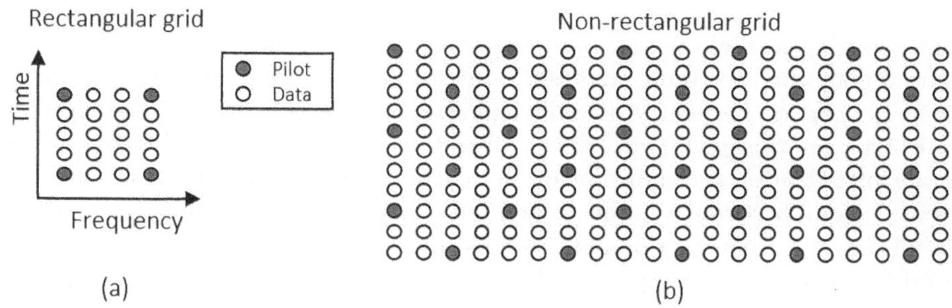

FIGURE 3.42
Illustration of the scatter-type pilot symbols.

the same subcarrier, accurately capturing time-domain channel variations but exhibiting limitations in handling frequency-domain channel variations. Consequently, comb-type pilot symbols are better suited for flat fading channels. Mixed type pilot symbols combine characteristics of both block and comb types, as shown in Figure 3.41. By extensively placing pilot symbols in both time and frequency domains, they enable accurate estimation of channel behaviors in both domains. However, this approach also incurs higher system overhead. Scatter-type pilot symbols distribute pilot symbols simultaneously in the time and frequency domains, as shown in Figure 3.42, aiming to strike a better balance between system overhead and channel behavior estimation in both domains.

Single-carrier transmission technique based on cyclic prefix: OFDM systems offer excellent transmission efficiency and ease of implementation. However, they suffer from the high PAPR problem, which limits the use of low-power transceivers. To address this issue, a single-carrier transmission technique based on the cyclic prefix has been developed as an extension of OFDM systems. This technique inherits the advantages of OFDM systems while mitigating the PAPR problem. The single-carrier transmission technique based on the cyclic prefix incorporates the cyclic prefix mechanism and frequency-domain equalization to achieve similar benefits as OFDM systems.

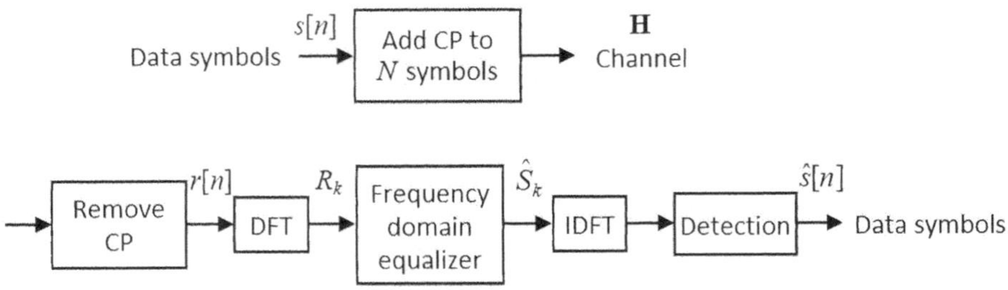

FIGURE 3.43
Illustration of the SC-FDE architecture based on CP.

By adopting a single-carrier approach, the PAPR problem is significantly reduced compared to traditional OFDM systems. This technique often employs a single-carrier frequency-domain equalizer (SC-FDE), as depicted in Figure 3.43. In this technique, the transmit data symbol $s[n]$ of length N is transmitted after the addition of the cyclic prefix. At the receiver, the received signal first removes the cyclic prefix. Then, DFT operations are applied to obtain the frequency-domain signals for channel equalization. The equalized frequency-domain signal is then transformed back into the time-domain for symbol detection. With the introduction of the cyclic prefix mechanism, the received signals after removing the cyclic prefix can be expressed as the convolution product of the transmitted signal and the channel:

$$r[n] = s[n] \odot_N h[n] + n[n]. \tag{3.70}$$

After performing the DFT operations, we obtain

$$R_k = \mathrm{DFT}\{r[n]\} = S_k H_k,\ k = 0, \cdots, N-1, \tag{3.71}$$

where

$$S_k = \mathrm{DFT}\{s[n]\},\ k = 0, \cdots, N-1. \tag{3.72}$$

Eq. (3.72) is similar to the aforementioned OFDM model, where frequency-domain equalization can be utilized to remove the channel effects. It can be expressed as:

$$\hat{S}_k = R_k H_k^{-1},\ k = 0, \cdots, N-1. \tag{3.73}$$

Finally, the IDFT operations will be utilized to recover the time-domain symbol estimates from the frequency-domain signal after equalization. This can be expressed as:

$$\hat{s}[n] = \mathrm{IDFT}\{\hat{S}_k\},\ n = 0, \cdots, N-1. \tag{3.74}$$

The architecture of SC-FDE is very similar to that of OFDM, as it also involves the use of a cyclic prefix, DFT/IDFT operations, and a frequency-domain equalizer. As a result, the complexity and performance of these two systems

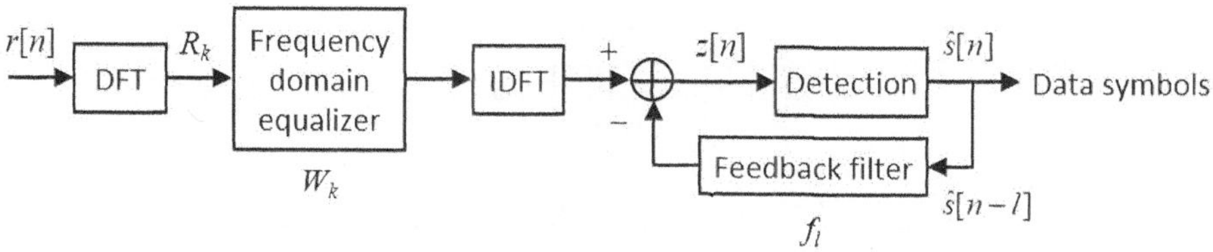

FIGURE 3.44
Illustration of the SC-FDE architecture based on CP with the decision feedback mechanism.

are comparable. However, SC-FDE offers the advantage of lower PAPR. The performance of SC-FDE can be further improved by introducing additional decision feedback, as depicted in Figure 3.44. In the single-carrier decision feedback equalizer (SC-DFE), time-domain feedback is utilized to mitigate postcursor Intersymbol Interference (ISI), while frequency-domain feedback is employed to eliminate precursor ISI. The output signal of the decision feedback equalizer, denoted as $z[n]$ in Figure 3.44, can be expressed as:

$$z[n] = \frac{1}{N}\sum_{k=0}^{N-1} W_k R_k e^{j2\pi\frac{kn}{N}} - \sum_{l=1}^{K_b} f_l^* \hat{s}[n-l], \tag{3.75}$$

where W_k is the equalization coefficient of k subcarrier, f_l is the coefficient of decision feedback filter, and K_b is the order of the decision feedback filter. Considering the utilization of the MMSE rule to determine W_k and f_l, we can express it as:

$$\min_{W_k, f_l} \mathbb{E}\{|z[n]-s[n]|^2\} \equiv \frac{1}{N}\sum_{k=0}^{N-1} |W_k H_k - F_k|^2 + \frac{\sigma_n^2}{N}\sum_{k=0}^{N-1} |W_k|^2, \tag{3.76}$$

where

$$F_k = 1 + \sum_{l=1}^{K_b} f_l^* e^{-j2\pi\frac{kl}{N}},\ k = 0, \cdots, N-1. \tag{3.77}$$

and σ_n^2 stands for the noise power. Assuming that $\{f_l\}$ is given, the frequency-domain equalization coefficient based on MMSE criterion can be expressed as:

$$W_k = \frac{H_k^* F_k}{\sigma_n^2 + |H_k|^2}. \tag{3.78}$$

By substituting Eq. (3.78) into Eq. (3.76), we can formulate the MMSE optimization equation for $\{F_k\}$ as:

$$\min_{F_k} \frac{\sigma_n^2}{N}\sum_{k=0}^{N-1} \frac{|F_k|^2}{\sigma_n^2 + |H_k|^2}. \tag{3.79}$$

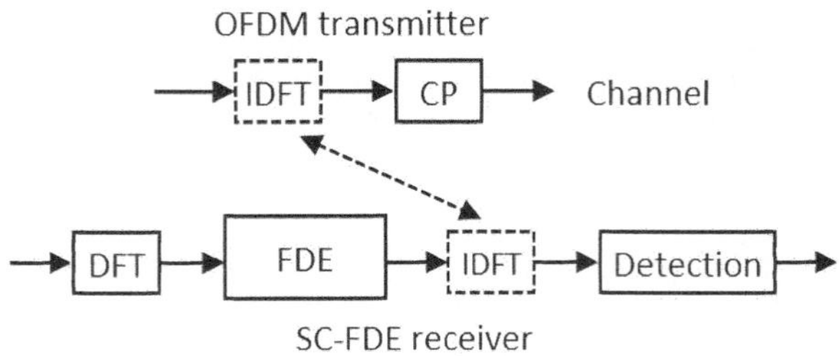

FIGURE 3.45
Illustration of the relationship between SC-FDE and OFDM systems.

The optimal solution of the above optimization problem can be obtained by adjusting f_l.[11] It can be expressed as:

$$\boldsymbol{f} = [f_1, \cdots, f_{K_b}]^T = \boldsymbol{V}^{-1}\boldsymbol{v}, \tag{3.80}$$

where

$$\begin{aligned} \mathbf{v} &= [v_1, \cdots, v_{K_b}]^T \\ \mathbf{V} &= \begin{bmatrix} v_0 & v_{-1} & \cdots & v_{1-K_b} \\ v_1 & v_0 & \cdots & v_{2-K_b} \\ \vdots & \vdots & \vdots & \vdots \\ v_{K_b-1} & v_{K_b-2} & \cdots & v_0 \end{bmatrix}. \\ v_l &= \frac{\sigma_n^2}{N} \sum_{k=0}^{N-1} \frac{e^{-2\pi \frac{kl}{N}}}{\sigma_n^2 + |H_k|^2}, \quad l = 0, \cdots, K_b - 1. \end{aligned} \tag{3.81}$$

Finally, we can substitute Eqs. (3.77) and (3.80) into Eq. (3.78) to obtain the frequency-domain equalization coefficient. The main difference between SC-FDE and OFDM systems lies in the position of the IDFT operations, as illustrated in Figure 3.45. In OFDM systems, the IDFT operations are performed in the transmitter to achieve multi-carrier transmission. On the other hand, SC-FDE systems employ IDFT operations in the receiver to restore symbols after frequency-domain equalization. By exploiting the similarity between SC-FDE and OFDM systems, communication systems can perform OFDM transmission in the uplink (e.g., from base stations) and SC-FDE reception in the downlink (e.g., in mobile stations) without the need for additional DFT or IDFT circuits. For example, in 4G mobile communication systems, base stations utilize OFDM systems for signal transmission in the uplink, while mobile stations employ single-carrier transmission in the downlink, as shown in Figure 3.46. This configuration offers two main advantages: (1) Complex computations are concentrated in the base stations, resulting in computational resource savings and reduced battery power consumption in mobile stations.

[11] Please refer to ref. [87] for more details.

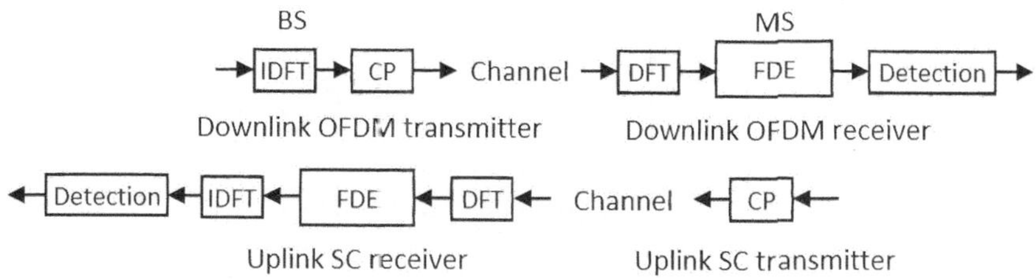

FIGURE 3.46
Illustration of the transceiver configuration strategy in mobile communication systems.

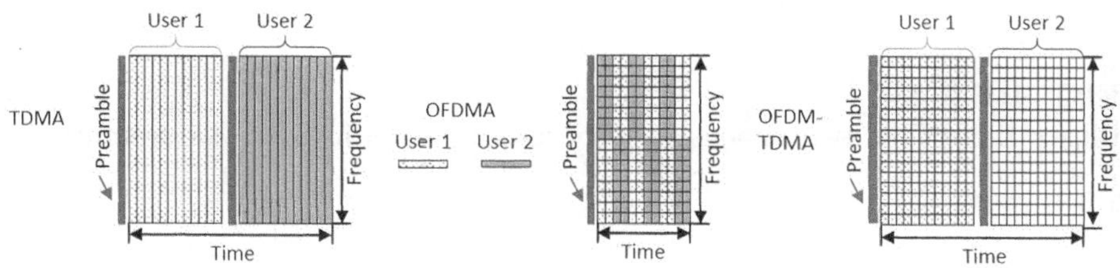

FIGURE 3.47
Illustration of the resource allocation in TDMA, OFDM-TDMA, and OFDMA systems.

(2) By utilizing single-carrier signals in mobile stations, the lower PAPR leads to improved transmission efficiency and energy savings. It is worth noting that in the 5G New Radio (NR) mobile communication system, the downlink transmission also utilizes OFDMA due to the availability of high-performance amplifiers for OFDMA transmission in most user equipment. However, in scenarios with challenging channel conditions, SC-FDMA may be employed to further lower PAPR and achieve better transceiver efficiency.

OFDMA/SC-FDMA introduction: In the previous discussions, we focused on single-user systems for point-to-point transmission. However, both OFDM systems and single-carrier systems can also support multi-user transmission, enabling multiple access techniques such as OFDMA and SC-FDMA, respectively. As shown in Figure 3.47, time division multiple access (TDMA) is used to divide the time-domain resources into different timeslots, which are then assigned to different users. OFDM-TDMA systems further assign all OFDM subcarriers to a single user in each timeslot. OFDMA assigns different sets of subcarriers to different users in each timeslot, combining the benefits of TDMA and frequency division multiple access (FDMA) systems. The flexibility of OFDMA allows for the allocation of time-domain and frequency-domain resources to different users, as illustrated in Figure 3.48. By taking advantage of selective fading in the channel across both time-domain and frequency-domain, OFDMA systems can allocate appropriate resources to

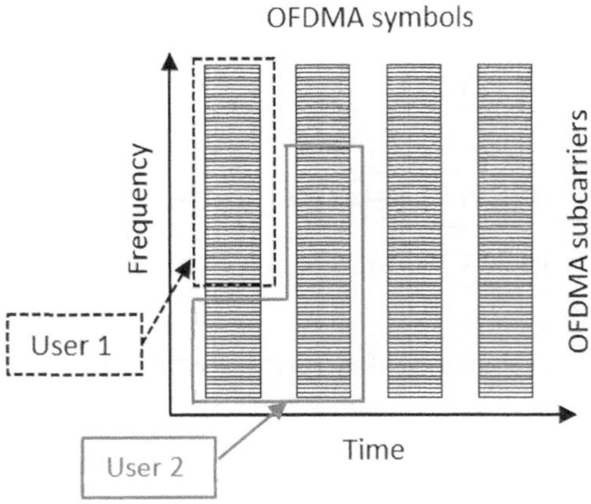

FIGURE 3.48
Illustration of the two-dimension resource allocation in OFDMA systems.

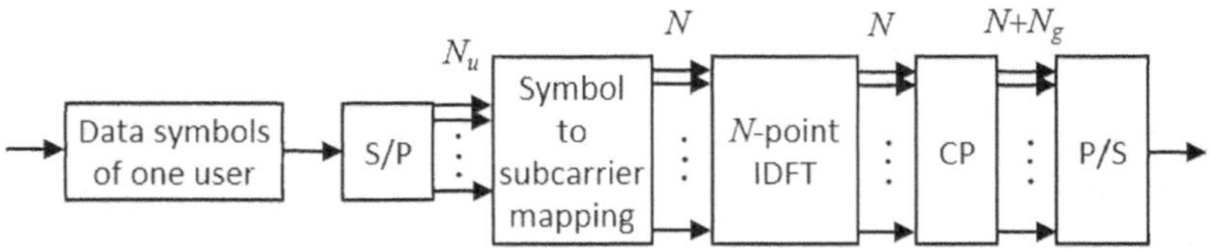

FIGURE 3.49
Illustration of the transmitter structure in OFDMA systems.

different users, thus improving overall system spectrum efficiency. This concept is known as multi-user diversity. For example, if the channel response is better for user 1 than user 2 in a particular timeslot and subcarrier, and vice versa in another timeslot and subcarrier, the system can assign the favorable resources to each user accordingly to enhance the overall system performance. The architecture of OFDMA systems is similar to that of OFDM systems, with the main difference being the mapping of data symbols from different users to assigned subcarriers prior to the IDFT operations, as shown in Figure 3.49. Each user maps their data symbols to their assigned subcarriers out of the total available subcarriers. After the DFT operations at the receiver, the corresponding receiver extracts information from the assigned subcarriers and performs equalization and detection to complete the reception. Similar to OFDM systems, OFDMA systems also face PAPR issues. SC-FDMA is a single-carrier multiple access technique based on OFDM systems, as depicted in Figure 3.50. It shares similarities with OFDM systems, with the main difference being the additional DFT operation used to transform time-domain data symbols to the frequency domain. Subsequently, the procedures for SC-FDMA are the same as those for OFDMA systems, where frequency-domain symbols from different users are placed in different subcarriers. The reason for em-

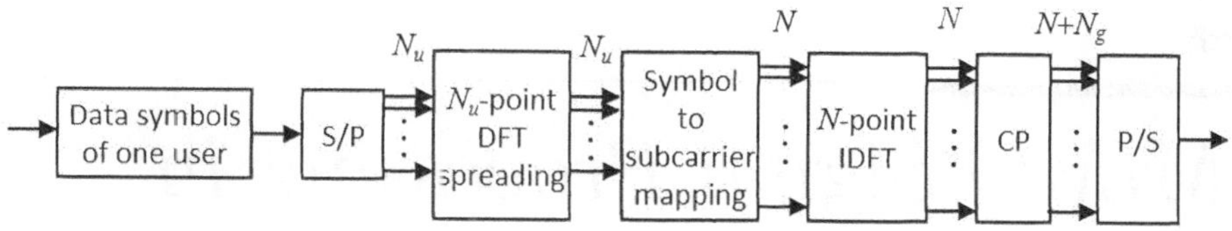

FIGURE 3.50
Illustration of the transmitter structure in SC-FDMA systems.

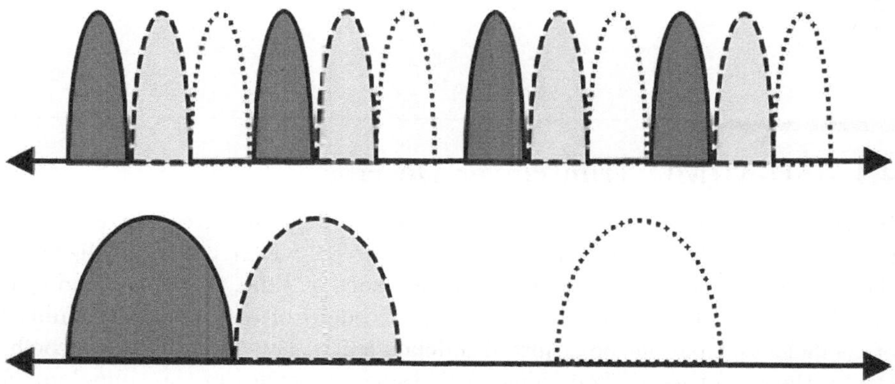

FIGURE 3.51
Illustration of the subcarrier allocation in SC-FDMA systems.

ploying the extra DFT operation in SC-FDMA is to allow each time-domain symbol to utilize the maximum bandwidth for transmission, leveraging the single-carrier property. The subcarrier allocation in SC-FDMA can be either distributed or localized, as illustrated in Figure 3.51. Distributed allocation assigns frequency-domain symbols from the same user to non-contiguous subcarriers, while localized allocation assigns frequency-domain symbols from the same user to contiguous subcarriers for data transmission.

4

MIMO Signal Processing in Communication Systems

4.1 SU-MIMO Transceiver Designs

In wireless communication system designs, the degree of freedom is an important idea, referring to the effective resources, which can be utilized by the systems for performance improvements, such as time, frequency, and spatial domains. For example, the degree of freedom of a filter is the number of coefficients, that can be adjusted independently. By adjusting those coefficients, the performance and signal processing properties in the time-domain and frequency-domain can be designed accordingly. In the spatial domain, a degree of freedom can be obtained by employing multi-antenna systems. The degree of freedom of a multi-antenna system is the number of antennas, that can be utilized independently. By adjusting the antenna weights, multi-antenna systems can also adjust their performance and signal processing properties in the time-domain and frequency-domain. In current wireless communications, the multi-antenna technique is an effective solution to boost system performance. The idea is to configure multiple antennas in the transmitter and/or receiver so that signal transmission/reception can take place in multiple spatial locations simultaneously. Due to the excellent performance improvements brought by the multi-antenna technique, almost all advanced wireless and mobile communication standards have included it in key technology options. Multi-antenna systems need to work with corresponding signal processing solutions to effectively utilize the available spatial degree of freedom to improve communication system performance. Generally speaking, there are two goals regarding multi-antenna signal processing solutions: (1) Improve signal quality by strengthening signal quality or mitigating interference. (2) Improve data rate by increasing system capacity. To achieve the above goals, multi-antenna systems can utilize the spatial degree of freedom flexibly by performing signal processing to obtain the following gains: array gain, diversity gain, interference suppression gain, and spatial multiplexing gain. As shown

DOI: 10.1201/9781003408239-4

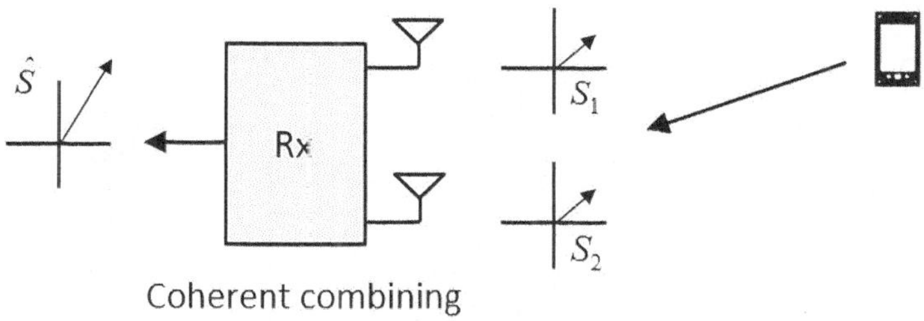

FIGURE 4.1
Illustration of array gain.

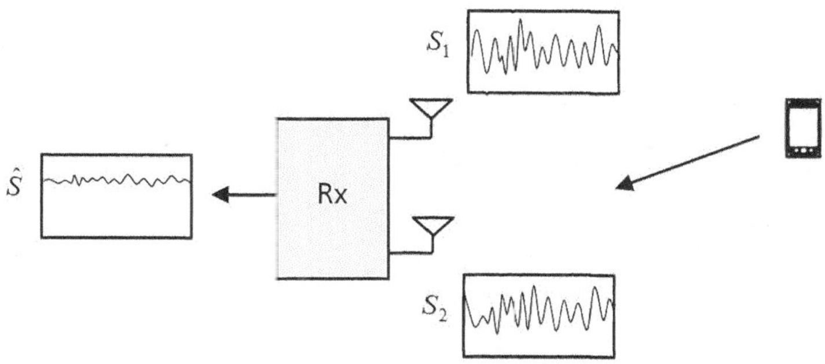

FIGURE 4.2
Illustration of receiver diversity gain.

in Figure 4.1, array gain can be obtained by performing coherent processing[1] to signals from multiple antennas to improve the signal-to-noise ratio (SNR). Specifically, if channel information is available in the transmitter, the transmit signals from different antennas can be processed in advance so that the received signals are coherent and consequently with an improved SNR, as shown in Figures 4.2 and 4.3. Utilizing the transmission/reception of multi-antenna systems, the attenuation phenomenon of received signals in fading channels can be mitigated, providing the so-called diversity gain.

Furthermore, by performing destructive interference in multi-antenna systems to mitigate interference signals, the system signal-to-interference-plus-noise ratio (SINR) can also be improved, being the so-called interference suppression gain, as shown in Figure 4.4. Finally, in the case that the transmitter and receiver are both with multiple antennas, multiple data streams can be transmitted simultaneously for the increased data rate, providing the spatial multiplexing gain, as shown in Figure 4.5. By employing the above techniques, multi-antenna signal processing can improve the coverage, link quality, trans-

[1]The coherent processing aims to adjust the phase of different signals to a consistent one.

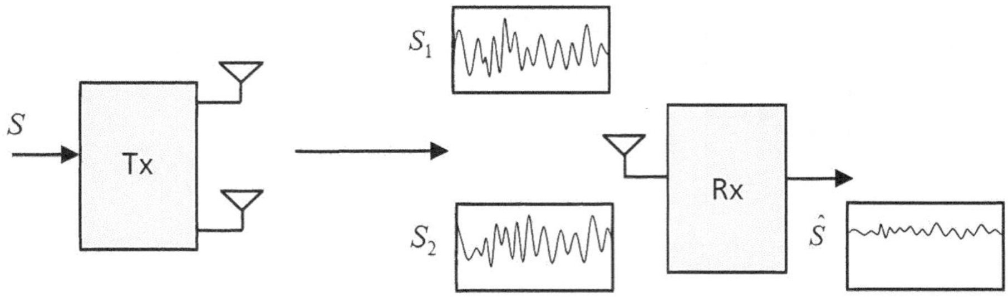

FIGURE 4.3
Illustration of transmitter diversity gain.

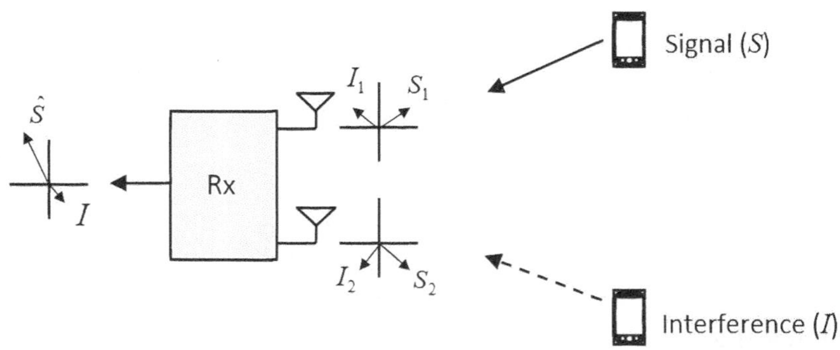

FIGURE 4.4
Illustration of interference suppression gain.

mission rate, and system capacity (or spectrum efficiency) of communication systems. In this chapter, we aim to introduce the above signal processing techniques and the corresponding transceiver designs in different scenarios. We will first cover a basic application of multi-antenna systems: beamforming, then introduce the theoretical background of multi-antenna systems, including diversity gain, channel capacity, and multiple-input multiple-output-orthogonal frequency division multiplexing (MIMO-OFDM) system application. Furthermore, we will discuss different multi-antenna signal processing techniques, including transmitter/receiver diversity, space-time code, and MIMO detection. Finally, in the case that channel information is available in the transmitter, the MIMO precoding technique will be introduced to conclude the content of this chapter.

Beamforming techniques: Being a basic application of multi-antenna systems, beamforming is a multi-antenna technique, utilizing an antenna array to transmit or receive signals to achieve the goal of improving SNR or SINR to consequently enhance communication performance or reliability. The beamforming technique can be employed in the transmitter or receiver. The idea of

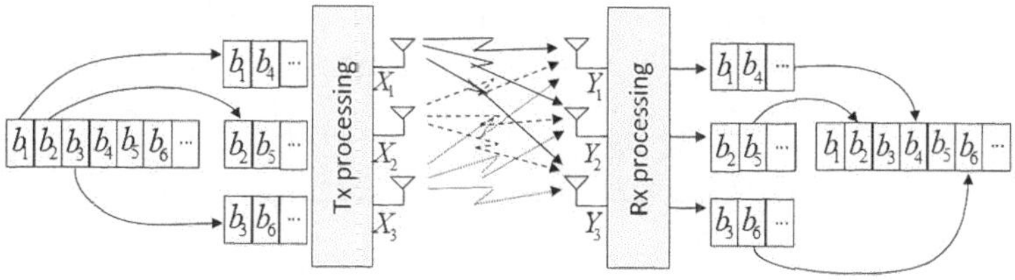

FIGURE 4.5
Illustration of spatial multiplexing gain.

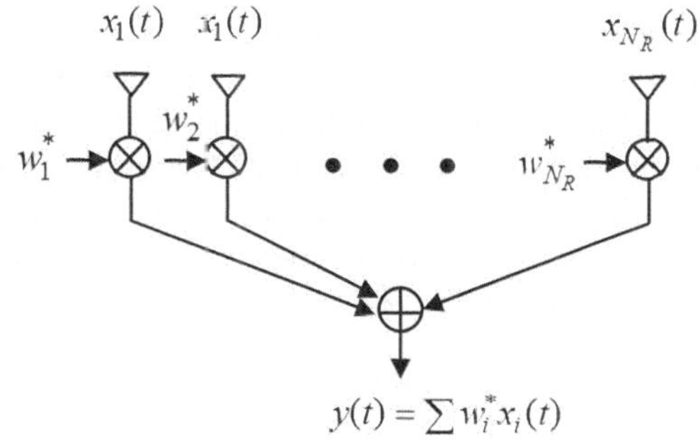

FIGURE 4.6
Illustration of the receiver beamforming architecture.

receiver beamforming is to adjust the amplitude or phase of received signals from different antennas to enable coherent superposition for improved SNR. On the other hand, the idea of transmitter beamforming is also aiming to adjust the amplitude or phase of transmitted signals to enhance the strength of transmitted signals in the direction of the receiver. It is noteworthy that the reciprocity exists between transmitter beamforming and receiver beamforming. In other words, in the same beamforming architecture, the spatial response of the transmitted signal and received signal is the same. As a result, the design of transmitter beamforming and receiver beamforming is similar. The architecture of receiver beamforming is shown in Figure 4.6. Assuming there are N_{R} antennas in the antenna array of the receiver, the equivalent baseband received signal can be expressed as:

$$y(t) = \sum_{i=1}^{N_{\mathrm{R}}} w_i^* x_i(t) = \mathbf{w}^H \mathbf{x}(t), \tag{4.1}$$

where $\mathbf{x}(t) = [x_1(t), x_2(t), \cdots, x_{N_{\mathrm{R}}}(t)]^T$; $\mathbf{w} = [w_1, w_2, \cdots, w_{N_{\mathrm{R}}}]^T$. $x_i(t)$ is the received signal in ith antenna and w_i is the weight in the antenna.[2] Hence,

[2]Each weight is a complex-value number, presenting the amplitude and phase adjustments to the baseband signals.

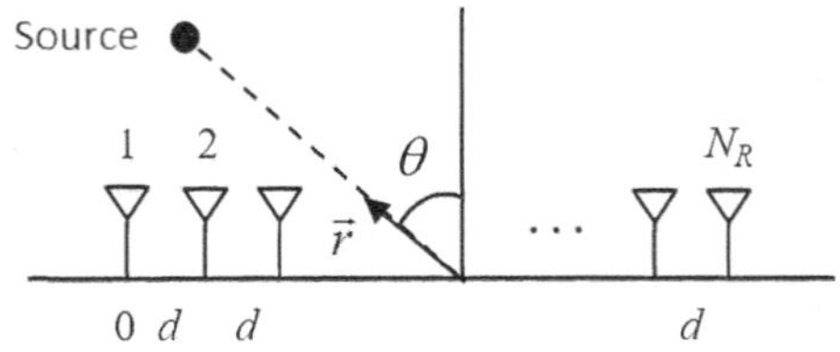

FIGURE 4.7
Illustration of the uniform linear array.

$\mathbf{x}(t)$ is the received signal vector, and $\mathbf{w}$ is the weight vector. The weight vector is the key parameter of the beamformer so that designing weight vector is equal to determining the properties of the beamformer. Next, we will introduce three representative received beamforming designing methodologies: matched beamforming, minimum variance distortionless response (MVDR) beamforming,[3] and minimum mean square error (MMSE) beamforming. Assuming a uniform linear array (ULA) is adopted as shown in Figure 4.7. The received signal vector can be expressed as:

$$\mathbf{x}(t) = \mathbf{a}(\theta_1)u(t) + \mathbf{i}(t) + \mathbf{n}(t), \tag{4.2}$$

where $\mathbf{a}(\theta)$ is the steering vector of the antenna array, θ_1 is the direction of the received signal, and $u(t)$ is the received signal, $\mathbf{i}(t)$ is the interference signal vector, and $\mathbf{n}(t)$ is the noise vector. As a result, the output signal from the beamformer and the average power can be expressed as:

$$y(t) = \mathbf{w}^H\mathbf{x}(t) = \mathbf{w}^H\mathbf{a}(\theta_1)u(t) + \mathbf{w}^H\mathbf{i}(t) + \mathbf{w}^H\mathbf{n}(t), \tag{4.3}$$

$$\begin{aligned} \mathbb{E}\{|y(t)|^2\} &= \mathbb{E}\{|\mathbf{w}^H\mathbf{x}(t)|^2\} \\ &= \mathbf{w}^H\mathbf{R}_{xx}\mathbf{w} \\ &= \mathbb{E}\{|u(t)|^2\}\,\mathbf{w}^H\mathbf{a}(\theta_1)\mathbf{a}^H(\theta_1)\mathbf{w} + \mathbf{w}^H\mathbf{R}_{ii}\mathbf{w} + \sigma_n^2\mathbf{w}^H\mathbf{w} \end{aligned} \tag{4.4}$$

where $\mathbf{R}_{xx}$ is the correlation matrix of the antenna array received signal, being expressed as:

$$\mathbf{R}_{xx} = \mathbb{E}\{\mathbf{x}(t)\mathbf{x}^H(t)\} = \mathbb{E}\{|u(t)|^2\}\mathbf{a}(\theta_1)\mathbf{a}^H(\theta_1) + \mathbf{R}_{ii} + \sigma_n^2\mathbf{I}_{N_R}. \tag{4.5}$$

In Eq. (4.5), $\mathbf{R}_{ii}$ is the correlation matrix of interference signal, σ_n^2 is the noise power. Since the noise in different antennas are independent, the noise correlation matrix can be expressed as $\sigma_n^2\mathbf{I}_{N_R}$.[4]

The goal of matched beamforming is to maximize the SNR of the received signal direction and the idea is similar to the commonly used matched filter in communication systems. The design of matched beamforming can be

[3] MVDR is the abbreviation of minimum variance distortionless response. Please refer to ref. [88].

[4] $\mathbf{I}_n$ is the unit matrix with the dimension $n \times n$.

expressed as the constrained optimization problem below:

$$\begin{aligned} &\max_{\mathbf{w}_{\mathrm{MF}}} \ \mathbf{w}_{\mathrm{MF}}^H \mathbf{a}(\theta_0)\mathbf{a}^H(\theta_0)\mathbf{w}_{\mathrm{MF}} \\ &\text{subject to} \ \ \mathbf{w}_{\mathrm{MF}}^H \mathbf{w}_{\mathrm{MF}} = \frac{1}{N_R}, \end{aligned} \tag{4.6}$$

where θ_0 is the assumed direction of the received signal (look direction). In Eq. (4.6), the goal is to maximize the received response to the θ_0 direction while fixing the received noise power. Hence, if θ_0 is close to the real direction of the received signal θ_1[5], the matched beamforming design will maximize SNR to achieve our goal. The solution can be obtained via the Cauchy inequality, which can be expressed as:

$$\mathbf{w}_{\mathrm{MF}} = \frac{1}{N_R}\mathbf{a}(\theta_0). \tag{4.7}$$

The above result is similar to the matched filter. Specifically, the matched filter adjusts its filter impulse response to match the received signal in time-domain while matched beamforming designs its beamforming weight to match the characteristic (steering vector) of the received signal in spatial-domain. When matched beamforming is utilized in a ULA, the equivalent received signal can be expressed as:

$$\mathbf{w}_{\mathrm{MF}}\mathbf{x}(t) = \frac{1}{N_R}\sum_{i=1}^{N_R} x_i(t) e^{-j2\pi\frac{d}{\lambda_c}(i-1)\sin\theta_0}. \tag{4.8}$$

One can notice that the result in Eq. (4.8) is also the Fourier transform result of the antenna array received signal $\mathbf{x}(t)$. As a result, matched beamforming is also called Fourier beamforming. Matched beamforming can also be utilized in the transmitter, where θ_0 presents the transmitting direction instead. In the above discussions, one can notice that the goal of matched beamforming is only to maximize the SNR of the look direction and fails to consider the existence of interference. Hence, although matched beamforming can improve received signal quality significantly in an interference-free environment, the achieved performance may drop due to the existence of interference. Thus, we further introduce MVDR beamforming and MMSE beamforming, which can handle interference effectively.

The idea of MVDR beamforming is to minimize the total received power while fixing the spatial response of the look direction as 1. Since the total received signal actually includes the desired signal, interference signal, and noise, minimizing the total received power while maintaining the strength of the desired signal is equal to minimizing the power of interference and noise,

[5]In an ideal case, the look direction should be equal to the real direction. However, in actual cases, there may be a slight mismatch between those two angles.

providing effective interference suppression. The design of MVDR beamforming can be formulated as the constraint optimization problem as:

$$\begin{aligned} &\min_{\mathbf{w}_{\mathrm{MV}}} \ \mathbf{w}_{\mathrm{MV}}^H \mathbf{R}_{xx}(\theta_0)\mathbf{w}_{\mathrm{MV}} \\ &\text{subject to} \ \ \mathbf{w}_{\mathrm{MV}}^H \mathbf{a}(\theta_0) = 1. \end{aligned} \tag{4.9}$$

The solution can be obtained via the Lagrange multiplier, which can be expressed as:

$$\begin{cases} \nabla_{\mathbf{w}_{MV}} \mathbf{w}_{\mathrm{MV}}^H \mathbf{R}_{xx}\mathbf{w}_{\mathrm{MV}} - \lambda \nabla_{\mathbf{w}_{MV}} [\mathbf{w}_{\mathrm{MV}}^H \mathbf{a}(\theta_0) - 1] = 0 \\ \mathbf{w}_{\mathrm{MV}}^H \mathbf{a}(\theta_0) = 1 \end{cases} \tag{4.10}$$

$$\begin{cases} \mathbf{R}_{xx}\mathbf{w}_{\mathrm{MV}} = \lambda \mathbf{a}(\theta_0) \\ \mathbf{w}_{\mathrm{MV}}^H \mathbf{a}(\theta_0) = 1 \end{cases} \tag{4.11}$$

$$\mathbf{w}_{\mathrm{MV}} = \frac{1}{\mathbf{a}^H(\theta_0)\mathbf{R}_{xx}^{-1}\mathbf{a}(\theta_0)} \mathbf{R}_{xx}^{-1}\mathbf{a}(\theta_0). \tag{4.12}$$

Since the design of $\mathbf{w}_{\mathrm{MV}}$ is related to the received signal, MVDR beamforming is an adaptive signal processing technique by automatically adjusting $\mathbf{w}_{\mathrm{MV}}$to suppress the interference signal and noise. For example, in the case of larger interference than noise, the MVDR beamformer will focus more on interference suppression. On the other hand, in the case of larger noise than interference, the MVDR beamformer will concentrate on noise suppression and devolve to the matched filter. The main advantage of MVDR beamforming is the above automatic adjusting mechanism. However, if there is a mismatch between the angles of the look direction and the actual signal direction, the performance of the MVDR beamformer will drop since the actual signal will be misidentified as an interference signal and be suppressed consequently. Also, in the case that the desired signal and interference signal is coherent,[6] the performance of the MVDR beamformer will also drop since destructive interference between the desired signal and interference signal will be conducted to minimize the total signal power.

The idea of MMSE beamforming is to minimize the mean square error between desired signal $u(t)$ and the actual received signal, expressed as:

$$\min_{\mathbf{w}_{\mathrm{MS}}} \ \mathbb{E}\{|\mathbf{w}_{\mathrm{MS}}^H \mathbf{x}(t) - u(t)|^2\}. \tag{4.13}$$

This is a typical MMSE problem with the solution as:

$$\mathbf{w}_{\mathrm{MS}} = \mathbf{R}_{xx}^{-1}\mathbf{r}_{xs}; \quad \mathbf{r}_{xs} = \mathbb{E}\{\mathbf{x}(t)u_1^*(t)\}. \tag{4.14}$$

The design of the MMSE beamformer does not require the direction of the desired signal, and thus will not be impacted by the mismatch between the

[6]The definition of "coherent" here is that two signals are proportional to each other. This will happen when signals arrive at the receiver via multi-path mirror reflections.

lookup direction and the actual signal direction. However, the extra training signal is required to obtain $\mathbf{r}_{xs}$, which will also increase the system overhead. The goal of MMSE beamforming is to restore the desired signal as precisely as possible and utilize training signals to guide the beamforming design. As a result, it can be performed appropriately no matter the faced environment, being the main difference between MVDR beamforming and MMSE beamforming.

4.2 MIMO Principles

The most common performance indicators for evaluating communication systems are reliability and capacity. Reliability generally refers to the error rate of transmission, and the factors that cause errors mainly come from channels, interference, and noise effects. Capacity refers to the amount of effective data transmission within a fixed time period, or the effective data transmission rate. Similar to reliability, the main factors that limit the capacity of communication systems also come from channels, interference, and noise effects. For example, in a quiet environment, the two parties can speak at a faster speed without difficulty or misunderstanding, indicating that the system has higher reliability and capacity. However, if the environment becomes noisy, the two parties must slow down their speaking speed, otherwise, they may misunderstand each other. From a communication perspective, the reliability and capacity of the system decrease at this time. The main goal of designing communication systems is to overcome the effects of channels, interference, and noise in the environment to improve the system's reliability and capacity. In communication systems, reliability is generally expressed as an error rate. The commonly used methods to represent error rate are symbol error rate (SER) and bit error rate (BER). The former refers to the probability of symbol errors, while the latter refers to the probability of transmitting bit errors. For example, if the transmitter carries 100 bits on 25 16-QAM symbols for transmission and 2 symbols have errors at the receiver, resulting in 3 bit errors after conversion, then the symbol error rate is 8% (2/25), and the bit error rate is 3% (3/100). The error rate of a communication system is generally represented by SNR, which represents the environmental effects and takes into account the transmission power, noise intensity, and channel effects. Interference is often included in noise owing to the fact that, from the receiver's perspective, the effects of interference and noise are basically similar. Therefore, the method of evaluating the performance of a transmission and reception link is to calculate the variation curve of its error rate under different SNR.

Reliability of MIMO communication systems: In Chapter 2, it was mentioned that the channel effects can mainly be divided into path loss, shadow fading, and multipath fading. The former two represent the impact of the transmission and reception locations on the channel. Once the positions of the transmitter and receiver are fixed, the path loss and shadow fading become fixed as well. Under the influence of the channel effects, the definition of SNR needs to take into account the large-scale average effects of path loss and shadow fading, expressed as:

$$\mathrm{SNR} = \frac{P_t G_{\mathrm{GL}} G_{\mathrm{SF}}}{\sigma_n^2}, \tag{4.15}$$

where P_t represents transmission power, and G_{GL} and G_{SF} respectively represent the average gain of path loss and shadowing fading. At this point, the channel effect is characterized by small-scale multipath fading. Generally, the value of multipath fading is random, and it is composed of many independent paths, each with the same statistical characteristics. In this case, assuming slow and flat fading, the multipath fading effect can be represented as a Gaussian random variable based on the central limit theorem, as expressed as:

$$y(t) = hx(t) + n(t), \tag{4.16}$$

where $y(t)$ is the received signal, $x(t)$ is the transmitted signal, and h is the complex Gaussian random variable representing multipath fading, with its amplitude following the Rayleigh distribution. Therefore, Eq. (4.16) is also known as the Rayleigh fading signal model. In this model, there are mainly two reasons for receiving errors: the channel is too poor, or the SNR is too low. When the SNR is relatively high, the channel is the main cause of errors, and the relationship between error rate p_e and SNR is given by Eq. (4.17).

$$p_{\mathrm{e}} \propto \frac{1}{\mathrm{SNR}}. \tag{4.17}$$

As shown in Eq. (4.17), on a logarithmic scale, the error rate curve against SNR approaches a straight line with a slope of -1 in the high SNR region. This indicates that the error rate decreases slowly as SNR increases, which is a highly undesirable situation. The most effective way to improve this phenomenon is to use multiple-antenna techniques. If one or both of the transmission and reception ends have multiple antennas, and the antenna spacing is large enough, then any link between any pair of transmitting and receiving antennas can be regarded as an independent channel. For example, if there are four antennas at the transmitting end and two antennas at the receiving end, there are eight independent channels.[7] If a communication system uses d independent channels to transmit the same signal, the relationship between error rate and SNR in the high SNR region becomes:

$$p_{\mathrm{e}} \propto \frac{1}{\mathrm{SNR}^d}. \tag{4.18}$$

[7]This is an ideal assumption. In reality, the number of effective independent channels is usually smaller than the product of the number of transmitting and receiving antennas.

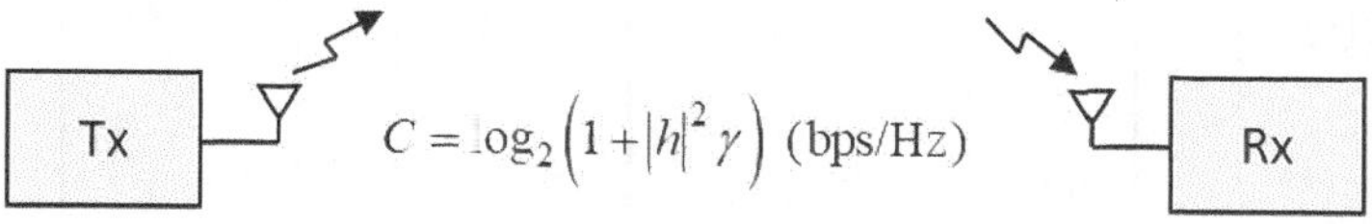

FIGURE 4.8
Illustration of the SISO system and corresponding channel capacity.

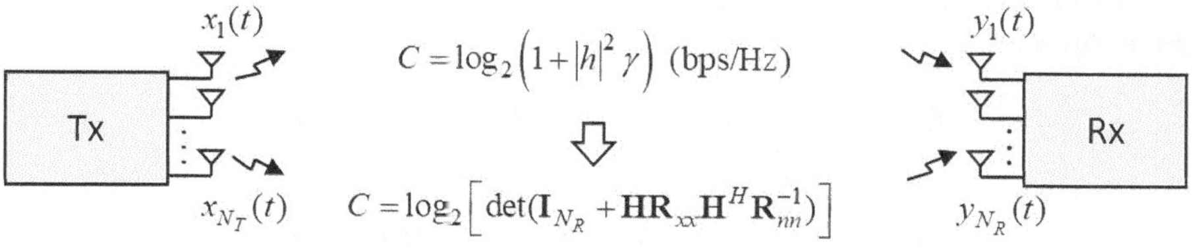

FIGURE 4.9
Illustration of the MIMO system and corresponding channel capacity.

In other words, in the logarithmic scale, the error rate versus SNR curve tends to a straight line with a slope of $-d$ in the high SNR region, indicating that a larger value of d can provide a greater improvement in system reliability. The parameter d in Eq. (4.19) represents the diversity gain,[8] which refers to the effective number of independent channels transmitting the same signal. The maximum diversity gain in a multi-antenna system is equal to the product of the number of transmitting and receiving antennas.

Multi-antenna communication system capacity: In communication systems, capacity is another important performance metric, referring to the amount of effective data transmitted per unit time, with units of bits per second (bps). Capacity is proportional to the system bandwidth, and is therefore often expressed as the amount of data transmitted per second per hertz of bandwidth, with units of bits per second per hertz (bps/Hz), similar to the concept of spectrum efficiency. In this section, we will discuss the impact of multiple antenna techniques on channel capacity. First, let us consider the simplest single-input single-output (SISO) system, as shown in Figure 4.8, where the transmit–receive relationship can be represented by Eq. (4.16). According to Shannon's formula, the channel capacity of a SISO system can be expressed as:

$$C = \log_2(1 + |h|^2\gamma)(\text{bps/Hz})\ ;\ \gamma = \frac{P}{\sigma_n^2}, \tag{4.19}$$

where P is the transmitted power and γ is SNR. Next, let us consider a MIMO system with N_T transmit antennas and N_R receive antennas, as shown in Figure 4.9. When the channel is slow and experiences flat fading, the signal model of the system can be represented as:

[8]In different literature, d is also referred to as the diversity order, but this book adopts the term diversity gain for consistency.

$$\begin{bmatrix} y_1(t) \\ \vdots \\ y_{N_\mathrm{R}}(t) \end{bmatrix} = \begin{bmatrix} h_{11} & \cdots & h_{1N_\mathrm{T}} \\ \vdots & \ddots & \vdots \\ h_{N_\mathrm{R}1} & \cdots & h_{N_\mathrm{R}N_\mathrm{T}} \end{bmatrix} \begin{bmatrix} x_1(t) \\ \vdots \\ x_{N_\mathrm{R}}(t) \end{bmatrix} + \begin{bmatrix} n_1(t) \\ \vdots \\ n_{N_\mathrm{R}}(t) \end{bmatrix}, \tag{4.20}$$

leading to

$$\mathbf{y}(t) = \mathbf{H}\mathbf{x}(t) + \mathbf{n}(t), \tag{4.21}$$

where $\mathbf{H}$ is the channel matrix. Then, the capacity of the MIMO system[9] can be expressed as:

$$C = \log_2[\det(\mathbf{I}_{N_\mathrm{R}} + \mathbf{H}\mathbf{R}_{xx}\mathbf{H}^H\mathbf{R}_{nn}^{-1})](\mathrm{bps/Hz}). \tag{4.22}$$

$$\mathbf{R}_{xx} = \mathbb{E}\{\mathbf{x}(t)\mathbf{x}^H(t)\} \ ; \ \mathbf{R}_{nn} = \mathbb{E}\{\mathbf{n}(t)\mathbf{n}^H(t)\}. \tag{4.23}$$

In Eqs. (4.22) and (4.23), $\det(\cdot)$ denotes the determinant operation and $\mathbf{R}_{xx}$ and $\mathbf{R}_{nn}$ represent the correlation matrices of the transmitted signal and noise, respectively. To ensure fairness, it is generally assumed that the total transmission power of the MIMO system is the same as that of the SISO system, both of which are P. Therefore, it is required that $\mathbf{R}_{xx}$ satisfies the following equation:

$$\mathrm{tr}(\mathbf{R}_{xx}) = P. \tag{4.24}$$

That is to say, the total transmission power of the N_T transmission antennas of the system is equal to P. Figure 4.9 shows the variation in capacity from SISO to MIMO systems, which although different, still exhibit similar structures.

The above discussion is on the channel capacity given the channel $\mathbf{H}$. In reality, $\mathbf{H}$ is random, and therefore the channel capacity is also random. Therefore, in general analysis, the expected value of $\mathbf{H}$ is taken into account as

$$C = \mathbb{E}\{\log_2[\det(\mathbf{I}_{N_\mathrm{R}} + \mathbf{H}\mathbf{R}_{xx}\mathbf{H}^H\mathbf{R}_{nn}^{-1})]\}(\mathrm{bps/Hz}). \tag{4.25}$$

If the transmitted signals and noises from different antennas are statistically independent and follow the same probability distribution (i.e., independent, identically distributed, i.i.d.), then $\mathbf{R}_{xx}$ and $\mathbf{R}_{nn}$ can be expressed as follows:

$$\mathbf{R}_{xx} = \frac{P}{N_\mathrm{T}}\mathbf{I}_\mathrm{T} \ ; \ \mathbf{R}_{nn} = \sigma_n^2\mathbf{I}_\mathrm{R}. \tag{4.26}$$

Then, Eq. (4.26) can be simplified to Eq. (4.28):[10]

$$\begin{aligned} C &= \mathbb{E}\{\log_2[\det(\mathbf{I}_{N_\mathrm{R}} + \frac{\gamma}{N_\mathrm{T}}\mathbf{H}\mathbf{H}^H)]\} \\ &= \mathbb{E}\{\log_2[\det(\mathbf{I}_{N_\mathrm{R}} + \frac{\gamma}{N_\mathrm{T}}\mathbf{H}^H\mathbf{H})]\}(\mathrm{bps/Hz}) \ ; \ \gamma = \frac{P}{\sigma_n^2}. \end{aligned} \tag{4.27}$$

[9] Please refer to refs. [89, 90] for more details.

[10] The property $\det(\mathbf{I} + \mathbf{AB}) = \det(\mathbf{I} + \mathbf{BA})$ is utilized in this simplifying process.

By assuming $N_\mathrm{R} \leq N_\mathrm{T}$, we perform an eigenvalue decomposition (EVD) on $\mathbf{H}\mathbf{H}^H$ in Eq. (4.27) to obtain $\mathbf{U}\Lambda\mathbf{U}^H$, and simplifying it yields to

$$\begin{aligned} C &= \mathbb{E}\{\log_2[\det(\mathbf{I}_{N_\mathrm{R}} + \frac{\gamma}{N_\mathrm{T}}\mathbf{U}\Lambda\mathbf{U}^H)]\} \\ &= \mathbb{E}\{\log_2[\det(\mathbf{U}\mathbf{U}^H) + \frac{\gamma}{N_\mathrm{T}}\mathbf{U}\Lambda\mathbf{U}^H)]\}(\mathrm{bps/Hz}) \\ &= \mathbb{E}\{\log_2[\det(\mathbf{I}_{N_\mathrm{R}} + \frac{\gamma}{N_\mathrm{T}}\Lambda)]\} \\ &= \mathbb{E}\left\{\log_2\left[\prod_{i=1}^{N_\mathrm{R}}(1+\frac{\gamma}{N_\mathrm{T}}\lambda_i)\right]\right\} \\ &= \sum_{i=1}^{N_\mathrm{R}} \mathbb{E}\{\log_2[(1+\frac{\gamma}{N_\mathrm{T}}\lambda_i)]\}(\mathrm{bps/Hz}). \end{aligned} \tag{4.28}$$

In Eq. (4.28), $\lambda_1, \cdots, \lambda_{N_\mathrm{R}}$ are the eigenvalues of $\mathbf{H}^H\mathbf{H}$, and the derivation quotes the equation $\mathbf{U}\mathbf{U}^H = \mathbf{I}_{N_\mathrm{R}}$. Using this result, one can notice that the channel capacity of the MIMO system can be regarded as the sum of N_R sets of parallel SISO channel capacities.[11] Furthermore, according to Jensen's inequality, when the norm of $\mathbf{H}$ is fixed, the maximum value of Eq. (4.28) occurs when $\lambda_1 = \lambda_2 = \cdots = \lambda_{N_\mathrm{R}}$. When N_R is sufficiently large and the elements in the channel $\mathbf{H}$ are independent and identically distributed (i.i.d.), it can be proven that $\lambda_1, \cdots, \lambda_{N_\mathrm{R}}$ are approximately equal to $N_\mathrm{T}\mathbb{E}\{|h_{11}|^2\}$. Then we assume $N_\mathrm{R} \geq N_\mathrm{T}$, we also perform EVD on $\mathbf{H}^H\mathbf{H}$ in Eq. (4.27) and conduct similar derivation. A similar conclusion can be obtained: when N_R is sufficiently large and the elements in the channel $\mathbf{H}$ are i.i.d., it can be proven that $\lambda_1, \cdots, \lambda_{N_\mathrm{T}}$ are approximately equal to $N_\mathrm{R}\mathbb{E}\{|h_{11}|^2\}$. In summary, when the channel $\mathbf{H}$ is full rank and its elements tend to be i.i.d., the following approximate expression for the maximum channel capacity can be obtained:

$$\begin{aligned} C &= N_\mathrm{R}\log_2(1+\mathbb{E}\{|h_{11}|^2\}\gamma)\ (\mathrm{bps/Hz})\ N_\mathrm{R} \leq N_\mathrm{T} \\ &= N_\mathrm{T}\log_2(1+\frac{N_\mathrm{R}}{N_\mathrm{T}}\mathbb{E}\{|h_{11}|^2\}\gamma)\ (\mathrm{bps/Hz})\ N_\mathrm{R} \geq N_\mathrm{T}. \end{aligned} \tag{4.29}$$

Eq. (4.29) has several important implications: (1) The capacity of MIMO channels increases linearly with the smaller number of antennas at the transmitter or receiver, and this increase is called the multiplexing gain, which is $\min\{N_\mathrm{R}, N_\mathrm{T}\}$ in Eq. (4.29). (2) For the effect of multiplexing gain to be significant, $\mathbf{H}$ must be full rank and its elements must approach i.i.d. This condition's physical significance is that the spatial channel must have sufficient rich multipath. From the above discussion, it can be inferred that multipath fading, which is considered a negative factor in traditional communication systems, instead becomes the driving force of channel capacity in MIMO systems. The rich multipath makes the MIMO channel equivalent to multiple

[11] In the case that $\mathbf{H}$ is not full rank, a portion of eigenvalues of $\mathbf{H}\mathbf{H}^H$ will be zero. As a result, the number of parallel SISO channels will be less than N_R.

independent SISO sub-channels, and MIMO systems can use this effect to transmit multiple data streams at the same time and frequency, increasing the data transmission rate. The multiplexing gain effect of MIMO systems is undoubtedly one of the greatest highlights in the wireless communication field in the past 20 years. It can increase system capacity without increasing bandwidth or transmission power, which is particularly attractive for commercial mobile communication with extremely high spectrum costs. MIMO technology has become a necessary item in current mainstream wireless and mobile communication standards.

The above discussion is based on the assumption that the transmitter has no knowledge of the channel $\mathbf{H}$. Therefore, in Eqs. (4.28) and (4.29), it is assumed that the transmitted signals from each antenna and each subchannel have the same power. However, when the transmitter has knowledge of the channel $\mathbf{H}$, the optimal strategy for power allocation is no longer to evenly distribute the power. In this case, the channel capacity should be expressed as follows:

$$C = \log_2[\det(\mathbf{I}_{N_{\mathrm{T}}} + \frac{1}{\sigma_n^2}\mathbf{R}_{xx}\mathbf{H}^H\mathbf{H})] = \sum_{i=1}^{Q} \log_2(1 + \frac{P_i}{\sigma_n^2}\lambda_i)(\mathrm{bps/Hz}), \quad (4.30)$$

where P_i is the power allocated to the ith subchannel, λ_i is the ith eigenvalue of $\mathbf{H}^H\mathbf{H}$, and Q is the rank of $\mathbf{H}$ subject to the total power constraint at the transmitting end:

$$\sum_{i=1}^{Q} P_i = P. \quad (4.31)$$

The optimal set of P_i values that maximizes the capacity in Eq. (4.30) can be obtained through a water-filling procedure, as given below:

$$P_i = (\lambda^{-1} - \lambda_i^{-1})^+, \ i = 1, \cdots, Q. \quad (4.32)$$

and λ is determined by:[12]

$$\sum_{i=1}^{Q} P_i = \sum_{i=1}^{Q} (\lambda^{-1} - \lambda_i^{-1})^+ = P. \quad (4.33)$$

Given the EVD as:

$$\frac{1}{\sigma_n^2}\mathbf{H}^H\mathbf{H} = \mathbf{V}\mathbf{\Lambda}\mathbf{V}^H, \quad (4.34)$$

the covariance matrix of the transmitted signal can be constructed by the following equation:

$$\mathbf{R}_{xx} = \mathbf{V}_Q\mathbf{P}\mathbf{V}_Q^H = \mathbf{V}_Q\mathrm{diag}([P_1, \cdots, P_Q]^T)\mathbf{V}_Q^H. \quad (4.35)$$

In Eq. (4.35), $\mathbf{V}_Q$ is the matrix formed by the first Q rows of $\mathbf{V}$, $\mathbf{P}$ is a $Q \times Q$ diagonal matrix with diagonal elements $\{P_1, \cdots, P_Q\}$. Eq. (4.35) suggests a

[12] $(x)^+ = x$ if $x \geq 0$, $(x)^+ = 0$ if $x \leq 0$.

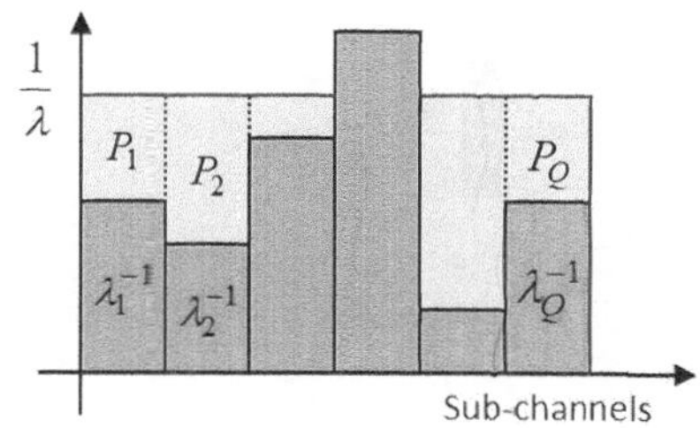

FIGURE 4.10
Illustration of the water-filling method for transmit power allocation.

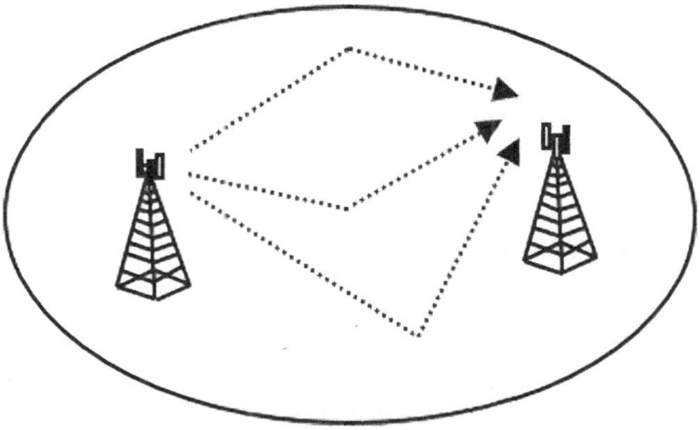

FIGURE 4.11
Illustration of the uncorrelated high-rank model.

signal generation procedure for transmission as follows: (1) Generate a set of independent signals with power levels $P_1, \cdots, P_Q$; (2) Multiply this set of signals by $\mathbf{V}_Q$ and transmit them from each transmitting antenna. The power allocation scheme expressed in Eq. (4.32) can be illustrated using the water-filling analogy, as shown in Figure 4.10: imagine pouring water into a bottom-up uneven pool, where the total amount of water represents the total transmission power, and different heights in the pool represent different sub-channels. The lower heights correspond to poorer sub-channels. When all the water has been poured into the pool, the water level in each area represents the power allocated to the corresponding sub-channel. With this allocation rule, better sub-channels are allocated more power, while some extremely poor sub-channels may not receive any power, such as the most protruding sub-channel in Figure 4.10. Therefore, the spirit of the water-filling method can be said to be "the rich get richer, and the poor get poorer."

MIMO channel models: We conclude this section with a brief classification of MIMO channel models based on the multipath structure. First, the uncorrelated high-rank (UHR) model is shown in Figure 4.11, which assumes

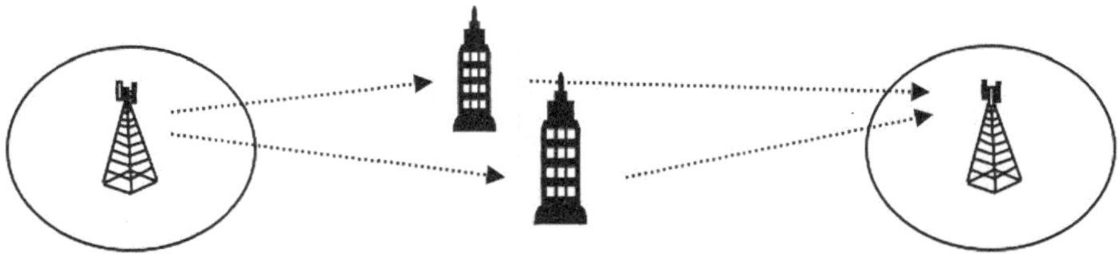

FIGURE 4.12
Illustration of the uncorrelated low-rank model.

FIGURE 4.13
Illustration of the correlated low-rank model.

that the overall multipath structure is rich, and the elements of the channel matrix $\mathbf{H}$ are i.i.d. $\mathcal{CN}(0,1)$ complex Gaussian random variables. In this case, $\mathbf{H}$ has the maximum rank, and the system can achieve the maximum diversity gain $N_\mathrm{R}N_\mathrm{T}$ and the maximum multiplexing gain $\min\{N_\mathrm{R}, N_\mathrm{T}\}$. Second, the uncorrelated low-rank (ULR) model is shown in Figure 4.12, which assumes that the local multipath structure is rich around the transmitting and receiving ends, but the multipath structure between the two ends is sparse. The channel matrix $\mathbf{H}$ can be expressed as Eq. (4.36), where it has i.i.d. channel vector structure $\mathbf{h}_\mathrm{T}$ and $\mathbf{h}_\mathrm{R}$ around the transmitting and receiving ends, which are $\mathcal{CN}(\mathbf{0},\mathbf{I})$ multivariate complex Gaussian distributions, but the line-of-sight (LoS) model is assumed between the two ends with only one path. In this case, the rank of $\mathbf{H}$ is 1, and the system's diversity gain is equal to $\min\{N_\mathrm{R}, N_\mathrm{T}\}$ and the multiplexing gain is 1.

$$\mathbf{H} = \mathbf{h}_\mathrm{R}\mathbf{h}_\mathrm{T}^H,\ \mathbf{h}_\mathrm{R} \sim \mathcal{CN}(\mathbf{0}, \mathbf{I}_{N_\mathrm{R}}); \mathbf{h}_\mathrm{T} \sim \mathcal{CN}(\mathbf{0}, \mathbf{I}_{N_\mathrm{T}}). \tag{4.36}$$

Finally, the correlated low-rank (CLR) model, as shown in Figure 4.13, assumes that the local multipath structures around the transmitter and receiver, as well as between them, are sparse. The channel matrix $\mathbf{H}$ can be expressed as Eq. (4.37), where $\mathbf{u}_\mathrm{T}$ and $\mathbf{u}_\mathrm{R}$ are the spatial feature vectors of the transmitter and receiver, respectively, and h_T and h_R are the fading factors at both ends, with the overall channel being a simple LoS model. In this case, the rank of $\mathbf{H}$ is 1, and both the diversity gain and the multiplexing gain of the system are equal to 1, providing only beamforming gain.

$$\mathbf{H} = h_\mathrm{R}h_\mathrm{T}^*\mathbf{u}_\mathrm{R}\mathbf{u}_\mathrm{T}^H,\ h_\mathrm{R} \sim \mathcal{CN}(0,1); h_\mathrm{T} \sim \mathcal{CN}(0,1). \tag{4.37}$$

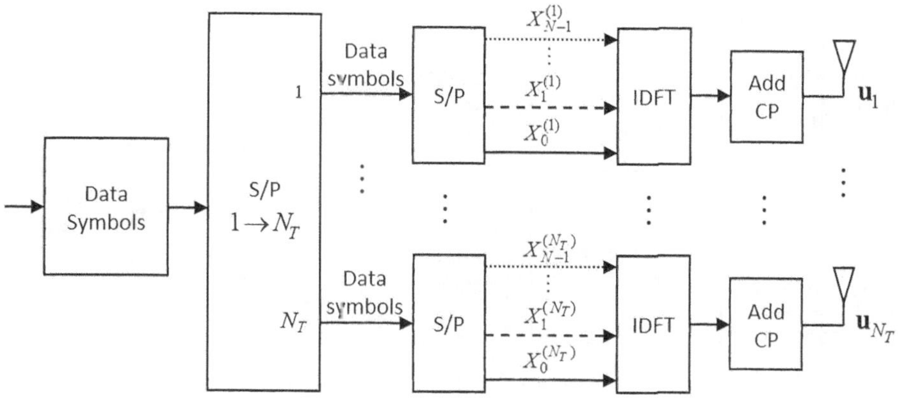

FIGURE 4.14
Illustration of the MIMO-OFDM transmitter architecture.

4.3 MIMO-OFDM System Model

Chapter 3 introduced the principles and advantages of the OFDM system, while this chapter presents the principles and advantages of MIMO. The combination of these two technologies results in the MIMO-OFDM system. MIMO-OFDM has become the mainstream transmission technology in advanced wireless and mobile communication systems. The primary reason for this is that OFDM exhibits a narrowband subcarrier structure, where each subcarrier experiences flat fading. In contrast, MIMO technology is relatively easier to analyze and implement under the assumption of flat fading. MIMO-OFDM can be regarded as a multi-antenna version of OFDM, applying MIMO technology to each subcarrier of OFDM. As shown in Figures 4.14 and 4.15, assuming the system has N subcarriers, the transmitter has $N_{\rm T}$ antennas, and the receiver has $N_{\rm R}$ antennas. In this framework, the transmitter divides the data symbols into $N_{\rm T}$ data streams and performs the same OFDM modulation processing at each transmitting antenna. At the receiver, the signals from each receiving antenna undergo OFDM demodulation processing, resulting in $N_{\rm R}N$ received signals. At this point, a MIMO signal can be observed on each subcarrier, consisting of $N_{\rm T}$ transmitted signal components and $N_{\rm R}$ received signal components. For each subcarrier, the $N_{\rm R}$ received signals can be detected using MIMO detection, thereby recovering the NT data streams and original data symbols transmitted. Before delving into the detailed discussion of the MIMO-OFDM system model, let's first explain the general MIMO system model, taking into account channel effects as depicted in Figure 4.16. In this model, the received signal can be expressed as Eq. (4.38), where $\mathbf{y}[n]$, $\mathbf{H}[n]$, $\mathbf{x}[n]$, and $\mathbf{n}[n]$ represent the received signal, channel, transmitted signal,

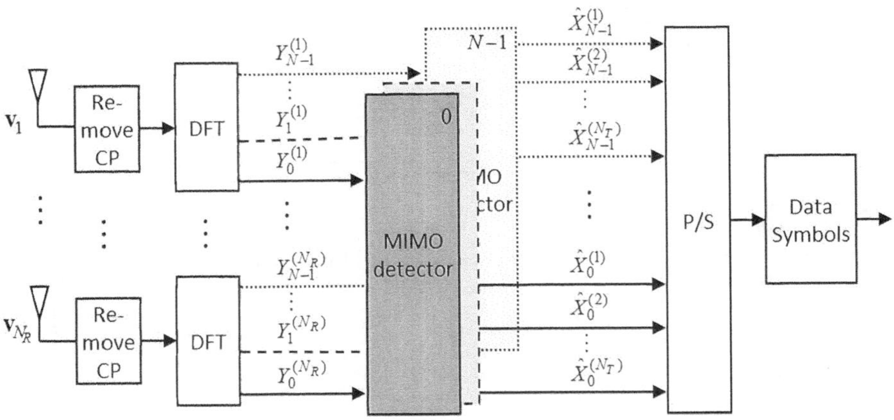

FIGURE 4.15
Illustration of the MIMO-OFDM receiver architecture.

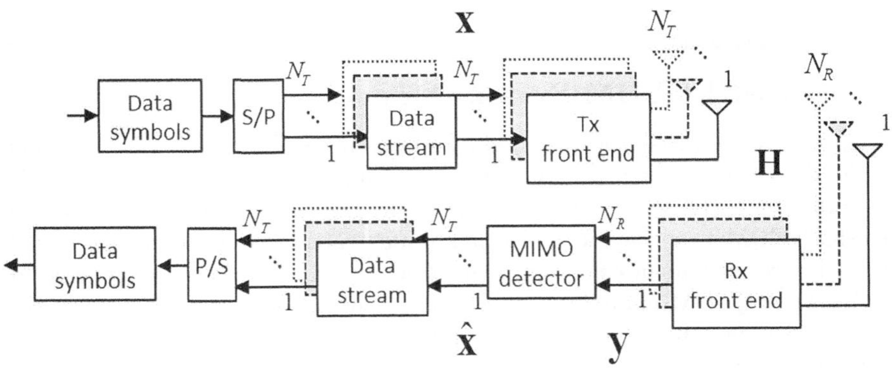

FIGURE 4.16
Illustration of the general architecture of MIMO systems.

and noise, respectively.[13] The specific details are described in Eqs. (4.39) and (4.40):

$$\mathbf{y}[n] = \sum_{i=0}^{L_h-1} \mathbf{H}[l]\mathbf{x}[n-l] + \mathbf{n}[n]. \tag{4.38}$$

$$\begin{aligned} &\mathbf{y}[n] = [y^{(1)}[n], \cdots, y^{(N_\mathrm{R})}[n]]^T; \mathbf{x}[n] = [x^{(1)}[n], \cdots, x^{(N_\mathrm{R})}[n]]^T \\ &\mathbf{n}[n] = [n^{(1)}[n], \cdots, n^{(N_\mathrm{R})}[n]]^T. \end{aligned} \tag{4.39}$$

[13]We consider the digital signals after sampling in this section.

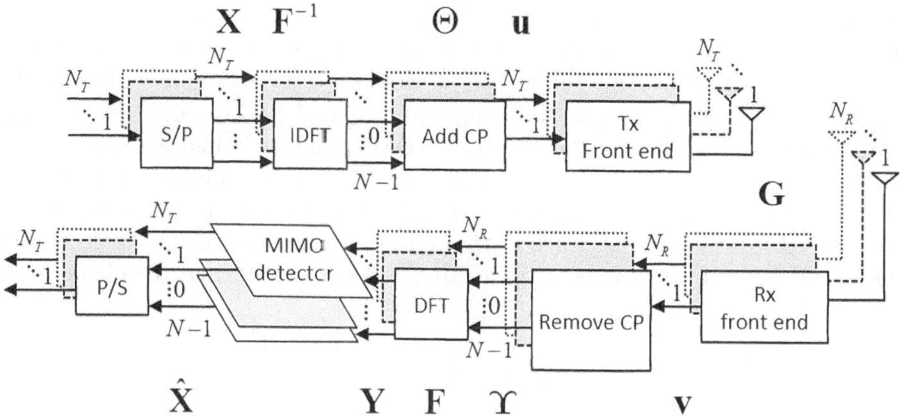

FIGURE 4.17
Illustration of the architecture of MIMO-OFDM systems.

$$\mathbf{H}[n] = \begin{bmatrix} h_{11}[n] & h_{12}[n] & \cdots & h_{1N_\mathrm{T}}[n] \\ h_{21}[n] & h_{22}[n] & \cdots & h_{2N_\mathrm{T}}[n] \\ \vdots & \vdots & \ddots & \vdots \\ h_{N_\mathrm{R}1}[n] & h_{N_\mathrm{R}2}[n] & \cdots & h_{N_\mathrm{R}N_\mathrm{T}}[n] \end{bmatrix} \tag{4.40}$$

It is worth noting that in this context, the channel is assumed to exhibit frequency-selective fading, where L_h represents the length of the time-domain channel response. In Eq. (4.40), $h_{ij}[n]$ denotes the SISO time-domain channel response between the ith receiving antenna and the jth transmitting antenna, while $\mathbf{H}[n]$ can be viewed as the nth coefficient of the MIMO time-domain channel response. In the case of a flat fading channel, the aforementioned MIMO system can be simplified to Eq. (4.41):

$$\mathbf{y}[n] = \mathbf{H}\mathbf{x}[n] + \mathbf{n}[n]. \tag{4.41}$$

In this scenario, the analysis and processing of the system become more straightforward, and the benefits of MIMO are easier to realize. However, in most communication systems today, there is a higher demand for bandwidth, and the channels are predominantly characterized by frequency-selective fading. To achieve a flat fading channel environment, the most direct and effective approach is to divide the system's bandwidth into multiple smaller subbands, known as subcarriers, and implement MIMO technology on each subcarrier. Based on this concept, the combination of MIMO and OFDM becomes a natural choice.

In this section, we will introduce the MIMO-OFDM signal model, depicted in Figure 4.17. In the model, $\mathbf{X}$ represents the frequency-domain transmitted signal (data symbols) with a length of NN_T, u represents the time-domain transmitted signal (OFDM symbols) with a length of $(N+N_{\mathrm{g}})N_{\mathrm{T}}$, including the CP, $\mathbf{v}$ represents the time-domain received signal with a length of $(N+N_{\mathrm{g}})N_{\mathrm{R}}$, $\mathbf{Y}$ represents the frequency-domain received signal with a length of NN_{R}, and $\hat{\mathbf{X}}$ represents the estimated value of $\mathbf{X}$. Due to the temporal and spatial dimensions of the MIMO-OFDM signal, it is highly complex. To achieve clarity and conciseness, we represent it in matrix and vector forms. In Figure 4.17, the original transmitted signal (data symbols) is first divided into N_{T} parallel data streams. Each data stream is then divided into code blocks of length N, and these signals are arranged in a vector format, resulting in $\mathbf{X}$ given as:

$$\mathbf{X} = [\mathbf{X}_0^T \mathbf{X}_1^T \cdots \mathbf{X}_{N-1}^T]^T; \mathbf{X}_k = [X_k^{(1)} X_k^{(2)} \cdots X_k^{(N_{\mathrm{T}})}]^T. \tag{4.42}$$

Next, each block of $\mathbf{X}$ undergoes an inverse discrete Fourier transform (IDFT) operation, which is equivalent to multiplying by the IDFT matrix $\mathbf{F}^{-1}$. This yields N_{T} time-domain signals of length N. Each time-domain signal is then appended with a CP by multiplying it with the $\boldsymbol{\Theta}$ matrix, and the processed signals are arranged in a vector format, resulting in the OFDM symbols $\mathbf{u}$, where the relationship between $\mathbf{u}$ and $\mathbf{X}$ is expressed as:

$$\mathbf{u} = (\boldsymbol{\Theta} \otimes \mathbf{I}_{N_{\mathrm{T}}})(\mathbf{F}^{-1} \otimes \mathbf{I}_{N_{\mathrm{T}}})\mathbf{X} = (\boldsymbol{\Theta}\mathbf{F}^{-1} \otimes \mathbf{I}_{N_{\mathrm{T}}})\mathbf{X}, \tag{4.43}$$

where $\otimes$ represents the Kronecker product, defined as:

$$\mathbf{A} = \begin{bmatrix} a_{11} & a_{12} \\ a_{21} & a_{22} \end{bmatrix} \Rightarrow \mathbf{A} \otimes \mathbf{B} = \begin{bmatrix} a_{11}\mathbf{B} & a_{12}\mathbf{B} \\ a_{21}\mathbf{B} & a_{22}\mathbf{B} \end{bmatrix}, \tag{4.44}$$

and the structure of the CP matrix $\boldsymbol{\Theta}$ is given by

$$\boldsymbol{\Theta} = \begin{bmatrix} \mathbf{O}_{N_{\mathrm{g}} \times (N-N_{\mathrm{g}})} & \mathbf{I}_{N_{\mathrm{g}}} \\ \multicolumn{2}{c}{\mathbf{I}_N} \end{bmatrix}_{(N+N_{\mathrm{g}}) \times N}. \tag{4.45}$$

It is worth noting that the structure of $\mathbf{X}$ involves grouping by subcarriers as sub-vectors $\{\mathbf{X}_0, \mathbf{X}_1, ..., \mathbf{X}_{N-1}\}$, and within each sub-vector, elements are grouped by transmitting antennas as $\{X_k^{(1)}, ..., X_k^{(N_{\mathrm{T}})}\}$. This differs from the previous procedure, hence the need to represent it using Kronecker product notation. The advantage of the notation in Eq. (4.42) is that it treats the MIMO-OFDM signal as a vector version of the OFDM signal. This allows for a focus on the processing at the OFDM level before addressing the MIMO aspect.

When the CP is shorter than the delay spread, the received signal corresponding to a particular OFDM symbol is only interfered by the previous OFDM symbol, as shown in Figure 4.18. Therefore, the received signal $\mathbf{v}$ after passing through the channel can be represented as:

$$\mathbf{v} = \mathbf{G} \cdot \mathbf{u} + \mathbf{G}^{(-)} \cdot \mathbf{u}^{(-)} + \mathbf{n}, \tag{4.46}$$

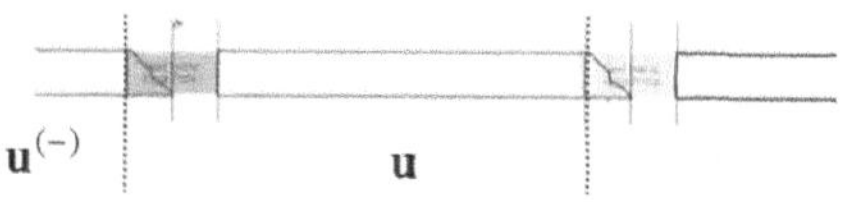

FIGURE 4.18
Illustration of the inter-symbol interference effect in OFDM systems.

where $\mathbf{u}$ and $\mathbf{u}^{(-)}$ are the current and previous OFDM symbols, respectively, $\mathbf{n}$ is the noise, and $\mathbf{G}$ and $\mathbf{G}^{(-)}$ are the channels corresponding to $\mathbf{u}$ and $\mathbf{u}^{(-)}$, respectively, given as:

$$\mathbf{G} = \begin{bmatrix} \mathbf{H}[0] & \mathbf{O} & \cdots & \mathbf{O} & \mathbf{O} & \mathbf{O} & \mathbf{O} \\ \mathbf{H}[1] & \mathbf{H}[0] & \ddots & \vdots & \mathbf{O} & \mathbf{O} & \mathbf{O} \\ \vdots & \mathbf{H}[1] & \ddots & \mathbf{O} & \vdots & \mathbf{O} & \mathbf{O} \\ \mathbf{H}[L_h-1] & \vdots & \ddots & \mathbf{H}[0] & \mathbf{O} & \cdots & \mathbf{O} \\ \mathbf{O} & \mathbf{H}[L_h-1] & \vdots & \mathbf{H}[1] & \mathbf{H}[0] & \mathbf{O} & \vdots \\ \vdots & \vdots & \ddots & \vdots & \mathbf{H}[1] & \ddots & \mathbf{O} \\ \mathbf{H}[0] & \mathbf{O} & \cdots & \mathbf{H}[L_h-1] & \cdots & \mathbf{H}[1] & \mathbf{H}[0] \end{bmatrix}; \tag{4.47}$$

and

$$\mathbf{G}^{(-)} = \begin{bmatrix} \mathbf{O} & \vdots & \mathbf{H}[L_h-1] & \vdots & \mathbf{H}[1] \\ \vdots & \ddots & \vdots & \ddots & \vdots \\ \mathbf{O} & \cdots & \mathbf{O} & \vdots & \mathbf{H}[L_h-1] \\ \vdots & \ddots & \vdots & \ddots & \vdots \\ \mathbf{O} & \cdots & \mathbf{O} & \vdots & \mathbf{O} \end{bmatrix}. \tag{4.48}$$

Next, the receiving end reverses the signal processing steps performed at the transmitting end. Specifically, the CP is first removed, followed by applying the discrete Fourier transform (DFT) operation to obtain $\mathbf{Y}$, expressed as:

$$\begin{aligned} \mathbf{Y} &= (\mathbf{F} \otimes \mathbf{I}_{N_\mathrm{R}})(\boldsymbol{\Upsilon} \otimes \mathbf{I}_{N_\mathrm{R}})\mathbf{v} \\ &= (\mathbf{F} \otimes \mathbf{I}_{N_\mathrm{R}})(\boldsymbol{\Upsilon} \otimes \mathbf{I}_{N_\mathrm{R}})\mathbf{G}(\boldsymbol{\Theta} \otimes \mathbf{I}_{N_\mathrm{T}})(\mathbf{F}^{-1} \otimes \mathbf{I}_{N_\mathrm{T}})\mathbf{X} + (\mathbf{F}\boldsymbol{\Upsilon} \otimes \mathbf{I}_{N_\mathrm{R}})\mathbf{n} \\ &= [\mathbf{Y}_0^T \mathbf{Y}_1^T \cdots \mathbf{Y}_{N-1}^T]^T; \mathbf{Y}_k = [Y_k^{(1)} Y_k^{(2)} \cdots Y_k^{(N_\mathrm{T})}]^T, \end{aligned} \tag{4.49}$$

where $\boldsymbol{\Upsilon}$ is given as:

$$\boldsymbol{\Upsilon} = [\mathbf{O}_{N \times N_\mathrm{g}} \mathbf{I}_{N \times N}]. \tag{4.50}$$

By combining the removal of the CP and the channel effects, an equivalent channel matrix $\mathbf{H}_e$ can be formed as:

$$\mathbf{H}_e = (\boldsymbol{\Upsilon} \otimes \mathbf{I}_{N_\mathrm{R}})\mathbf{G}(\boldsymbol{\Theta} \otimes \mathbf{I}_{N_\mathrm{T}})$$
$$= \begin{bmatrix} \mathbf{H}[0] & 0 & 0 & 0 & \mathbf{H}[L_h-1] & \cdots & \mathbf{H}[1] \\ \mathbf{H}[1] & \mathbf{H}[0] & 0 & \vdots & 0 & 0 & \mathbf{H}[2] \\ \vdots & \mathbf{H}[1] & \ddots & 0 & \vdots & 0 & \vdots \\ \mathbf{H}[L_h-1] & \vdots & \ddots & \mathbf{H}[0] & 0 & \vdots & \mathbf{H}[L_h-1] \\ 0 & \mathbf{H}[L_h-1] & \vdots & \mathbf{H}[1] & \mathbf{H}[0] & 0 & \vdots \\ \vdots & \vdots & \ddots & \vdots & \ddots & \ddots & 0 \\ 0 & 0 & \cdots & \mathbf{H}[L_h-1] & \mathbf{H}[L_h-2] & \cdots & \mathbf{H}[0] \end{bmatrix} \tag{4.51}$$

It can be observed that $\mathbf{H}_e$ is a block circulant matrix. Utilizing the properties of matrix algebra, the block circulant matrix $\mathbf{H}_e$, after undergoing block IDFT and block DFT operations, becomes a block diagonal matrix $\mathbf{H}$, given as:

$$\mathbf{H} = (\mathbf{F}\boldsymbol{\Upsilon} \otimes \mathbf{I}_{N_\mathrm{R}})\mathbf{H}_e(\mathbf{F}^{-1}\boldsymbol{\Upsilon} \otimes \mathbf{I}_{N_\mathrm{T}}) = \begin{bmatrix} \mathbf{H}_0 & \mathbf{O} & \cdots & \mathbf{O} \\ \mathbf{O} & \mathbf{H}_1 & \ddots & \vdots \\ \vdots & \ddots & \ddots & \mathbf{O} \\ \mathbf{O} & \cdots & \mathbf{O} & \mathbf{H}_{N-1} \end{bmatrix} \tag{4.52}$$

By the above, Eq. (4.49) can then be simplified as:

$$\mathbf{Y} = (\mathbf{F}\boldsymbol{\Upsilon} \otimes \mathbf{I}_{N_\mathrm{R}})(\boldsymbol{\Upsilon} \otimes \mathbf{I}_{N_\mathrm{R}})\mathbf{G}(\boldsymbol{\Theta} \otimes \mathbf{I}_{N_\mathrm{T}})(\mathbf{F}^{-1} \otimes \mathbf{I}_{N_\mathrm{T}})\mathbf{X} + \mathbf{N} = \mathbf{H}\mathbf{X} + \mathbf{N}, \tag{4.53}$$

where the relationship between the transmitted and received signals for the kth subcarrier is expressed as:

$$\mathbf{Y}_k = \mathbf{H}_k\mathbf{X}_k + \mathbf{N}_k, \tag{4.54}$$

where

$$\mathbf{Y}_k = \begin{bmatrix} Y_k^{(1)} \\ \vdots \\ Y_k^{(N_\mathrm{R})} \end{bmatrix} = \begin{bmatrix} H_k^{(1,1)} & H_k^{(1,2)} & \cdots & H_k^{(1,N_\mathrm{T})} \\ H_k^{(2,1)} & \ddots & \ddots & \vdots \\ \vdots & \ddots & \ddots & \vdots \\ H_k^{(N_\mathrm{R},1)} & \cdots & \cdots & H_k^{(N_\mathrm{R},N_\mathrm{T})} \end{bmatrix} \begin{bmatrix} X_k^{(1)} \\ \vdots \\ X_k^{(N_\mathrm{T})} \end{bmatrix} + \begin{bmatrix} N_k^{(1)} \\ \vdots \\ N_k^{(N_\mathrm{R})} \end{bmatrix} \tag{4.55}$$

It can be seen from Eq. (4.55) that although MIMO-OFDM appears complex, it can be simplified as treating different subcarriers as independent flat-fading MIMO systems on the corresponding subcarriers.

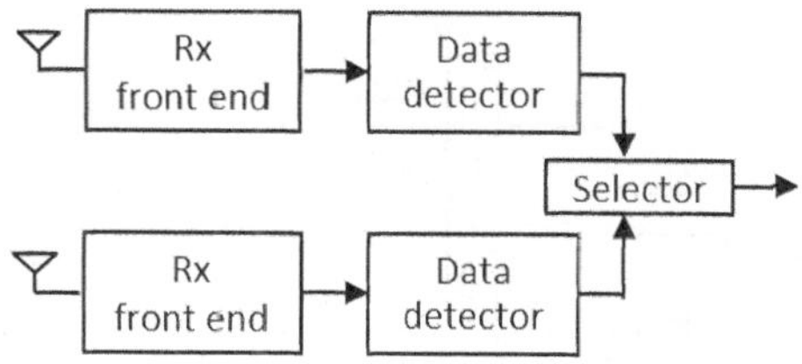

FIGURE 4.19
Illustration of the selection diversity.

In the following subsection, based on the signal model described in Eq. (4.55), we will explore the encoding, detection, and precoding techniques for MIMO systems. Before discussing MIMO encoding, let us introduce several spatial diversity approaches.

Receive diversity: The concept of receive diversity aims to improve the reliability of signal reception by exploiting the characteristics of different received signals experiencing diverse channel fading. Specifically, in scenarios where both the transmitter and receiver have a single antenna, channel fading may lead to unstable reception at the receiver. However, if the receiver is equipped with multiple antennas, and the antennas are sufficiently spaced apart, the channels between the receiver and transmitter antennas can be considered uncorrelated. In this case, as long as one of the channels provides good reception, the receiver can reliably capture the signal, thereby enhancing reception reliability. Receive diversity can be achieved through mechanisms such as selection diversity, switched diversity, and linear combining. Selection diversity involves choosing the best-quality signal among all received signals, as illustrated in Figure 4.19. Switched diversity entails switching to another receiving antenna when the quality of the received signal falls below a predefined threshold, as shown in Figure 4.20. Linear combining involves combining the signals from multiple receiving antennas linearly to obtain the final received signal, as depicted in Figure 4.21.[14] The commonly used linear combining techniques are equal gain combining (EGC) and maximum ratio combining (MRC). Assuming there are two receiving antennas at the receiver, and each antenna receives a signal as follows: $r_1[n] = A_1 e^{j\phi_1} x[n] + n_1[n], r_2[n] = A_2 e^{j\phi_2} x[n] + n_2[n]$. After linear combining, we obtain:

$$y[n] = \alpha_1 r_1[n] + \alpha_2 r_2[n] \tag{4.56}$$

The mechanism of EGC is to sum the received signals from each antenna after compensating for their phases without adjusting their amplitudes. The corresponding weights for EGC are $\alpha_1 = e^{-j\phi_1}, \alpha_2 = e^{-j\phi_2}$. On the other hand, the mechanism of MRC is to compensate for the phase of each received signal and multiply it by its amplitude before summing them. The corresponding

[14]For the sake of simplicity, let's consider the explanation using two receiving antennas.

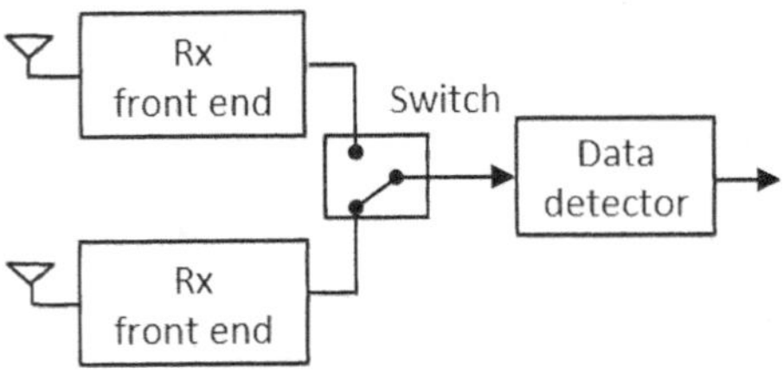

FIGURE 4.20
Illustration of the switched diversity.

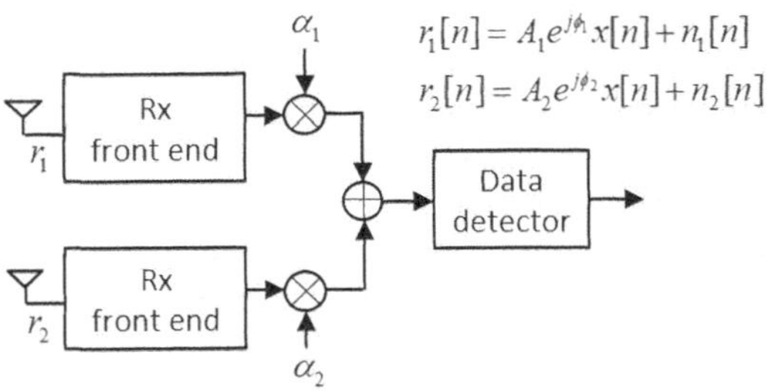

FIGURE 4.21
Illustration of the linear combining.

weights for MRC are $\alpha_1 = A_1 e^{-j\phi_1}, \alpha_2 = A_2 e^{-j\phi_2}$. As the name suggests, the objective of MRC is to maximize the SNR after combining. Its mechanism is equivalent to a matched filter or a matched beamformer, where the receiver matches the spatial response of the received signals.

Transmit diversity: The concept of transmit diversity in wireless communication systems aims to enhance reception reliability by utilizing signals transmitted from different antennas, which may experience diverse channel fading characteristics. Similar to receive diversity, if the transmitter has multiple antennas with sufficient spacing between them, the channels between each transmit antenna and the receive antenna can be considered uncorrelated. Therefore, as long as any one of the channels is good enough, the receiver can stably receive the signal and improve reception reliability. At first glance, transmit diversity may seem similar to receive diversity, where the signal processing at the receiver is moved to the transmitter. However, there is a fundamental difference between them, which is the presence or absence of channel state information (CSI). In most wireless communication systems, coherent demodulation is employed at the receiver, allowing it to estimate the channel response based on reference signals or pilot signals. However, the transmitter does not have access to this channel information. In time division duplex (TDD) systems, the reciprocity of the channel allows the transmitter to directly obtain CSI. In frequency division duplex (FDD) systems, the transmitter needs to rely on feedback from the receiver to obtain CSI, but the accuracy of CSI at the transmitter may be compromised if the channel changes rapidly. Considering this practical factor, the design of transmit diversity strategies needs to take into account whether the transmitter has CSI or not. When the transmitter has CSI, it can utilize the reverse operation of receive diversity to achieve diversity gain at the receiver. However, in the case where the transmitter lacks CSI, the diversity mechanism is referred to as open-loop transmit diversity, as the system does not receive CSI feedback from the receiver. One representative example of open-loop transmit diversity is cyclic delay diversity (CDD), which is commonly used in OFDM systems.[15] The relationship between the transmitted and received signals in CDD can be expressed as follows:

$$r[n] = h_1[n] * s[n] + h_2[n] * s[((n - n_d))_N], \tag{4.57}$$

where $h_1[n]$ and $h_2[n]$ represent the channels between the two transmit antennas and the receive antenna. $(())_N$ denotes the cyclic shift with a period of N (number of subcarriers). In an OFDM system, the time-domain cyclic delay is equivalent to a phase shift in the frequency domain. Assuming the original frequency-domain symbol for the kth subcarrier is X_k, the relationship between $s[n]$ and X_k is given by the IDFT:

$$s[n] = \frac{1}{N}\sum_{k=0}^{N-1} X_k e^{j2\pi\frac{kn}{N}}, \; n = 0, 1, \cdots, N-1. \tag{4.58}$$

In this case, the frequency-domain received signal can be represented as follows:

$$Y_k = H_k^{(1)} X_k + H_k^{(2)} e^{j2\pi\frac{kn_d}{N}} X_k = (H_k^{(1)} + H_k^{(2)} e^{j2\pi\frac{kn_d}{N}})X_k, \tag{4.59}$$

where the relationship between Y_k and $r[n]$ can be expressed as the DFT:

$$Y_k = \sum_{n=0}^{N-1} r[n] e^{j2\pi\frac{kn}{N}}, \; k = 0, 1, \cdots, N-1. \tag{4.60}$$

$H_k^{(1)}$ and $H_k^{(2)}$ are the frequency responses of $h_1[n]$ and $h_2[n]$, respectively. From Eq. (4.59), it can be observed that CDD modifies the frequency response of the channel, transforming the original channel H_k into an equivalent composite channel $H_k^{(c)}$:

$$H_k^{(c)} = H_k^{(1)} + H_k^{(2)} e^{j2\pi\frac{kn_d}{N}}. \tag{4.61}$$

The key point of the above discussion is that CDD introduces delay, causing additional phase variations on different subcarriers at the receiver. This is analogous to deliberately adding frequency-selective fading effects to the channel, allowing the system to achieve frequency diversity gain at the receiver without requiring CSI at the transmitter.

[15] We continue with the example of two transmit antennas. The CDD technique has been employed in the 3GPP LTE-A (Long Term Evolution-Advanced) standard.

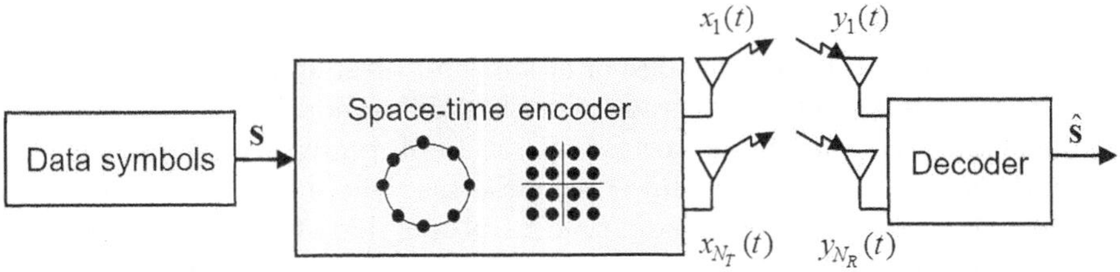

FIGURE 4.22
Illustration of space-time codes.

Another method to achieve open-loop transmit diversity is through space-time codes, as shown in Figure 4.22. Space-time codes encode the transmitted symbols in the time and spatial domains, utilizing the redundancy provided by these domains to achieve diversity gain at the receiver. Similar to channel coding, space-time codes can be classified into space-time trellis codes (STTC) and space-time block codes (STBC). When there is sufficient space-time redundancy, space-time codes can also provide limited error correction capabilities. In practice, STTCs are more complex and difficult to implement, so this section will focus on STBCs. The STBC encodes the input symbol block $\mathbf{s}$ into a space-time codeword $\mathbf{X}$, as shown in the following equation:

$$\mathbf{s} = [s_1, s_2, \cdots, s_L]^T \rightarrow \mathbf{X} = \begin{bmatrix} x_1(1) & x_1(2) & \cdots & x_1(K) \\ x_2(1) & x_2(2) & \cdots & x_2(K) \\ \vdots & \vdots & \ddots & \vdots \\ x_{N_\mathrm{T}}(1) & x_{N_\mathrm{T}}(2) & \cdots & x_{N_\mathrm{T}}(K) \end{bmatrix}, \tag{4.62}$$

where L is the time-domain length of the input symbol block, K is the time-domain length of the output space-time codeword, and N_T is the number of transmit antennas.[16] The equation demonstrates that the space-time block code encodes an $L \times 1$ symbol into an $N_\mathrm{T} \times K$ codeword, providing significant redundancy. The code rate in the time domain, denoted as R, is given by $R = L/K$, which can theoretically reach 1. The Alamouti code is a representative space-time block code, as shown in Figure 4.23. The encoding mechanism of the Alamouti code is represented as follows:

$$\mathbf{s} = [s_1, s_2]^T \rightarrow \mathbf{X} = \begin{bmatrix} s_1 & -s_2^* \\ s_2 & s_1^* \end{bmatrix}. \tag{4.63}$$

Hence, the code rate is $R = 1$. Alamouti code encodes two symbols into a 2×2 space-time codeword. At the transmitter, in the first time slot, the two antennas transmit s_1 and s_2 respectively, while in the second time slot, they transmit $-s_2^*$ and s_1^*. At the receiver, let y_1 and y_2 be the received signals in

[16]Commonly used space-time block codes typically satisfy the condition $K \geq N_\mathrm{T}$.

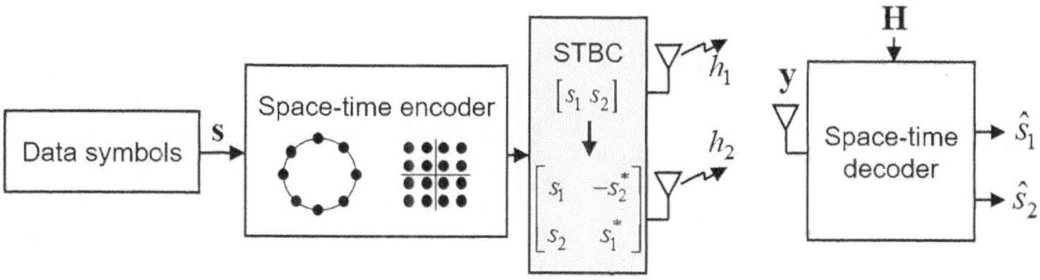

FIGURE 4.23
Illustration of the Alamouti codes.

the first and second time slots, respectively. As shown in

$$\begin{aligned} y_1 &= h_1 s_1 + h_2 s_2 + n_1 \\ y_2 &= -h_1 s_2^* + h_2 s_1^* + n_2, \end{aligned} \tag{4.64}$$

taking the complex conjugate of y_2 and representing it in vector form yields:

$$\mathbf{y} = \begin{bmatrix} y_1 \\ y_2^* \end{bmatrix} = \begin{bmatrix} h_1 & h_2 \\ h_2^* & -h_1^* \end{bmatrix} \begin{bmatrix} s_1 \\ s_2 \end{bmatrix} + \begin{bmatrix} n_1 \\ n_2^* \end{bmatrix} = \mathbf{Hs} + \mathbf{n} \tag{4.65}$$

In Eqs. (4.64) and (4.65), h_1 and h_2 represent the channels between the transmitting antennas and the receiving antennas, n_1 and n_2 are the noise terms, and $\mathbf{H}$ can be seen as the equivalent channel matrix between the transmitted symbols and the received signals. Therefore, the decoding at the receiver can utilize the channel information $\mathbf{H}$ and the received signal $\mathbf{y}$ to recover the transmitted symbols $\mathbf{s}$. The theoretically optimal decoding method is the maximum likelihood (ML) decoding, as shown in Eq. (4.66):

$$\hat{\mathbf{s}} = \arg\min_{\mathbf{s}\in\mathcal{S}} ||\mathbf{y} - \mathbf{Hs}||^2, \tag{4.66}$$

where $\mathcal{S}$ represents the set of symbols. ML decoding requires a multidimensional search, which is impractical in practice, so suboptimal solutions are often used instead. By observation, it can be seen that $\mathbf{H}$ satisfies the condition $\mathbf{H}^H\mathbf{H} = (|h_1|^2 + |h_2|^2)\mathbf{I}$. Therefore, multiplying both sides of Eq. (4.65) by $\mathbf{H}^H\mathbf{H}$ yields:

$$\mathbf{H}^H\mathbf{y} = (|h_1|^2 + |h_2|^2) \begin{bmatrix} s_1 \\ s_2 \end{bmatrix} + \mathbf{H}^H\mathbf{n} \tag{4.67}$$

and

$$\begin{bmatrix} \hat{s}_1 \\ \hat{s}_2 \end{bmatrix} = \frac{1}{|h_1|^2 + |h_2|^2}\mathbf{H}^H\mathbf{y} = \begin{bmatrix} s_1 \\ s_2 \end{bmatrix} + \frac{1}{|h_1|^2 + |h_2|^2}\mathbf{H}^H\mathbf{n}. \tag{4.68}$$

Eq. (4.68) shows that s_1 and s_2 can be directly obtained through normalized $\mathbf{H}^H\mathbf{y}$ decision, making it a relatively simple decoding process. The information suggested in Eq. (4.67) is that the Alamouti code achieves a diversity

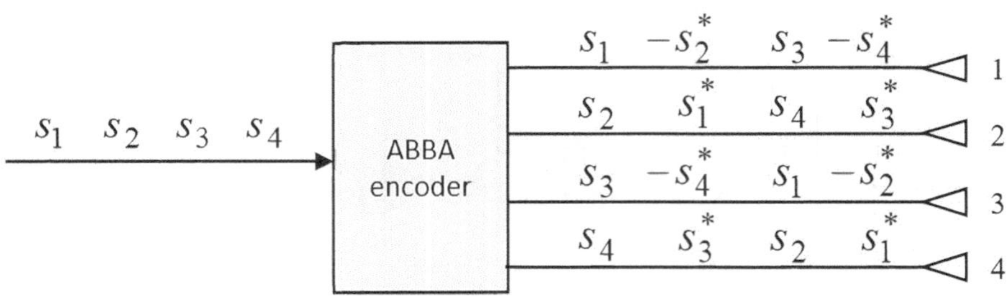

FIGURE 4.24
Illustration of ABBA codes.

gain of 2. Both s_1 and s_2 are transmitted through two channels, h_1 and h_2, to the receiver, which combines the two received signals using maximum ratio combining. In other words, Alamouti code cleverly transforms the scenario of two transmitters and one receiver into an equivalent scenario of one transmitter and two receivers, allowing even a single-antenna receiver to benefit from full diversity gain. Alamouti code is also applicable in the frequency domain, where y_1 and y_2 represent received signals in different frequency bands, and the derivation is similar to the time domain. Due to its excellent performance, simple decoding process, and flexible application, Alamouti code has been widely used in mainstream wireless and mobile communication system standards. However, it is limited to using only two transmit antennas. Alamouti code possesses the property of orthogonal design, expressed as:

$$\mathbf{X}\mathbf{X}^H = (|s_1|^2 + |s_2|^2)\mathbf{I}. \tag{4.69}$$

Orthogonal design is the key factor that enables Alamouti code to possess the aforementioned advantages. In scenarios with more than two antennas, there is no simple design method for STBCs. To maintain diversity gain and simple decoding, a trade-off must be made by reducing the code rate. If the code rate is to be increased, orthogonal design needs to be abandoned. ABBA code is a representative example of a non-orthogonal STBC, as shown in Figure 4.24. Eqs. (4.70) and (4.71) provide the equations of the procedure shown in the figure:

$$\mathbf{s} = [s_1, s_2, s_3, s_4]^T \to \mathbf{X} = \begin{bmatrix} s_1 & -s_2^* & s_3 & -s_4^* \\ s_2 & s_1^* & s_4 & s_3^* \\ s_3 & -s_4^* & s_1 & -s_2^* \\ s_4 & s_3^* & s_2 & s_1^* \end{bmatrix}; \tag{4.70}$$

$$\mathbf{X} = \begin{bmatrix} \mathbf{A} & \mathbf{B} \\ \mathbf{B} & \mathbf{A} \end{bmatrix}; \mathbf{A} = \begin{bmatrix} s_1 & -s_2^* \\ s_2 & s_1^* \end{bmatrix}; \mathbf{B} = \begin{bmatrix} s_3 & -s_4^* \\ s_4 & s_3^* \end{bmatrix}. \tag{4.71}$$

Therefore, the code rate is $R = 1$. ABBA code encodes four symbols into a 4×4 space-time codeword, transmitted over four antennas in four time slots. The principle behind ABBA code is to treat two symbols as a pair and construct Alamouti space-time codewords. These Alamouti codewords are then interleaved to form the final space-time codeword matrix, which gives ABBA code its distinctive structure. Similar to Eq. (4.65), the received signals in the four time slots for the ABBA code can be expressed as:

$$\mathbf{y} = \begin{bmatrix} y_1 \\ y_2^* \\ y_3 \\ y_4^* \end{bmatrix} = \begin{bmatrix} h_1 & h_2 & h_3 & h_4 \\ h_2^* & -h_1^* & h_4^* & -h_3^* \\ h_3 & h_4 & h_1 & h_2 \\ h_4^* & -h_3^* & h_2^* & -h_1^* \end{bmatrix} \begin{bmatrix} s_1 \\ s_2 \\ s_3 \\ s_4 \end{bmatrix} + \mathbf{n} = \mathbf{Hs} + \mathbf{n}. \quad (4.72)$$

Multiplying both sides of the equation by $\mathbf{H}^H$ yields:

$$\mathbf{H}^H\mathbf{y} = \begin{bmatrix} \rho & 0 & \beta & 0 \\ 0 & \rho & 0 & \beta \\ \beta & 0 & \rho & 0 \\ 0 & \beta & 0 & \rho \end{bmatrix} \begin{bmatrix} s_1 \\ s_2 \\ s_3 \\ s_4 \end{bmatrix} + \mathbf{H}^H\mathbf{n},$$
$$\rho = \sum_{n=1}^{4} |h_n|^2; \beta = 2\text{Re}\{h_1^* h_3 + h_2^* h_4\} \quad (4.73)$$

Unlike Alamouti code, the $\mathbf{H}^H\mathbf{H}$ matrix in Eq. (4.73) is not diagonal, indicating inter-symbol interference. Therefore, the receiver needs to perform additional signal processing or even ML decoding to eliminate this interference, which is the cost paid by ABBA code to achieve higher code rates.

Block codes can also be implemented in both spatial and frequency domains, which is known as the concept of space-frequency block code (SFBC). In scenarios with more than two antennas, SFBC can be combined with frequency shift transmit diversity (FSTD) to achieve higher code rates. This combination is referred to as SFBC-FSTD.[17] Let's take the example of Alamouti SFBC to illustrate, as shown in Figure 4.25. The encoding mechanism of SFBC-FSTD can be represented as follows:

$$\mathbf{s} = [s_1, s_2, s_3, s_4]^T \rightarrow \mathbf{X} = \begin{bmatrix} s_1 & s_2 & 0 & 0 \\ -s_2^* & s_1^* & 0 & 0 \\ 0 & 0 & s_3 & s_4 \\ 0 & 0 & -s_4^* & s_3^* \end{bmatrix}. \quad (4.74)$$

The SFBC-FSTD code encodes four symbols into a 4×4 space-frequency codeword, transmitted over four frequencies and one time slot, resulting in a code rate of $R = 4$. The encoding mechanism involves treating two symbols as a group to construct an Alamouti space-frequency codeword, and then arranging two Alamouti space-frequency codewords diagonally to form the final space-frequency codeword matrix. Similar to Eq. (4.72), the received

[17]The SFBC-FSTD technique has been applied in the 3GPP LTE-A standard.

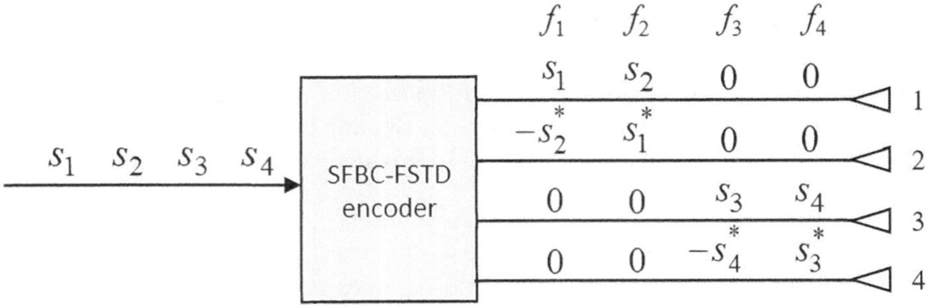

FIGURE 4.25
Illustration of SFBC-FSTD codes.

signals for the four frequencies in SFBC-FSTD can be represented as:

$$\mathbf{y} = \begin{bmatrix} y_1 \\ y_2^* \\ y_3 \\ y_4^* \end{bmatrix} = \begin{bmatrix} h_1 & -h_2^* & 0 & 0 \\ h_2 & h_1^* & 0 & 0 \\ 0 & 0 & h_3 & -h_4^* \\ 0 & 0 & h_4 & h_3^* \end{bmatrix} \begin{bmatrix} s_1 \\ s_2 \\ s_3 \\ s_4 \end{bmatrix} + \mathbf{n} = \mathbf{Hs} + \mathbf{n} \tag{4.75}$$

Then, by multiplying both sides of the equation by $\mathbf{H}^H$, we obtain:

$$\mathbf{H}^H\mathbf{y} = \begin{bmatrix} \rho_1 & 0 & 0 & 0 \\ 0 & \rho_1 & 0 & 0 \\ 0 & 0 & \rho_2 & 0 \\ 0 & 0 & 0 & \rho_2 \end{bmatrix} \begin{bmatrix} s_1 \\ s_2 \\ s_3 \\ s_4 \end{bmatrix} + \mathbf{H}^H\mathbf{n}; \quad \rho_1 = |h_1|^2 + |h_2|^2; \rho_2 = |h_3|^2 + |h_4|^2. \tag{4.76}$$

Unlike the ABBA code, Eq. (4.76) has $\mathbf{H}^H$ as a diagonal matrix, indicating no interference between symbols. This advantage is achieved at the cost of consuming more frequency resources. Similar to the Alamouti code, s_1, s_2, s_3, s_4 can be obtained directly through normalized $\mathbf{H}^H\mathbf{y}$ decision.

4.4 MIMO Signal Detection

In this section, we will discuss the detection techniques for MIMO systems in the scenario of spatial multiplexing without channel information at the transmitter. We continue to assume that the MIMO channel exhibits slow fading and flat fading, as depicted in Figure 4.26. The signal model can be represented as:

$$\mathbf{y}[n] = \mathbf{H}\mathbf{x}[n] + \mathbf{n}[n], \tag{4.77}$$

where $\mathbf{x}[n]$ is the transmitted signal vector, $\mathbf{y}[n]$ is the received signal vector, N_{T} and N_{R} are the numbers of transmit and receive antennas, respectively.

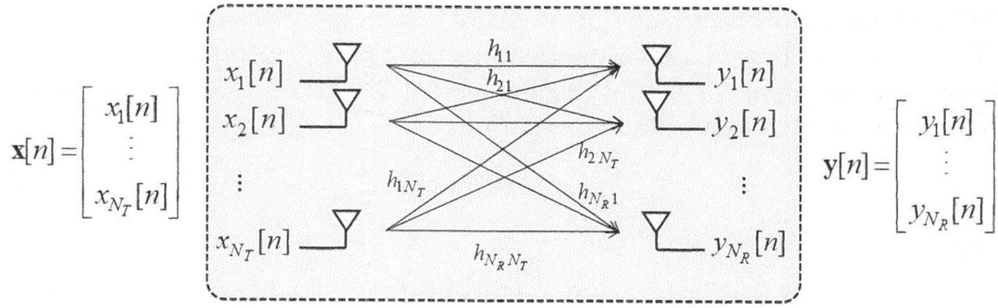

FIGURE 4.26
Illustration of the signal model regarding MIMO systems exhibiting slow fading and flat fading.

$\mathbf{H} = [h_{ij}]$ is the $N_\text{R} \times N_\text{T}$ channel matrix, where h_{ij} denotes the i.i.d. complex Gaussian fading gains following $\mathcal{CN}(0,1)$ distribution. $\mathbf{n}[n]$ is the noise vector with elements being i.i.d. complex Gaussian random variables following $\mathcal{CN}(0,\sigma^2)$, and σ^2 represents the noise power. Finally, we let P as the average transmit power of the transmit antennas and define $\gamma = P/\sigma_n^2$ as the SNR without considering the channel effects.[18] As shown in Figure 4.26 and Eq. (4.77), the original transmitted signal in a MIMO system is divided into N_T sub-signals, which are transmitted from different antennas. After passing through the channel, these sub-signals become intertwined at the receiving antennas, resulting in inter-antenna interference (IAI). Therefore, the receiver needs to employ specific processing techniques to separate the sub-signals and achieve spatial multiplexing. This is the objective of MIMO detection, which aims to recover the transmitted signal $\mathbf{x}[n]$ using the received signal $\mathbf{y}[n]$ and the channel information $\mathbf{H}$. The ML method provides the optimal solution for MIMO detection. Assuming that the transmitted signal comes from a finite set of symbols, ML MIMO detection can be formulated as the following problem:

$$\hat{\mathbf{x}}[n] = \arg\min_{\mathbf{x}[n]\in S} ||\mathbf{y}[n] - \mathbf{H}\mathbf{x}[n]||^2, \tag{4.78}$$

where S represents the set of transmitted symbols. The ML solution is obtained by substituting all possible symbol combinations into Eq. (4.78) and selecting the one with the minimum Euclidean distance. However, ML detection requires an exhaustive search in a high-dimensional space, resulting in high complexity that is generally impractical for real-world communication systems. Therefore, alternative suboptimal solutions are needed. In the following sections, we will introduce three methods: linear detection, nonlinear detection, and sphere decoding. It is assumed that $N_\text{T} \leq N_\text{R}$, meaning that the receiving end has sufficient degrees of freedom to recover the transmitted signal.

[18] To ensure a fair comparison, regardless of the number of transmitting antennas, we assume a fixed total transmit power in MIMO systems.

MIMO linear detection: Linear MIMO detection consists of two steps: (1) suppressing inter-antenna interference through signal processing and (2) recovering the transmitted data using symbol detection. These steps can be represented by the following two equations:

$$\mathbf{z}[n] = \mathbf{W}^H \mathbf{y}[n]; \tag{4.79}$$

and

$$\hat{\mathbf{x}}[n] = \text{Dec}\{\mathbf{z}[n]\}, \tag{4.80}$$

where $\mathbf{W}$ is an $N_{\text{R}} \times N_{\text{T}}$ interference suppression matrix, $\mathbf{z}[n]$ is the received signal after interference suppression, and Dec represents symbol decision. The problem in linear MIMO detection is to find an appropriate $\mathbf{W}$. The most common linear MIMO detection technique is zero-forcing (ZF) detection, which aims to completely eliminate inter-antenna interference while simultaneously accurately recovering all transmitted signals without distortion. The solution for the interference suppression matrix $\mathbf{W}$ in ZF detection is obtained as follows:

$$\mathbf{W} = \mathbf{H}(\mathbf{H}^H\mathbf{H})^{-1} = (\mathbf{H}^+)^H, \tag{4.81}$$

where $\mathbf{H}^+$ represents the pseudo-inverse of $\mathbf{H}$. ZF detection completely suppresses inter-antenna interference, and the processed signal is given by:

$$\mathbf{z}[n] = \mathbf{x}[n] + (\mathbf{H}^H\mathbf{H})^{-1}\mathbf{H}^H\mathbf{n}[n]. \tag{4.82}$$

In the absence of noise, ZF detection is an optimal solution. However, in situations where the channel is ill-conditioned and $\mathbf{H}$ is close to rank deficiency, the pseudo-inverse $\mathbf{H}^+$ may amplify the noise power, thus affecting the performance of the symbol decision. Another commonly used linear MIMO detection technique is MMSE detection. The principle of MMSE detection is to minimize the mean square error (MSE) between the transmitted and received signals. Mathematically, it can be expressed as:

$$\mathbf{W} = \arg\min_{\mathbf{W}} \mathbb{E}\{||\mathbf{x}[n] - \mathbf{W}^H\mathbf{y}[n]||^2\}, \tag{4.83}$$

with the solution as

$$\mathbf{W} = \mathbf{H}(\mathbf{H}^H\mathbf{H} + \frac{N_{\text{T}}}{\gamma}\mathbf{I}_{N_{\text{T}}})^{-1}. \tag{4.84}$$

The processed signal is given by:

$$\mathbf{z}[n] = (\mathbf{H}^H\mathbf{H} + \frac{N_{\text{T}}}{\gamma}\mathbf{I}_{N_{\text{T}}})^{-1}\mathbf{H}^H\mathbf{H}\mathbf{x}[n] + (\mathbf{H}^H\mathbf{H} + \frac{N_{\text{T}}}{\gamma}\mathbf{I}_{N_{\text{T}}})^{-1}\mathbf{H}^H\mathbf{n}[n]. \tag{4.85}$$

MMSE detection takes into account both the channel and noise effects, making it more robust than ZF detection in adverse channel conditions. By examining Eqs. (4.82) and (4.85), we can observe that MMSE detection approximates ZF detection at high SNR and approaches matched filtering at low SNR. Although

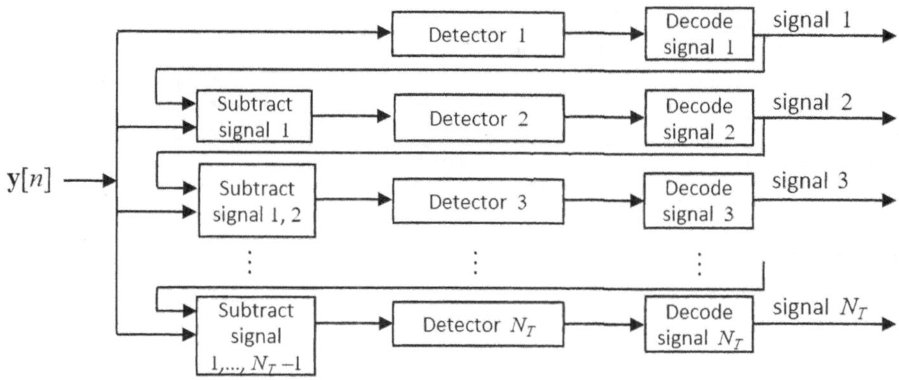

FIGURE 4.27
Illustration of the successive interference cancellation mechanism.

linear detection techniques have lower complexity, their performance may not be sufficient to achieve high-quality transmission in communication systems.
MIMO nonlinear detection: A promising approach to improve linear detection is the introduction of interference cancellation mechanisms. Unlike interference suppression, interference cancellation aims to restore the interfering signals and subtract them from the received signals, thereby purifying the received signals before performing detection. This process leads to enhanced detection performance. Interference cancellation can be broadly classified into two categories: successive interference cancellation (SIC) and parallel interference cancellation (PIC). In the following, we will focus on MIMO detection techniques that combine linear detection with SIC interference cancellation. Since interference restoration and cancellation involve nonlinear operations, these detection techniques are considered nonlinear. The concept of SIC is illustrated in Figure 4.27. When a particular signal is successfully detected, it is subtracted from the received signal in the subsequent detection stage, reducing the interference component in the received signal and facilitating the detection of the next signal. Specifically, the first linear detector detects the first signal, and then the second linear detector subtracts the detected signal from its received signal before detecting the second signal. At this stage, the detector only needs to handle the interference formed by the third to N_{T}th signals, thereby having more degrees of freedom to combat noise and improve detection performance. This process continues until the last linear detector, at which point all interferences have been subtracted. The application of the aforementioned detection process to MIMO detection results in a linear SIC-MIMO detection architecture, as depicted in Figure 4.28. The corresponding detection process is summarized in Figure 4.29. In this context, $z_l[n]$ represents the processed signal of the lth detector, $\mathbf{w}_l$ is the weight vector of the lth detector, $\hat{x}_l[n]$ denotes the decision signal of the lth detector, and $\mathbf{h}_l$ corresponds to the channel vector of the lth signal. The linear detectors in the

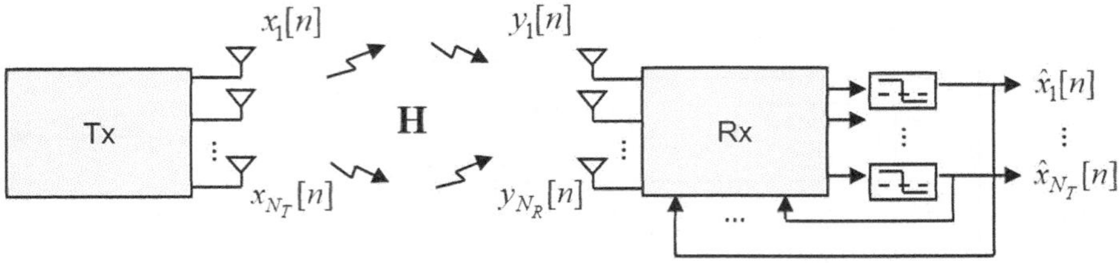

FIGURE 4.28
Illustration of the linear SIC-MIMO detector architecture.

$$
\begin{aligned}
&z_1[n] = \mathbf{w}_1^H \mathbf{y}[n] \\
&\hat{x}_1[n] = \mathrm{Dec}\{z_1[n]\} \\
&\mathbf{y}_{(1)}[n] = \mathbf{y}[n] - \mathbf{h}_1 \hat{x}_1[n] \\
\\
&z_2[n] = \mathbf{w}_2^H \mathbf{y}_{(1)}[n] \\
&\hat{x}_2[n] = \mathrm{Dec}\{z_2[n]\} \\
&\mathbf{y}_{(2)}[n] = \mathbf{y}_{(1)}[n] - \mathbf{h}_2 \hat{x}_2[n] \\
&\vdots \\
&z_{N_T}[n] = \mathbf{w}_{N_T}^H \mathbf{y}_{(N_T-1)}[n] \\
&\hat{x}_{N_T}[n] = \mathrm{Dec}\{z_{N_T}[n]\}
\end{aligned}
$$

FIGURE 4.29
Illustration of the linear SIC-MIMO detection procedure.

aforementioned process can be ZF or MMSE detectors, or any detector that linearly extracts signals from interference. In the SIC detection algorithm described above, it is assumed that the signals eliminated at each stage are correct. However, in practical scenarios, signal detection is not error-free, and detection errors can lead to the propagation of erroneous interference, resulting in more severe interference. This phenomenon is known as error propagation. To mitigate error propagation, a commonly used approach is to adjust the order of signal detection in advance, known as ordered SIC (OSIC). The principle of ordering can be determined based on the system's performance requirements. One intuitive idea is to detect stronger signals earlier since they are less prone to errors. Therefore, the order can be determined based on the SNR values of the signals, i.e., $\mathrm{SNR}_{o1} > \mathrm{SNR}_{o2} > \cdots > \mathrm{SNR}_{oN_\mathrm{T}}$, where SNR_{ol} represents the SNR of the lth original signal after sorting. According to this ordering principle, apart from detecting signals with lower error rates first, it also avoids severe interference caused by high SNR signals to other signals. Another approach to mitigate error propagation is to use channel coding. When a signal is received with errors, channel coding can be employed to correct the errors before feeding the corrected signal into the next stage, reducing the probability of error propagation. The MIMO detection scheme that combines linear interference suppression and OSIC is commonly referred to as V-BLAST (vertical Bell Lab Layered space-time).[19]

MIMO sphere detection: As previously mentioned, the optimal MIMO detection method is the ML method, which seeks the best solution using Eq. (4.78). However, an exhaustive search in practice is infeasible due to its computational complexity. Moreover, linear and SIC) detection methods do not fully achieve the desired performance. Therefore, the exploration of alternative low-complexity nonlinear detection methods, which approach the performance of the ML method, has become a crucial research topic in the field of MIMO.

One such algorithm that fulfills the requirements of high performance and low complexity in MIMO detection is sphere decoding (SD) [92]. SD, also known as the sphere decoder, has gained significant attention in recent years and is widely utilized in the development of practical communication systems. The fundamental idea behind SD is to set a finite search region to reduce the complexity of ML detection. It centers the search around the received signal, within a multidimensional sphere of a given radius D, to replace the global search of ML. In other words, SD only performs ML search for codewords that satisfy the following condition:

$$||\mathbf{y}[n] - \mathbf{H}\mathbf{x}[n]||^2 \leq D^2. \tag{4.86}$$

The concept of sphere decoding may seem simple, but a critical issue in its implementation is how to determine the parameter D. If D is chosen too large, the complexity increases, while selecting it too small may lead to the possibility of missing the true optimal solution. To further reduce complexity, SD utilizes QR decomposition on the channel matrix $\mathbf{H}$, transforming the high-dimensional search into multiple low-dimensional searches. The QR decomposition is represented as Eq. (4.87):

$$\mathbf{H} = [\mathbf{Q}_1\mathbf{Q}_2]\begin{bmatrix} \mathbf{R} \\ \mathbf{O} \end{bmatrix}, \tag{4.87}$$

where $\mathbf{Q}_1$ is an $N_\mathrm{R} \times N_\mathrm{T}$ matrix, $\mathbf{Q}_2$ is an $N_\mathrm{R} \times (N_\mathrm{R} - N_\mathrm{T})$ matrix, $[\mathbf{Q}_1\mathbf{Q}_2]$ is a unitary matrix, and $\mathbf{R}$ is an $N_\mathrm{T} \times N_\mathrm{T}$ upper triangular matrix with the structure as shown below

$$\mathbf{R} = [\mathbf{Q}_1\mathbf{Q}_2]\begin{bmatrix} r_{11} & r_{12} & \cdots & r_{1N_\mathrm{T}} \\ 0 & r_{22} & \cdots & r_{2N_\mathrm{T}} \\ \vdots & \vdots & \ddots & \vdots \\ 0 & 0 & \cdots & r_{N_\mathrm{T}N_\mathrm{T}} \end{bmatrix}, \tag{4.88}$$

The matrix $\mathbf{O}$ is an $(N_\mathrm{R} - N_\mathrm{T}) times N_\mathrm{T}$ all-zero matrix. By substituting Eq. (4.87) into Eq. (4.86) and multiplying the left-hand side by $[\mathbf{Q}_1\mathbf{Q}_2]^H$, we

[19]Please refer to ref. [91] for more details.

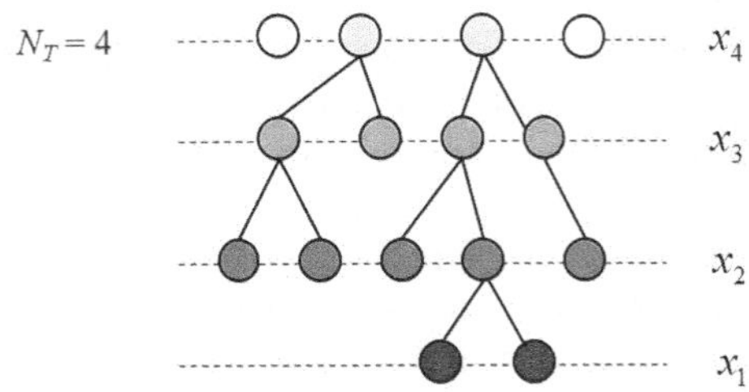

FIGURE 4.30
Illustration of the tree-based search sphere decoding method.

obtain Eq. (4.89) as follows:[20]

$$||\mathbf{Q}_1^H \mathbf{y}[n] - \mathbf{R}\mathbf{x}[n]||^2 \leq D^2 - ||\mathbf{Q}_2^H \mathbf{y}[n]||^2 = d^2. \tag{4.89}$$

Let $\mathbf{z}[n] = \mathbf{Q}_1^H \mathbf{y}[n]$, then Eq. (4.90) can be rewritten as follows:

$$||\mathbf{z}[n] - \mathbf{R}\mathbf{x}[n]||^2 = || \begin{bmatrix} z_1 \\ \vdots \\ z_{N_\mathrm{T}} \end{bmatrix} - \begin{bmatrix} r_{11} \; r_{12} & \cdots & r_{1N_\mathrm{T}} \\ 0 \; r_{22} & \cdots & r_{2N_\mathrm{T}} \\ \vdots \; \vdots & \ddots & \vdots \\ 0 \; 0 & \cdots & r_{N_\mathrm{T}N_\mathrm{T}} \end{bmatrix} \begin{bmatrix} x_1 \\ \vdots \\ x_{N_\mathrm{T}} \end{bmatrix} ||^2 \leq d^2. \tag{4.90}$$

Eq. (4.90) can be expanded to Eq. (4.91) as follows:

$$\sum_{i=1}^{N_\mathrm{T}} |z_i - \sum_{j=1}^{N_\mathrm{T}} r_{ij} x_j|^2 = \sum_{i=1}^{N_\mathrm{T}} B_i \leq d^2. \tag{4.91}$$

where B_i represents the branch metric, indicating the measurement of errors, and the accumulation of branch metrics is known as the path metric, defined as follows:

$$P_l = \sum_{i=1}^{N_\mathrm{T}} P_i. \tag{4.92}$$

From Eq. (4.91), it is evident that solving for x_i only depends on $x_{i+1}, x_{i+2}, \cdots, x_{N_\mathrm{T}}$, thus transforming the high-dimensional sphere search problem into a tree-based search to enhance search efficiency. The representation of the tree-based search in Figure 4.30 depicts each layer performing a search for a single codeword, with different nodes representing different possible values for that codeword. The mathematical description of the tree-based search in Figure 4.30 is as follows: First, in N_Tth layer, the search for x_{N_T} is

[20]By leveraging the property of unitary transformations, the norm of vectors remains unchanged.

performed as follows:

$$|z_{N_{\rm T}} - z_{N_{\rm T}N_{\rm T}} x_{N_{\rm T}}|^2 \leq d^2 - \sum_{i=1}^{N_{\rm T}-1} B_i \approx d^2. \tag{4.93}$$

The approximation used here implies that the currently unknown values $B_1, \cdots, B_{N_{\rm T}-1}$ are discarded and not considered during the search process. After completing the search the $x_{N_{\rm T}}$ (node) satisfying Eq. (4.93) is retained in the $N_{\rm T}$th layer of the tree diagram. Subsequently, for each retained $x_{N_{\rm T}}$, in the $N_{\rm T} - 1$th layer, a search is performed for $x_{N_{\rm T}-1}$ as follows:

$$\begin{aligned} &|z_{N_{\rm T}-1} - r_{(N_{\rm T}-1)N_{\rm T}} x_{N_{\rm T}} - r_{(N_{\rm T}-1)(N_{\rm T}-1)} x_{N_{\rm T}-1}|^2 \\ &\leq d^2 - |z_{N_{\rm T}} - r_{(N_{\rm T})N_{\rm T}} x_{N_{\rm T}}|^2 - \sum_{i=1}^{N_{\rm T}-2} B_i \\ &\approx d^2 - |z_{N_{\rm T}} - r_{(N_{\rm T})N_{\rm T}} x_{N_{\rm T}}|^2. \end{aligned} \tag{4.94}$$

The mentioned approximation implies that the currently unknown values $B_1, \cdots, B_{N_{\rm T}-2}$ are discarded and not considered during the search process. After completing the search, the $x_{N_{\rm T}-1}$ (node) satisfying Eq. (4.94) is recorded in the $N_{\rm T} - 1$th layer of the tree diagram. This process is repeated iteratively until reaching the first layer. After executing the aforementioned search procedure, the optimal solution is determined as the one having the minimum total path metric P_1 among all complete path combinations.

4.5 MIMO Precoding

MIMO precoding is a signal processing technique that applies suitable transformations to the transmitted signals before transmission. This allows the receiving end to achieve the desired system performance. During the design of a precoder, the transmitting end often needs to utilize channel information, as the quality of the channel and the accuracy of its information significantly impact the performance of the precoder. This section will introduce several MIMO precoding design methods under different system configurations and various forms of channel information. From a system architecture perspective, MIMO precoding can be categorized into single-user and multi-user precoding. Regarding channel information, it can be further classified into precoding based on Channel State Information at the Transmitter (CSIT) and precoding based on a codebook.

Single-user MIMO precoding: The concept of SU-MIMO precoding system is depicted in Figure 4.31. In this system, the transmitting end aims to transmit Q signals, and the precoded transmitted signal vector is denoted as:

$$\mathbf{x}[n] = \mathbf{F}\mathbf{s}[n], \tag{4.95}$$

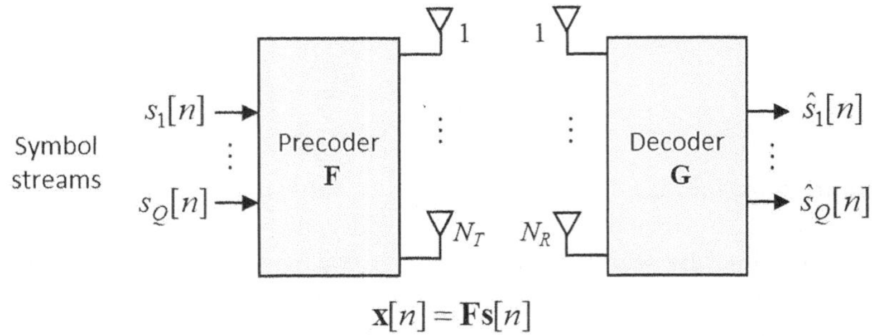

FIGURE 4.31
Illustration of the SU-MIMO precoding systems.

where $\mathbf{s}[n] = [s_1[n], ..., s_Q[n]]^T$ represents the original transmitted signal vector, and $\mathbf{F}$ is the $N_\mathrm{T} \times Q$ precoding matrix. After passing through the channel $\mathbf{H}$, the received signal vector is given by:

$$\mathbf{y}[n] = \mathbf{H}\mathbf{x}[n] + \mathbf{n}[n] = \mathbf{H}\mathbf{F}\mathbf{s}[n] + \mathbf{n}[n]. \tag{4.96}$$

To facilitate the derivation, we assume that $\{s_i[n]\}$ are mutually independent and have unit power, and $\mathbf{n}[n]$ is a noise vector with i.i.d. $\mathcal{CN}(0, \sigma^2)$ elements. Therefore, their respective correlation matrices can be represented as:

$$\mathbf{R}_{ss} = \mathbb{E}\{\mathbf{s}[n]\mathbf{s}^H[n]\} = \mathbf{I}_Q; \mathbf{R}_{nn} = \mathbb{E}\{\mathbf{n}[n]\mathbf{n}^H[n]\} = \sigma_n^2 \mathbf{I}_{N_\mathrm{R}}. \tag{4.97}$$

Finally, after undergoing the corresponding MIMO decoding process, the received signal becomes:

$$\hat{s}[n] = \mathbf{G}\mathbf{y}[n] = \mathbf{G}\mathbf{H}\mathbf{F}\mathbf{s}[n] + \mathbf{G}\mathbf{n}[n], \tag{4.98}$$

where $\mathbf{G}$ is the $Q \times N_\mathrm{R}$ decoding matrix. The purpose of the aforementioned precoding system is to design appropriate matrices $\mathbf{F}$ and $\mathbf{G}$ to ensure that the decoded signal $\hat{s}[n]$ closely approximates the original transmitted signal $s[n]$. It should be noted that the number of original transmitted signals Q must not exceed the smaller value between the number of transmitting antennas and receiving antennas, i.e., $Q \leq \min N_\mathrm{R}, N_\mathrm{T}$, and should not exceed the rank of the channel matrix $\mathbf{H}$ to adhere to the system's multiplexing gain constraint.

Assuming that the transmitting end has complete CSIT, the most common approach for precoding design is to maximize the system transmission rate. This entails designing a precoding matrix that maximizes the MIMO channel capacity, as represented in Eq. (4.99):

$$\max_{\mathbf{F}} \log_2[\det(\mathbf{I}_Q + \frac{1}{\sigma_n^2}\mathbf{F}^H\mathbf{H}^H\mathbf{H}\mathbf{F})](\text{bps/Hz}) \text{ subject to } \mathrm{tr}(\mathbf{F}\mathbf{F}^H) = \mathrm{tr}(\mathbf{R}_{xx}) \leq P \tag{4.99}$$

where $\mathbf{R}_{xx} = \mathbb{E}\{\mathbf{x}[n]\mathbf{x}^H[n]\} = \mathbf{F}\mathbf{F}^H$, as obtained from Eq. (4.97), and P is the total power constraint on the transmitted signals. The design method to maximize the transmission rate, as described previously, resembles the water-filling algorithm, and its optimal solution is given by:

$$\mathbf{R}_{xx}^{\text{opt}} = \mathbf{V}_Q \text{diag}([P_1, \cdots, P_Q]^T)\mathbf{V}_Q^H \tag{4.100}$$

where $\mathbf{V}\Lambda\mathbf{V}^H$ is the eigenvalue decomposition of $(1/\sigma^2)\mathbf{H}^H\mathbf{H}$, $\mathbf{V}_Q$ is the matrix consisting of the first Q rows of $\mathbf{V}$, and $P_1, P_2, ..., P_Q$ represents the power allocated to each transmitted signal. Finally, by comparing Eq. (4.100) with the structure of $\mathbf{R}_{xx} = \mathbf{F}\mathbf{F}^H$, we obtain the optimal precoder as:

$$\mathbf{F}^{\text{opt}} = \mathbf{V}_Q \text{diag}([\sqrt{P_1}, \cdots, \sqrt{P_Q}]^T) \tag{4.101}$$

It is noteworthy that the maximum transmission rate precoding only depends on the transmitting end, hence the decoder can be any arbitrary MIMO detector.

The purpose of MMSE precoding is to minimize the mean square error between the decoded signal and the transmitted signal, as shown in Eq. (4.102):

$$\min_{\mathbf{G},\mathbf{F}} \mathbb{E}\{||\hat{\mathbf{s}}[n] - \mathbf{s}[n]||^2\} \text{ subject to } \text{tr}(\mathbf{F}\mathbf{F}^H) = \text{tr}(\mathbf{R}_{xx}) \leq P \tag{4.102}$$

where $\hat{\mathbf{s}}$ is the estimated transmitted signal. The error vector is given by:

$$\hat{\mathbf{s}}[n] - \mathbf{s}[n] = \mathbf{GHFs}[n] - \mathbf{s}[n] + \mathbf{Gs}[n] \tag{4.103}$$

The solution to Eq. (4.101) can be represented as follows:[21]

$$\mathbf{F} = \mathbf{V}\mathbf{\Phi_f} \ ; \ \mathbf{G} = \mathbf{\Phi_g}\mathbf{V^H}\mathbf{H^H}/\sigma_{\mathbf{n}}^2, \tag{4.104}$$

where $\mathbf{V\Lambda V^H}$ is the eigenvalue decomposition of $\mathbf{H}^H\mathbf{H}$, and the expressions for $\mathbf{\Phi_f}$ and $\mathbf{\Phi_g}$ are given by:

$$\mathbf{\Phi_f} = (\lambda^{-1/2}\mathbf{\Lambda}^{-1/2} - \mathbf{\Lambda}^{-1})_+^{1/2} \ ; \ \mathbf{\Lambda_g} = (\lambda^{1/2}\mathbf{\Lambda}^{-1/2} - \lambda\mathbf{\Lambda}^{-1})_+^{1/2}\mathbf{\Lambda}^{-1/2}, \tag{4.105}$$

Here, λ represents the Lagrange multiplier, which must satisfy the total transmit power constraint, and $(.)_+$ denotes the operation of setting negative elements of a matrix to zero. By substituting Eq. (4.104) into Eq. (4.103) and rearranging, it can be observed that **GHF** becomes a diagonal matrix. This implies that the system, after precoding, channel, and decoding, has an equivalent channel that has been diagonalized. Consequently, the transmitted signals are separated and decoded individually.

Codebook-based MIMO precoding: The precoder designs introduced above are based on the assumption that the transmitter has complete channel information, which is difficult to implement in practice due to the large feedback bandwidth requirement.[22] A more practical approach is to only feedback

[21]Please refer to ref. [93] for more details.

[22]This holds in the context of an FDD system. If a TDD system is employed, the transmitter can independently estimate the channel.

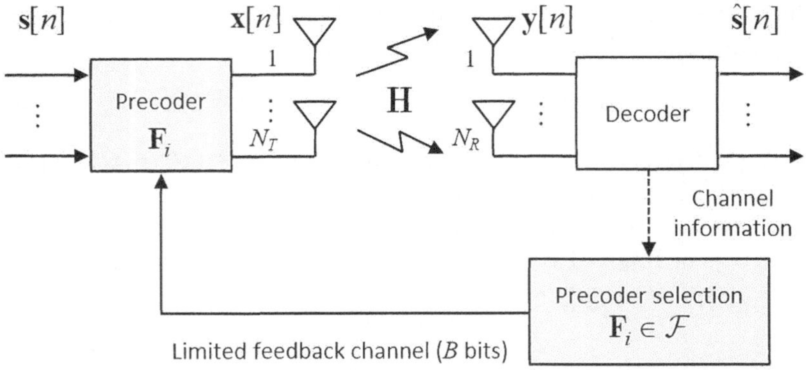

FIGURE 4.32
Illustration of the codebook-based MIMO precoding systems.

limited channel or precoder information. A common method for limited feedback is channel quantization and precoder quantization. Currently, in many mainstream communication systems, the adopted scheme falls under precoder quantization, known as codebook-based precoding. The concept of codebook-based precoding is as follows: A set of precoders is designed in advance and integrated into a codebook. Each precoder in the codebook is assigned a unique index number. Both the transmitter and the receiver possess complete information about this codebook. Before transmitting the signal, the receiver performs channel estimation and selects an appropriate precoder from the codebook based on the channel information. The index of the selected precoder is then fed back to the transmitter, which uses the chosen precoder to transmit the signal. Assuming the codebook contains no more than 2^B precoders, the receiver only needs to send back B bits of information, significantly reducing the feedback data volume. The architecture of the codebook-based precoding system is depicted in Figure 4.32, where $\mathcal{F}$ represents the codebook, and $\mathbf{F}_i$ is the ith precoder in the codebook. The most crucial issues in codebook-based precoding are as follows: (1) how to select the optimal precoder from the codebook; (2) how to design the codebook to maximize the system performance while minimizing the feedback overhead. These considerations play a pivotal role in determining the overall efficiency and performance of codebook-based precoding systems. Among them, the more well-known methods include Grassmannian packing, vector quantization (VQ), and random vector quantization (RVQ).[23] After providing the codebook and channel information, the receiver can select different criteria for precoder selection. The minimum singular value criterion aims to maximize the minimum singular value of the equivalent channel $\mathbf{HF}_i$, expressed mathematically as follows:

$$\mathbf{F}^{\text{opt}} = \arg \max_{\mathbf{F}_i \in \mathcal{F}} \lambda_{\min}\{\mathbf{HF}_i\} \text{ subject to } \text{tr}(\mathbf{F}_i\mathbf{F}_i^H) \leq P. \tag{4.106}$$

[23]Please refer to ref. [94] for more details.

Here, $\lambda_{\min}\{\mathbf{HF}_i\}$ denotes the minimum singular value of $\mathbf{HF}_i$. As $\mathbf{HF}_i$ represents the equivalent channel after combining the precoder, increasing its minimum singular value can result in a smaller condition number for the equivalent channel, leading to improved transmission performance. The maximum capacity criterion seeks to maximize the capacity of the equivalent channel $\mathbf{HF}i$ and is mathematically expressed as follows:

$$\mathbf{F}^{\text{opt}} = \arg\max_{\mathbf{F}_i \in \mathcal{F}} \log_2[\det(\mathbf{I}_Q + \frac{1}{\sigma_n^2}\mathbf{F}_i^H\mathbf{H}^H\mathbf{HF}_i)] \text{ subject to } \text{tr}(\mathbf{F}_i\mathbf{F}_i^H) \leq P. \tag{4.107}$$

The MMSE criterion aims to minimize the mean-square error between the decoded signal and the transmitted signal. It is mathematically expressed as follows:

$$\mathbf{F}^{\text{opt}} = \arg\min_{\mathbf{F}_i \in \mathcal{F}} \text{tr}((\mathbf{I}_Q + \frac{1}{\sigma_n^2}\mathbf{F}_i^H\mathbf{H}^H\mathbf{HF}_i)^{-1}) \text{ subject to } \text{tr}(\mathbf{F}_i\mathbf{F}_i^H) \leq P. \tag{4.108}$$

Here, the MMSE criterion assumes that the receiver decoding utilizes the MMSE detection based on the equivalent channel $\mathbf{HF}_i$. Apart from the MMSE criterion, the aforementioned codebook-based precoding methods do not involve decoder design. Users may choose an appropriate MIMO detection strategy according to their specific requirements.

Multi-mode MIMO precoding based on codebook: MIMO systems can provide varying degrees of diversity gain or multiplexing gain based on their channel conditions. Thus, under different channel conditions, a careful tradeoff should be made to select the most appropriate transmission mode. Precoding in conjunction with the chosen transmission mode is referred to as multimode precoding, where "mode" indicates the number of transmitted signals or data streams. Multi-mode precoding simultaneously selects the precoder and the mode, enabling the system to comprehensively respond to channel conditions and transmission requirements. The system architecture is illustrated in Figure 4.33, where Q represents the number of transmitted data streams (i.e., modes), $\mathcal{M}$ is the set of all modes, and $\mathcal{F}_Q$ corresponds to the codebook used for mode Q. In multi-mode precoding, the selection of the optimal precoder from all the modes and their corresponding precoders is based on the selection criteria described in the previous section. Taking the maximum capacity criterion as an example, its mathematical expression is as follows:

$$\begin{aligned}
&\mathbf{F}^{\text{opt}} = \mathbf{F}_Q^{\text{opt}} \\
&Q = \arg\max_{q \in \mathcal{M}} \log_2[\det(\mathbf{I}_q + \frac{1}{\sigma_n^2}(\mathbf{F}_q^{\text{opt}})^H\mathbf{H}^H\mathbf{HF}_q^{\text{opt}})] \\
&\mathbf{F}_Q^{\text{opt}} = \arg\max_{\mathbf{F}_i \in \mathcal{F}_q} \log_2[\det(\mathbf{I}_q + \frac{1}{\sigma_n^2}\mathbf{F}_i^H\mathbf{H}^H\mathbf{HF}_i)] \\
&\text{subject to } \text{tr}(\mathbf{F}_i\mathbf{F}_i^H) \leq P.
\end{aligned} \tag{4.109}$$

In this formulation, the total transmission power constraint ensures that the output power at the transmitter remains constant across different modes to

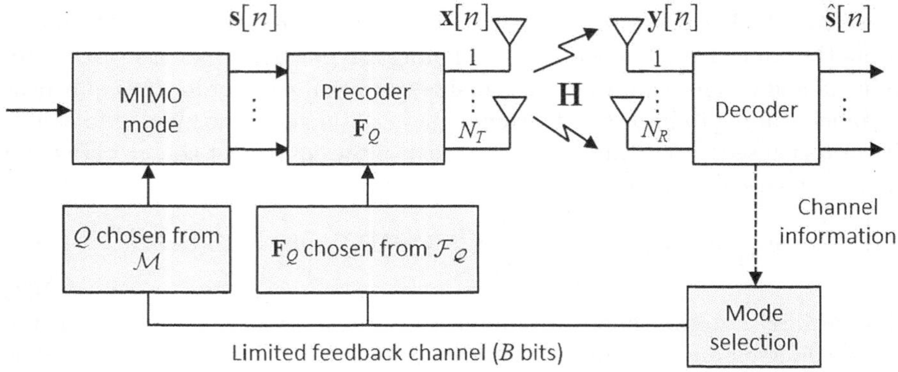

FIGURE 4.33
Illustration of the codebook-based multi-mode MIMO precoding systems.

achieve fairness. Eq. (4.109) indicates that the selection criterion aims to find the best precoder from each mode, and then from this set of candidates, the final selected precoder is determined.

5

Interference Management and Multiple Access in Communication Systems

5.1 Overview of MU-MIMO

In the previous chapter, we discussed the mainstream signal processing techniques to support multiple-input multiple-output (MIMO) communication in single-user scenarios with dedicated time/frequency resources. However, in practical scenarios, a base station will often need to serve a group of users simultaneously. Moreover, in the coverage of the base station, different communications, such as WiFi, Bluetooth, and so on, might also take place at the same time. While a straightforward idea is to provide dedicated time/frequency resources to ensure interference-free transmissions, we will discuss how modern communication systems handle mutual interference issues by joint transceiver designs to allow minimum cross-interference when sharing time/frequency/spatial resources. Note that this consideration is increasingly important due to the exponentially increasing number of devices, making it almost impossible to have delicate time/frequency/spatial resources for each device. Also, note that the discussed solutions in this chapter can be applied to both aforementioned cases, no matter the geo-located scenario (e.g., a base station aims to serve a group of users simultaneously) or geo-separated scenario (e.g., several transceiver pairs in each other's coverage) as long as the transmitters have a certain level of joint design capability. In the following sections, we will divide different solutions based on the main considered resource domain, that is, time/frequency/spatial domains into different categories to let readers understand the mainstream resource management solutions specialized for interference mitigation purposes.

5.2 Beamforming/Precoding Techniques

In the most normal scenario, a group of users will be served using the same time/frequency resources. In this case, a lot of research is conducted to

DOI: 10.1201/9781003408239-5

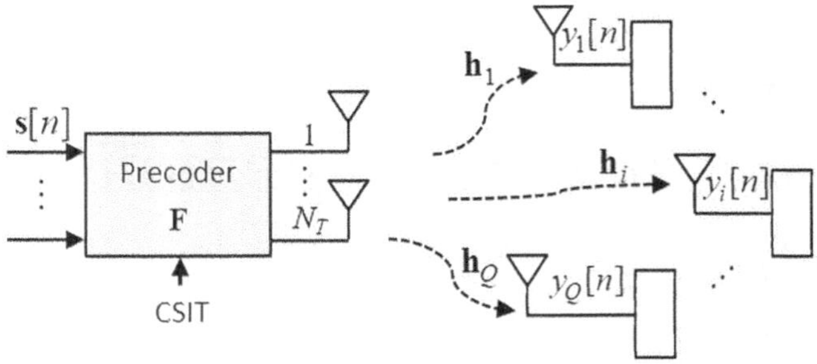

FIGURE 5.1
Illustration of the multi-user MIMO precoding systems.

discuss how to separate those transmissions in the spatial domain to minimize the cross-interference with each other. In this sense, different beamforming and precoding solutions are proposed with different trade-offs between achieved performance and implementation complexity, so that they can be employed in different scenarios. However, to perform this type of design, the assumption that all channel state information (CSI) to different users is known on the transmitter side is required. Thus, we can provide corresponding transceiver designs to the spatial profile of each user (i.e., CSI). Following each introduced solution, there are some works discussing how to relax this assumption, providing enhancements to work with imperfect CSI and statistical CSI. Interested readers are encouraged to check those recent designs [95–98], which will not be covered in this section owing to page limitations. The concept of multi-user MIMO (MU-MIMO) precoding involves equipping the transmitter with multiple antennas to transmit Q signals to Q different receivers, each equipped with a single antenna. The system concept is illustrated in Figure 5.1. In this scenario, the transmitter applies precoding to the transmitted signals, while each receiver individually receives its corresponding signal. The received signal is represented as follows:

$$\mathbf{y}[n] = \mathbf{H}\sum_{q=1}^{Q} \mathbf{f}_q s_q[n] + \mathbf{n}[n] = \mathbf{HFs}[n] + \mathbf{n}[n], \tag{5.1}$$

where $\mathbf{y}[n] = [y_1[n], ..., y_Q[n]]^T$ denotes the vector composed of Q received signals, and $\mathbf{s}[n] = [s_1[n], ..., s_Q[n]]^T$ denotes the vector composed of Q transmitted signals. The matrix $\mathbf{F} = [\mathbf{f}_1, ..., \mathbf{f}_Q]$ represents the $N_\mathrm{T} \times Q$ precoding matrix, where $\mathbf{f}_i$ is the precoding vector designed for the ith receiver. The $Q \times N_\mathrm{T}$ channel matrix $\mathbf{H} = [\mathbf{h}_1, ..., \mathbf{h}_Q]$ consists of all channels and $\mathbf{h}_i$ denotes the channel from the transmitter to the ith receiver. Under the aforementioned assumptions, $s_i[n]$ are mutually independent and have unit power, and $\mathbf{n}[n]$ is a noise vector with independent and identically distributed (i.i.d.) $\mathcal{CN}(0, \sigma_n^2)$ elements. In the context of a multi-user MIMO precoding system, the receiver is equipped with a single antenna, which severely limits its ability to counteract interference. Therefore, the primary objective of precoding is to

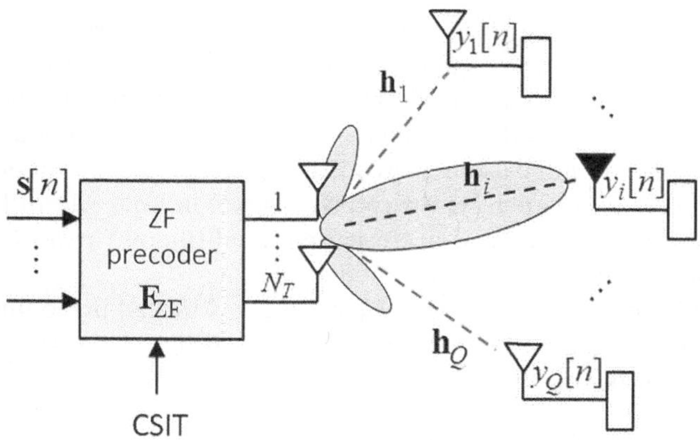

FIGURE 5.2
Illustration of the zero-forcing precoding systems.

design the precoding matrix $\mathbf{F}$ in such a way that mutual interference among the receivers is minimized, enabling each receiver to recover its intended signal. To achieve this goal, the transmitter needs to possess channel information beforehand. The direct approach to eliminate mutual interference among the receivers is the zero-forcing (ZF) precoding method, similar to ZF MIMO detection. The objective of ZF precoding is to direct individual transmitted signals toward their respective receivers while ensuring no impact on other receivers, as depicted in Figure 5.2. In this context, the precoding vector $\mathbf{f}_i$ acts as a transmitting beamformer specifically designed for the ith receiver. It aligns the beam toward the ith receiver while nulling the signals toward other receivers, i.e.,

$$\mathbf{h}_i^T \mathbf{f}_{\mathrm{ZF},q} = 0, \ q \neq i. \tag{5.2}$$

The solution for the ZF precoder is given as follows:

$$\mathbf{f}_{\mathrm{ZF},q} = \frac{[\mathbf{H}^+]_i}{k\|[\mathbf{H}^+]_i\|}, \tag{5.3}$$

where $\mathbf{H}^+ = \mathbf{H}^H(\mathbf{H}\mathbf{H}^H)^{-1})$ is the pseudo-inverse of $\mathbf{H}$, and $[\mathbf{H}^+]_i$ represents the ith row of $\mathbf{H}^+$. The parameter k is the normalization coefficient, ensuring that the precoder complies with the total transmission power constraint. Based on Eq. (5.2), the received signal at the ith receiver can be expressed as:

$$\begin{aligned} y_i[n] &= \mathbf{h}_i^T \sum_{q=1}^{Q} \mathbf{f}_{\mathrm{ZF},q} s_q[n] + n_i[n] \\ &= \mathbf{h}_i^T \mathbf{f}_{\mathrm{ZF},i} s_i[n] + \sum_{q=1,q\neq i}^{Q} \mathbf{h}_i^T \mathbf{f}_{\mathrm{ZF},q} s_q[n] + n_i[n] \\ &= \frac{1}{k\|[\mathbf{H}^+]_i\|} s_i[n] + n_i[n]. \end{aligned} \tag{5.4}$$

From Eq. (5.4), it can be observed that for all other receivers, the combination effect of precoding and the channel is zero, meaning they will not receive any interference from the transmitted signal $s_i[n]$. Similar to ZF detection, the performance of ZF precoding is affected in adverse channel conditions, but not due to noise amplification. Instead, it is caused by the amplification of the norm of $\mathbf{H}^+$. Consequently, the received signal in Eq. (5.1) will be divided by a large value, leading to a decrease in signal-to-noise ratio (SNR). This phenomenon is referred to as power reduction.

The objective of minimum mean-square error (MMSE) precoding is to minimize the mean squared error between the received signal and the transmitted signal, as shown below:

$$\min_{\mathbf{F}} \mathbb{E}\{||\mathbf{y}[n] - \mathbf{s}[n]||^2\} \text{ subject to } \mathrm{tr}(\mathbf{F}\mathbf{F}^H) \leq P, \tag{5.5}$$

where the error vector is defined as:

$$\mathbf{y}[n] - \mathbf{s}[n] = \mathbf{H}\mathbf{F}\mathbf{s}[n] - \mathbf{s}[n] + \mathbf{n}[n]. \tag{5.6}$$

The solution to Eq. (5.4) can be expressed as:

$$\mathbf{F} = \frac{1}{k}\mathbf{H}^H(\mathbf{H}\mathbf{H}^H + \frac{Q\sigma_n^2}{P}\mathbf{I}_Q)^{-1}, \tag{5.7}$$

where k is the normalization coefficient to ensure that the precoder complies with the total transmission power constraint. The MMSE precoder strikes a balance between the effects of interference and noise. At high SNR, it behaves similarly to the ZF precoder, while at low SNR, it resembles the matched precoder.[1]

The concept of SINR-based precoding involves setting the minimum SINR value that each receiver needs to satisfy. Then, from all the solutions that meet this condition, the precoding matrix that minimizes the total transmission power is selected. This design concept can be formulated as an optimization problem:[2]

$$\begin{aligned} &\min_{\mathbf{F}} \mathrm{tr}(\mathbf{F}\mathbf{F}^H) \\ &\text{subject to } \frac{|\mathbf{h}_i^T\mathbf{f}_i|^2}{\sum_{q=1,q\neq i}^{Q} |\mathbf{h}_i^T\mathbf{f}_q|^2 + \sigma_n^2} \geq \mathrm{SINR}_{\mathrm{T},i}, \\ &\mathrm{tr}(\mathbf{F}\mathbf{F}^H) \leq P, \end{aligned} \tag{5.8}$$

where $\mathrm{SINR}_{\mathrm{T},i}$ represents the target SINR for the ith receiver, and the optimal solution still needs to satisfy the total transmission power constraint. Eq. (5.8) represents a common non-convex quadratic problem, and obtaining the optimal solution is challenging. In practice, the problem is often tackled using semi-definite relaxation (SDR) to find the best approximate solution that meets the given constraints.

[1] The matched precoder directly matches the channel by employing precoder $\mathbf{F} = \mathbf{H}^H$.
[2] Please refer to ref. [99] for more details.

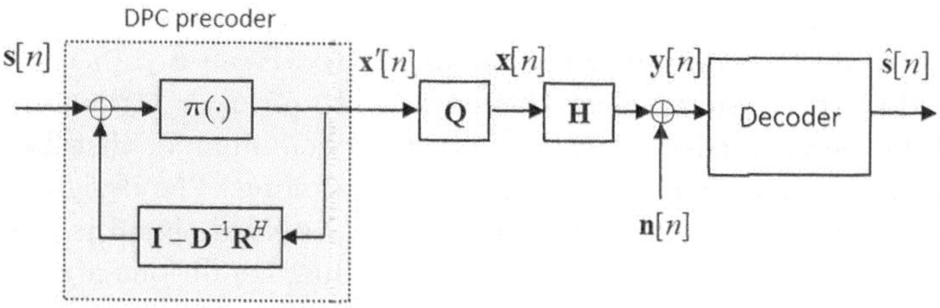

FIGURE 5.3
Illustration of the DPC precoding and decoding systems.

The concept of dirty paper coding (DPC) aims to handle interference effectively by utilizing precoding techniques when the transmitter has prior knowledge of the interference.[3] The design is based on a specific modulo function defined as follows:

$$\pi(y) = y - \lfloor \frac{y + \tau/2}{\tau} \rfloor \tau, \tag{5.9}$$

where $\lfloor . \rfloor$ denotes the floor function, which only takes the integer part of the input and τ is a design parameter. If the signal source is s and the known interference is i, then the precoded signal x is given by:

$$x = \pi(s - i) = s - i - \tau k. \tag{5.10}$$

The variable k represents the integer quotient resulting from the modulo function operation. Utilizing the modulo function offers the advantage of restricting the encoded signal to the interval $[-\tau/2, \tau/2]$, with its magnitude remaining constant regardless of variations in interference. At the decoding end, the received signal (including the signal source, interference, and noise) only needs to undergo the original modulo function operation to obtain the decoded signal, as shown in Eq. (5.11):

$$\begin{aligned} \hat{s} &= \pi(x + i + n) = \pi(s - i - \tau k + i + n) \\ &= \pi(s + n). \end{aligned} \tag{5.11}$$

From Eq. (5.11), it is evident that the decoder can completely eliminate interference. As long as $|s + n|$ does not exceed $\tau/2$, the decoded signal will appear as if there was no interference

A multi-user MIMO precoding system incorporating the concept of dirty paper coding is illustrated in Figure 5.3. Assuming $N_{\mathrm{T}} \geq N_{\mathrm{R}}$, $\mathbf{s}[n] = [s_1[n], ..., s_{N_{\mathrm{R}}}[n]]^T$ represents the original transmitted signal vector, and the MIMO transmission-reception relationship is given as:

$$\begin{aligned} \mathbf{y}[n] &= \mathbf{H}\mathbf{x}[n] + \mathbf{n}[n] = \mathbf{R}^H \mathbf{x}'[n] + \mathbf{n}[n], \\ \mathbf{x}[n] &= \mathbf{Q}\mathbf{x}'[n], \end{aligned} \tag{5.12}$$

[3]Please refer to refs. [100, 101] for more details.

where $\mathbf{H}^H = \mathbf{QR}$ represents the QR decomposition of $\mathbf{H}^H$, and $\mathbf{Q}$ satisfies $\mathbf{Q}^H\mathbf{Q} = \mathbf{I}$, while $\mathbf{R}$ is an upper triangular matrix. From Eq. (5.12), it can be observed that the channel effect transforms into an upper triangular matrix $\mathbf{R}^H$, and interference effects have been partially eliminated, thereby reducing the complexity of the front-end dirty paper precoding. The design premise of the dirty paper precoder is to ensure that the received signal is not affected by interference. Therefore, the following precoding conditions are considered:

$$\mathbf{R}^H\mathbf{x}^{'}[n] = \mathbf{Ds}[n], \tag{5.13}$$

where

$$\mathbf{R}^H = \begin{bmatrix} r_{11} & 0 & \cdots & 0 \\ r_{21} & r_{22} & \cdots & 0 \\ \vdots & \vdots & \ddots & \vdots \\ r_{N_\mathrm{R}1} & r_{N_\mathrm{R}2} & \cdots & r_{N_\mathrm{R}N_\mathrm{R}} \end{bmatrix}, \tag{5.14}$$

$$\mathbf{D} = \mathrm{diag}(r_{11}, \cdots, r_{N_\mathrm{R}N_\mathrm{R}}). \tag{5.15}$$

Expanding Eq. (5.13) results in the following set of equations:

$$\begin{bmatrix} r_{11} & 0 & \cdots & 0 \\ r_{21} & r_{22} & \cdots & 0 \\ \vdots & \vdots & \ddots & \vdots \\ r_{N_\mathrm{R}1} & r_{N_\mathrm{R}2} & \cdots & r_{N_\mathrm{R}N_\mathrm{R}} \end{bmatrix} \begin{bmatrix} x_1^{'}[n] \\ x_2^{'}[n] \\ \vdots \\ x_{N_\mathrm{R}}^{'}[n] \end{bmatrix} = \begin{bmatrix} r_{11}s_1[n] \\ r_{22}s_2[n] \\ \vdots \\ r_{N_\mathrm{R}N_\mathrm{R}}s_{N_\mathrm{R}}[n] \end{bmatrix}. \tag{5.16}$$

The below set of equations can be obtained:

$$\begin{aligned} x_1^{'}[n] &= s_1[n] \\ x_2^{'}[n] &= s_2[n] - \frac{r_{21}}{r_{22}}x_1^{'}[n] \\ &\vdots \\ x_{N_\mathrm{R}}^{'}[n] &= s_{N_\mathrm{R}}[n] - \frac{r_{N_\mathrm{R}(N_\mathrm{R}-1)}}{r_{N_\mathrm{R}N_\mathrm{R}}}x_{N_\mathrm{R}-1}^{'}[n] - \cdots - \frac{r_{N_\mathrm{R}1}}{r_{N_\mathrm{R}N_\mathrm{R}}}x_1^{'}[n]. \end{aligned} \tag{5.17}$$

In Eq. (5.17), one can notice that the ith equation depends only on the first $i-1$ equations, so the solution should be obtained in a top-down manner. This set of equations can be simplified into matrix form as follows:

$$\mathbf{x}^{'}[n] = \mathbf{s}[n] + (\mathbf{I} - \mathbf{D}^{-1}\mathbf{R}^H)\mathbf{x}^{'}[n]. \tag{5.18}$$

Note that the encoded signal should undergo the modulo function operation to maintain the transmission signal power constraint.

5.3 Scheduling Techniques

After visiting how the current communication systems utilize spatial diversity for multi-user interference mitigation, we want to introduce existing scheduling techniques in this section to let readers understand how time resource management is scheduled when serving multiple users. The introduced resource scheduling techniques have two main benefits, making the resource scheduling mechanism also a popular and practical solution in modern communication systems. First, compared with the precoding/beamforming solutions to improve user experience, resource scheduling solutions can be implemented in no CSI or partial CSI scenarios (e.g., resource scheduling can only be based on channel quality indicator (CQI), instead of full CSI). On the other hand, precoding/beamforming solutions often require precise CSI to carefully design spatial responses to serve different users using the same time/frequency resources. Second, even with precise CSI, resource scheduling techniques can be implemented and work with precoding/beamforming solutions for further performance enhancement. In fact, in modern communication systems, this procedure is often done in the MAC layer by performing user grouping; then precoding/beamforming solutions are developed in the PHY layer to serve a user group using the same time/frequency resources by means of utilizing spatial diversity. Mathematically speaking and without losing generality, we consider a MU-MIMO system containing a single base station with M antennas and K serving users, which has N antennas. If the previously discussed linear precoding solutions are employed to support independent data streams for each antenna in each user, the system has a fundamental constraint to serve at most $K_{\max} = \lfloor M/N \rfloor$ users using the same time/frequency resource. Thus, one can define the search space as $\mathcal{K}^{\mathrm{SS}} = \{\mathcal{K} \subseteq \{1, 2, ..., K\} : 1 \leq |\mathcal{K}| \leq K_{\max}\}$ containing all possible user scheduling results. Accordingly, we can form the following optimization problem for the joint design of the precoding/beamforming problem and user scheduling problem:

$$\max_{\mathcal{K} \subseteq \mathcal{K}^{\mathrm{SS}}} \max_{\mathcal{P}} \sum_{k \in \mathcal{K}} w_k \log_2(1 + \mathrm{SINR}_k), \tag{5.19}$$

where the outer optimization aims to find the optimal user grouping result, and then the inner optimization can allow available spatial resources to serve each user in the group simultaneously. This problem is interesting because of the considered trade-off between a number of serving users and the allocation of spatial diversity. On the one hand, trying to serve as many users as possible, or even all users, at the same time can lead to more terms from different users (i.e., $\log_2(1 + \mathrm{SINR}_k)$) to be summed up. On the other hand, the underlying channel from the transmitter to all users might become ill-conditioned, and a lot of terms will consequently be close to 0 and do not contribute to a higher sum-rate. Thus, this problem is worthy of consideration to obtain the optimal trade-off between those factors. In Eq. (5.19), $\mathcal{P}$ is the set containing all precoding/beamforming weightings to affect the SINR of each user and w_k is the predefined user weights based on service or application type. Depending on the considered scenarios, different types of constraints can be added to emphasize beamforming constraint, power constraint, and fairness (e.g., the minimum achievable rate when serving each user) for the considered scenarios. In this section, we assume that mature precoding/beamforming solutions are employed to solve the inner optimization, letting us only concentrate on the outer optimization problem first. Obviously, brute-force solutions can be employed to solve the outer optimization problem by searching over each possible solution exhaustively as $|\mathcal{K}^{\text{SS}}| = \sum_{m=1}^{K_{\max}} C_m^k$. However, this is often computationally prohibited, especially when $K >> K_{\max}$ holds. Thus, different suboptimal solutions are developed, trying to achieve a better trade-off between performance and complexity and being the main content of the rest of this section. In light of this direction, we introduce two mainstream solutions, direct solution and indirect solution, for the interested user scheduling problem.

Direct solution: Direct solutions aim to solve the original optimization problem in Eq. (5.19) without any relaxation. For example, the idea of greedy algorithms is widely used to do so. Given a fixed precoding structure and CSIT assumption, the set $\mathcal{K}$ is constructed by a sequence of decisions. In each round, every newly selected user finds a local maximum for an objective function to ensure the improvement of the objective function without heavy computations. Thus, the greedy scheduling rule requires low computational complexity and is easy to implement. However, it does not guarantee either performance or convergence to the optimal solution since a lot of possible combinations are skipped and will never be considered in the following rounds. An example of a greedy user scheduler is ref. [102]. This example shows the usage of the combination of a greedy solution (for outer optimization) and ZF precoding solution (for inner optimization) to provide comprehensive designs to Eq. (5.19). Note that the transmitter must have channel information so that every iteration the precoders are recalculated to solve the inner problem. Also, note that the maximum number of iterations is at most $K_{\max}$. Specifically, since the transmitter knows the precoders and the transmit powers, the achievable rates, SINR, BERs or other desired performance indicators are known for each iteration, allowing the scheduler to identify when it is worth adding more users to $\mathcal{K}$. This type of solution is especially efficient when the candidacy space is too large to be searched exhaustively by providing an acceptable solution with low complexity. However, also inheriting the drawback of greedy-based algorithms, the obtained solution might show a considerable performance gap with the optimal solution since a lot of possible combinations are never considered during the whole process. Nonetheless, this is still a mainstream solution when it comes to MIMO scheduling; interested readers are also encouraged to read refs. [103, 104] for more details.

Indirect solution: Instead of solving the original problem directly, this direction of solutions tries to decouple Eq. (5.19) as two subproblems to allow a sequential solving process, showing as:

$$\begin{cases} \mathcal{K}^{\text{map}} = \arg\max_{\mathcal{K}\subseteq\bar{\mathcal{K}}:|\mathcal{K}|\leq K_{\max}} f(\mathbf{H}(\mathcal{K})), \\ \max\limits_{\mathcal{P}} : \sum\limits_{k\in\mathcal{K}^{\text{map}}} w_k \log_2(1+\text{SINR}_k). \end{cases} \tag{5.20}$$

The first equation models the combinational user grouping problem, where $f(\mathbf{H}(\mathcal{K}))$ stands for the selected performance indicator. One can notice that the user grouping problem is an NP-C problem while the second problem can be solved using convex optimization efficiently. Thus, different literature in this direction develops various algorithms to solve user grouping problems with acceptable complexity or relax the problem by introducing assumptions. Interested readers are also encouraged to read refs. [105,106] for more details.

In the above works, the direct solution utilizes greedy user scheduling and ZF precoding to iteratively solve the interested problem, while the indirect solution decouples the original problem into two sub-problems to be solved independently. Now, we aim to take a closer look at how existing indirect solutions can be applied to finish the joint design of user scheduling and precoding/detection matrix simultaneously. Let us consider a downlink scenario, where a base station with N_{T} antennas aims to serve a set of available users $\mathcal{K}_{\text{user}}$ at the same time. Assuming that kth user is with $N_{\text{R},k}$ antennas for data reception. At tth timeslot, only a subset of users (i.e., $\mathcal{K}(t) \subseteq \mathcal{K}_{\text{user}}$) can be served simultaneously due to the limited spatial diversity to serve multiple users at the same time and we denote the antenna index as $k = 1, 2, ..., \mathcal{K}(t)$ to refer to the assigned antenna at the moment. Thus, all the antennas can be assigned to a specific user as

$$\sum_{k=1}^{\mathcal{K}(t)} d_k(t) = N_{\text{T}}, \tag{5.21}$$

where $d_k(t)$ is the number of antennas scheduled for kth user, satisfying $d_k(t) \leq N_{\text{R},k}$. Specifically, the transmitted signal for kth user can be expressed as $\mathbf{x}_k(t) \in \mathbb{C}^{d_k(t)\times 1}$ will be processed by linear precoding matrix $\mathbf{M}_k(t) \in \mathbb{C}^{N_{\text{T}}\times d_k(t)}$ so that the received signal $\mathbf{y}_k(t) \in \mathbb{C}^{N_{\text{R},k}\times 1}$ at the kth user can be expressed as:

$$\mathbf{y}_k(t) = \mathbf{H}_k(t)\mathbf{M}_k(t)\mathbf{x}_k(t) + \sum_{i\neq k, i\in\mathcal{K}(t)} \mathbf{H}_k(t)\mathbf{M}_i(t)\mathbf{x}_i(t) + \mathbf{n}_k(t), \tag{5.22}$$

where $\mathbf{H}_k(t) \in \mathbb{C}^{N_{\text{R},k}\times N_{\text{T}}}$ is the channel matrix and $\mathbf{n}_k(t) \in \mathbb{C}^{N_{\text{R},k}\times 1}$ is the noise vector. To decode the desired data, kth user can employ a detection matrix $\mathbf{B}_k(t) \in \mathbb{C}^{N_{\text{R},k}\times d_k(t)}$ as:

$$\tilde{\mathbf{y}}_k(t) = \mathbf{B}_k(t)^H \mathbf{y}_k(t) = \tilde{\mathbf{H}}_k(t)\mathbf{M}_k(t)\mathbf{x}_k(t) + \sum_{i\neq k, i\in\mathcal{K}(t)} \tilde{\mathbf{H}}_k(t)\mathbf{M}_i(t)\mathbf{x}_i(t) + \tilde{\mathbf{n}}_k(t), \tag{5.23}$$

where $\tilde{\mathbf{H}}_k(t) = \mathbf{B}_k(t)^H \mathbf{H}_k(t)$ represents the equivalent channel matrix and $\tilde{\mathbf{n}}_k(t) = \mathbf{B}_k(t)^H \mathbf{n}_k(t)$ stands for the equivalent noise vector. With the above formulations, the joint scheduling and precoding design problem can be expressed as:

$$\{\mathcal{B}, \mathcal{M}\} = \arg\max_{\mathcal{B},\mathcal{M}} \sum_{i=1}^{\mathcal{K}_{\text{user}}} C_i(\mathcal{B}, \mathcal{M}), \tag{5.24}$$

where $\mathcal{B} = \{\mathbf{B}_i\}_{i=1}^{\mathcal{K}_{\text{user}}}$, $\mathcal{M} = \{\mathbf{M}_i\}_{i=1}^{\mathcal{K}_{\text{user}}}$, $C_i(\mathcal{B}, \mathcal{M}) = \log_2 |\mathbf{I}_{d_i} + \mathbf{R}_i^{-1}(\tilde{\mathbf{H}}_i \mathbf{M}_i \mathbf{M}_i^H \tilde{\mathbf{H}}_i^H)|$ is the capacity for ith user and $\mathbf{R}_i$ is the interference and noise term. Note that the unit power constraint should be applied to Eq. (5.24) by allowing $\sum_{i=1}^{\mathcal{K}_{\text{user}}} \text{tr}(\mathbf{M}_i \mathbf{M}_i^H) \leq P_{\max}$ where $P_{\max}$ is the predefined power limit. To solve the above optimization problem, a simple but effective solution is provided below for the reader's reference.

An intuitive solution to the above optimization is to schedule users with better channel conditions to maximize the system sum-rate. Specifically, for all candidate user channel $\mathbf{H}_i$, we perform singular value decomposition as $\mathbf{H}_i = \mathbf{U}_i \mathbf{\Lambda}_i \mathbf{V}_i^H$, then the proposed algorithm selects the N_{T} largest singular values of the set $\{\lambda_{i,j} | i = 1, ..., \mathcal{K}_{\text{user}}, j = 1, ..., \text{rank}(\mathbf{H}_i)\}$ to build the detection matrices $\mathbf{B}_k$ with the corresponding left singular vectors $\mathbf{v}_{i,j}$. This is equivalent to conducting the optimization of the sub-problem below:

$$\mathcal{B} = \arg\max_{\mathcal{B}} \sum_{i=1}^{\mathcal{K}_{\text{user}}} ||\mathbf{B}_i^H \mathbf{H}_i||^2. \tag{5.25}$$

Note that the user scheduling is also done in the above optimization since only users with better channel conditions will be allowed to transmit and receive desired data in each timeslot; other users remain silent and wait for the scheduling in the following timeslots. After the user scheduling and detection matrices are designed in the above step, the precoding matrices $\mathbf{M}_k$ can also be derived by maximizing the sum-rate with the additional zero interference constraint among the users. This will lead to optimal performance in the high-SNR region, since interference will be the major bottleneck here, while the performance still has some room for further improvement in the low-SNR region. Specifically, the optimization problem below is considered:

$$\mathcal{M} = \arg\max_{\mathcal{M}} \sum_{i=1}^{\mathcal{K}_{\text{user}}} C_i(\mathcal{B}, \mathcal{M}). \tag{5.26}$$

In Eq. (5.26), besides the power constraint $\sum_{i=1}^{\mathcal{K}_{\text{user}}} \text{tr}(\mathbf{M}_i \mathbf{M}_i^H) \leq P_{\max}$, an additional constraint $\tilde{\mathbf{H}}_i \mathbf{M}_j = 0$, if $i \neq j$ will be considered to ensure zero interference. The resulting precoding matrix $\mathbf{M}_k$ can be expressed as

$$\mathbf{M}_k = \mathbf{\Theta}_k \mathbf{P}_k = \bar{\mathbf{V}}_k^0 \tilde{\mathbf{V}}_k \mathbf{P}_k, \tag{5.27}$$

where $\bar{\mathbf{V}}_k^0 \in \mathbb{C}^{N_{\text{T}} \times d_k}$ can be obtained by choosing the basis of the d_k-dimensional null space of the $N_{\text{T}} \times N_{\text{T}} - d_k$ matrix $\bar{\mathbf{H}}_k^H$ with the singular value decomposition:

$$
\begin{aligned}
\bar{\mathbf{H}}_k^H &= [\tilde{\mathbf{H}}_1^H ... \tilde{\mathbf{H}}_k^H ... \tilde{\mathbf{H}}_K^H]^H \\
&= [\bar{\mathbf{U}}_i \bar{\mathbf{U}}_i^0] \begin{bmatrix} \bar{\mathbf{\Lambda}}_i & \mathbf{O} \\ \mathbf{O} & \mathbf{O} \end{bmatrix} [\bar{\mathbf{V}}_i \bar{\mathbf{V}}_i^0]^H.
\end{aligned} \tag{5.28}
$$

In Eq. (5.27), selecting matrix $\tilde{\mathbf{V}}_k \in \mathbb{C}^{d_k \times d_k}$ (i.e., the range space of $\tilde{\mathbf{H}}_k^H$) guarantees the zero inter-user interference constraint. Moreover, the diagonal power allocation matrix $\mathbf{P}_k \in \mathbb{C}^{d_k \times d_k}$ can be obtained by the multi-user waterfall criterion for optimal power allocations. Simulation results confirm that improved performance can be obtained compared to benchmark designs in the high-SNR region. We suggest readers interested in this direction to read refs. [107–112] for more details and the recent developments.

5.4 NOMA Transceiver Designs

In the idea of resource scheduling, non-overlapping time-domain resources are specified to different users for interference-free transmissions. Similarly, in the orthogonal frequency division multiplexing chapter, we also introduce how to design transceivers to allow multiple users to transmit over different frequencies simultaneously. While those orthogonal transmission schemes provide great performance and enable serving multiple users simultaneously, it is almost impossible to serve the increasingly growing number of end users using limited time/frequency resources. Thus, researchers also investigate the usage of non-orthogonal data transmissions. In this direction, non-orthogonal multiple access (NOMA) transmissions have caught a lot of attention in recent years. Specifically, instead of letting multiple users occupy time/frequency resources exclusively to utilize orthogonal resources, NOMA allows signals from multiple users to be superposed in the power-domain in a non-orthogonal manner (i.e., multiple users will utilize the same time-frequency domain for data transmission). On the one hand, the spectral utilization can be further improved by doing so. On the other hand, the transceivers also need to be specially designed to tackle the messy messages containing overlapping signals from multiple users, which has been a popular research topic recently.

In this direction, MIMO-NOMA represents a cutting-edge innovation in wireless communications, combining the strengths of MIMO systems and NOMA to address the escalating demands for higher spectral efficiency, massive connectivity, and enhanced user experience in next-generation networks like 5G and beyond. The rapid proliferation of mobile devices and the burgeoning Internet of Things (IoT) ecosystem necessitate an exponential increase in network capacity and efficiency. Traditional orthogonal multiple access (OMA) techniques such as time division multiple access (TDMA), frequency division multiple access (FDMA), and orthogonal frequency division

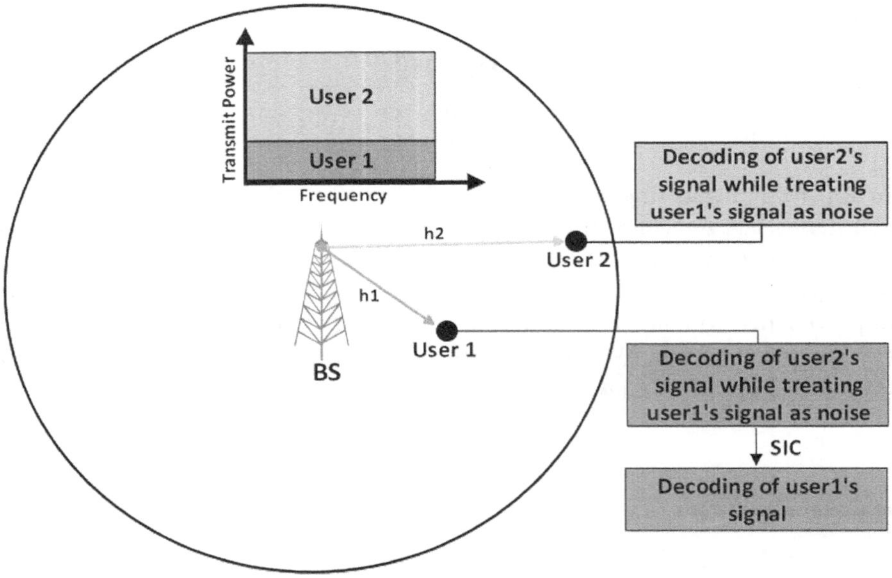

FIGURE 5.4
An illustration of a two-user downlink power-domain NOMA scheme with superposition coding and successive interference cancellation decoding (copyright from ref. [113]).

multiple access (OFDMA) allocate distinct time, frequency, or code resources to each user, inherently limiting the number of simultaneous users and leading to suboptimal spectral utilization. These methods struggle to meet the high data rate requirements and low latency expectations of modern applications. NOMA addresses these limitations by allowing multiple users to share the same time-frequency resources but distinguishes them in the power domain, as shown in Figure 5.4. This approach leverages differences in users' channel conditions, enabling superposition coding at the transmitter and successive interference cancellation (SIC) at the receiver. By allocating higher power levels to users with weaker channel conditions and lower power levels to those with stronger channels, NOMA can significantly enhance spectral efficiency and user connectivity. This power-domain multiplexing allows NOMA to serve more users simultaneously within the same bandwidth, substantially improving overall network throughput and accommodating the diverse service requirements of modern communication systems. The integration of NOMA with MIMO technology further amplifies these benefits. MIMO systems, which utilize multiple antennas at both the transmitter and receiver, exploit spatial diversity and multiplexing gains to enhance link reliability and increase data rates. By combining MIMO with NOMA, the system can simultaneously exploit both spatial and power domains, achieving unprecedented levels of spectral efficiency and capacity. MIMO-NOMA can support a larger number of users with higher data rates, improved reliability, and better quality of service (QoS). This is particularly advantageous in scenarios with a high density of users and varying channel conditions, such as urban environments and indoor hotspots. The motivation behind MIMO-NOMA is multifaceted. First, there is a pressing need to accommodate the explosive growth of connected devices and the IoT, which traditional OMA methods cannot efficiently support. Second, the diverse application scenarios of next-generation networks require flexible and scalable solutions that can dynamically allocate resources based on real-time conditions and user requirements. MIMO-NOMA's ability to dynamically adjust power allocation and leverage spatial multiplexing makes it ideal for such environments. Third, the demand for higher data rates and lower latency in applications such as virtual reality (VR), augmented reality (AR), and autonomous driving necessitates more efficient use of the available spectrum. MIMO-NOMA's enhanced spectral efficiency directly addresses this need. However, the integration of MIMO and NOMA also introduces new challenges. Signal processing complexity increases, particularly in the design of effective SIC algorithms and power allocation strategies. Resource allocation and user grouping become more complex due to the interplay between spatial and power domains. Additionally, managing interference in dense user environments and ensuring fairness among users with diverse channel conditions are critical issues that require innovative solutions. Overall, MIMO-NOMA is a sophisticated and promising technology that represents a significant advancement in wireless communications. It offers a pathway to meeting the stringent requirements of future wireless networks, including higher capacity, better spectral efficiency, and enhanced user experience.

In the following example, we will introduce the NOMA system model and theoretical rate to let readers better understand this innovative and emerging data transmission technique. To intuitively see the provided benefits, a comparison between NOMA and OMA will also be elaborated mathematically in this section. We consider the downlink scenario, where a base station with M antennas utilizes NOMA to serve multiple users with the same time-frequency resources simultaneously. We assume that those users are randomly grouped into M clusters with two users in each group, and each user has N antennas ($N > M$) for data reception, as shown in Figure 5.5. Thus, for the kth user in the mth cluster (i.e., user (m,k), $m \in \{1,2,...,M\}$ and $k \in \{1,2\}$), the received signal of the user (m,k) $\mathbf{y}_{m,k}$ can be expressed as:

$$\mathbf{y}_{m,k} = \mathbf{H}_{m,k}\mathbf{P}\tilde{\mathbf{s}} + \mathbf{n}_{m,k}, \tag{5.29}$$

where $\mathbf{H}_{m,k} \in \mathbb{C}^{N\times M}$ is the downlink channel matrix, $\mathbf{P} \in \mathbb{C}^{M\times M}$ is the precoding matrix applied by the base station to the symbol vector $\tilde{s} \in \mathbb{C}^{M\times 1}$, and $\mathbf{n}_{m,k} \in \mathbb{C}^{N\times 1}$ is the effective noise.

$$\mathbf{v}_{m,k}^H\mathbf{y}_{m,k} = \mathbf{v}_{m,k}^H\mathbf{H}_{m,k}\mathbf{P}\tilde{\mathbf{s}} + \mathbf{v}_{m,k}^H\mathbf{n}_{m,k}. \tag{5.30}$$

In the content below, we assume $|\mathbf{v}_{m,k}^H|^2 = 1$ and $\mathbf{P} = \mathbf{I}_M$, thus $\mathbf{v}_{m,k}^H\mathbf{H}_{m,k}\mathbf{p}_n =$

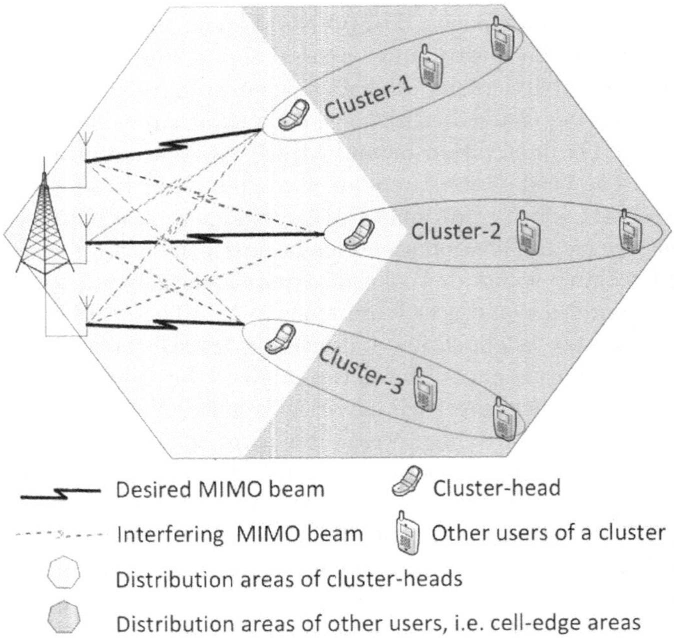

FIGURE 5.5
Illustrations of the simulation model for a 3-user MIMO-NOMA system, where the number of the transmit antennas, receive antennas and UEs are 3, 9, and 9, respectively (reproduced from ref. [114]).

0 for any $m \neq n$ to simplify the derivation process. Specifically, for MIMO-NOMA transmission, the signals transmitted from the base station to different users utilize the same time-frequency resources by means of power domain division, that is:

$$\tilde{\mathbf{s}} = \begin{bmatrix} \lambda_{1,1}s_{1,1} + \lambda_{1,2}s_{1,2} \\ \vdots \\ \lambda_{M,1}s_{M,1} + \lambda_{M,2}s_{M,2} \end{bmatrix}, \tag{5.31}$$

where $s_{m,k}$ is the information for user (m, k) and $\lambda_{m,k}$ is the power allocation coefficient for user (m, k). By denoting the mth column of $\mathbf{P}$ by $\mathbf{p}_m$, Eq. (5.22) can be rewritten as:

$$\begin{aligned} \mathbf{v}_{m,k}^H \mathbf{y}_{m,k} &= \mathbf{v}_{m,k}^H \mathbf{H}_{m,k} \mathbf{p}_m (\lambda_{m,1}s_{m,1} + \lambda_{m,2}s_{m,2}) \\ &\quad + \sum_{n=1, n\neq m}^{M} \mathbf{v}_{m,k}^H \mathbf{H}_{m,k} \mathbf{p}_n \tilde{s}_n + \mathbf{v}_{m,k}^H \mathbf{n}_{m,k}. \end{aligned} \tag{5.32}$$

We further assume that the channel gain of users in the same group is placed in order, that is:

$$|\mathbf{v}_{m,1}^H \mathbf{H}_{m,1} \mathbf{p}_m|^2 \geq |\mathbf{v}_{m,2}^H \mathbf{H}_{m,2} \mathbf{p}_m|^2, \ m \in 1, 2, ..., M. \tag{5.33}$$

In MIMO-NOMA, in the same group, the transmitter (i.e., base station) will assign more power to the user with worse channel gain, as expressed as:

$$\lambda_{m,1} \leq \lambda_{m,1}. \ \lambda_{m,1} + \lambda_{m,2} = 1 \text{ and } m \in 1, 2, ..., M. \tag{5.34}$$

Thus, on the receiver side, SIC decoding can be applied directly to resolve the superposition message effectively. To do so, user $(m, 2)$ will decode its own information $s_{m,2}$ by treating signal $s_{m,1}$ as interference. Thus, the achievable rate of user $(m, 2)$ can be expressed as:

$$\begin{aligned} & R_{m,2}^{\text{MIMO-NOMA}} \\ & \leq \log_2\left(1 + \frac{\rho \lambda_{m,2} |\mathbf{v}_{m,2}^H \mathbf{H}_{m,2} \mathbf{p}_m|^2}{\rho(1-\lambda_{m,2})|\mathbf{v}_{m,2}^H \mathbf{H}_{m,2} \mathbf{p}_m|^2 + \rho \sum_{n=1, n\neq m}^{M} |\mathbf{v}_{m,2}^H \mathbf{H}_{m,2} \mathbf{p}_n|^2 + 1}\right) \\ & = \log_2\left(1 + \frac{\rho \lambda_{m,2} |\mathbf{v}_{m,2}^H \mathbf{H}_{m,2} \mathbf{p}_m|^2}{\rho(1-\lambda_{m,2})|\mathbf{v}_{m,2}^H \mathbf{H}_{m,2} \mathbf{p}_m|^2 + 1}\right), \end{aligned} \tag{5.35}$$

where ρ is the transmit signal-to-noise ratio. Then, as for the user $(m, 1)$, the user will need to decode the message for user $(m, 2)$ first, and then the user can reconstruct and remove it from the overall received signal for decoding the desired message with improved signal quality. Thus, the achievable rate $(m, 1)$ can be expressed as:

$$\begin{aligned} R_{m,1}^{\text{MIMO-NOMA}} & \leq \log_2\left(1 + \frac{\rho(1-\lambda_{m,2})|\mathbf{v}_{m,1}^H \mathbf{H}_{m,1} \mathbf{p}_m|^2}{\rho \sum_{n=1, n\neq m}^{M} |\mathbf{v}_{m,1}^H \mathbf{H}_{m,1} \mathbf{p}_n|^2 + |\mathbf{v}_{m,1}|^2}\right) \\ & = \log_2(1 + \rho(1-\lambda_{m,2})|\mathbf{v}_{m,1}^H \mathbf{H}_{m,1} \mathbf{p}_m|^2). \end{aligned} \tag{5.36}$$

In contrast, MIMO-OMA allocates delicate time-frequency resources to two users in the same group for interference-free data transmission. Assuming that α of the degree of freedom is allocated to user $(m, 2)$ and the remaining $1-\alpha$ to user $(m, 1)$. Thus, we define the transmit SNR allocated to user $(m, 2)$ as $\gamma\rho/\alpha$ and the SNR allocated to user $(m, 1)$ as $(1-\gamma)\rho/(1-\alpha)$, where γ is the power factor. With the above prior, the achievable rate for user $(m, 1)$ can be expressed as:

$$R_{m,1}^{\text{MIMO-OMA}} \leq (1-\alpha)\log_2\left(1 + \frac{(1-\gamma)\rho|\mathbf{v}_{m,1}^H \mathbf{H}_{m,1} \mathbf{p}_m|^2}{(1-\alpha)}\right). \tag{5.37}$$

Similarly, the achievable rate for user $(m, 2)$ can be expressed as:

$$R_{m,2}^{\text{MIMO-OMA}} \leq \alpha \log_2\left(1 + \frac{\gamma\rho|\mathbf{v}_{m,2}^H \mathbf{H}_{m,2} \mathbf{p}_m|^2}{\alpha}\right). \tag{5.38}$$

According to Jensen's inequality and concavity of log(.), we obtain

$$\begin{aligned} &R_{m,1}^{\text{MIMO-OMA}} + R_{m,2}^{\text{MIMO-OMA}} \\ &\leq \log_2(1 + (1-\alpha)\frac{(1-\gamma)\rho|\mathbf{v}_{m,1}^H \mathbf{H}_{m,1}\mathbf{p}_m|^2}{(1-\alpha)} + \alpha\frac{\gamma\rho|\mathbf{v}_{m,2}^H \mathbf{H}_{m,2}\mathbf{p}_m|^2}{\alpha}) \\ &= \log_2(1 + (1-\gamma)\rho|\mathbf{v}_{m,1}^H \mathbf{H}_{m,1}\mathbf{p}_m|^2 + \gamma\rho|\mathbf{v}_{m,2}^H \mathbf{H}_{m,2}\mathbf{p}_m|^2), \end{aligned} \tag{5.39}$$

where the equality holds if $\frac{(1-\gamma)\rho|\mathbf{v}_{m,1}^H \mathbf{H}_{m,1}\mathbf{p}_m|^2}{(1-\alpha)} = \frac{\gamma\rho|\mathbf{v}_{m,2}^H \mathbf{H}_{m,2}\mathbf{p}_m|^2}{\alpha}$. This can be achieved by letting optimal allocation as:

$$\alpha^* = \frac{\gamma|\mathbf{v}_{m,2}^H \mathbf{H}_{m,2}\mathbf{p}_m|^2}{\gamma|\mathbf{v}_{m,2}^H \mathbf{H}_{m,2}\mathbf{p}_m|^2 + (1-\gamma)|\mathbf{v}_{m,1}^H \mathbf{H}_{m,1}\mathbf{p}_m|^2}. \tag{5.40}$$

By utilizing this optimal allocation factor, the optimal sum-rate shown in Eq. (5.31) can be achieved when equality holds. On the other hand, from Eqs. (5.27) and (5.28), the sum-rate of MIMO-NOMA of a group can be expressed as:

$$\begin{aligned} R_{m,1}^{\text{MIMO-NOMA}} + R_{m,2}^{\text{MIMO-NOMA}} &= \log_2(1 + \rho(1-\lambda_{m,2})|\mathbf{v}_{m,1}^H \mathbf{H}_{m,1}\mathbf{p}_m|^2) \\ &+ \log_2(1 + \frac{\rho\lambda_{m,2}|\mathbf{v}_{m,2}^H \mathbf{H}_{m,2}\mathbf{p}_m|^2}{\rho(1-\lambda_{m,2})|\mathbf{v}_{m,2}^H \mathbf{H}_{m,2}\mathbf{p}_m|^2 + 1}). \end{aligned} \tag{5.41}$$

Given that $\frac{\rho(1-\lambda_{m,1})|\mathbf{v}_{m,1}^H \mathbf{H}_{m,1}\mathbf{p}_m|^2+1}{\rho(1-\lambda_{m,2})|\mathbf{v}_{m,2}^H \mathbf{H}_{m,2}\mathbf{p}_m|^2+1} \geq 1$ holds, we also have

$$\begin{aligned} &R_{m,1}^{\text{MIMO-NOMA}} + R_{m,2}^{\text{MIMO-NOMA}} \\ &= \log_2(1 + \rho(1-\lambda_{m,2})|\mathbf{v}_{m,1}^H \mathbf{H}_{m,1}\mathbf{p}_m|^2 \\ &+ \rho\lambda_{m,2}|\mathbf{v}_{m,2}^H \mathbf{H}_{m,2}\mathbf{p}_m|^2 \frac{\rho(1-\lambda_{m,2})|\mathbf{v}_{m,1}^H \mathbf{H}_{m,1}\mathbf{p}_m|^2 + 1}{\rho(1-\lambda_{m,2})|\mathbf{v}_{m,2}^H \mathbf{H}_{m,2}\mathbf{p}_m|^2 + 1}) \\ &\geq \log_2(1 + \rho(1-\lambda_{m,2})|\mathbf{v}_{m,1}^H \mathbf{H}_{m,1}\mathbf{p}_m|^2 + \rho\lambda_{m,2}|\mathbf{v}_{m,2}^H \mathbf{H}_{m,2}\mathbf{p}_m|^2). \end{aligned} \tag{5.42}$$

Finally, if we align the power allocation of the MIMO-NOMA and MIMO-OMA by letting $\lambda_{m,2} = \gamma$ above, we obtain

$$R_{m,1}^{\text{MIMO-NOMA}} + R_{m,2}^{\text{MIMO-NOMA}} \geq R_{m,1}^{\text{MIMO-OMA}} + R_{m,2}^{\text{MIMO-OMA}}. \tag{5.43}$$

In conclusion, we prove that for any channel matrix $\mathbf{H}_{m,k}$, the optimal achievable of MIMO-NOMA can always outperform MIMO-OMA system, being the main motivation to develop such systems by better tackling mutual interference when allowing multiple users to utilize the same time-frequency resources. However, one should also note that the drawback is the additional complexity burden of performing extra steps, such as user grouping and SIC, in the NOMA transceiver compared to the OMA transceiver. In light of this direction, a great body of literature has already worked on efficient NOMA transceiver designs. We suggest readers interested in this direction to read refs. [115–117] for further study.

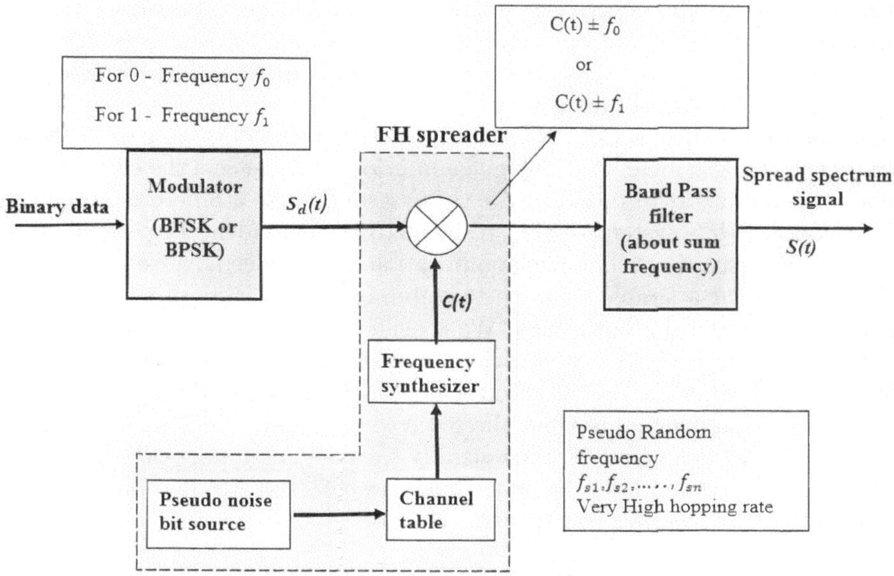

FIGURE 5.6
Frequency hopping spread spectrum subsystem transmitter (copyright from ref. [118]).

5.5 FHSS Techniques

Besides the previously discussed methodologies, frequency hopping spread spectrum (FHSS) is another technique, that can be used to improve the robustness and reliability of wireless communications by rapidly switching the carrier among many frequency channels. This method is highly effective in mitigating interference and enhancing security. Thus, in the context of modern wireless communications, FHSS is also expected to play a crucial role in managing interference, which is a significant challenge due to the dynamic and dense nature of practical scenarios. Specifically, as shown in Figure 5.6, the block diagram of a typical FHSS system involves several key components: A pseudo-random number generator to create the hopping sequence, frequency synthesizers to rapidly change the carrier frequency, and synchronization mechanisms to ensure that the transmitter and receiver stay in synchronization status. The process begins with the pseudo-random number generator producing a sequence of frequencies that both the transmitter and receiver agree upon before communication starts. The transmitter then modulates the data onto a carrier signal and hops between the frequencies in the sequence. The receiver, aware of the same sequence, tunes to the corresponding frequencies at the correct times, thus demodulating the received signal correctly. The effectiveness of FHSS in mitigating interference can be analyzed using concepts from information theory and signal processing. In a noisy environment where interference sources are unpredictable, FHSS's ability to spread the signal over a wide frequency range reduces the probability of significant signal degradation. From a signal processing perspective, the rapid frequency changes mean that any given interferer only affects the communication for a very short duration before the system hops to a different frequency. This makes FHSS systems highly resilient to both accidental and deliberate interference, ensuring reliable communication even under adverse conditions. Here we present a simple example to elaborate the benefits provided by the FHSS for interference mitigation. We consider binary frequency shift keying (BPSK) with the FHSS system, where noncoherent matched filtering is utilized for demodulation on the receiver. Thus, the receiver will make an error if the transmitted signal is 1 but the received signal falls into the decision region of -1 and vice versa. Mathematically speaking, assuming that the signal energy is E_b and the Gaussian noise power is $\frac{N_0}{2}$, the error probability of a common AWGN channel can be expressed as:

$$P_e = \frac{1}{2} e^{-\frac{E_b}{2N_0}}. \tag{5.44}$$

Now we consider a jammer with jamming power J while a transmitter with transmit power P. The transmitter utilizes a bandwidth $W = R_b$, where R_b is the transmission rate in (bits/second). The jammer also allocates its total power over the whole bandwidth thus we have $N_0 = J/W$ in the case that the jamming power is significantly larger than the noise power. In this case, we have error probability as:

$$P_e = \frac{1}{2} e^{-\frac{E_b}{2N_0}} = \frac{1}{2} e^{-\frac{P/W}{2J/W}} = \frac{1}{2} e^{-\frac{P}{2J}}. \tag{5.45}$$

Now, we assume the FHSS system provides a frequency degree of freedom so that we have a total bandwidth W_{total} for secure transmission. In this case, we have $\eta = \frac{W_{\text{total}}}{W}$ non-overlapping channel to select for data transmission. To interfere the legitimate transmission, a jammer without any prior knowledge can choose to spread its jamming power to the whole spectrum W_{total}, thus we have $N_0 = J/W_{\text{total}}$ and the error rate of the legitimate transmission can be expressed as:

$$P_e = \frac{1}{2} e^{-\frac{E_b}{2N_0}} = \frac{1}{2} e^{-\frac{P/W}{2J/W_{\text{total}}}} = \frac{1}{2} e^{-\frac{P\eta}{2J}}. \tag{5.46}$$

In the above example, one can see that $10 \log_{10} \eta$ (dB) gain is provided by the additional frequency degree of freedom against interference. Thus, η is also named as spreading gain in FHSS-related research literature. As an alternative, the jammer can also choose to only jam a fraction ρ of the whole bandwidth to concentrate its power for interfering with legitimate transmission. Then, if the jammer and transmitter utilize the same channel, the error

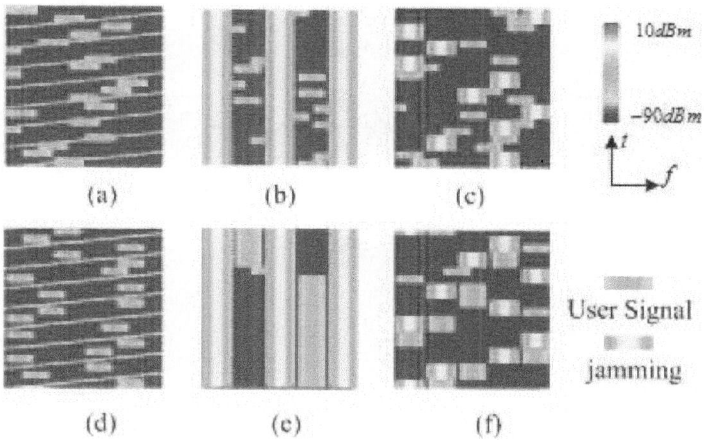

FIGURE 5.7
Environmental states at initial and convergent stages under different jamming patterns (copyright from ref. [119]).

probability will be higher due to the increased noise power $N_0 = J/(\rho W_{\text{total}})$ while the error probability remains the same as Eq. (5.44). Thus, the error rate can be expressed as:

$$P_e = \frac{\rho}{2} e^{-\frac{P\eta\rho}{2J}} + \frac{1-\rho}{2} e^{-\frac{E_b}{2N_0}}. \tag{5.47}$$

From the above discussion, one can notice that by providing more channels for the FHSS selection, the system performance can be further improved due to the provided spreading gain. Furthermore, to avoid the jammer learning the frequency-hopping pattern, the transmitter often adopts a random frequency-hopping pattern so that the probability of a channel being selected by the jammer and transmitter simultaneously can be approximated as uniform, as shown in Figure 5.7. One of the key benefits of FHSS is its ability to provide secure communication. By hopping frequencies in a predetermined pattern known only to the transmitter and receiver, FHSS makes it difficult for unauthorized parties to intercept or jam the signal. This intrinsic security feature is particularly valuable in military and secure communication applications. Moreover, FHSS is less susceptible to narrowband interference, which affects only a small portion of the spectrum at any given time. As a result, the impact of such interference on FHSS systems is significantly reduced compared to fixed-frequency systems. This characteristic makes FHSS an ideal choice for environments with high levels of electromagnetic interference or in scenarios where multiple devices operate simultaneously within the same frequency band. In industrial and commercial applications, FHSS contributes to enhanced performance and reliability of wireless networks. It allows for efficient spectrum utilization, reducing the potential for congestion and im-

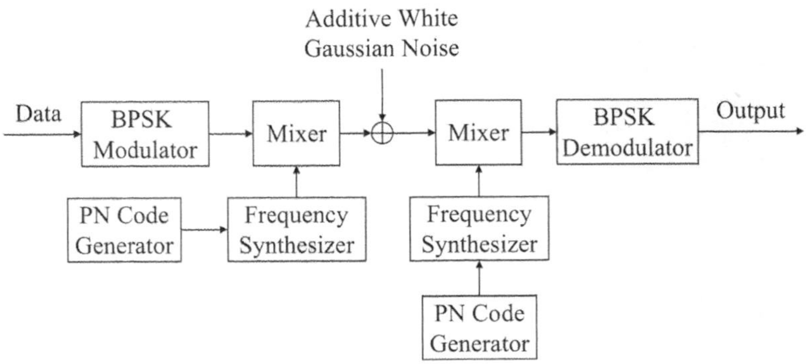

FIGURE 5.8
Illustration of the implementation of FHSS solutions.

proving overall network capacity. The dynamic frequency hopping mechanism ensures that the communication link remains stable, even in the presence of potential interference sources. Specifically, FHSS operates by dividing the available bandwidth into multiple frequency channels. The transmitter and receiver hop between these channels in a synchronized manner according to a pseudorandom sequence, as shown in Figure 5.8. This hopping pattern reduces the likelihood of interference affecting the communication, as any single channel is used for only a short period. By frequently changing frequencies, FHSS minimizes the impact of narrowband interference and makes the system more resistant to jamming or interference introduced by simultaneous usage of the same bands. The pseudo-random hopping sequence also makes it difficult for eavesdroppers to intercept the communication without knowledge of the hopping pattern. FHSS is well-suited for the highly dynamic and mobile nature of vehicular communications, where the network topology frequently changes. It provides a robust solution for managing interference in vehicular ad-hoc networks (VANETs), enhancing both security and communication reliability. Protocols like hopping reservation multiple access (HRMA) and multi-band MAC exemplify effective implementations of FHSS, demonstrating significant performance gains in terms of throughput and interference mitigation. As VANETs continue to evolve, the integration of FHSS with emerging technologies will be critical to addressing future challenges and realizing the full potential of intelligent transportation systems. Originally developed for military applications, FHSS has found widespread use in various civilian communication systems, including Wi-Fi, Bluetooth, and other wireless technologies. The fundamental principle behind FHSS is to spread the transmitted signal over a wide range of frequencies by rapidly switching, or "hopping," between different frequency channels according to a predetermined pseudo-random sequence. This hopping sequence is known to both the transmitter and receiver, allowing them to stay synchronized despite the frequent changes in frequency. The applications of FHSS extend beyond simple interference mitigation. In the context of wireless networks, FHSS can improve the overall network performance by reducing the probability of collisions and allowing multiple users to share the same frequency band without significant mutual interference. This makes FHSS an important technique for modern wireless communication systems, which must operate in increasingly congested spectral environments.

Part II

Radar Sensing Basics

6

Waveform Designs and Basic Signal Processing in Radar Systems

6.1 Overview of Radar Sensing Systems

A radar system refers to a transceiver system that transmits radiofrequency (RF) electromagnetic (EM) waves toward a region of interest and then receives and detects these EM waves when reflected from objects in the region. The history of the radar concept can be traced back to the late 19th century when Heinrich Hertz began experimenting and demonstrated that transmitted EM waves could be reflected by metallic objects. However, it wasn't until the early 20th century that the first radar system was invented by German inventor Christian Hülsmeyer, demonstrating its capacity to detect ships in fog and thereby avoid collisions. At present, radar systems are not only used to detect targets such as vehicles, ships, and aircraft but can also sense activities and soil composition. As a result, radar has become indispensable not only in military applications but also in various aspects of social and economic development (such as weather forecasting, resource detection, environmental monitoring, etc.) and scientific research (including astronomical research, atmospheric physics, and ionospheric structure research). Different radar systems employ EM waves with different frequency bands to utilize their characteristics for various applications. Radar bands can range from 3 MHz to 300 GHz, with the majority of radar systems operating between 300 MHz and 35 GHz, as shown in Table 6.1. While the allocation shown in Table 6.1 follows the radar band allocation defined by the International Telecommunications Union (ITU), different government agencies, such as the Federal Communications Commission (FCC) in the United States, provide more specific regulations by limiting the range of frequencies for different radar applications.

The current frequency range used in automotive radar is typically between 24 and 100 GHz, with 24 GHz and 77 GHz being the mainstream frequencies. Research has indicated that at higher frequencies, signal attenuation increases, thereby limiting the radar's detection range. The key functions of radar include the detection, localization, and tracking of objects of interest. In

DOI: 10.1201/9781003408239-6

TABLE 6.1
ITU radar band allocation.

Frequency Range	Wavelength Range	Band Name	Usage
3-30 MHz	10-100 m	HF	Coastal radar systems
30-300 MHz	1-10 m	VHF	Very long range
300-1000 MHz	0.3-1 m	UHF	Very long range
1-2 GHz	15-30 m	L-band	Long range
2-4 GHz	7.5-15 cm	S-band	Terminal air traffic control, marine radar
4-8 GHz	3.75-7.5 cm	C-band	Satellite transponders, synthetic aperture radar
8-12 GHz	2.5-3.75 cm	X-band	Marine radar, weather, ground surveillance, synthetic aperture radar
12-18 GHz	1.67-2.5 cm	Ku-band	Satellite transponders
18-24 GHz	1.11-1.67 cm	K-band	Satellite transponders, radar guns, weather
24-40 GHz	0.75-1.11 cm	Ka-band	Mapping, surveillance

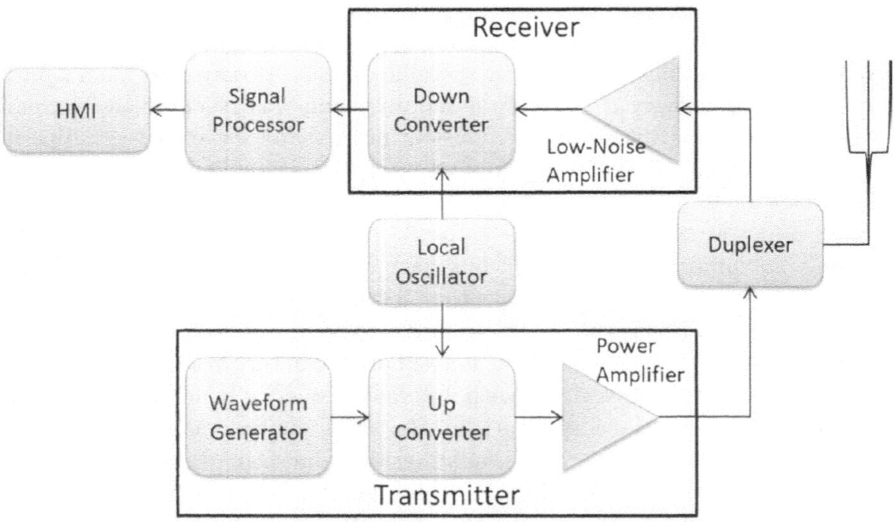

FIGURE 6.1
Illustration of the general block diagram of radar systems.

automotive applications, these objects include vehicles, pedestrians, bicycles, motorcycles, and obstacles found on or along the road. Radar systems generally consist of three main subsystems: the transmitter, the receiver, and the signal processing subsystem, as depicted in Figure 6.1. The antenna serves as the external electromagnetic wave receiver, and the transmitter subsystem is responsible for generating the transmitted signal. The ranging capability of radar is primarily determined by the design of the transmitter, considering factors such as the generated transmission power and associated costs. Additionally, the power radiated by the antenna must be adjusted based on the detection requirements, further influencing the transmitter design. For instance, the maximum detectable range is directly proportional to the fourth root of the transmission power. This implies that to double the detectable range, the power must increase by 16 times (2^4 times). The transmitter typically comprises a waveform generator, an upconverter, and a power amplifier. Various waveforms used in radar systems will be discussed in later sections. The main components of the receiver are the low-noise amplifier (LNA) and the downconverter. While the role of the receiving subsystem is to capture known reflected signals, it must also address clutter and other unwanted signals (such as noise and interference) to maximize the signal-to-noise ratio (SNR) of the desired signal. The signal processing subsystem utilizes various radar signal processing algorithms to extract useful information from received signals, determining target location, tracking, and target identification. Robust signal processing techniques are necessary due to clutter and noisy environments. However, constraints such as hardware resources and installation space limit the complexity of achievable algorithms, especially in real-time automotive and navigation applications. Thus, a comprehensive consideration is required to balance various indicators.

The development of automotive radar sensors encounters challenges in both hardware and software aspects. Effective signal processing algorithms play a crucial role in the software domain, without which the utility of radar sensors would be significantly curtailed. On the hardware side, considerations such as size and weight are pivotal factors influencing whether Original Equipment Manufacturers (OEMs) would embrace radar sensors as an option for vehicle models. Since it is not mandatory to install radar sensors in vehicles currently, minimizing sensor costs and the complexity of vehicle integration becomes paramount. One of the radar's advantages over other sensors is its ability to be discreetly installed behind the bumper fascia without compromising the vehicle's aesthetics. While radar sensors enhance safety, preserving the vehicle's appearance is preferable to avoid additional design efforts. Consequently, radar sensors must be designed to fit within the limited space behind the bumper. For passenger cars, the width of the bumper is typically constrained to 40–50 centimeters. However, not the entire length of the bumper can be utilized, and most vehicles target the bumper's edge near the headlights. Therefore, radar designers must aim to make the sensor smaller than the bumper's width. Although weight imposes relatively minor restrictions on radar, the prevailing trend is to reduce it to below 200 g to minimize installation costs. Calibration, tied to radar size, can be simplified based on the radar's dimensions. The subsequent challenge lies in radar waveform design and signal processing, as radar applications are broadly categorized into short-range, medium-range, and long-range. This led to the development of short-range radar (SRR), medium-range radar (MRR), and long-range radar (LRR). After delineating the radar application, the subsequent step is to define constraints and resolutions for distance, velocity, and angle. This necessitates waveform design, ultimately influencing the radar antenna's design, which is integral to the overall radar system design. Waveform design encompasses determining the modulation scheme, frequency bandwidth, and the transmission and reception of signals. Detailed elucidation on this subject will be provided in subsequent articles. During the waveform design process, iterative decisions must be made for radar waveforms, antenna configurations, and signal processing algorithms until a combination satisfying the target application is achieved. This iterative process is time-consuming, halting only when critical performance indicators are met, necessitating close collaboration among hardware developers, software developers, and system testing personnel. Since radar perception of targets invariably involves target tracking, striking a balance between available radar device resources and tracking performance to select a suitable tracking algorithm is not a trivial task. This spans options ranging from simple one-step predictive filters to complex nonlinear filters such as extended Kalman filters and particle filters. These choices will be explored in detail in subsequent sections.

6.2 Radar Detection Fundamentals

In most radar applications, radar systems must meet specific performance criteria, encompassing parameters such as maximum range, range resolution, maximum velocity, velocity resolution, angle coverage, and other specifications. Similarly, radar applications in automobiles require adherence to distinct performance benchmarks, with fundamental characteristics governed by the radar equation. The antenna acts as the intermediary between the radar system and the EM wave propagation medium. The radar equation delineates the correlation among transmitted signal power, received signal power, reflected target distance, reflected target characteristics, and antenna properties. This equation can be expressed as follows:

$$P_{\mathrm{r}} = \frac{P_{\mathrm{t}} G_{\mathrm{t}} G_{\mathrm{r}} \lambda^2 \sigma_{\mathrm{s}}}{(4\pi)^3 R^4} = \frac{P_{\mathrm{t}} G_{\mathrm{t}} A_{\mathrm{e}} \sigma_{\mathrm{s}}}{(4\pi)^2 R^4}, \tag{6.1}$$

where P_{r} is the received signal power, P_{t} is the transmitted signal power. The antenna characteristics are represented by the transmit gain G_{t} and receive gain G_{r}. The parameter $A_{\mathrm{e}} = G_{\mathrm{r}}\lambda^2/4\pi$ is the effective aperture of the receiving antenna. The characteristics of the reflecting object at distance R are represented by σ_{s}, known as the radar cross-section (RCS). RCS is an indicator of the target's ability to reflect radar signals in the direction of the radar receiver and is generally challenging to estimate. λ represents the wavelength of the emitted electromagnetic signal, which can be considered constant for any given radar system since the wavelength is only dependent on the frequency of the electromagnetic wave. The radar equation, as outlined above, can be expressed in terms of distance R as follows:

$$R = \sqrt[4]{\frac{P_t G_t A_e \sigma_s}{P_r (4\pi)^2}}. \tag{6.2}$$

From the equation provided, it becomes apparent that if we assume all other variables remain constant, the maximum detection range of the radar is directly proportional to the fourth root of the transmitted power. However, when accounting for losses associated with signal reception, an estimate of the maximum detectable range can be derived. To extend the coverage range of the radar system, it's essential to recognize that increasing transmitted power arbitrarily is not a feasible solution. High power usage is not energy-efficient, and adhering to the principle that lower energy consumption results in greater stability, controlling high-power devices becomes challenging. Consequently, achieving the reception of echo signals necessitates higher receiver sensitivity. By further simplifying Eq. (6.1), we can assume that the transmit gain and receive gain are equal and constant. Additionally, since λ is constant, the main variables influencing received power become the transmit power, distance, and RCS. Moreover, if we define the radar equation using the minimum detectable power of the receiver $S_{\min}$, Eq. (6.2) can be expressed as:

$$R = \sqrt[4]{\frac{P_t G_t A_e \sigma_s}{S_{\min} (4\pi)^2}}. \tag{6.3}$$

The effect of propagation losses can be integrated into the radar equation to better align with real-world scenarios. The minimum received power can be articulated in relation to the SNR:

$$S_{\min} = kT_0 B F_n \mathrm{SNR}. \tag{6.4}$$

In this equation, kT_0B denotes the thermal noise of an ideal Ohmic conductor, where k represents the Boltzmann constant, T signifies the standard temperature or absolute temperature (measured in Kelvin), and B stands for the receiver bandwidth (expressed in Hz) or effective noise bandwidth. The noise figure F_n elucidates the nonlinear characteristics introduced by the non-ideal receiver circuit, and it is a dimensionless value defined as the ratio of the receiver input to output SNR. Additionally, the loss factor denoted by L can be incorporated into the radar equation, resulting in the following revised expression:

$$R = \sqrt[4]{\frac{P_t G_t A_e \sigma_s}{kT_0 B F_n \mathrm{SNR} (4\pi)^2 L}}. \tag{6.5}$$

Furthermore, factors such as pulse accumulation and radar signal propagation losses can be incorporated into the radar equation. However, for the purpose of our discussion, Eq. (6.5) suffices to demonstrate the impact of losses on the maximum detectable range. Depending on the functional requirements of the radar, the radar equation can assume various forms, facilitating fine-tuning for specific applications. For instance, it can be adapted for surveillance radar and radar jamming systems. In most scenarios, crucial radar parameters such as antenna gain, wavelength, and noise figure remain constant. Given a fixed target distance (e.g., vehicles), determining the necessary transmit power to double the detectable range is straightforward. Doubling the detectable range affords the system or driver ample time to react and prevent collisions. The distance is directly proportional to the fourth root of the transmit power, denoted as $R \propto \sqrt[4]{P_\mathrm{t}}$. Thus, to double the detectable range, we require $2R \propto 2\sqrt[4]{P_\mathrm{t}} = \sqrt[4]{16P_\mathrm{t}}$. Consequently, with all other factors held constant, increasing the transmit power by a factor of 16 is necessary. However, it's worth noting that simply boosting transmit power can impose significant limitations on system requirements, as mentioned earlier.

In automotive applications, the radar equation serves to compute the maximum detectable distance. Nevertheless, estimating the RCS of a car poses a challenge due to variations in shape, size, and texture among automotive targets. Consequently, accurately estimating the RCS of a vehicle remains a complex task. According to the ITU, for automotive radars operating within the 77.5–78 GHz frequency range, the detection distance in meters is represented as:

$$R = \sqrt[4]{\frac{P_\mathrm{t} G_\mathrm{A}^2 \lambda^2 \sigma_\mathrm{tg}}{S_\mathrm{min}(4\pi)^3}}. \tag{6.6}$$

The provided formula assumes equal transmit and receive gains. Here, P_t denotes the transmitter power (in watts), G_A signifies the antenna gain, σ_tg represents the effective target area, which is set to 1 m^2, λ stands for the wavelength, calculated as 3.859×10^{-3} m at a frequency of 77.75 GHz, and S_min represents the receiver sensitivity (in watts). However, in practice, overall system losses L_s, encompassing polarization losses, atmospheric propagation losses, and antenna pattern losses, are commonly taken into account. In this scenario, the maximum detection distance can be expressed as:

$$R = \sqrt[4]{\frac{P_\mathrm{t} G_\mathrm{A}^2 \lambda^2 \sigma_\mathrm{s}}{S_\mathrm{min}(4\pi)^3 L_\mathrm{s}}}. \tag{6.7}$$

Alternatively, we can rewrite Eq. (6.5) in the form of Eq. (6.8):

$$R = \sqrt[4]{\frac{P_\mathrm{t} G_\mathrm{A}^2 \lambda^2 \sigma_\mathrm{s}}{kT_0 B F_\mathrm{n} \mathrm{SNR}_\mathrm{min}(4\pi)^3 L_\mathrm{s}}}, \tag{6.8}$$

TABLE 6.2
Typical radar parameters within the 76–81 GHz frequency range.

Parameters	Typical Range	Remarks
Detection Range (m)	0 ~ 250	Depends on the expected application, such as Short Range Radar (SRR), Medium Range Radar (MRR), and Long Range Radar (LRR)
Transmit Power (dBm)	10 ~ 13	Depends on regulations
Transmit/Receive Antenna Gain (dBi)	10 ~ 25	Depends on the azimuth and elevation angles of the field of view
Receiver Noise Figure (dB)	10 ~ 20	Depends on hardware implementation
Radar Cross Section (RCS) (dBsm)	-10 ~ 20	Pedestrians, trucks, buses
Receiver Sensitivity (dBm)	-120 ~ -115	Depends on hardware implementation
Minimum Detection Signal-to-Noise Ratio (dB)	10 ~ 20	Depends on hardware implementation

where $\mathrm{SNR}_{\min}$ is the minimum corresponding to the minimum received power. For most automotive applications, typical radar parameters within the 76–81 GHz frequency range are as shown in Table 6.2.

6.3 Range and Doppler Estimation

It is reasonable to assume that the signal form of the linear frequency modulated continuous wave (LFMCW) radar transmission waveform manifests as a frequency-modulated continuous sawtooth wave. Linear frequency modulation implies that the frequency of the modulated signal changes linearly with time. In the time domain, this manifests as a waveform with frequency linearly varying over time. Conversely, in the frequency domain, the transmission signal's frequency is directly proportional to time, as depicted in Figure 6.2. Based on this, the transmission signal model of LFMCW can be represented by the following formula:

$$s_{\mathrm{t}}(t) = A\cos\left(2\pi(f_0 t + ut^2/2) + \phi_0\right), t \in [0, T]. \tag{6.9}$$

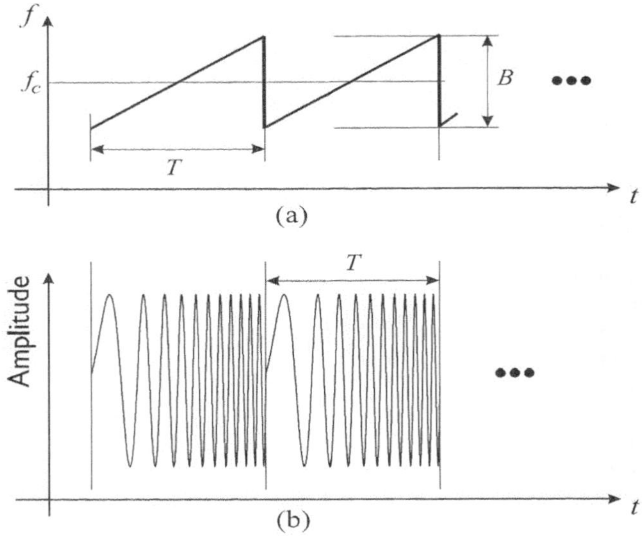

FIGURE 6.2
Frequency-modulated continuous-wave (FMCW) signal: (a) frequency variation over time, (b) instantaneous chirp signals.

In this representation, the sweep bandwidth of the transmission signal is designated as B, the pulse width during transmission is denoted as T, and the frequency modulation slope is expressed as B/T, identified as u. Consequently, the phase of the single-cycle LFMCW radar transmission signal model can be articulated in the following manner:

$$p_{\mathrm{t}}(t) = 2\pi(f_0 t + ut^2/2), t \in [0, T], \tag{6.10}$$

Range measurement in a stationary scenario: Under the assumption that the distance from the stationary target to the radar is R, and the speed of electromagnetic wave propagation in the air is c, the delay τ of the received signal compared to the transmitted signal is given by $\tau = 2R/c$. Hence, in the ideal scenario, the model of the received echo signal by the receiving antenna from the target can be represented as follows:

$$s_{\mathrm{r}}(t) = KA\cos\left(2\pi(f_0(t-\tau) + u(t-\tau)^2/2) + \phi_0\right), t \in [0, T], \tag{6.11}$$

where K is the attenuation loss during the transmission process. From the aforementioned formula, it's apparent that the echo signal maintains the same signal form as the transmitted signal, albeit with a fixed time delay τ relative to the transmitted signal. Consequently, the phase of the echo signal can be expressed as:

$$p_{\mathrm{r}}(t) = 2\pi(f_0(t-\tau) + u(t-\tau)^2/2), t \in [0, T]. \tag{6.12}$$

Combining the received echo signal $s_\mathrm{r}(t)$ with the transmitted signal $s_\mathrm{t}(t)$ and subsequently passing it through a low-pass filter yields a single-frequency sinusoidal signal. Through the derivation of the formula, the phase expression of the signal can be obtained as:

$$\begin{aligned} p_\mathrm{t}(t) - p_\mathrm{r}(t) &= 2\pi(f_0 t + ut^2/2) + \phi_0 - \left(2\pi(f_0(t-\tau) + u(t-\tau)^2/2) + \phi_0\right) \\ &= 2\pi f_0 \tau + 2\pi u\tau t + \pi u \tau^2. \end{aligned} \tag{6.13}$$

At this point, it is evident that the frequency difference between the transmitted signal and the echo signal from a single target is a single-frequency signal. According to the above formula, the frequency of the intermediate frequency signal, f_m, can be obtained as follows:

$$f_\mathrm{m} = u\tau = \frac{u2R}{c} = \frac{2BR}{cT}. \tag{6.14}$$

Sampling the intermediate frequency signal with an ADC, and then performing fast Fourier transform (FFT) to extract the frequency information from the spectrum, assuming that the frequency corresponding to the peak value of the FFT spectrum is f_m, the distance information of the target can be represented as follows:

$$R = \frac{cTf_\mathrm{m}}{2B}. \tag{6.15}$$

Range measurement in the case of moving targets: Assuming that, within the coverage area of the EM wave, a target is at a distance R_0 from the transmitting antenna at time t_0, moving away from the antenna with a radial velocity v, and considering the direction away from the antenna as the positive direction, the model formula for the received echo signal from the target remains consistent with that of a single target, as shown below:

$$s_\mathrm{r}(t) = KA\cos\left(2\pi(f_0(t-\tau) + u(t-\tau)^2/2) + \phi_0\right), t \in [0, T]. \tag{6.16}$$

However, the delay τ changes, as shown below:

$$\tau = \frac{2R_0}{c} = \frac{2(R+vt)}{c}. \tag{6.17}$$

At this moment, the phase of the intermediate frequency signal obtained through heterodyning is as follows:

$$\begin{aligned} p_\mathrm{t}(t) - p_\mathrm{r}(t) &= 2\pi(f_0 t + ut^2/2) + \phi_0 - \left(2\pi(f_0(t-\tau) + u(t-\tau)^2/2) + \phi_0\right) \\ &= 2\pi f_0 \tau + 2\pi u\tau t + \pi u \tau^2, \end{aligned} \tag{6.18}$$

which is exactly the same as Eq. (6.13). The equation after substituting τ

becomes:

$$\begin{aligned}p_{\rm t}(t) - p_{\rm r}(t) &= \frac{4\pi f_0(R+vt)}{c} + \frac{4\pi u(R+vt)t}{c} - \frac{4\pi u(R+vt)}{c^2}\\ &= 2\pi\left(\frac{2f_0 v}{c} + \frac{2uR}{c} - \frac{4uRv}{c^2}\right) + \pi\left(\frac{4uv}{c} - \frac{4uv^2}{c^2}\right)t^2 \\ &+ \left(\frac{4\pi R f_0}{c} - \frac{4\pi u R^2}{c^2}\right).\end{aligned} \tag{6.19}$$

Clearly, according to the above formula, it can be seen that the intermediate frequency signal of the motion target's signal remains a linear frequency-modulated signal. Therefore, the frequency modulation slope $u_{\rm m}$, carrier frequency $f_{\rm m}$, and initial phase $\phi_{\rm m}$ are as follows:

$$\begin{aligned}u_{\rm m} &= \frac{4uv}{c} - \frac{4uv^2}{c^2},\\ f_{\rm m} &= \frac{2f_0 v}{c} + \frac{2uR}{c} - \frac{4uRv}{c^2},\\ \phi_{\rm m} &= \frac{4\pi R f_0}{c} - \frac{4\pi u R^2}{c^2}.\end{aligned} \tag{6.20}$$

Because the speed of light, c, is equal to 3×10^8 m/s, the quadratic term of c can be neglected. Approximately written as:

$$\begin{aligned}u_{\rm m} &= \frac{4uv}{c} - \frac{4uv^2}{c^2} \sim \frac{4uv}{c},\\ f_{\rm m} &= \frac{2f_0 v}{c} + \frac{2uR}{c} - \frac{4uRv}{c^2} \sim \frac{2f_0 v}{c} + \frac{2uR}{c},\\ \phi_{\rm m} &= \frac{4\pi R f_0}{c} - \frac{4\pi u R^2}{c^2} \sim \frac{4\pi R f_0}{c}.\end{aligned} \tag{6.21}$$

Then if the influence of the Doppler frequency on the intermediate frequency signal frequency is neglected, then:

$$\begin{aligned}f_{\rm m} &= \frac{2uR}{c},\\ \phi_{\rm m} &= \frac{4\pi R f_0}{c}.\end{aligned} \tag{6.22}$$

The approximate formula neglects the dependence of the intermediate frequency signal frequency on the object's velocity. In fast FMCW radar, this influence is typically very small. Moreover, after completing the Doppler FFT processing, it can be easily further corrected through Doppler phase compensation. Hence, the range information can still be extracted from the intermediate signal, being similar to the stationary scenario. Also, from Eqs. (6.14) and (6.22), we can also conclude that

$$R_{\max} = \frac{f_{\rm m} c}{2u}. \tag{6.23}$$

As a result, given a fixed analog-to-digital converter (ADC) capability, the maximized estimation distance can be adjusted by adjusting frequency modulation slope $u = B/T$. Also, we have

$$\Delta f = \frac{u2\Delta R}{c}. \tag{6.24}$$

The Fourier transform theory states that the observation window (T) can distinguish frequency components with intervals exceeding $1/T$ Hz. It can be represented by the following formula.

$$\Delta f > \frac{1}{T}. \tag{6.25}$$

Hence, with u fixed, the range resolution can be considered equivalent to the radar system's frequency resolution and we have

$$d_{\text{res}} = \Delta d > \frac{c}{2uT} = \frac{c}{2B}. \tag{6.26}$$

From the above formula, it can be concluded that the wider the signal bandwidth (B), the higher the range resolution. Therefore, adjusting the signal's bandwidth (B) also adjusts the radar system's range resolution.

From the principle of range, it can be understood that emitting a single chirp can provide distance information about an object. However, a single chirp cannot achieve velocity measurement; at least two chirps need to be emitted for velocity measurement. If multiple objects need to be detected, a greater number of chirps is required, denoted as N. N chirps constitute a data frame. Similar to how distance information is concealed in the intermediate frequency, the velocity information of an object is concealed in the phase difference of the signal. For two adjacent cycles of signals, due to the short interval time T, and limited distance resolution, the peak position in the distance dimension of the FFT spectrum within the two cycles remains almost unchanged. However, because phase is more sensitive than distance, even small changes in distance between cycles can cause a change in the initial phase of the intermediate frequency signal. According to the properties of the Fourier transform, the initial phase of the signal is reflected in the phase corresponding to the complex value at the peak. By calculating the phase difference between adjacent cycles, the velocity of the target can be obtained:

$$\Delta\phi = \frac{4\pi\Delta R}{\lambda} = \frac{4\pi vT}{\lambda}, \tag{6.27}$$

where T is the period interval time. So the velocity between two pulses is:

$$v = \frac{\lambda\Delta\phi}{4\pi T}. \tag{6.28}$$

Due to the phase difference inherent in velocity measurements, such measurements are ambiguous. This kind of measurement is only unambiguous under certain conditions, that is, $|\Delta\phi| < \pi$. Thus, we also have

$$v_{\text{max}} = \frac{\lambda}{4T}, \tag{6.29}$$

as the maximized velocity estimation constraint. Similarly, according to Fourier Transform theory, two discrete frequencies can only be distinguished under specific conditions, that is $\Delta w = w_2 - w_1 > 2\pi/N_{\text{fft}}$, where N_{fft} is the number of FFT points. We also have

$$v_{\text{res}} > \frac{\lambda}{2N_{\text{fft}}T}, \tag{6.30}$$

as the velocity resolution constraint.

6.4 Target Detection

After having an idea of how modern radar systems estimate the range and Doppler velocity of a target, we aim to discuss another important operation in current radar systems, target detection. The purpose of target detection is typically to distinguish real target echo signals from noise and clutter. Thus, if we skip this operation in the radar receiver, one can imagine that we will get a lot of ghost targets from noise and clutter besides real targets to degrade the radar system performance. To do so, the availability of statistical characteristics of the target radar cross-section (RCS) can significantly enhance the performance of target detection algorithms. In this direction, Swerling introduced the Swerling model, which describes the statistical characteristics of an object's RCS using different degrees of freedom chi-square distributions. Take the Swerling model I as an example, the reflection of targets in a single scan has a constant RCS magnitude, but it varies with the scan according to the PDF of a chi-square distribution with two degrees of freedom. The PDF is given by the following expression:

$$f(\sigma) = \frac{1}{\sigma_{\text{avg}}} e^{-\frac{\sigma}{\sigma_{\text{avg}}}}, \sigma \geq 0, \tag{6.31}$$

where σ_{avg} is the average value of RCS. The Swerling model can be incorporated into the probability of target detection as follows. If the input signal to the threshold detector consists of signal components with amplitudes A from Gaussian noise with embedded variance ξ^2, denoted as $r(t)$, then the PDF of $r(t)$ can be represented as:

$$f(r(t)) = \frac{r(t)}{\xi^2} I_0\left(\frac{r(t)A}{\xi^2}\right) e^{-\frac{r(t)^2+A^2}{2\xi^2}}, \tag{6.32}$$

where $I_0(.)$ is the first-type zero-order modified Bessel function. If only noise is considered, then:

$$f(r(t)) = \frac{r(t)}{\xi^2} I_0 \left(\frac{r(t)A}{\xi^2} \right) e^{-\frac{r(t)^2}{2\xi^2}}. \tag{6.33}$$

For a given detection threshold P_{thr}, the false alarm probability P_{fa} is represented by the following formula:

$$P_{\text{fa}} = \int_{P_{\text{thr}}}^{\infty} f(r(t)) = \frac{r(t)}{\xi^2} I_0 \left(\frac{r(t)A}{\xi^2} \right) e^{-\frac{r(t)^2+A^2}{2\xi^2}} d(r(t)) = e^{-\frac{P_{\text{thr}}^2}{2\xi^2}}. \tag{6.34}$$

From the above formula, the threshold can be expressed in terms of the false alarm probability as:

$$P_{\text{thr}} = \sqrt{2\xi^2 \ln \left(\frac{1}{P_{\text{fa}}} \right)}. \tag{6.35}$$

From the PDF of $r(t)$, we can define the detection probability P_{d} as:

$$P_{\text{d}} = \int_{P_{\text{thr}}}^{\infty} f(r(t)) = \frac{r(t)}{\xi^2} I_0 \left(\frac{r(t)A}{\xi^2} \right) e^{-\frac{r(t)^2+A^2}{2\xi^2}} d(r(t)). \tag{6.36}$$

Many approximate values of P_{fa} and P_{d} are possible, and they can be obtained from tables for a given required SNR. From an automotive perspective, insights into the bump target model can be obtained from surveys of ground vehicles. Ground vehicles consist of large trucks, heavy-duty trucks, medium-sized cars, and medium-sized trucks as described above. The target behavior of Swerling I ground vehicles has been shown to be the most common case across all datasets and parameter variations, despite limited changes in angle.

In a practical scenario with the presence of noise and clutter, peak detection is required to select valid targets from the received reflection signals. The spectrum typically consists of multiple peaks. For automotive applications, peak detection becomes more challenging due to the reflection signals received from roadside obstacles and ground reflections. Different strategies can achieve varying degrees of effectiveness in performing peak detection tasks, with distance-Doppler spectrum being commonly used in most cases. The above expressions of P_{d} and P_{fa} are correct, but they become ineffective when there is a mixture of different targets in the radar field of view. Moreover, the reflection characteristics of targets are also influenced by factors such as distance, angle, size, and shape of the target. Therefore, multiple unit thresholds will be required to describe such scenarios. Adaptive detection thresholds have long been a subject of research to address this challenge. To address the issues associated with fixed threshold and multi-unit threshold approaches, we have applied the constant false alarm rate (CFAR) method and achieved some success. It should be noted that this method is not without cost (computational cost). In automotive radar applications, besides radar size, the increase in computational cost is also a significant concern.

The requirement of CFAR detection algorithms increases computational speed and device memory with each processing, thus necessitating a trade-off between performance and cost. In this section, we briefly outline some important CFAR algorithms. Specifically, maintaining a constant false alarm rate is desirable because detection algorithms are highly sensitive to the noise and clutter that almost always exist in received radar echoes. Therefore, the overall goal of all radar detection schemes is to ensure that false alarms do not fluctuate arbitrarily. During the detection process, each unit uses a threshold to assess the presence or absence of a target. It is advantageous to be able to detect both high and low fidelity targets while maintaining a constant false alarm rate. This requires the use of adaptive thresholding methods, which are commonly employed in modern radar systems. The most common forms of basic CFAR methods include cell averaging CFAR (CA-CFAR), cell averaging greatest of CFAR (CAGO-CFAR), and cell averaging least of CFAR (CALO-CFAR). The CFAR principle dates back to the late 1960s. Methods for addressing false alarm issues include implementing CFAR schemes, which adjust the detection threshold as a function of the perceived environment. Although there are numerous types of CFAR models, they typically revolve around the concept of a "background average" (sometimes referred to as cell averaging CFAR). This model estimates the interference (noise or clutter) level from radar cells on either side of the range cell of interest and utilizes this estimate to determine whether a target is present in the central cell of interest. This process iterates through the range, progressively extracting one cell at a time until all range cells have been examined. The basic idea behind the model is that when noise is present, the cells surrounding the cell of interest will contain a good estimate of the noise for that cell; assuming that the noise or interference is temporarily uniform spatially. In theory, the model produces a constant false alarm rate that is independent of the noise or clutter level, as long as the noise follows a Rayleigh distribution within all distances considered by the model. In the CA-CFAR method, as shown in Figure 6.3, the threshold is calculated by taking the average power of the cells surrounding the cell under test (CUT) rather than using a single fixed value. The CUT is the cell being examined to determine the presence or absence of a target. To ensure that the CUT does not influence the threshold calculation, the cells immediately surrounding the CUT are excluded from the calculation, and these cells are referred to as the guard cells. For one-dimensional CFAR, the guard cells are located to the left and right of the CUT, while for two-dimensional CFAR, the guard cells form a ring around the CUT. To determine whether a target exists in the CUT, the power in the CUT must be greater than the power in all guard cells and also greater than the calculated average power level. Other CFAR methods use different approaches to determine thresholds, achieving different trade-offs. Nevertheless, their basic principles are very similar to CA-CFAR, so they will not be individually introduced here. We refer interested readers to ref. [120] for more details regarding the comparison of different CFAR algorithms.

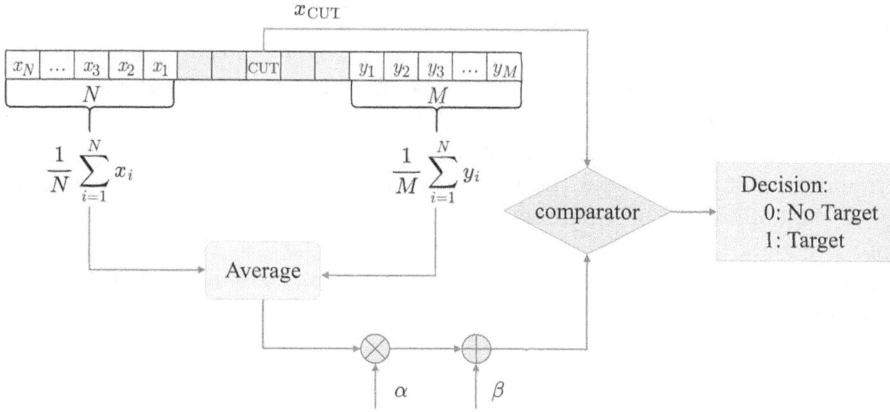

FIGURE 6.3
Illustration of the CA-CFAR algorithm.

6.5 Common Radar Waveforms

In previous sections, we introduced different functionalities of radar systems, mainly considering FMCW waveform. In summary, FMCW radar operates by transmitting continuous waveforms with a linearly modulated frequency. This modulation creates a frequency-swept waveform that is continuously transmitted and received by the radar system. By comparing the transmitted and received signals, FMCW radar can measure the range and velocity of targets based on the frequency difference, known as the Doppler shift. Moreover, FMCW radar offers several advantages, including high-range resolution, which allows for the accurate detection and localization of targets, particularly in short-range applications. Additionally, FMCW radar is relatively immune to interference and clutter due to its narrowband nature and linear frequency modulation. This makes it well-suited for applications such as automotive radar for collision avoidance, range measurement, and adaptive cruise control systems.

Recently, research on orthogonal frequency division multiplexing (OFDM) radar has become increasingly popular because such radar systems have great potential to serve as the foundation of integrated sensing and communication systems by providing both applications. From the communication standpoint, this achieves high spectral efficiency as well as simple extraction of communication data. Meanwhile, from the radar standpoint, it enables efficient digital demodulation of the radar waveform. OFDM not only enables favorable modulation for both applications, but also combines both functionalities via a single waveform. This initially motivated research on OFDM radar. Currently, OFDM is often studied as a means for efficient implementation of digital, software-defined radar—independent of the communication aspect. This concept differs from FMCW radar in terms of generating waveforms digitally and performing demodulation in the digital domain. Broadly speaking, this is equivalent to operating with arbitrary digitally generated waveforms and matched filter-based processing at the receiver. For OFDM radar, this large degree of flexibility in the waveform choice enables communication and radar capabilities to be combined by embedding communication information into the radar waveform. Specifically, OFDM radar, on the other hand, is based on a technique widely used in communication systems for transmitting data over multiple subcarriers. That is, the resultant signal model can be expressed as:

$$x(t) = \sum_{m=0}^{M-1} \sum_{n=0}^{N-1} D(mN+n) e^{j2\pi n \Delta f t} \mathrm{rect}_T(t - mT), \tag{6.37}$$

where N represents the number of subcarriers, M is the number of consecutive symbols evaluated, Δf is the OFDM subcarrier spacing, $T = T_{\mathrm{cp}} + T_{\mathrm{sym}}$ is the OFDM symbol duration consisting of the cyclic shift (CP) duration T_{cp} and the symbol duration T_{sym}, and $D(n)$ is the modulated symbol. One can notice that the signal model is similar to the OFDM symbol for communication usage. Thus, the signal processing procedure in the receiver is also similar: (1) remove CP, (2) conduct Fourier transform, and (3) perform element-wise complex division by the transmit symbols to get the desired information. After the above procedure, the information from the nth subcarrier of the mth OFDM symbol can be expressed as:

$$y_{m,n} = \gamma e^{-j2\pi n \Delta f \tau} e^{j2\pi f_c \eta m T} + z_{m,n}, y_{m,n} = \gamma e^{-j2\pi n \Delta f \tau} e^{j2\pi f_c \eta m T} + z_{m,n}, \tag{6.38}$$

where γ is the radar channel gain, $\tau = 2R/c$ is the round-trip delay associated with the range R, and $\eta = 2v/c$ is the normalized Doppler shift associated with the velocity v. Then, following the same procedure as FMCW radar by taking a two-dimensional Fourier transform in the time-frequency domain will provide the range and Doppler estimates to realize radar functionality. In the above equation, the radar waveform is divided into multiple orthogonal subcarriers, each transmitting data simultaneously. By using multiple subcarriers, OFDM radar can achieve high data rates and spectral efficiency. The block diagram of OFDM radars is shown in Figure 6.4 for reference. Additionally, the orthogonal nature of the subcarriers enables robustness against multipath fading and interference, making OFDM radar well-suited for applications in complex environments where traditional radar systems may struggle. OFDM radar is particularly advantageous for applications requiring high-resolution imaging, target classification, and radar-based surveillance, as it offers enhanced signal processing capabilities and improved performance in challenging scenarios. We recommend interested readers to refer to ref. [121] for the detailed OFDM radar introduction, including signal model, signal processing procedure, and achieved performance. In summary, while both FMCW radar

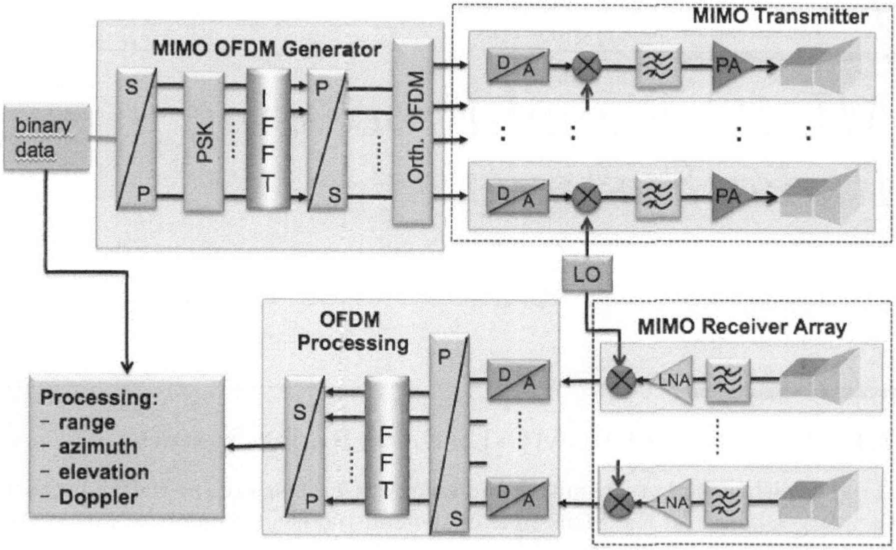

FIGURE 6.4
Block diagram of the OFDM radars (copyright from ref. [122]).

and OFDM radar have their unique features and advantages, FMCW radar excels in short-range applications with high-range resolution and interference robustness, while OFDM radar offers high data rates and robust performance in complex environments, making it suitable for applications requiring advanced signal processing and imaging capabilities.

7

MIMO Signal Processing in Radar Systems

7.1 Overview of MIMO-Enabled Radar Systems

A radar with a single transmitter and receiver can measure the radial distance to a target based on the flight time of the emitted electromagnetic waves. Although it provides distance estimation capability, a single-transmitter, single-receiver antenna typically cannot directly measure the angle of the target. The first solution to enable angle estimation with a single antenna radar system is to use antennas with very narrow beamwidths with mechanical rotating scanning mechanisms. However, this method also comes with additional requirements, such as requiring large antennas to achieve narrow beamwidths. Additionally, rotating large antennas requires a significant amount of space and cannot simultaneously cover the entire field of view. The second solution involves using a radar with multiple fixed antennas to obtain the angle of the target, rather than rotating large antenna arrays. This type of radar can utilize multiple antennas to detect the target without physical rotation. Moreover, because all antennas can operate simultaneously, the refresh rate of radar target detection is faster. We refer to this type of radar with multiple transmitter and receiver antennas as multiple-input multiple-output (MIMO) radar. When more antennas are used, the radar's angular resolution improves.

In the previous chapter, we introduced the principle of radar operation by transmitting electromagnetic waves from a transmitting antenna, which are then reflected by the target and received by a receiving antenna. By calculating the time delay between transmission and reception, the distance to the target can be determined. In MIMO radar, to determine the angle of the target, multiple antennas are used. Specifically, when the positions of each antenna are slightly different, the time it takes to receive the reflected signal also varies slightly. By analyzing the subtle time differences between each antenna, the function of determining the target angle can be achieved. In practical applications, the phase of the received signal is more important than absolute time differences because absolute time differences require extremely high time resolution. Under the assumption that the difference in distance between antennas to the target is less than the radar's range resolution (i.e., near-field

DOI: 10.1201/9781003408239-7

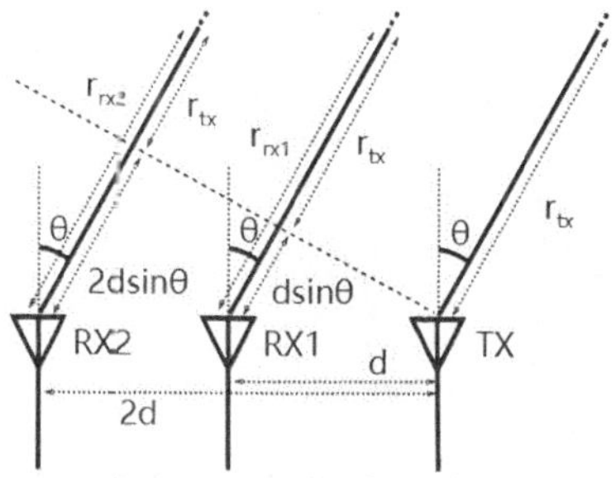

FIGURE 7.1
Illustration of the radar signal traveling distance in a 1×2 radar system.

assumption), the distance traveled by the radar waveform is the sum of the distance from the transmitting antenna to the target and the distance from the target to the receiving antenna. The calculation is as follows:

$$R_{\mathrm{Tx}} + R_{\mathrm{Rx}} = 2R_{\mathrm{Tx}} + d \sin\theta, \tag{7.1}$$

as shown in Figure 7.1. Because the multiple receiving antennas are at different distances from the transmitter, the radar waveform travels slightly different distances, which can be used to solve for the target's angle. Specifically, the signal received from a target at a receiving antenna is in the form:

$$r(t) = As(t) \exp\left(\frac{2\pi j}{\lambda} d \sin\theta\right), \tag{7.2}$$

where A is the complex amplitude of the received signal, $s(t)$ is the transmitted signal. The exponential term represents displacement and is related to the wavelength of the transmitted signal. As indicated by the equation above, by comparing the phase difference of the signals received by two antennas, it is possible to easily determine the angle of the target's echo. This is known as the phase difference angle measurement method. For example, for two antennas spaced d apart from each other, the phase difference of the received signals is:

$$\Delta\phi = \exp\left(\frac{2\pi j}{\lambda} d \sin\theta\right). \tag{7.3}$$

Therefore, the target angle θ can be easily obtained from the equation above. In order to obtain a single solution, the spacing between array antennas must be less than half of the wavelength $\frac{\lambda}{2}$. Otherwise, due to the periodicity of the sine wave, it is possible to obtain the same phase at two or more angles, resulting in grating lobes. However, the aforementioned phase difference angle measurement method can only be used to measure a single target angle, and estimating multiple target angles requires a special design. Then, when it

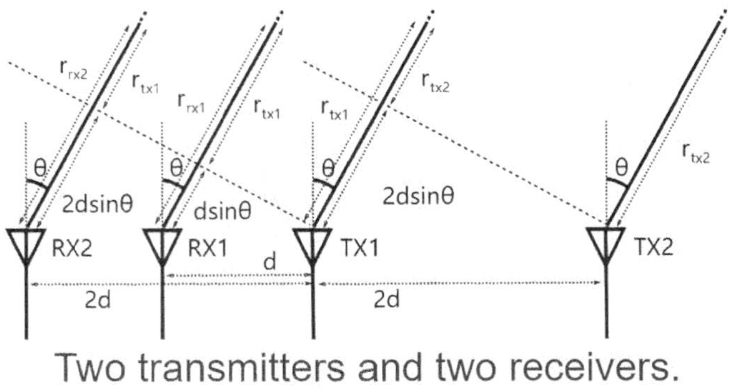

FIGURE 7.2
Illustration of the radar signal traveling distance in a 2×2 radar system.

comes to the angle resolution, we are interested in the minimal angle $\Delta\theta$ we can distinguish after processing, that is:

$$\Delta\omega = \frac{2\pi d}{\lambda}(\sin(\theta + \Delta\theta)) > \frac{2\pi}{N}, \tag{7.4}$$

where N is the number of received antennas from the fast Fourier transform (FFT) properties. Utilizing the property

$$\frac{\sin(\theta + \Delta\theta) - \sin(\theta)}{\Delta\theta} = \cos(\theta), \tag{7.5}$$

from the concept of derivations, we obtain

$$\frac{2\pi d}{\lambda}\cos(\theta)\Delta\theta > \frac{2\pi}{N}. \tag{7.6}$$

Thus,

$$\theta_{\text{res}} = \Delta\theta = \frac{\lambda}{Nd\cos(\theta)}. \tag{7.7}$$

In the best case ($\theta = 0$ and $d = \frac{\lambda}{2}$), we have $\theta_{\text{res}} = \frac{2}{N}$, being a motivation to employ multiple antenna radars for the improved angle estimation resolution.

Next, we aim to demonstrate other benefits of MIMO radar and several widely used designs to fully utilize the spatial degree of freedom when we have multiple transmit antennas and multiple received antennas at the same time. We are now considering a similar antenna configuration to the one before. However, a second transmitting antenna is placed at a distance of $2d$ from the first transmitting antenna, as shown in Figure 7.2. When the signals between Tx1-Rx1 and Tx1-Rx2 are exactly the same as before, signals between Tx2-Rx1 and Tx2-Rx2 are also received. In fact, if we write down all four pairs of transmitting-receiving antenna signal propagation distances, we cannot find any difference in the length between the 2Tx-2Rx array with four antennas and the 1Tx-4Rx array with five antennas. This implies that their topological structures are completely identical. Thus, if Tx2 is moved to the same position as Rx2, the absolute signal propagation distance measured will be exactly the same as the result obtained from the 1Tx-4Rx array, as shown in Figure 7.3. From the above examples, we can draw an important conclusion: by using two transmitting antennas, we can save the cost of one antenna. If we further consider larger arrays with more antennas, the number of antennas that can be saved will also increase accordingly. Moreover, each arbitrary pair of transmitting antenna and receiving antenna combination in MIMO radar can independently perform measuring tasks, while an antenna array with only one transmitting antenna can only perform measurements equal to the number of receiving antennas. That is, in a MIMO radar with 32 transmitting antennas and 32 receiving antennas, the total number of measurements is $32 \times 32 = 1024$. This is equivalent to the measurement accuracy (angular resolution) that can be achieved by a radar with only one transmitting antenna but 1024 receiving antennas. Note that to make the principle of MIMO radar possible, we must assume that the receiving antennas can separate the signals from different transmitting antennas. If they transmit the same waveform at the same time, we will see superimposed received signals at the receiving end that cannot be separated. In practice, transmitting antennas can ensure that received signals can be separated by various means, such as transmitting signals at different times or frequencies, or using waveforms that are orthogonal to each other. By doing so, receiving antennas can receive signals and separate the signals from different transmitting antennas from each other, which is known as MIMO radar waveform design. Also, in the example above, we found that the performance of the 2TX-2RX radar array is similar to that of the 1TX-4RX radar array, but if the antenna spacing is not specially designed, there is no way to match as shown in the example above. While it is relatively easy to analyze radar arrays with a single transmitting antenna, how to generalize this process becomes very important in radar system designs. In practice, to simplify the analysis, we often introduce the concept of virtual elements to design radar arrays, as shown in Figure 7.4. The so-called virtual element refers to an overlapping transmitting antenna and receiving antenna, and this

Antenna pair	Traveled distance
TX2-RX1	$2r_{tx1} - d\sin\theta$
TX2-RX2	$2r_{tx1}$
TX1-RX1	$2r_{tx1} + d\sin\theta$
TX1-RX2	$2r_{tx1} + 2d\sin\theta$

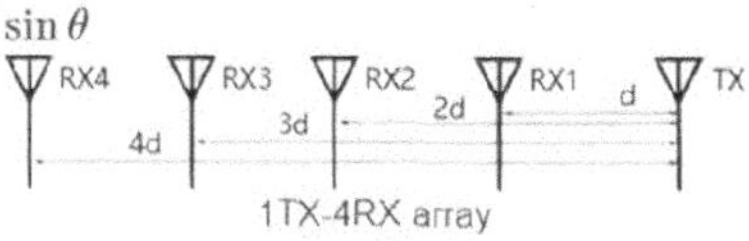

FIGURE 7.3
Illustration of the radar signal traveling distance in a 1×4 radar system.

antenna can only receive its own transmitted signal. For example, in the figure below, we place a centered virtual element in the TX-RX pair at a distance of d. When we consider the signal travel distance of the original TX-RX pair, we can obtain $2r_{\text{TX}} + d\sin\theta$, which is exactly equal to the signal travel distance from the virtual element to the receiver $2(r_{\text{TX}} + (d/2)\sin\theta) = 2r_{\text{TX}} + d\sin\theta$, we can use this property for radar array design.

7.2 Direction-of-Arrival (DOA) Estimation

In addition to achieving cost savings in antennas through equalization calculations, another advantage of MIMO radar is that it can provide degrees of freedom inherent in multi-antenna systems to further enhance radar performance. Below are several commonly used algorithms for angle estimation in MIMO radar. We first consider an array radar system composed of M receiving antennas and K transmitting antennas. Then, the received signal can be represented as:

$$\mathbf{x}(t) = \mathbf{A}(\theta)\mathbf{s}(t) + \mathbf{n}(t), \tag{7.8}$$

where $\mathbf{x}(t) \in \mathbb{C}^{M\times 1}$ is the receiving signal, $\mathbf{A}(\theta) \in \mathbb{C}^{M\times K} = [\mathbf{a}(\theta_1), ..., \mathbf{a}(\theta_K)]$ is the direction matrix, $\mathbf{s}(t) \in \mathbb{C}^{K\times 1}$ is the transmitting signal, and $\mathbf{n}(t) \in \mathbb{C}^{M\times 1}$ is the noise vector. The elements of the direction matrix are composed of the defined steering vectors, which are defined as:

$$\mathbf{a}(\theta_i) = \left[1, e^{-\frac{j2\pi d\sin(\theta_i)}{\lambda}}, ..., e^{-\frac{j2\pi d(M-1)\sin(\theta_i)}{\lambda}}\right], \tag{7.9}$$

where d is the antenna element spacing, using a uniform linear array (ULA), λ is the wavelength of the propagated signal, and θ_i is the arrival direction

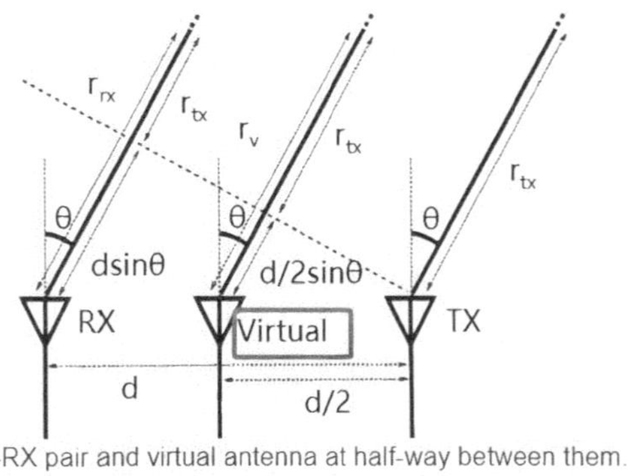

FIGURE 7.4
Illustration of the virtual element in MIMO radar system.

(angle) of the signal from the ith transmitting antenna. In some algorithms that require optimization of weights, it is necessary to consider the weighted output of sensors. In this case, the weighted linear combination of sensor outputs can be represented as follows:

$$y(t) = \sum_{m=1}^{M} w_m^* x_m(t) = \mathbf{w}^H \mathbf{x}(t), \tag{7.10}$$

where w_m^* is the conjugate weighting of the mth receiving antenna.
Delay-and-sum method: The delay-and-sum method calculates the DoA by measuring the signal strength at each possible arrival angle (scanning) and selecting the arrival angles at power peaks. In the case of weights $\mathbf{w}$, according to Eq. (7.10), equal to the steering vector, it will introduce a power peak in the beam. The highest power point corresponds to the estimated angle of arrival. The output mean power of the beamformer using this method is given by:

$$P_{\mathrm{DS}}(\theta) = \mathbb{E}\left[|y(t)|^2\right] = \mathbf{w}^H \mathbb{E}\left[\mathbf{x}(t)\mathbf{x}^H(t)\right]\mathbf{w} = \mathbf{w}^H \mathbf{R}\mathbf{w}. \tag{7.11}$$

Let $s(t)$ arriving with steering angle θ_0, of all the possible weight vectors, the receiving antenna will have the biggest gain in the direction θ_0, when $\mathbf{w} = \mathbf{a}(\theta_0)$. This is because $\mathbf{w} = \mathbf{a}(\theta_0)$ aligns the phases of the components of the arrival signal of θ_0 in the sensors. In the DS method, a scan is performed on all possible angles of arrival and the power measurement is performed on all of them. The mean power of the steering angle is:

$$P_{\mathrm{DS}}(\theta) = \mathbf{w}^H \mathbf{R}\mathbf{w} = \mathbf{a}^H(\theta)\mathbf{R}\mathbf{a}(\theta). \tag{7.12}$$

Hence, the arrival angles θ are determined by evaluating the power peaks. Despite being computationally simpler, the width and height of the side lobes limit the performance (i.e., discrimination capability) and effectiveness of the DS method when signals from multiple directions/sources are involved, implying poor resolution. One way to improve it consists of increasing the number of sensors, thus increasing the elements of vector $\mathbf{a}(\theta)$, which increases the delay-sum signal processing and complexity.
MVDR method: Capon's minimum variance distortionless response or MVDR is similar to the delay-and-sum technique, since it evaluates the power of the received signal in all possible directions. The power from the DoA with angle θ is measured by constraining the beamformer gain to be 1 in that direction and using the remaining degrees of freedom to minimize the output power contributions of signals coming from all other directions. The optimization problem can be stated mathematically as a constrained minimization problem. The idea is that for each possible angle θ, the power in the cost function must be minimized w.r.t. $\mathbf{w}$ subject to a single constraint:

$$\begin{aligned} &\min_{\mathbf{w}} \mathbf{w}^H \mathbf{R}\mathbf{w} \\ &\text{subject to } \mathbf{w}^H \mathbf{a}(\theta) = 1 \end{aligned} \tag{7.13}$$

resulting in the MVDR received power solution:

$$P_{\text{MVDR}}(\theta) = \frac{1}{\mathbf{a}^H(\theta)\mathbf{R}^{-1}\mathbf{a}(\theta)}. \tag{7.14}$$

The disadvantage of this method is that an inverse matrix computation is required, which may become poorly conditioned if highly correlated signals are present. This method, however, provides higher resolution than the delay-and-sum method.

Multiple signal classifier (MUSIC) DoA estimator: The steering vectors corresponding to the incoming signals lie in the signal subspace; therefore, they are orthogonal to the noise subspace. One way to estimate the DoAs of multiple signal sources is to search through the set of all possible steering vectors and find those that are orthogonal to the noise subspace. MUSIC DoA estimator implements such a strategy. If $\mathbf{a}(\theta)$ is the steering vector corresponding to one of the incoming signals, then $\mathbf{a}(\theta)^H\mathbf{Q}_{\text{n}} = 0$, where $\mathbf{Q}_{\text{n}}$ is the noise subspace matrix. In practice, $\mathbf{a}(\theta)$ will not be precisely orthogonal to the noise subspace due to errors in estimating $\mathbf{Q}_{\text{n}}$. However, the function

$$P_{\text{MUSIC}}(\theta) = \frac{1}{\mathbf{a}^H(\theta)\mathbf{Q}_{\text{n}}\mathbf{Q}_{\text{n}}^H\mathbf{a}(\theta)}, \tag{7.15}$$

implies a very large value when θ is equal to the DoA related to one of the signals. $P_{\text{MUSIC}}(\theta)$ function is known as a pseudo "spectrum" provided by MUSIC. In terms of implementation, the MUSIC-DoA first estimates a basis for the noise subspace, $\mathbf{Q}_{\text{n}}$, and then determines the L peaks in Eq. (7.15); the associated angles provide the DoA estimates.

Estimation of signal parameters via rotational invariance techniques (ESPRIT): The ESPRIT-based DoA estimates are obtained neither from nonlinear optimization nor from the search of any spectral measure. Hence, it results in a computational complexity lower than the extrema-searching methods, scanning for all possible angles of arrival. Specifically, the ESPRIT operates under an array of antennas with M elements, divided into sensor doublets as shown in Figure 7.5. Each sensor is distant d from its respective pair, and each doublet is distant Δ from one another. The doublets can be separated to form 2 subarrays with m elements in each. The distance d may be different from Δ as shown in Figure 7.5(b), which makes it quite dynamic in cases of non-uniform arrays. However, the most commonly used antenna arrays possess sensors uniformly spaced, as depicted in Figure 7.5(a), being the considered configuration in this section. The subarrays are represented by $\mathbf{x}_1$ and $\mathbf{x}_2$. The output of the $\mathbf{x}_1$ and $\mathbf{x}_2$ subarrays is expressed as:

$$\begin{aligned} \mathbf{x}_1[n] &= \sum_{l=0}^{L} s_l[n]\mathbf{a}(\theta_l) + \mathbf{n}_{\mathbf{x}_1}[n], \\ \mathbf{x}_2[n] &= \sum_{l=0}^{L} s_l[n]e^{j\frac{2\pi}{\lambda}\Delta\sin(\theta_l)}\mathbf{a}(\theta_l) + \mathbf{n}_{\mathbf{x}_2}[n], \end{aligned} \tag{7.16}$$

for $n = 1, 2, ..., S$ samples. In the above equations, $\mathbf{x}_1$ and $\mathbf{x}_2$ are $m \times 1$ vectors, $\mathbf{n}_{\mathbf{x}_1}$ and $\mathbf{n}_{\mathbf{x}_2}$ are noise vectors with the same dimensions as signal vectors. In matrix form, Eq. (7.16) can be represented as:

$$\begin{aligned} \mathbf{x}_1[n] &= \mathbf{A}\mathbf{s} + \mathbf{n}_{\mathbf{x}_1}[n], \\ \mathbf{x}_2[n] &= \mathbf{A}\mathbf{\Phi}\mathbf{s} + \mathbf{n}_{\mathbf{x}_2}[n], \end{aligned} \tag{7.17}$$

where $\mathbf{\Phi} = \text{diag}\{e^{j\frac{2\pi}{\lambda}\Delta\sin(\theta_0)}, ..., e^{j\frac{2\pi}{\lambda}\Delta\sin(\theta_{L-1})}\}$ is a $L \times L$ diagonal matrix relating the signals received by the two subarrays, named the rotational operator to present the extra delay caused by Δ on the second subarray. The total array vector can be further expressed as:

$$\mathbf{x}[n] = \begin{bmatrix} \mathbf{x}_1[n] \\ \mathbf{x}_2[n] \end{bmatrix} = \begin{bmatrix} \mathbf{A} \\ \mathbf{A}\mathbf{\Phi} \end{bmatrix} \mathbf{s}[n] + \begin{bmatrix} \mathbf{n}_{\mathbf{x}_1}[n] \\ \mathbf{n}_{\mathbf{x}_2}[n] \end{bmatrix} = \mathbf{Q}_\text{s}\mathbf{s}[n] + \mathbf{n}[n]. \tag{7.18}$$

The $\mathbf{Q}_\text{s}$ structure is exploited to estimate the diagonal elements of $\mathbf{\Phi}$ without knowing $\mathbf{A}$. The $\mathbf{Q}_\text{s}$ columns span the signal subspace of the concatenated subarrays. Hence, by performing eigen-decomposition of $\mathbf{R} = \mathbb{E}\{\mathbf{x}(t)\mathbf{x}^H(t)\}$, we will obtain $\mathbf{Q} = [\mathbf{Q}_\text{s}\mathbf{Q}_\text{n}]$ containing the signal subspace and noise subspace. If $\mathbf{E}_\text{s}$ is a matrix whose columns form a basis for the subspace of the signal corresponding to the data vector $\mathbf{x}$, then $\mathbf{Q}_\text{s}$ and $\mathbf{E}_\text{s}$ are related by a $L \times L$ transformation $\mathbf{T}$ expressed by:

$$\mathbf{E}_\text{s} = \mathbf{Q}_\text{s}\mathbf{T} = \begin{bmatrix} \mathbf{A}\mathbf{T} \\ \mathbf{A}\mathbf{\Phi}\mathbf{T} \end{bmatrix} = \begin{bmatrix} \mathbf{E}_1 \\ \mathbf{E}_2 \end{bmatrix}. \tag{7.19}$$

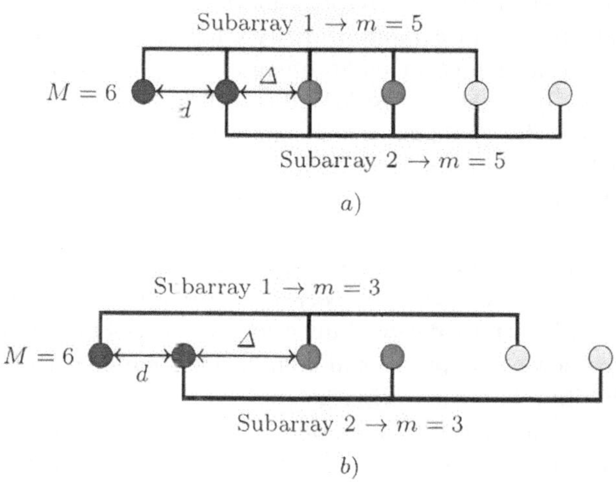

FIGURE 7.5
Two examples of ESPRIT subarrays formation using $M = 6$ antenna elements: (a) Equidistant array with 3 equidistant identical doublets and $d = \Delta$. (b) Array with 3 non-equidistant identical doublets and $d = \Delta$.

It can be seen that the subspace of $\mathbf{E}_1$, $\mathbf{E}_2$, and $\mathbf{A}$ are the same. So $\mathbf{E}_1$, $\mathbf{E}_2$, and $\mathbf{A}$ have the same range. As a result, a nonsingular $L \times L$ matrix $\boldsymbol{\Psi}$ can be defined as:

$$\mathbf{E}_1 \Psi = \mathbf{E}_2, \tag{7.20}$$

hence Ψ can be defined by:

$$\begin{aligned} \mathbf{A T}\boldsymbol{\Psi} &= \mathbf{A}\boldsymbol{\Phi}\mathbf{T}, \\ \mathbf{A T}\boldsymbol{\Psi}\mathbf{T}^{-1} &= \mathbf{A}\boldsymbol{\Phi}, \\ \Psi &= \mathbf{T}^{-1}\boldsymbol{\Phi}\mathbf{T}. \end{aligned} \tag{7.21}$$

As a result, the eigenvalues of $\boldsymbol{\Psi}$ must be equal to the diagonal elements of the $\boldsymbol{\Phi}$, and $\mathbf{T}$ columns are the eigenvectors of $\boldsymbol{\Psi}$. This is the key relationship in the development of ESPRIT and their properties. The signal parameters are obtained as nonlinear functions of the eigenvalues of the operator that maps $\boldsymbol{\Psi}$ one set of vectors ($\mathbf{E}_1$) spanning an m-dimensional signal subspace into another ($\mathbf{E}_2$). Then, since the L eigenvalues ϕ_l of $\boldsymbol{\Phi}$ are calculated, the angles of arrival can be computed as:

$$\begin{aligned} \phi_l &= e^{j\frac{2\pi}{\lambda}\Delta\theta_l}, l = 1, ..., L, \\ \theta_l &= \arcsin\left(\frac{\lambda \arg(\phi_l)}{2\pi\Delta}\right), \end{aligned} \tag{7.22}$$

where $\arg(\theta) = \arctan\left(\frac{\mathrm{Im}(\theta)}{\mathrm{Re}(\theta)}\right)$.

In conclusion, DOA estimation is a critical task in radar systems, and various methods exhibit distinct advantages and limitations. The Delay-and-Sum method, while simple and computationally efficient, suffers from low resolution and poor performance in environments with closely spaced signals or high noise levels. In contrast, the MVDR method provides improved resolution and robustness by optimizing the beamforming process, though it requires accurate knowledge of the signal covariance matrix and is computationally intensive. The MUSIC method offers superior resolution and accuracy by leveraging eigenspace decomposition, making it effective even in scenarios with closely spaced sources; however, it is computationally expensive and sensitive to model mismatches and noise. Finally, the ESPRIT method, based on subspace decomposition and rotational invariance, achieves high-resolution DOA estimation with lower computational complexity compared to MUSIC but is constrained by its reliance on array geometry and assumptions of source coherence. Thus, while each method caters to specific operational needs, their practical application depends on trade-offs between computational resources, resolution requirements, and robustness to noise and environmental factors.

7.3 Target Tracking

After obtaining target information using the methods introduced in the above sections, we can further use filtering mechanisms to capture and track the moving status of the targets, which is the main focus of this section. Being the most used tracking solution, Kalman filter is a computationally efficient recursive filter based on the theory of linear dynamic systems. It estimates the state of a discrete-time linear dynamic system from noisy measurements. Since its introduction, the Kalman filter has become a fundamental tool in modern estimation and control systems, finding widespread applications across diverse fields. In aerospace engineering, it is extensively used for navigation and guidance, particularly in applications such as spacecraft attitude estimation and autonomous drone control. In robotics, it enables real-time localization and mapping (SLAM) by integrating noisy sensor data. The Kalman filter also plays a crucial role in finance, where it is used to estimate market trends and predict stock prices from noisy datasets. Moreover, in medical technology, it aids in physiological signal processing, such as filtering electrocardiograms (ECGs) or estimating metabolic rates from sensor data. Beyond these domains, the filter is widely employed in signal processing, weather forecasting, and even augmented reality systems, showcasing its versatility in fusing noisy measurements and predicting system states in dynamic environments. Its ability to provide optimal estimates in linear systems with Gaussian noise makes it indispensable in many modern technologies.

Specifically, Kalman filter considers a discrete-time linear dynamic system. The process model is described as follows:

$$\mathbf{x}(k) = \mathbf{F}(k-1)\mathbf{x}(k-1) + \mathbf{G}(k-1)\mathbf{u}(k-1) + \mathbf{v}(k-1), \tag{7.23}$$

where $\mathbf{x}(k)$ is the state vector at time k, $\mathbf{F}(k-1)$ is the state transition matrix, $\mathbf{G}(k-1)$ is the input control matrix, $\mathbf{u}(k-1)$ is the control input, and $\mathbf{v}(k-1)$ is zero-mean Gaussian white noise. The covariance matrix of the process noise at time k is defined by $\mathbb{E}\{\mathbf{v}(k)\mathbf{v}(k)^T\} = \mathbf{Q}(k)$. In the case that the state vector $\mathbf{x}(k)$ cannot be observed, we further define the measurement equation as follows:

$$\mathbf{z}(k) = \mathbf{H}(k)\mathbf{x}(k) + \mathbf{w}(k), \tag{7.24}$$

where $\mathbf{z}(k)$ is the measurement vector at time k, $\mathbf{H}(k)$ is the observation matrix, and $\mathbf{w}(k)$ is the zero-mean Gaussian white noise. Similarly, the covariance matrix of the measurement noise at time k is defined by $\mathbb{E}\{\mathbf{w}(k)\mathbf{w}(k)^T\} = \mathbf{R}(k)$. In the scenario of automotive radar processing, the ground truth of $\mathbf{x}(k)$ remains unknown and can only be observed. Thus, $\mathbf{z}(k)$ is obtained from measurements of target states, namely distance, velocity, and DOA. For filtering and tracking purposes, the state vector $\mathbf{x}(k)$ includes the position, velocity, and acceleration of the target in both range and azimuth directions, correspondingly. The Kalman filter operation requires prior $\mathbf{P}(k|k-1)$ and posterior $\mathbf{P}(k|k)$ estimates of the error covariance matrix based on respective state estimates. The prior state estimate $\hat{\mathbf{x}}(k|k-1)$ is the estimate of the state at time k based solely on available measurements up to time $k-1$, without considering the current measurement $\mathbf{z}(k)$. On the other hand, the posterior state estimate $\hat{\mathbf{x}}(k|k)$ is the estimate of the system at time k taking into account the current measurement $\mathbf{z}(k)$. Based on these values, the estimated error covariance can be further expressed as follows:

$$\begin{aligned}\mathbf{P}(k|k-1) &= \mathbb{E}\left[(\mathbf{x}(k) - \hat{\mathbf{x}}(k|k-1))(\mathbf{x}(k) - \hat{\mathbf{x}}(k|k-1))^T\right],\\ \mathbf{P}(k|k) &= \mathbb{E}\left[(\mathbf{x}(k) - \hat{\mathbf{x}}(k|k))(\mathbf{x}(k) - \hat{\mathbf{x}}(k|k))^T\right].\end{aligned} \tag{7.25}$$

Below, we introduce the typical procedure to execute the Kalman filter, including (i) initialization, (ii) prediction, and (iii) update steps.

Initialization: In common automotive radar scenarios, the first step is to initialize $\mathbf{F}, \mathbf{G}, \mathbf{Q}, \mathbf{H}, \mathbf{R}$. Then, the state estimate and covariance estimate can be computed as $\hat{\mathbf{x}}(0)$ and $\mathbf{P}(0)$ for further processing.

Prediction: Based on the initial value, the prediction step projects the state and its uncertainty forward in time using the system model. Specifically, the prior estimation of status can be expressed as:

$$\hat{\mathbf{x}}(k|k-1) = \mathbf{F}(k-1)\hat{\mathbf{x}}(k-1|k-1), \tag{7.26}$$

and for error covariance as:

$$\mathbf{P}(k|k-1) = \mathbf{F}(k-1)\mathbf{P}(k-1|k-1)\mathbf{F}(k-1)^T + \mathbf{Q}(k-1). \tag{7.27}$$

Update: Then, the update step refines the predicted state by incorporating the measurement. In this step, a critical parameter, Kalman gain, will be computed to determine the weight of the measurement in updating the state. Assuming the measurement $\mathbf{z}(k)$ is available, the Kalman gain is calculated as follows:

$$\mathbf{K}(k) = \mathbf{P}(k|k-1)\mathbf{H}(k)^T \left(\mathbf{H}(k)\mathbf{P}(k|k-1)\mathbf{H}(k)^T + \mathbf{R}(k)\right)^{-1}. \tag{7.28}$$

Then we can update the posterior estimation of status as:

$$\hat{\mathbf{x}}(k|k) = \hat{\mathbf{x}}(k|k-1) + \mathbf{K}(k)\mathbf{z}(k) + \mathbf{H}(k)\hat{\mathbf{x}}(k-1|k-1), \tag{7.29}$$

and for error covariance as:

$$\mathbf{P}(k|k) = \mathbf{P}(k|k-1) + \mathbf{K}(k)\mathbf{H}(k)\mathbf{P}(k-1|k-1). \tag{7.30}$$

After multiple iterations, the error covariance $\mathbf{P}$ will converge to the minimized posterior error covariance with optimal Kalman gain $\mathbf{K}(k)$ to achieve automotive tracking. For interested readers, one can refer to refs. [123, 124] for the detailed derivations.

Although widely deployed in automotive radar tracking applications, the Kalman filter algorithm discussed above has several drawbacks when applied to radar measurements:

Linearity assumption: The above discussion is based on the assumption that the underlying processes are linear. The Kalman filter provides accurate estimates only for linear process and measurement models. In nonlinear scenarios, it may not achieve optimal estimation. To set a linear environment, we assumed a constant velocity model for our process, which may not hold true in practical applications where both process and measurement models are nonlinear.
Gaussian noise assumption: Another assumption that both process and measurement noise are Gaussian may not always hold true. While the Kalman filter works well when noise has known statistical properties, such as white noise, its performance deteriorates when dealing with colored noise or when the system has uncertainties. Moreover, the Kalman filter may struggle to effectively track rapidly accelerating or decelerating targets, highlighting the need to improve upon the underlying assumptions.

To address these limitations, various extensions and modifications to the Kalman filter, such as the extended Kalman filter (EKF) or unscented Kalman filter (UKF), have been proposed. These approaches aim to relax the linearity assumption and handle non-Gaussian noise characteristics more effectively, thereby enhancing the filter's performance in real-world scenarios. The EKF models the state transition and measurements as differentiable functions of the state. These functions are not explicitly linear, as shown below:

$$\mathbf{x}(k) = f\left(\mathbf{x}(k-1), \mathbf{u}(k)\right) + \mathbf{v}(k), \tag{7.31}$$

$$\mathbf{z}(k) = f\left(\mathbf{x}(k)\right) + \mathbf{w}(k). \tag{7.32}$$

In the above equation, instead of using a linear state transition matrix $\mathbf{F}(k)$, more general functions $f\left(\mathbf{x}(k-1), \mathbf{u}(k)\right)$ are utilized. The assumptions about noise are similar to those in the Kalman Filter. The prediction and update steps of the EKF are carried out as follows: **(1) Prediction**: Perform prior estimation for status as:

$$\hat{\mathbf{x}}(k|k-1) = f\left(\hat{\mathbf{x}}(k-1|k-1), \mathbf{u}(k)\right), \tag{7.33}$$

and for error covariance as:

$$\mathbf{P}(k|k-1) = \mathbf{F}(k)\mathbf{P}(k-1|k-1)\mathbf{F}(k)^T + \mathbf{Q}(k-1), \tag{7.34}$$

where $\mathbf{F}(k)$ is defined as

$$\mathbf{F}(k) = \left.\frac{\partial f}{\partial x}\right|_{\hat{\mathbf{x}}(k-1|k-1)}. \tag{7.35}$$

(2) Update: Assuming the measurement $\mathbf{z}(k)$ is available, the Kalman gain is calculated as follows:

$$\mathbf{K}(k) = \mathbf{P}(k|k-1)\mathbf{H}(k)^T\left(\mathbf{H}(k)\mathbf{P}(k|k-1)\mathbf{H}(k)^T + \mathbf{R}(k)\right)^{-1}, \tag{7.36}$$

where $\mathbf{H}(k)$ is defined as

$$\mathbf{H}(k) = \left.\frac{\partial h}{\partial x}\right|_{\hat{\mathbf{x}}(k|k-1)}. \tag{7.37}$$

Then we can update the posterior estimation of status as:

$$\mathbf{P}(k|k) = \mathbf{P}(k|k-1) - \mathbf{K}(k)\mathbf{H}(k)\mathbf{P}(k-1|k-1). \tag{7.38}$$

The optimal Kalman gain $\mathbf{K}(k)$ in the above formula also minimizes the posterior error covariance, and detailed derivations can be found in refs. [125,126]. Similar to the Kalman filter case, posterior processing can be applied to $\hat{\mathbf{x}}(k|k)$ as well to obtain smoothed estimates. Although already providing improved results than Kalman filter by introducing nonlinearity, EKF still faces several issues for further improvements:

Sub-optimal on MMSE: The state estimation provided by Kalman filter is not optimal in terms of minimum mean square error (MMSE). There might be other estimation techniques that yield better results in terms of error minimization.

Sensitivity to initial state estimation: Incorrect initial state estimation can lead to filter divergence, where the estimated state diverges significantly from the true state over time.

Computational complexity: For many applications, especially in automotive contexts, the computational complexity associated with Jacobian matrix calculations can be high. This computational burden may become prohibitive, particularly in real-time applications where rapid processing is required.

Gaussian assumption: The Gaussian assumption for both process and measurement models may not always hold true in real-world scenarios. Extended Kalman filters' performance might be limited when dealing with non-Gaussian noise or when the system deviates significantly from the assumed Gaussian distribution. To address these limitations, researchers have developed various extensions and alternatives to the EKFs, such as the UKF and particle filter. These techniques aim to mitigate the shortcomings of the Kalman filters and improve its performance in diverse and challenging environments.

For interested readers, we refer to this survey paper [127] for more details and recent developments in this direction.

7.4 MIMO Radar Waveform Designs

As discussed in the previous section, waveform diversity empowers MIMO radar with tremendous potential, offering greater flexibility in waveform parameter design and corresponding processing methods compared to traditional radars. However, the waveform design process of MIMO radar signal processing is also more complex than that of traditional radars. This section will introduce the primary forms and application characteristics of orthogonal waveforms in MIMO radar. In an orthogonal waveform MIMO radar system, let the transmitted waveforms from each channel be $s_1(t), s_2(t), ..., s_{N_t}(t)$. If the waveforms satisfy the orthogonality condition:

$$\int s_m(t)s_n^*(t)dt = \begin{cases} E \text{ if } m = n, \\ 0 \text{ if } m \neq n, \end{cases} \tag{7.39}$$

where E represents the waveform energy, the transmitted waveforms are considered mutually orthogonal. When the orthogonality condition is met, the received signals can be separated through matched filtering at the receiver, resulting in a data matrix that preserves the angular information of the target relative to the transmitting array. This property forms the foundation for the improved angle estimation accuracy in centralized MIMO radar systems. Specifically, methods such as frequency division, code division, or a hybrid combination of frequency and code division can be employed to construct orthogonal waveforms in MIMO radar. By modulating the same signal envelope onto several uniformly spaced carrier frequencies, frequency-division stepped-frequency orthogonal waveforms for MIMO radar can be constructed. In this case, the detection signal for the nth transmission channel can be expressed as:

$$s_n(t) = u(t)e^{j2\pi(f_0+n\Delta f)t}, \quad n = 0, 1, \ldots, N_t, \tag{7.40}$$

where f_0 is the starting frequency, Δf is the frequency interval between transmission channels, and $u(t)$ represents the envelope signal. If the envelope signal has a bandwidth B, the total bandwidth of the frequency-division orthogonal signals is:

$$B_{\text{total}} = B + (M - 1)\Delta f. \tag{7.41}$$

The orthogonality of the waveforms depends on the frequency interval Δf; generally, the larger Δf, the better the orthogonality, but also occupying increasingly more system bandwidth. The envelope signal $u(t)$ can take various forms, such as linear frequency modulation (LFM) signals, nonlinear frequency modulation signals, pseudo-random phase-coded signals, or other possible signal types.

If the same carrier frequency is used across different transmission channels but the waveform codes are mutually orthogonal, code-division orthogonal signals for MIMO radar can be constructed. In this case, the modulation envelope of the nth transmission channel can be expressed as:

$$u_n(t) = \sqrt{\frac{1}{K}} \sum_{k=0}^{K-1} e^{j\phi_{n,k}} \operatorname{rect}\left(\frac{t - kT_z}{T_z}\right), \tag{7.42}$$

where rect(.) is the rectangular window function, T_z is the width of a single code element, and K is the length of the coded signal. To ensure orthogonality, the following condition must be satisfied:

$$\sum_{k=0}^{K-1} e^{j\phi_{m,k}} e^{-j\phi_{n,k}} = \begin{cases} K, & m = n, \\ 0, & m \neq n. \end{cases} \tag{7.43}$$

Here, $\phi_{n,k}$ represents the phase of the k th element in the code of the n th channel, and its value can fall within the range $[0, 2\pi)$. Depending on the values of $\phi_{n,k}$, the coded signals can be binary-phase, quadrature-phase, octal-phase, or higher-order discrete phase codes, collectively referred to as orthogonal uniform discrete phase-coded signals. Alternatively, if $\phi_{n,k}$ takes arbitrary values in $[0, 2\pi)$, the resulting signals are called orthogonal continuous phase-coded signals. The design of code-division orthogonal waveforms often employs optimization techniques, with criteria such as minimizing peak sidelobes or the total energy of auto-correlation and cross-correlation sidelobes. Common optimization methods include simulated annealing, genetic algorithms, and sequential quadratic programming. Additionally, greedy algorithms and other local search methods have been used to further improve the performance metrics of the coded signals.

In addition to orthogonal frequency-division and code-division signals, other specialized signal types, referred to as hybrid signals, have also been investigated. Taking the coded-linear frequency modulation (coded-LFM) signal as an example, it replaces the simple pulse in orthogonal coded signals with a linear frequency modulation waveform, resulting in a fundamental hybrid signal waveform. It can be expressed as:

$$u_n(t) = \sqrt{\frac{1}{KT_z}} \sum_{k=0}^{K-1} \left(\text{rect}\left(\frac{t - kT_z}{T_z} \right) e^{j\pi u(t-kT_z)^2} \right) e^{j\phi_{n,k}}. \tag{7.44}$$

Here, $\text{rect}(\cdot)$ represents the rectangular window function, T_z is the width of a single code element, and K is the length of the coded signal. The term $e^{j\pi u(t-kT_z)^2}$ introduces the linear frequency modulation component, while $e^{j\phi_{n,k}}$ applies the phase coding to each element. In addition to phase-coded linear frequency modulation hybrid signals, other forms of hybrid signals have been explored, such as frequency-division phase-coded signals, discrete frequency-coded linear frequency modulation signals, and more complex hybrid signal formats. These advanced waveforms provide increased flexibility and improved performance in various MIMO radar applications.

Besides the orthogonal MIMO radar waveform we discussed above, MIMO radar waveform can also be further designed to achieve different goals. For example, by introducing the concept of beamforming, MIMO radar waveform can be designed to minimize the beam pattern matching error criterion. This involves designing the transmit or receive beam pattern to approximate a target beam pattern as closely as possible. As a result, the resulting weight vector minimizes the error between the designed beam pattern and the desired beam pattern, ensuring better spatial resolution and interference rejection. On the other hand, minimizing the integrated sidelobe-to-mainlobe ratio can also be treated as the optimization goal for the design of MIMO radar waveform. This approach focuses on reducing the energy of the sidelobe region relative to the energy of the mainlobe, thereby improving the spatial resolution and suppressing interference. Thus, by minimizing this criterion, the beam pattern achieves a sharper mainlobe with suppressed sidelobes, enhancing target detection and interference rejection capabilities. If the purpose of waveform design in radar systems is to maximize the target detection probability while maintaining a fixed false alarm probability. The maximum signal-to-interference-plus-noise ratio (SINR) criterion can be used to do so. It ensures that the desired signal is optimally received in the presence of interference and noise, leading to better detection, estimation, and communication quality. For interested readers, we refer to refs. [128–131] for the representative works in this direction.

8

Interference Mitigation in Radar Systems

8.1 Overview of Interference Mitigation Techniques in Radar Systems

The electromagnetic spectrum, as discussed in previous chapters, is a vital resource shared by communication and sensing systems, thereby constraining system capacity in both domains. As a result, future sensing environments will increasingly encounter complex electromagnetic interference due to the coexistence of radar and communication applications. To address these challenges, the concept of cognitive radar, characterized by a closed-loop architecture encompassing transmission, reception, and processing was proposed recently for next-generation radar designs with interference mitigation capabilities. Unlike conventional radar systems, cognitive radar can dynamically adapt its transmission waveforms and reception processing methods in response to changes in the external environment, allowing it to function effectively within more intricate electromagnetic conditions. In recent years, advancements in cognitive radar technology and artificial intelligence have enabled radar systems to acquire cognitive and intelligent evolutionary capabilities. This progress has given rise to a novel paradigm referred to as cognitive intelligent radar. This paradigm is distinguished by a closed-loop architecture with self-evolution and learning capabilities. According to ref. [132], cognitive intelligent radar incorporates adaptive environmental sensing, flexible transmission and reception design, transceiver co-design, and resource scheduling. Such features empower cognitive intelligent radar to operate efficiently in complex and dynamic electromagnetic countermeasure environments. A brief overview of the underlying radar transmitter and receiver designs is presented below.

Detect-and-avoid at the transmitter: This primarily involves leveraging prior knowledge of radar targets and electronic interference, historical radar echo data, and real-time cognitive information of the underlying environment to optimally design the radar's transmission waveforms, beams, and power at the transmission end. In a complex electromagnetic countermeasure environment, this enhances the radar's anti-interference capabilities. Specifically, from the game-theoretic perspective of radar versus interference, the

DOI: 10.1201/9781003408239-8

radar's advantage lies in its ability to control the transmission. Thus, cognitive transmission is one of the core components of cognitive intelligent radar, and its essence lies in fully utilizing active control over transmission through various strategies such as agility, randomness, deception, and collaboration, thereby enhancing the radar's adaptability and survivability in complex electromagnetic environments.

Detect-and-suppress at the receiver: On the other hand, at the radar's reception end, cognitive optimization design is also implemented to filter the received echo signals across the spatial, temporal, frequency, and polarization domains. The core objective is to prevent interference energy from mixing with the radar target echoes as much as possible, providing relatively clean echo signals for subsequent signal and data processing. Depending on specific application needs, cognitive transmission, and cognitive reception can be optimized as an integrated design, such as through joint design of transmission waveforms and reception filters [133], to further enhance the radar system's anti-interference and anti-clutter capabilities.

In the above discussions, one can notice that interference sensing (i.e., interference identification and estimation) plays an important role in enabling consequent transmitter and receiver operations effectively. Moreover, besides the solely transmitter or receiver operations, centralized coordination of one or several radar transceivers, including transceiver co-design, and resource scheduling, is also a promising way to enhance radar interference mitigation capabilities. In the following discussions, we will elaborate on the basic idea of each scheme and provide several representative works using different schemes to demonstrate its practical usage. For the recent development in this direction, we refer interested readers to refs. [132, 134–136] for more details.

8.2 Interference Identification and Estimation

Interference sensing involves the estimation of interference characteristics to provide essential informational support for anti-interference decision-making. According to ref. [132], they can be categorized into shallow and deep features based on the dimensionality of these characteristics. Shallow features are physically interpretable characteristics, such as interference spectra, Doppler shifts, and intra-pulse modulation properties. In contrast, deep features are semantic representations within a high-dimensional feature space generated by deep neural networks, typically lacking direct physical interpretability. This chapter focuses exclusively on shallow features, which are estimable using traditional statistical methods. The exploration of deep features will be addressed in Chapter 15. Shallow feature estimation primarily employs techniques such as probabilistic statistics, time-frequency transformations, and higher-order matrix transformations. These methods facilitate the extraction of physical interference characteristics, enabling the identification of interference types and the estimation of parameters across various modulation domains, including time, frequency, and Doppler domains, thereby achieving effective interference characterization. For example, regarding range deception interference identification, refs. [137, 138] have analyzed the generation mechanisms of range deception interference signals, revealing that flaws in digital radio-frequency memory (DRFM) jamming technology can cause frequency shifts in the frequency domain center of range deception interference signals, providing a reference for identifying this type of interference. Follow-up literature [139] also developed a mathematical model for deception interference and proposed an interference type identification method that utilizes bispectral features.

When it comes to interference parameter estimation based on shallow features, this mainly involves using advanced matrix transformation techniques, such as short-time fractional Fourier transform, Radon transform, and Hilbert transform, to extract features of interference across time, frequency, and modulation domains, thus enabling the estimation of typical interference parameters. A representative work in this direction is ref. [140]. In this work, the interference caused by different modulated radar signals can be identified. Then, the interference cancellation can be performed for improved radar estimation capability in an interfered environment. Specifically, the normalized transmitted signal of a single chirp $s(\tau)$ can be expressed as:

$$s(\tau) = \cos[(w_0 + \Delta w \tau)\tau], \tag{8.1}$$

where τ is the relative time within the current chirp, w_0 is the fundamental angular frequency and Δw is the angular frequency deviation. On the receiver side, by mixing the transmitted signal of the m the frequency ramp and the received signal $x(t)$, we obtain

$$y_m(\tau) = x(mT + \tau)s(\tau), \tag{8.2}$$

where T is the duration of a single chirp. Furthermore, a low-pass filter is often applied to eliminate high-frequency mixing components, being expressed as:

$$y_m^{\mathrm{LP}}(\tau) = h(\tau)y_m(\tau), \tag{8.3}$$

where $h(\tau)$ is the frequency response of the low-pass filter with cut-off frequency w_c. In an interference-free environment, this term can be further expressed as:

$$y_m^{\mathrm{LP}}(\tau) = \frac{\alpha}{2}\cos(\Delta w \frac{2r}{c_0} + \phi_m), \tag{8.4}$$

where $\alpha << 1$ is the attenuation term, r is the distance, and ϕ_m is a constant for a stationary target and will reassemble the Doppler shift for a moving target. By sampling N times per ramp on the processed signal $y_m^{\mathrm{LP}}(\tau)$, a digital signal $\mathbf{A} \in \mathcal{C}^{M\times N}$ can be obtained and each element can be expressed as:

$$\mathbf{A}_{m,n} = y_m^{\mathrm{LP}}(\tau)(\frac{nT}{N}). \tag{8.5}$$

This matrix $\mathbf{A}$ represents a spectrogram or time-frequency image, providing important information for radar parameter estimations. However, if we further consider the practical environment with interference, this time-frequency image will be ruined. Specifically, assuming an interference signal $x_{\mathrm{I}}(t)$ with changing frequency $w_{\mathrm{I}}(t)$, the received signal can be expressed as:

$$x_{\mathrm{I}}(t) = \cos[w_{\mathrm{I}}(t)t]. \tag{8.6}$$

After the low-pass filter, the interference below cut-off frequency w_{c} will appear and corrupt the time-frequency image $\mathbf{A}_{m,n}$, being expressed as:

$$|w_0 + \Delta w\frac{nT}{N} - w_{\mathrm{I}}(mT + \frac{nT}{N})| < w_{\mathrm{c}}. \tag{8.7}$$

In this work, the authors notice that the interference caused by specific modulations shows a special pattern, motivating the authors to employ image processing schemes for detecting and removing those interference signals. To do so, the first step is to convert the time-frequency image to a typical image by means of quantization or other signal pre-processing methods. For example, by adjusting the scaling factor β and offset γ, the below conversion rule can be used to fulfill this need, expressed as:

$$\boldsymbol{I} = \beta \log_{10} |\mathbf{A}| + \gamma. \tag{8.8}$$

After this step, traditional image processing methods can be employed to perform pattern recognition purposes. Four example images containing the interference of one or two sensors, respectively, are shown in Figure 8.1. In this work, a maximally stable extreme regions (MSER) algorithm is employed to do so and shows good performance in both simulation and real tests. In the follow-up works, different image processing algorithms are also employed to further improve the interference identification performance in different scenarios. Notably, this is a pioneering work that considers the radar spectrogram as an image, and this concept is still widely used when introducing advanced learning-based achievements into radar signal processing domains. We will discuss this in Chapter 15.

After the success of the above work, researchers also aim to extend a similar concept to handle more types of radar interference signals, including interrupted sampling repeater (ISR) interference [141], chopping and interleaving (C&I) interference [142], and smeared spectrum interference [143]. Compared to full-pulse repeater interference, ISR interference has advantages such as minimal delay, fast response, and the ability to form false targets before the actual target, making it widely used. Especially, ISR interference has been improved since it was first proposed to overcome its original limitations, making it a powerful jamming attack to interfere with the target radar reception [144]. For example, in ref. [144], a novel ISR interference method based on joint subsection frequency shift and two-phase modulation is proposed. This approach leverages the coherent superposition of jamming signals and strategic modulation to enhance the interference effect for both linear-frequency-modulated

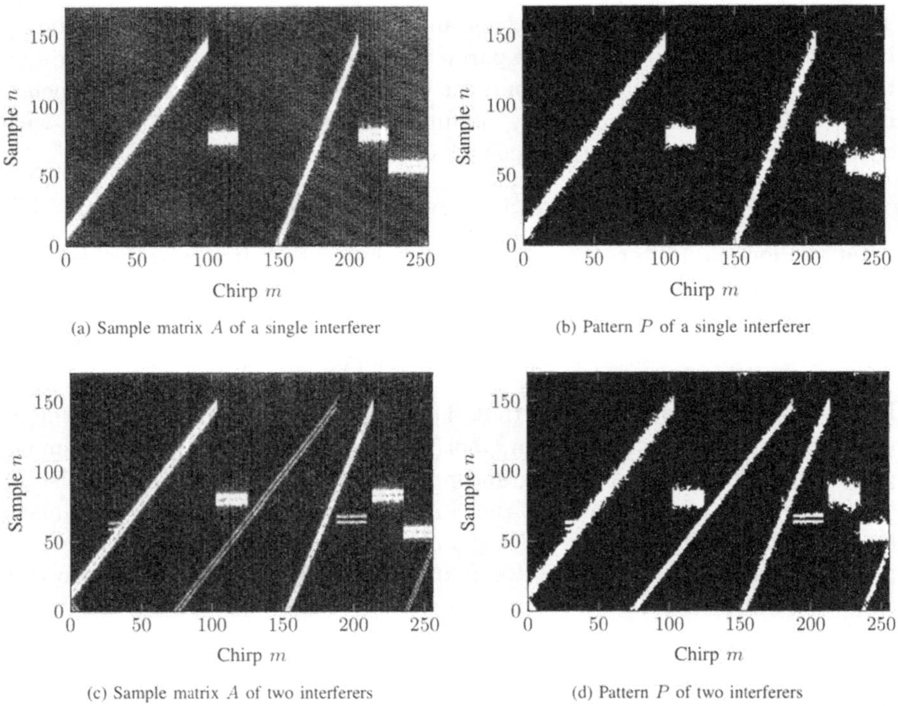

(a) Sample matrix A of a single interferer

(b) Pattern P of a single interferer

(c) Sample matrix A of two interferers

(d) Pattern P of two interferers

FIGURE 8.1
Examples of signals disturbed by FSK/LFM interference (copyright from ref. [140]).

(LFM) and phase-coded signals. Simulation results demonstrate that this joint modulation strategy effectively mitigates the inherent defects of Interrupted Sampling Repeater interference, enabling the formation of more robust and versatile jamming effects. This improvement enhances the ISR interference's applicability in countering advanced radar systems and modern anti-jamming measures. To tackle this type of interference, ref. [145] analyzed the cross ambiguity function (CAF) of ISR interference and proposed a Doppler compensation-based ISR interference identification method. Ref. [146] introduced an estimation method based on the short-time fractional Fourier transform, enabling the estimation of interference parameters such as the number of slices, slice width, and frequency modulation slope in ISR interference. Ref. [141] introduces an interference suppression method for ISR jamming based on the singular spectrum entropy function (SSEF), integrating principles of singular value decomposition (SVD) and information entropy theories. The method begins with an adaptive multi-scale segmentation (AMS) approach to enhance salient signal characteristics while smoothing similar features. AMS employs segmentation criteria based on average segment energy and minimum segmentation constraints, leading to the creation of an improved delay embedded matrix. Singular spectrum extraction is performed using SVD, followed by the introduction of SSEF, a novel characteristic derived from the Shannon entropy model, addressing the limitations of traditional singular spectrum analysis and entropy-based methods under low signal-to-noise ratio (SNR) conditions. The SSEF is then utilized in an entropy-based threshold detection process to filter and suppress interference. Experimental results demonstrate that this method significantly improves target detection probability and peak-to-sidelobe ratio (PSR) after pulse compression, validating its robustness against noise and jamming, particularly under low SNR scenarios. Ref. [147] also examined the characteristics of the CAF after matched filtering of the interference and proposed a parameter estimation method based on radon transform and least squares. Ref. [148] proposes an anti-jamming method using minimum variance distortionless response (MVDR) beamforming, which leverages small-sample pure jamming signals for covariance matrix estimation. The method begins with time–frequency analysis of the one-dimensional range profile of the radar echo, enabling the identification of target and jamming signals on a two-dimensional range–frequency graph based on their time–frequency differences. Pure jamming signals are isolated through snapshot sampling in this graph. Given the limited snapshot data, a covariance matrix for pure jamming is reconstructed using an iterative adaptive approach (IAA). This matrix serves as training data for the MVDR beamformer, which effectively suppresses the main-lobe Interrupted Sampling Repeater interference. The proposed approach is validated through simulations and measured results, demonstrating its effectiveness in mitigating main-lobe Interrupted Sampling Repeater Jamming. These methods are generally applicable to relatively ideal interference environments to identify an interference based on "typical" interference behaviors. That being said, when it comes to increasingly difficult scenarios, shallow feature-based interference identification and estimation face severe challenges and the performance will drop significantly. For example, when radar simultaneously receives different types of interference, the aforementioned "typical" interference behavior of each type of interference might no longer exist, and thus it is very challenging to still identify the underlying interference based on those shallow features. Additionally, other imperfections, such as clutter, can further complicate interference parameter estimation. The above two constraints of traditional radar interference identification and estimation motivate the introduction of advanced learning-based achievements, and we will talk more about this in Chapter 15.

8.3 Detect-and-Avoid at the Transmitter

After identifying the interference type and estimating interference parameters, subsequent interference handling mechanisms can be performed to realize interference mitigation. In cognitive intelligent radar systems, anti-interference at the transmitter primarily relies on the real-time perception of interference direction, power, modulation patterns, and retransmission rules to optimally design waveforms and beams. This enhances the radar's anti-interference detection capabilities in complex electromagnetic environments. For example, after identifying interference, a cognitive anti-interference strategy can be employed to select or adaptively generate a waveform from the waveform library to effectively suppress the interference. In this direction, ref. [149] proposed a comprehensive waveform optimization design framework for cognitive radar anti-interference, which maximizes mutual information under constraints of SNR, energy, and power spectral density, balancing output SNR and target parameter estimation performance. Similarly, several methods are proposed to suppress interference based on specialized waveform design. For known noise amplitude modulation interference, a joint design method of transmitted waveforms and receiving filters based on the suppression of interference and sidelobes, improves target detection performance by minimizing target parameter estimation errors. Moreover, we can further approach this from the ambiguity function perspective, proposing a two-dimensional modulation anti-jamming optimization criterion, solved using alternating iteration methods to suppress interference in range and Doppler domains.

Ref. [150] is a great work to show an example of cognitive transmitter design in radar systems for anti-interference purposes. Specifically, by cognitively adapting the chirp bandwidth and transmission time of the radar transmitter, the FMCW radar detection performance can be maintained even with the existence of interference. To elaborate, a FMCW transmit signal $s_{\mathrm{T}}(t)$ is a train of P chirps spaced by the interval T_p, being expressed as:

$$s_{\mathrm{T}}(t) = \sum_{p=0}^{P-1} s_{\mathrm{T},p}(t - pT_p), \ 0 \leq t \leq PT_p. \tag{8.9}$$

In the receiver end, the received signal can be expressed as below by considering the channel effect:

$$s_{\mathrm{R}}(t) = \sum_{p=0}^{P-1} \sum_{k=0}^{K-1} \sum_{l_k=0}^{L_k-1} \alpha_{lk} s_{\mathrm{T},p}(t - \tau_{\mathrm{r},lk} - pT_p) e^{-2j\pi p \frac{T_p}{\tau_{\mathrm{D},lk}}}, \tag{8.10}$$

where α_{lk} is the attenuation of the lk the scatterer of the k the target by considering an environment with K extended targets and each target with L_k point scatterers. In the above equation, the channel effect is shown by the propagation delay $\tau_{\mathrm{r},lk}$ and Doppler $\tau_{\mathrm{D},lk}$. Moreover, the velocity information can be extracted from those two terms as $\tau_{s,k} = \tau_{\mathrm{r},lk} + \tau_{\mathrm{D},lk} = \frac{2(\tau_{lk} - v_{lk} t)}{c_0}$, where v_{lk} is the relative velocity between the ego-vehicle and reflecting target. By integrating the linear frequency ramp, we can obtain a phase course $\phi(t) = \frac{\alpha}{2} t^2 + \phi_0$, where $\alpha = \frac{B}{t_{up}}$, B is the chirp bandwidth and t_{up} is the chirp duration. If we assume that the initial phase $\phi_0 = 0$, the transmit chirp chain can be expressed as:

$$s_{\mathrm{T}}(t) = \sum_{p=0}^{P-1} e^{j\pi\alpha(t-pT_p)^2}, \tag{8.11}$$

and the receive waveform can be expressed as:

$$s_{\mathrm{R}}(t) = \sum_{p=0}^{P-1} \sum_{k=0}^{K-1} \sum_{l_k=0}^{L_k-1} \alpha_{lk} e^{j\pi\alpha(t-pT_p-\tau_{s,lk})^2}. \tag{8.12}$$

If we further consider the frequency over time aggregation that equals the derivative of the signal phase, we obtain

$$f_{\mathrm{T}}(t) = \sum_{p=0}^{P-1} \alpha(t - pT_p)\mathrm{rect}(\frac{t - t_{up} - pT_p}{t_{up}}), \tag{8.13}$$

where rect is the rectangular window function. In contrast, the received frequency ramp with propagation time shift and Doppler shift can be expressed as:

$$f_{\mathrm{R}}(t) = \sum_{p=0}^{P-1} \sum_{k=0}^{K-1} \sum_{l_k=0}^{L_k-1} [\alpha(t - 2\tau_{\mathrm{r},lk} - pT_p) + f_{\mathrm{D},lk}]\mathrm{rect}(\frac{t - t_{up} - 2\tau_{\mathrm{r},lk} - pT_p}{t_{up}}). \tag{8.14}$$

Thus, the beat frequency equals the difference $f_{\mathrm{B}}(t) = f_{\mathrm{T}}(t) - f_{\mathrm{R}}(t)$ can be expressed as:

$$f_{\mathrm{B}}(t) = \sum_{p=0}^{P-1} \sum_{k=0}^{K-1} \sum_{l_k=0}^{L_k-1} [\alpha\tau_{\mathrm{r},lk} - f_{\mathrm{D},lk}]\mathrm{rect}(\frac{t - t_{up} - \tau_{\mathrm{r},lk} - pT_p}{t_{up}}). \tag{8.15}$$

Then, the range and Doppler components of the beat frequency can be extracted using DFTs in fast-time and slow-time domains. With the above modeling, the aim of this paper is to implement a cognitive radar system instead of simply pausing radar illumination in the presence of a second radar system. Specifically, in this work, the cognitive strategy is to allocate chirps on free center frequencies within the complete allocated frequency range while adjusting chirp bandwidth and length to available frequency and time allocations, thus allowing flexible and persistent radar usage during the presence of interfering radar systems. In the real field test, the measurement results show that chirps of a reduced bandwidth mitigate frequency bands that are occupied by other radar systems and allow a reduction of the chirp duration while increasing the radar perception range by a factor of 16 to achieve the aim.

Similarly, ref. [151] introduced a method combining constant modulus sequences and filters through iterative quadratic programming under the condition of sensing clutter and interference, reducing the impact of suppression jamming by minimizing mean square error in target parameter estimation. Ref. [152] presented a relaxed and cyclic optimization algorithm that uses low-rank parameters to design multi-phase waveforms for multiple-input multiple-output (MIMO) radar, ensuring good detection performance under cognitive interference conditions. Furthermore, deception jamming is a malicious interference occurring when a jamming system captures radar signals and generates interference signals similar to target echoes by applying delays or Doppler frequency modulation. These generated signals are then retransmitted according to specific patterns, creating multiple false targets with fake time delays or Doppler modulation, significantly increasing the radar's false alarm probability and reducing the detection probability of real targets. On the other hand, in the condition that the prior clutter and interference information is known, a waveform design method optimizing the pulse initial phase to suppress inter-pulse retransmission-based velocity deception interference by forming nulls within a specific range of the Doppler spectrum of real target echoes can be developed for anti-interference purposes. In this direction, a waveform design method optimizes multi-notch pulse initial phase to enhance multi-target detection capability by setting notches near each real target echo Doppler spectrum to address the issue of multi-target detection under velocity deception interference. That is, the following optimization problem can be considered as

$$\min_{\mathbf{y}_J,\alpha} C(\mathbf{y}_J, \alpha) = ||\mathbf{y}_J - \mathbf{G}\alpha||^2. \tag{8.16}$$

where C is the notch power, α is a auxiliary variable, $\mathbf{y}_J = [y_J(1)y_J(2)...y_J(N)]^T$ and $y_J(n)$ is the matched filter output of the interference signal containing n-th received signal. Ref. [153] addresses the suppression of such jammers using frequency diverse array (FDA) MIMO radar, leveraging its additional degrees-of-freedom in the range domain. Mainbeam jammers are modeled as false targets lagging several pulses behind true targets. Data-independent beamforming is initially employed to suppress false targets by nulling the equivalent transmit beampattern with a suitable frequency increment. However, performance degradation may occur due to transmit spatial frequency mismatches caused by quantization errors, angle estimation errors, or frequency increment errors. To mitigate this issue, the paper proposes a preset broadened nulling beamformer (PBNBF), which introduces artificial interferences with appropriate powers around the nulls to create broadened notches, ensuring robust suppression of deceptive jammers. Numerical analysis in a multi-unmanned aerial vehicle (UAV) scenario demonstrates the PBNBF's effectiveness, achieving improved signal-to-interference-plus-noise ratio compared to conventional beamforming techniques. Ref. [148] introduced an inter-pulse amplitude-phase joint design method that includes amplitude degrees of freedom and considers constraints such as discrete phase and peak-to-average power ratio (PAPR), minimizing interference energy and target sidelobe energy within the stopband using the inexact alternating direction penalty method (IADPM). The considered optimization problem is presented as:

$$\min_{\mathbf{s}} f(\mathbf{s}) = u_1\mathbf{s}^H\mathbf{R}_J\mathbf{s} + u_2(\boldsymbol{J}|\mathbf{s}|)^H\mathbf{R}_T(\boldsymbol{J}|\mathbf{s}|), \tag{8.17}$$

where $\mathbf{s} = [s_1 s_2 ... s_N]^T$ is the transmitted waveform with N dimension, u_1 and u_2 are weighting scalar satisfying $u_1 + u_2 = 1$ to consider the trade-off between minimizing interference and maintaining sidelobe waveform, $\mathbf{s}^H\mathbf{R}_J\mathbf{s}$ is the interference power in a specific Doppler band, and $(\boldsymbol{J}|\mathbf{s}|)^H\mathbf{R}_T(\boldsymbol{J}|\mathbf{s}|)$ is the sidelobe power of the desired signal in a specific Doppler band. Under the constraint that the transmitted waveform should satisfy PAPR limitation and with constant power, the above optimization can be used to design an optimal waveform for given u_1 and u_2 consideration for interference control purposes. Ref. [154] proposed a constant modulus waveform design method that minimizes spectral amplitude within the interference band, generating waveforms with low-range sidelobes and specified spectral power suppression. Specifically focusing on the radar performance degeneration in terms of degraded sensitivity and increased false alarm rates, particularly in scenarios involving synchronous and asynchronous operations of radars with similar parameters (e.g., carrier frequency and chirp slope), ref. [155] proposes novel waveform design strategies to mitigate mutual interference effectively. For single-input-single-output (SISO) radar systems, two slow-time coding schemes are introduced. The first scheme reduces interference by Doppler-shifting it away from the target region, while the second minimizes the discrete periodic CAF (PCAF) in a desired area. For the more complex MIMO radar systems, an efficient cyclic algorithm is proposed to design transmit waveforms, leveraging the waveform diversity inherent to MIMO radars to reduce interference power. The proposed solutions are computationally efficient, require minimal modifications to existing radar systems, and are adaptable for both offline and online implementation. This work establishes a foundational approach for mitigating mutual interference in radar-dense environments, preparing for future scenarios where shared protocols and regulations will be critical for multi-vehicle operations. Ref. [156] presented a constant modulus waveform design method based on minimizing the integral sidelobe criterion using a reduced-order fourth-order sequence iterative algorithm to solve the optimization problem. Simulation results demonstrated that the optimized waveform performs better in anti-jamming scenarios involving multiple targets and range deception.

Ref. [157] addresses a significant challenge in radar technology: the interference caused by the proliferation of wind turbines on ground-based medium-to-high pulse repetition frequency (PRF) pulsed-Doppler air surveillance radars. The traditional approach of using fixed waveforms and PRFs in pulsed-Doppler radar systems can lead to range ambiguity, particularly when attempting to achieve unambiguous velocity measurements, which necessitates multiple coherent processing intervals (CPIs) at varying PRFs. To mitigate these issues, the authors propose two innovative solutions: first, applying a random initial phase to each transmit pulse while maintaining a fixed PRF, and second, randomizing the pulse repetition interval (PRI). Both methods enable the design of filters that effectively suppress wind turbine interference and ground clutter while allowing for unambiguous range measurements within a single CPI. The results demonstrate that these techniques not only resolve range ambiguity but also enhance detection capabilities for airborne targets, including those with slow or zero radial velocity beyond the ground clutter horizon. The performance of the proposed algorithms is rigorously evaluated through simulations and experimental data analysis, focusing on key metrics such as the probability of target detection and SINR loss. The findings suggest that these approaches hold significant promise for improving radar performance in environments affected by wind turbine interference, making them particularly relevant for applications within the Federal Aviation Administration and coastal radar systems. Ref. [158] addresses a critical challenge in autonomous driving technology, specifically focusing on improving the performance of radar sensors used in advanced driver assistance systems (ADAS). The research tackles two main issues: interference management and ghost target detection in FMCW radar systems. The authors introduce a novel technique that employs an orthogonal frequency division multiplexing (OFDM) waveform to overcome these limitations. The OFDM approach leverages maximum length sequence (MLS), also known as m-sequence, to provide orthogonality between different transmitted waveforms. This orthogonality is crucial for reducing interference between multiple radar systems operating in close proximity. To further enhance the robustness of the system, the researchers incorporate a scrambled sequence into the proposed waveform. This additional feature aims to minimize the impact of interference and improve the reliability of target detection. The study conducted simulations to evaluate the performance of the proposed algorithm in various scenarios, including single-neighbor and multiple-neighbor environments. These simulations were designed to test the effectiveness of the OFDM-based approach in mitigating interference and reducing ghost target detections. The results of these simulations were then compared to the performance of conventional FMCW radar systems under similar conditions. This comparative analysis provides insights into the relative advantages of the proposed technique over existing methods. The advancements proposed in this paper have significant implications for the future of autonomous vehicles. By improving the reliability and accuracy of radar sensors, this research contributes to enhancing the overall safety and performance of self-driving cars. The ability to more effectively manage interference and reduce ghost target detections could lead to more robust ADAS systems, potentially accelerating the development and adoption of higher levels of vehicle automation. As the automotive industry continues to push toward fully autonomous vehicles, innovations in sensor technology, such as those presented in this paper, play a crucial role in overcoming existing challenges and paving the way for safer, more reliable self-driving systems. These studies highlight the development of advanced cognitive waveform design strategies that optimize radar performance against various interference techniques, contributing to the robustness and adaptability of cognitive intelligent radar systems in complex electromagnetic environments.

8.4 Detect-and-Suppress at the Receiver

Besides the detect and avoid strategy at the transmitter end, the receiver end can also perform the detect and suppress strategy for radar interference mitigation purposes. Specifically, the anti-interference capabilities of cognitive radar receivers primarily rely on the design of reception filters in spatial, temporal, frequency, and polarization domains to accumulate the energy of target echoes while suppressing the energy of interference signals. The related techniques can be broadly categorized into cognitive anti-interference reception beamforming (spatial filtering), as well as temporal filtering, Doppler domain (frequency domain) filtering, polarization domain filtering, and multi-domain joint filtering [132]. Spatial reception beamforming refers to utilizing the angular information of targets and interference by designing spatial filters to accumulate the energy of signals from the target direction while suppressing signals from the direction of interference. This can be further divided into static beamforming and adaptive beamforming. For static beamforming, windowing techniques are widely used in low-sidelobe reception beamforming to counteract omnidirectional interference, owing to their low computational complexity and strong real-time processing capabilities. Classic window functions include the Chebyshev window and the Taylor window. In recent years, substantial research has been conducted on the design of window functions [159, 160], thereby enhancing the flexibility of traditional window functions. However, most of traditional windowing techniques are still limited to uniform arrays, even with the aforementioned enhancements. Compared to static beamforming, adaptive beamforming further utilizes the provided spatial degree of freedom to design appropriate reception beamforming suitable for the underlying environment. Although also coming with increasingly more complexity, adaptive spatial filtering often provides better results by customizing waveform to a specific interference detected in the environment. For example, ref. [161] is a great work to develop an adaptive digital beamforming technique in this direction. Through the special design, the proposed adaptive beamforming can perform interference suppression without the exact knowledge of the interfering signal's Direction of Arrival (DoA) and consequently without the calibrations. Specifically, the bemafoming output $y(k)$ can be expressed as:

$$y(k) = \mathbf{w}^T \mathbf{x}(k), \tag{8.18}$$

where $\mathbf{w} \in \mathbb{C}^{M\times 1}$ is the weighting vector and $\mathbf{x} \in \mathbb{C}^{M\times 1}$ is the sampled signal from M antenna elements. In the proposed adaptive design, an error term $e(k)$ is defined to evaluate the deviation of the beamforming output and desired

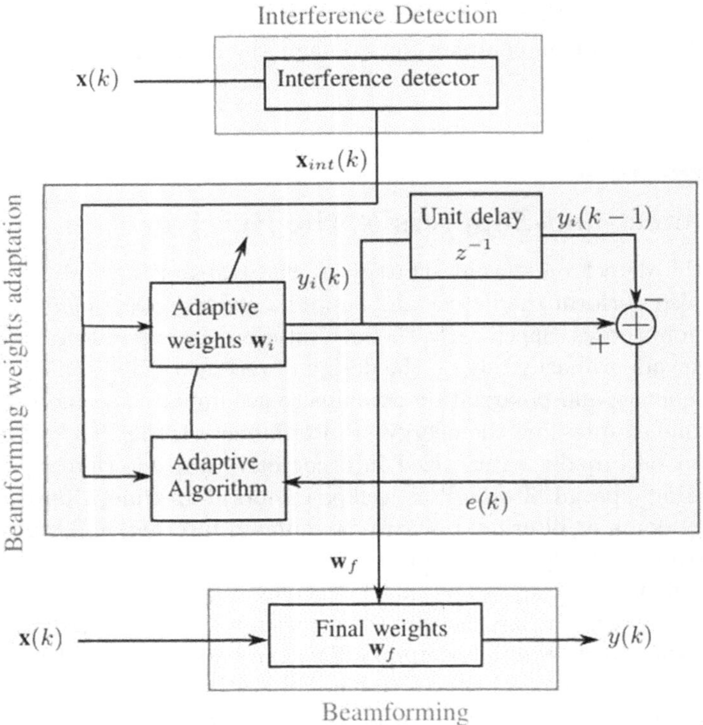

FIGURE 8.2
Block diagram of the complete interference cancellation system (copyright from ref. [161]).

signal $d(k)$, being expressed as:

$$e(k) = d(k) - y(k). \tag{8.19}$$

Thus, we can employ different optimization algorithms to adjust the beamforming weighting to achieve the above goal. In the case that the normalized least mean square algorithm is chosen, we can use the following recursive equation to update the beamforming weighting until convergence, as shown below:

$$\mathbf{w}(k+1) = \mathbf{w}(k) + u(k)e(k)\mathbf{x}^*(k), \tag{8.20}$$

where

$$u(k) = \frac{\beta}{\mathbf{x}^H(k)\mathbf{x}(k)}, \tag{8.21}$$

is the variable adaptation step size and $0 < \beta < 2$ is utilized to guarantee error convergence. Then, the proposed interference cancellation consists of three steps, as shown in Figure 8.2. In the interference detection step, the interfered part of the signal can be captured by evaluating the amplitude variation $v(k)$ of the signal, being expressed as

$$v(k) = x(k) - x(k-1), \tag{8.22}$$

where $x(k)$ is the received signal at a single channel. By calculating the highest radar cross section (RCS) that can be present in real road scenarios, a threshold can be defined in advance to filter all interfered signals for further processing. If we define the interference start and end as k_s and k_e respectively, the below equation expresses the signal output $\mathbf{x}_{\text{int}}(k)$ in this period:

$$\mathbf{x}_{\text{int}}(k) = \mathbf{x}_t(k) + \mathbf{x}_i(k) + \mathbf{n}(k), \ \ k_s < k < k_e. \tag{8.23}$$

where $\mathbf{x}_t(k)$ is the desired signal reflected from target, $\mathbf{x}_i(k)$ is the interference signal, and $\mathbf{n}(k)$ is the noise. In other words, the interference detector in Figure 8.2 identifies the samples of the input signal that are affected by the interference. With the above modeling, the aim of the adaptive beamforming design is to adaptively adjust the beamforming weighting to reduce the variation in the intermediate beamforming output $y_i(k)$– and bring it below the threshold. Thus, the error signal $e(k)$ is defined as below in this interfered period:

$$e(k) = y_i(k) - y_i(k-1), \ \ k_s < k < k_e. \tag{8.24}$$

In this setting, the aforementioned normalized least mean square algorithm will design a beamforming weighting, which can minimize the error caused by the interference signal by suppressing it to the noise level. Then, this beamforming vector can be used to perform interference cancellation by nulling the signal from the detected interference node. Simulation results confirm that up to 23 dB SINR improvement can be achieved after the proposed beamforming processing to achieve the design aim. Inspired by this work, techniques such as diagonal loading have been proposed based on the white noise gain constraint, iteratively calculating the appropriate diagonal loading parameter to further improve the robustness of adaptive beamforming. Furthermore, to reduce the computational complexity of the iterative model, methods based on the eigenspace robust adaptive beamformer design have been proposed, requiring accurate sensing of the number of jammers/desired signal sources. Similar ideas can also be used in temporal, Doppler, and polarization domains to separate target and interference signals in the receiver end.

Given the limitations of single-domain anti-jamming techniques, multi-dimensional joint filtering techniques, such as time-frequency, space-time, and space-time-frequency joint filtering, have gained increasing attention. Studies have examined the time-frequency distribution of intermittent sampling jamming and designed jamming suppression filters in the time-frequency domain. Further research combined signal reconstruction and time-frequency domain filtering to suppress various types of intermittent sampling jamming in multi-target environments. Ref. [162] tackles a significant challenge in passive bistatic radar systems: the effective cancellation of interference. The study begins by examining how three key factors—fractional time delay, channel linearity, and the SNR in the reference channel—impact the performance of interference cancellation techniques. The analysis reveals that these factors can significantly degrade the effectiveness of adaptive cancellation algorithms. To address these limitations, the authors propose a two-pronged approach. First, they introduce a signal reconstruction method designed to enhance the quality of the reference signal. This improvement is crucial for more accurate interference cancellation. Second, they develop a novel joint spatiotemporal domain interference suppression technique. This method is particularly effective in mitigating both direct-path and multi-path interference, resulting in a substantial improvement in target SNR. The paper's contributions are not merely theoretical; the researchers validate their proposed methods through real-world data experiments. These practical tests demonstrate the efficacy of their approach in actual radar operations, underscoring the potential for significant advancements in passive bistatic radar technology. By effectively addressing the interference cancellation problem, this research paves the way for more reliable and accurate passive radar systems, with potential applications in various fields requiring non-emitting surveillance capabilities. Ref. [163] addresses the challenge of spectrum sharing between radar systems and commercial services, focusing on temporal sharing techniques to reduce radar exclusion zones and improve spectral efficiency. The proposed approach is particularly relevant for scenarios involving multiple radars operating in close proximity within the same frequency band. The authors introduce signal processing techniques that enable a secondary user to transmit without exceeding a specified interference level at any radar. The method relies on the periodic behavior of radars and employs adaptive sensing to track radar behavior in real-time without requiring prior information. A crucial component of this approach is a pulse deinterleaving mechanism that separates multiple radar emissions in real-time, without the need for batch or offline processing. This real-time deinterleaving is essential for effectively managing the shared spectrum in dynamic environments. The temporal sharing approach presented in this paper is particularly suited for static or low mobility sharing scenarios, where the interference channel exhibits quasi-periodic features. By enabling more efficient use of the spectrum, this method can potentially increase the available bandwidth for commercial services while maintaining the operational integrity of radar systems. This research contributes to the ongoing efforts to improve spectrum utilization, addressing regulators' concerns about the underutilization of radar-reserved bands. By providing a framework for temporal sharing, the paper offers a practical solution that could lead to more flexible and efficient spectrum allocation policies, benefiting both radar operators and commercial service providers.

Ref. [164] introduces a novel real-time method for mitigating uncorrelated automotive radar interference, specifically targeting the challenges posed by mutual interference among frequency modulated continuous wave (FMCW) radar sensors in vehicles. As the deployment of these sensors increases, ensuring their reliable operation in complex vehicular environments becomes critical. The proposed approach utilizes the short time Fourier transform (STFT) in conjunction with L-Statistics to effectively address this issue. The method begins by computing the STFT of the beat signal and sorting each constant frequency line of the resulting spectrogram in ascending order, which strategically positions bins affected by interference toward the right side of the time-frequency plane. By coherently summing the first $q\%$ of time bins deemed unaffected by interference, the algorithm constructs an interference-free range profile. Notably, this method does not require a prior step for interference detection, simplifying the processing and enhancing efficiency. By setting the STFT step size to half of the window length, the algorithm achieves performance nearly equivalent to that of a unitary step size while significantly reducing computation time, thus facilitating real-time implementation. The effectiveness of this technique was validated through evaluations using synthetic automotive radar signals to explore various design parameters, including sliding window length and summation percentage. Subsequently, the method was successfully implemented on a Texas Instruments AWR1843 radar platform and tested in a controlled laboratory environment. The article provides a detailed account of the real-time processing flow, highlighting implementation specifics related to the hardware resources of the AWR1843 platform. Testing scenarios included utilizing the AWR1843 as a victim radar while employing up to three additional similar radar platforms as interference sources with varying parameters. This real-time solution fills a significant gap in existing interference mitigation strategies, which often rely on post-processing techniques, thereby offering a practical enhancement to the reliability and performance of automotive radar systems operating in environments with multiple sensors. Ref. [165] addresses the limitations inherent in single-domain filters used for interference suppression in antenna arrays, proposing a novel multi-domain collaborative oblique projection filter to enhance performance in challenging scenarios. Traditional single-domain filters, whether operating in the temporal, frequency, or spatial domain, often struggle when the differences between the target signal and interference are minimal within their respective domains. To overcome this issue, the authors theoretically derive multi-domain spaces that facilitate the distinction between target signals and interference. They introduce a multi-domain oblique projection operator designed to recover the original target while effectively suppressing interference. This innovative approach allows the filter to maintain excellent performance even when the differences in individual domains are insignificant, contrasting sharply with the significant degradation observed in both single-domain and cascade filters under similar conditions. As a practical demonstration of their method, the researchers present a space-polarization-frequency domain collaborative filter based on oblique projection. The study includes comprehensive performance analysis and simulation results that illustrate the superiority of the proposed filter for interference mitigation, marking a significant advancement in techniques for enhancing the reliability and effectiveness of antenna array systems in complex environments. There are also several representative works developing multi-domain radar filters to perform interference mitigation, we refer interested readers to refs. [166–170] for more details. Note that the above discussion also highlights that accurate identification of interference types and precise parameter estimation we discussed previously are crucial for effective interference mitigation.

8.5 Centralized Coordination

In the previous sections, we mainly focus on the interference mitigation operations in a single radar transceiver, either employing detect-and-avoid at the transmitter end or detect-and-suppress at the receiver end. Besides those approaches, joint decisions between multiple radar transceivers can also realize interference mitigation, named the centralized coordination approach. Centralized coordination for radar interference mitigation has made significant advancements by integrating diverse strategies and technologies to address increasingly sophisticated jamming threats. In this direction, cognitive intelligent radar anti-jamming has become a major research direction, incorporating intelligent scheduling algorithms that operate in a "perception-learning-decision-action" feedback loop. This approach allows radar systems to dynamically adjust waveform parameters and allocate multi-domain resources to counter agile, cluster, and intelligent jamming. Resource intelligent scheduling focuses on three primary areas: waveform, power, and frequency domain resource scheduling. Waveform parameter scheduling dynamically adjusts radar's transmission signals based on real-time environmental sensing, enhancing emission strategy unpredictability. Studies have utilized game theory to optimize beamforming and power distribution in multi-base radar scenarios [171, 172]. For example, several iterative beamforming and power distribution optimization algorithms for networked radars are proposed to minimize mutual interference from multiple radars. Ref. [172] proposed a potential game-based waveform allocation scheme, improving SNR under jamming conditions. Ref. [173] addressed deceptive jamming in multi-target tracking with a game model based on partially observable Markov decision processes, enhancing tracking accuracy. In this direction, ref. [174] is a great work to develop a centralized framework for multiple radars to achieve interference mitigation by means of frequency resource optimization. Different from the detect and avoid at the radar transmitter approach, v2x-communication is utilized to convey minimal information between multiple radars to realize centralization. Specifically, the radar interference between two opposite radars in different vehicles is analyzed based on the geometrical layout in Figure 8.3. To do so, the distance between two radars is decomposed into two directions, one is the extension direction of roads, denoted as $\pm H$, and the other is the vertical

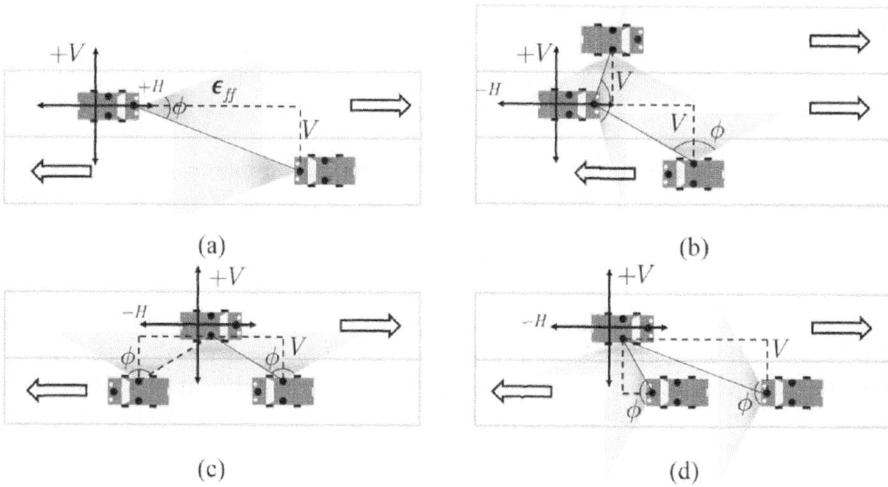

FIGURE 8.3
A simple layout of interfering automotive radars. (a) Front radar to front radar. (b) Side radars to front radar. (c) Side radars to side radar. (d) Front radars to side radar.

direction, denoted as $\pm V$. By doing so, given a target vehicle as the origin of the coordinate system, any surrounding vehicle can be represented as (H, V) for further analysis. Then, to quantize the front radar interference caused by the other front radars of the vehicles traveling in the opposite direction, the location of the interfering front radar will satisfy

$$\begin{cases} V > 0 \\ H > \epsilon_{ff} = \frac{V}{\tan \frac{\phi}{2}}, \end{cases} \tag{8.25}$$

where ϵ_{ff} is the minimum distance caused by the antenna beam width and ignored sidelodes. Similarly, we can also identify the interference source caused by the side radars in other vehicles to the front/side radar in the target vehicle. Then, in the proposed framework, for j the vehicle, its front radar, left side radar, and right side radar will be denoted as $3j-2$, $3j-1$, and $3j$, respectively. Thus, there are $M = 3 \times J$ radars in the consideration of the centralized framework. Assuming there are T orthogonal sub-bands with the same bandwidth, which are under the control of the centralized framework, $\mathrm{f}_m \in \{0,1\}^{1\times T}$ is used to denote the allocation result of a specific timeslot of radar m and $\mathbf{F} \in \{0,1\}^{M\times T}$ is used to collect the allocation results of all radars. In this framework, each vehicle is required to upload its parameters to a base station, including the locations and the orientations for each radar. Then given an allocation matrix $\mathbf{F}$, the base station calculates the potential interference between each radar pair that is utilizing the same sub-band for radar purposes based on the aforementioned geometrical modeling and collects the results as $\mathbf{P} \in \mathcal{R}^{M \times M}$ as

$$\mathbf{P}(m,n) = \begin{cases} 0, \text{ no potential interference} \\ P_t G^2 (\frac{\lambda}{4\pi})^2 R_{mn}^{-2}, \text{ otherwise,} \end{cases} \tag{8.26}$$

based on the radar equation. Based on the above modeling, the centralized radar resource allocation problem can be expressed as

$$\min_{\mathbf{F}} \sum (\mathbf{PF}), \tag{8.27}$$

satisfying $||\mathbf{f}_m||_0 = 1$, which means each radar can only be allocated to a sub-band. Different optimization algorithms can be used to solve the above optimization problem and obtain the optimal sub-band allocations. Moreover, more complex designs can be considered by modeling the potential interference with more factors, such as transmit power and timeslot. In this work, a greedy algorithm is employed to solve the considered minimization problem. Specifically, the allocation matrix will be initialized as a zero matrix first, then the base station allocates the sub-bands for radars in the order of their index. For each radar, the base station calculates the interference that will be introduced to all sub-bands and allocates the radar with the sub-band, ensuring that the introduced interference is minimal until convergence. Simulation results confirm the effectiveness of the proposed solution. Ref. [175] tackles the increasing challenge of interference among small, sophisticated wireless sensors sharing the electromagnetic spectrum, with a particular focus on automotive radars. As the number of these sensors is expected to grow rapidly over the next decade, effective interference mitigation becomes crucial. The authors propose a novel approach based on pseudo-random cyclic orthogonal sequences (PRCOS), which allows sensors to quickly learn their interference environment and avoid using overlapping frequency waveforms. This method ensures a minimum frequency separation between a radar's instantaneous transmitting frequencies, thereby creating a frequency guard that guarantees orthogonality between sequences. Consequently, this enables efficient spectrum sharing among radars in a decentralized manner, eliminating the need for centralized control systems and allowing radars to adaptively manage frequency separation according to the prevailing interference conditions. The paper makes several significant contributions, including the development of a mathematical model for mutual interference in automotive radar systems and an analytic framework to characterize effective interfering power based on frequency separation. Additionally, the authors introduce a new family of waveform sequences capable of mitigating multi-radar interference without requiring centralized coordination. They provide a new statistical characterization of mutual interference in an analytically tractable form, which is validated through simulations and experimental measurements on commercially available automotive radars. The results demonstrate considerable interference reduction when compared to traditional random stepped frequency waveform sequences (RSFWS), highlighting the effectiveness of the proposed model and framework in addressing mutual interference in automotive radar systems while offering insights applicable to a broader range of sensor scenarios. Ref. [176] addresses the critical issue of mutual interference among millimeter-wave automotive radars, which has become a significant limiting factor for radar detection performance as these systems are increasingly deployed in vehicles for ADAS. The authors present a comprehensive analysis of mutual interference in FMCW radars, considering both co-channel interference (CCI) and adjacent channel interference (ACI). They employ a stochastic geometry model to analyze CCI, while using deterministic analysis for ACI assessment. This dual approach allows for a more thorough understanding of the interference mechanisms in automotive radar systems. To mitigate the interference problem, the researchers propose a time-frequency division multiple access (TFDMA) scheme. This coordinated approach aims to reduce interference by efficiently allocating time and frequency resources among multiple radar systems. The performance of this TFDMA scheme is evaluated based on several metrics, including mitigation delay, probability of interference, effective detectable density, maximum number of interference-free radars, and control signaling overhead. Furthermore, the study explores power allocation strategies to enhance the effectiveness of the coordinated interference mitigation approach. This aspect is crucial for optimizing the overall performance of the radar systems in the presence of multiple interfering sources.

The authors validate their proposed framework for interference analysis using Monte Carlo simulations. The results demonstrate a significant performance improvement of 3.5 dB with the coordinated interference mitigation approach, highlighting its potential for enhancing the reliability and effectiveness of automotive radar systems in complex, multi-radar environments. This research contributes to the ongoing efforts to improve the performance of millimeter-wave automotive radars in the face of increasing mutual interference, which is essential for the continued development and deployment of advanced driver assistance systems and autonomous driving technologies. RadChat [177] is an innovative distributed networking protocol designed to mitigate interference among FMCW-based automotive radars, including self-interference, through the cooperative use of radar and communication systems. Operating in the 77 GHz radar band, RadChat integrates both radar and communication functionalities by employing different waveforms on the same hardware, requiring minimal modifications to standard FMCW-based automotive radar, which is widely recognized for its affordability and robustness in the automotive sector. The protocol includes a thorough interference analysis of FMCW radar and narrowband communication systems, revealing that radar-to-communication (R2C) and communication-to-radar (C2R) interference can significantly disrupt reliable concurrent communication and radar sensing. Consequently, it is essential that radar and communication signals with similar power levels do not share the same time-frequency resources. Additionally, the analysis indicates that radar-to-radar (R2R) interference can be effectively minimized if FMCW radar chirp sequences are staggered in their frequency sweeps. RadChat features a protocol for the physical (PHY) and medium access (MAC) layers that substantially reduces R2R radar interference while fulfilling automotive radar sensing requirements for both radars mounted on different vehicles and those on the same vehicle, achieving this in less than 80 milliseconds. The performance of RadChat has been rigorously analyzed and compared to standard FMCW systems within a single-hop dense vehicular ad-hoc network (VANET), demonstrating its effectiveness in significantly reducing mutual interference among automotive radars. This advancement addresses a critical challenge in enhancing the reliability and performance of advanced driver assistance systems (ADAS) and autonomous driving technologies. Ref. [178] addresses the pressing issue of mutual interference among automotive radar systems, particularly focusing on FMCW millimeter-wave (MMW) radars operating within the constrained frequency band of 77 to 81 GHz. As the number of vehicles equipped with these radars continues to surge, the limited spectrum availability leads to an increased risk of mutual interference, which can significantly hinder target detection and parameter estimation capabilities. To mitigate this interference, the authors propose two primary strategies. The first is a Sparse Interference Extraction (SIE) method that leverages the sparse characteristics of mutual interference in the two-dimensional (2D) time domain, allowing for effective extraction and removal of interfering signals from the received radar data. The second approach introduces a novel three-dimensional (3D) Tensor Decomposition (TD) method that capitalizes on the low-rank property of useful echoes across multiple channels, enabling the decomposition of received signals into distinct components of mutual interference and useful echoes. The effectiveness of the proposed TD method is evaluated through several numerical simulations, particularly in the context of MIMO systems operating under complex electromagnetic conditions. Furthermore, practical experiments are conducted to demonstrate the feasibility of this approach, comparing its performance against various state-of-the-art methods for interference mitigation in automotive radar systems. Overall, this research contributes significantly to enhancing the performance and reliability of automotive radar systems in increasingly crowded electromagnetic environments, addressing a critical challenge in the advancement of ADAS and autonomous driving technologies.

In the context of distributed radar systems, studies have exploited the high amplitude correlation of jamming signals across receiving nodes while preserving the independence of target echoes, enabling the detection and identification of range false target jamming. To further enhance detection, clustering methods have been proposed, utilizing amplitude ratio differences between true and false targets among receiving nodes. Additionally, spatial scattering characteristics, including Hermitian distances, have been used to identify false targets, while a homogeneity hypothesis method has been introduced to eliminate false targets on the localization plane through the spatial correlation of true targets. The suppression of range-velocity joint deception jamming has been achieved by exploiting differences in localization and Doppler characteristics between true and false targets. Radar polarization science has played a critical role in false target suppression, particularly through polarization identification, which distinguishes true targets from false ones by extracting polarization domain characteristics. Research has focused on analyzing scattering characteristics and polarization scattering matrices, and utilizing transient polarization projection vectors to differentiate target echoes from jamming signals. High-resolution time-division polarization measurement systems have addressed challenges related to both time-division and instantaneous polarization measurement systems, enabling the study of wideband target characteristics. Methods such as dual-station models have been proposed to combat polarization modulation jamming, extracting features for radar target identification and active decoy suppression. For fully polarized, complex-modulated false targets, techniques involving dynamic adjustments to the transmitted signal's carrier frequency have proven effective. In combat environments, the dynamic polarization characteristics of mid-course ballistic targets have been analyzed to improve polarization identification accuracy. By centralizing and coordinating these advanced techniques, modern radar systems have achieved robust performance against complex and dynamic electromagnetic threats. These efforts integrate innovative interference detection, resource optimization, and adaptive countermeasures, highlighting the effectiveness of cognitive intelligent radar anti-jamming strategies in real-world scenarios.

Part III

Integrated Sensing and Communication (ISAC) Systems

9

ISAC Systems and Architectures

9.1 Motivations and Possible Applications

With the establishment of the 5G standard, there has been a surge in interconnected devices and communication services, leading to rapid growth in the wireless communication industry. Consequently, the natural spectrum resources have become increasingly crowded. The scarcity of spectrum resources has resulted in resource competition among various service devices[1] and a significant increase in spectrum auction prices.[2] These developments have brought about a higher demand for spectrum usage in higher frequency bands.

Meanwhile, radar sensing, which has been significantly refined since its inception in the first half of the 20th century, has been deployed worldwide in various applications, including air traffic control (ATC), geophysical monitoring, weather observation, automotive safety, and national defense. Radar sensing operates in frequency bands with abundant wideband resources. As a result, the relevant frequency bands of radar sensing have become one of the optimal candidate bands for future communication system sharing. Currently, wireless communication systems such as 5G NR, LTE, and Wi-Fi are gradually expanding to higher frequencies, such as the millimeter-wave band. The upcoming 5G technology and future generations of wireless communication systems are also expected to coexist with radar sensing systems. However, with this coexistence, the interference in radar sensing bands is also increasing, making the interference problem in spectrum sharing between sensing and communication a critical area of research and development in sensor

[1]By 2025, the global number of interconnected devices is expected to reach 75 billion, further emphasizing the urgent need for additional spectrum resources.

[2]Since 2015, mobile network operators in the United Kingdom have been required to pay a total annual fee of £80.3 million for the use of the 900 MHz and 1800 MHz bands, which are utilized for voice and data services using a mix of 2G, 3G, and 4G technologies. Similarly, in Germany, the total amount raised from the auction of four frequency bands by mobile network operators exceeded 5 billion euros. The Federal Communications Commission in the United States completed its first 5G auction, selling 28 GHz spectrum licenses and raising $702 million.

DOI: 10.1201/9781003408239-9

fusion. Sensor and communication fusion can avoid the underutilization of spectrum resources, thereby improving operational efficiency. Consequently, spectrum sharing between sensor and communication systems has garnered significant attention. In addition, sensor and communication fusion technologies that share hardware resources have also become a major focus of interest. In fact, both radar sensing and communication systems are evolving toward higher frequencies, larger antenna arrays, and miniaturization. This trend makes their hardware architectures, communication channel characteristics, and signal processing increasingly similar. It presents an exciting opportunity to leverage wireless infrastructure for sensing purposes. Integrated hardware sensor fusion systems enable future networks to transcend traditional networks, providing ubiquitous sensing services for measuring and even imaging the surrounding environment. The vast amount of data acquired through such sensing capabilities is considered a powerful driving force for learning and establishing intelligence in the future smart world, finding extensive applications in many scenarios.

The application scenarios of sensor and communication fusion encompass various aspects of life. These include smart homes related to daily living, human-machine interaction in the domains of entertainment and education, in-vehicle sensing for connected cars and autonomous driving, intelligent factories in the manufacturing sector, traditional application environments in radar sensing for environmental monitoring and remote sensing, as well as topics related to base station sensing and drone surveillance in the context of 6G networks. In Table 9.1, we provide detailed explanations of the concepts and requirements of each application.

9.2 Fundamentals and Frameworks of ISAC

The development of sensing and communication technologies can be traced back to the military applications of the 1960s. Even before the birth of modern digital communication technologies, researchers at that time recognized that certain specific communication functionalities could be achieved through military radar sensing. Radar sensing, originally designed as a missile ranging instrument, embedded communication information into a series of radar sensing pulses using the pulse interval modulation (PIM) technique. In this section, the development of integrated sensing and communication (ISAC) technology will be outlined from a historical perspective, divided into four parts.

Early development of radar sensing: In the early stages of radar sensing, mechanical motors were used to drive the system, allowing periodic rotation of the antenna to search for targets in space. However, such radar sensing faced several key challenges, such as lack of functionality and flexibility, and susceptibility to interference. In view of these challenges, phased array

TABLE 9.1
The brief discussion of different ISAC application scenes.

Scene	Introduction	Application	Challenges
Smart Homes	• Wireless signal's amplitude/phase variations can be used to detect or identify human presence/proximity/falls/sleep/breathing/daily activities by extracting distance, Doppler, or micro-Doppler features during indoor movement. • Spatial positioning sensed by wireless signals can serve as prior knowledge to assist in location-aware control or wireless communication.	• Indoor positioning • Fall detection • Sleep monitoring • Breathing and heart rate detection • Intruder detection	• High range/time resolution of standard waveforms • Fixed pilot allocation • Clock-level synchronization • Narrowband signals with high Doppler resolution • Integration with other sensing devices (e.g., cameras)
Base Station Sensing	• IoT devices and cellular networks, with ISAC support, can provide sensing services for civilians, including enhanced network positioning, cooperative imaging within a given area, and mobile crowd sensing. • Dense deployment of ISAC transceivers can naturally form passive sensing networks to assist nearby devices. • Each base station can build a site-specific communication channel knowledge base by measuring the surrounding environment, expediting beam alignment.	• Drone monitoring • Cellular network positioning and tracking • Crowd identification and sensing • Area imaging • Channel information and sensing map construction • Passive sensing networks	• High range/time resolution of standard waveforms • Wireless resource allocation and optimization • Interference management • Cooperative sensing and image formation • Small-scale drone monitoring and tracking
Connected Vehicles	• ISAC-assisted connected vehicles can achieve high-speed communication and high-precision positioning simultaneously. • ISAC-assisted inter-vehicle communication provides environmental information to support rapid vehicle platooning, secure access, and synchronized positioning. • Roadside unit (RSU) networks can provide sensing services, extending the sensing range of passing vehicles beyond their line of sight.	• Raw data exchange and high-precision positioning • Hands-free communication security • Vehicle platooning • Simultaneous localization • Expanded sensor capabilities	• Full-duplex issues • Vehicle communication protocol design • Multisource sensor information fusion • Sensing-assisted vehicle communication
Remote Sensing	• A group of drones can collaborate to act as a moving antenna array. • Synthetic aperture imaging can be performed, achieving high-resolution all-weather day-night imaging	• Drone swarm synthetic aperture imaging • Satellite imaging and broadcasting	• Trajectory optimization for joint sensing and communication control • Resource allocation problems
Environmental Monitoring	• Changes in environmental features such as water vapor, air pollutants, and insects can be observed by analyzing millimeter-wave path loss. • Cellular networks with sensing capabilities can serve as built-in monitoring facilities, used for large-scale atmospheric observation networks.	• Weather forecasting • Pollution monitoring • Rainfall measurement • Insect monitoring	• Large-scale data analysis • Distributed beamforming and optimization
Smart Factories	• The goal of smart factories is to optimize manufacturing processes by continuously monitoring and adjusting them to achieve higher efficiency and personalization. This requires and relies on IoT machine technologies, including various computing, sensing, actuation, and machine-to-machine communication modes.	• Collaborative robots • Human-machine interaction • Digital twins • Video surveillance • Remote control • Predictive maintenance	• Current research focuses only on specific scenarios • Many issues regarding reliability, performance, and security remain unresolved.

technology, also known as electronic scanning array technology, was developed. Unlike mechanically rotating antennas, phased arrays generate spatial signal beams to electronically control the direction of sensing. FuMG 41/42 Mammut, the long-range phased array early warning radar system developed by GEMA company in Germany during the late stages of World War II, was capable of detecting targets flying at an altitude of 8,000 meters within a range of 300 kilometers.

Interaction between radar sensing and communication: Mammut was not only the first phased array radar sensing system but also the first multiple-antenna system, greatly influencing the development of multiple-input multiple-output (MIMO) systems. In 1994, Paula and Kailath obtained the first patent for MIMO communication, ushering in a new era for 3G, 4G, and 5G wireless networks. Triggered by MIMO communication technology, the Lincoln Laboratory at the Massachusetts Institute of Technology (MIT) presented the concept of centralized MIMO radar at the 2004 IEEE Radar Conference. Compared to phased array radar sensing, MIMO radar improved flexibility and sensing performance. Concepts such as degrees of freedom and diversity originated from MIMO communication theory and became the cornerstone of MIMO radar theory. The research on the integration of radar sensing and communication began in the early 2000s. In the 1990s, the Office of Naval Research (ONR) in the United States initiated the Advanced Multifunction Radio Frequency Concept (AMRFC) program, aiming to design integrated RF front-ends by dividing multiple antennas into different functional modules, such as radar sensing, communication, and electronic warfare modules. The research of ISAC emerged from the 1990s to the 2000s, motivated primarily by the AMRFC and subsequent projects like the ONR-sponsored InTop project. During this period, various ISAC schemes were proposed in the radar sensing community, with the overall idea of embedding communication information into commonly used radar sensing waveforms. For example, combining linear frequency modulation signals with phase-shift keying (PSK) modulation, which was the first ISAC waveform design using linear frequency modulation signals. Subsequently, many research works started focusing on modulating communication data by utilizing radar sensing waveforms such as linear frequency modulation signals and frequency/phase encoding waveforms as carriers. Orthogonal frequency division multiplexing (OFDM), a key technology in 4G and 5G wireless networks, was discovered for radar sensing purposes in the early 2010s. In OFDM radar sensing, the impact of random communication data generation can be mitigated by simply performing fast Fourier transform (FFT) and Inverse FFT (IFFT), while also separating the delay-Doppler. Two approaches based on chirp and OFDM signals respectively, are the design methods for "sensing waveform-based" and "communication waveform-based" radar sensing. In 2013, the Defense Advanced Research Projects Agency (DARPA) funded the Shared Spectrum Access for Radar and Communications (SSPARC) project, aiming to release a portion of the spectrum below 6 GHz from radar sensing bands for shared use by radar sensing and communication. This led to another interesting research topic within the cognitive radio framework, "radar sensing-communication coexistence," where radar sensing and communication systems coexist in the same frequency band without excessive mutual interference. In this context, radar sensing and communication systems coexist in the same frequency band without excessive mutual interference. In addition to spectrum coexistence and interference management addressed by RCC, ISAC further achieves a deeper integration of sensing and communication functions through shared infrastructure.

Parallel development of radar sensing and communication: In 2010, Marzetta proposed the concept of massive MIMO (mMIMO), which later became one of the core technologies for 5G and subsequent networks. In 2013, NYU WIRELESS published a groundbreaking paper on the feasibility of using millimeter-wave (mmWave) signals for mobile communications, which had a far-reaching impact. Since then, mmWave and mMIMO have become a complementary and synergistic pair. Due to the reduced signal wavelength, mMIMO antenna arrays can be physically smaller, and with the high beamforming gain provided by mMIMO arrays, mmWave signals can be transmitted over longer distances. However, a key challenge in the large-scale deployment of mMIMO mmWave technology is the significant hardware cost and energy consumption associated with the required large number of mmWave RF chains. This has compelled wireless researchers to rethink the RF front-end architecture of mMIMO systems. Among them, the hybrid analog digital (HAD) structure, which connects a large number of antennas with a small number of RF chains through carefully designed phase shifters, has emerged as a viable solution. Coincidentally, in the same year as the birth of mMIMO, the concept of Phased-MIMO radar emerged, attempting to strike a balance between phased array and MIMO radar. By transmitting separate waveforms on each antenna, MIMO radar favors increased degrees of freedom at the cost of limited array gain. In contrast, phased array radar sensing concentrates the transmitted power in the direction of the target, offering higher array gain but at the expense of reduced degrees of freedom. Similar to the case of the HAD structure used for communication, a natural idea is to design a system architecture that bridges the gap between the two by connecting multiple antennas with a limited number of RF chains through a phase shifter array. Through this approach, phased-MIMO radar achieves a flexible trade-off between phased array and MIMO radar. In the extreme case of having only one RF chain, phased-MIMO radar simplifies to phased array radar. On the other hand, if the number of RF chains equals the number of antennas, phased-MIMO radar is equivalent to MIMO radar. However, despite the parallel but largely independent developments of radar sensing and communication, there exist issues of device duplication, such as the devices used between phased array radar sensing and communication and between MIMO radar and MIMO communication. Multi-base radar sensing can be performed in parallel with cooperative communication. It is worth noting that there are similarities between radar sensing and communication signal processing, including beamforming used for both communication and radar sensing, hypothesis testing for target detection and symbol detection, as well as mmWave communication channel estimation and radar sensing target estimation.

Fusion of radar sensing and communication: The mentioned similarities have led to the integration of both technologies into systems and devices, providing a clear opportunity for simultaneous sensing and communication through a single transmission service. In fact, radar sensing and communication technologies have become so tightly intertwined that they have been evolving in the same direction over the past decade. In other words, high-frequency bands and massive antenna arrays inherently demand more spectrum and spatial resources. From a communication perspective, wide bandwidth and antenna arrays increase communication capacity and provide abundant connectivity. On the other hand, the increased bandwidth and number of antennas significantly enhance the performance of radar sensing in terms of distance and angular resolution, enabling more accurate perception of multiple targets or mapping complex environments. Radar sensing and communication also tend to exhibit similarities in channel characteristics and signal processing methods as they operate in the millimeter-wave frequency range. Especially due to the less abundant propagation paths compared to the sub-6 GHz frequency range, the mmWave communication channel is sparse and dominated by line-of-sight (LoS) paths. Therefore, the mmWave communication channel models align better with physical geometric shapes, and when combined with mMIMO, they have facilitated the development of beam-domain signal processing for mmWave communication. These techniques are not limited to beam training, beam alignment, beam tracking, and beam management, all of which can be based on the HAD structure. It is worth noting that communication in the beam domain to some extent emulates traditional radar signal processing, where beam training and tracking can be analogized to target search and tracking. As a result, the boundary between radar sensing and communication has become blurred, and sensing functionalities are not necessarily limited to radar sensing infrastructure. Wireless infrastructure and devices can also be perceived through radio transmission and signals, constituting the foundation and fundamental principles of ISAC technology.

Recently, the increasing demand for spectrum and the emergence of numerous applications in future 6G networks have made the collection of large amounts of physical world information an indispensable function. Considering hardware costs, communication system performance, and the capability of sensing information, the development of sensing and communication fusion technology is imminent. Currently, sensing and communication fusion technology is still in the academic development stage, and relevant standards have not been clearly defined. Many technological advancements are initially based on existing standards for further improvement, such as the use of IEEE 802.11p in vehicular networks. Industry organizations, such as Hexa led by Nokia in Europe and Huawei in China, have also proposed relevant white papers and conducted prototype testing. In fact, the sixth generation of mobile communication standards (6G) will build upon existing beyond 5G (B5G) network technologies and deepen the integration between networks and the physical world. Therefore, in addition to the continuous evolution of spectrum and standardization processes, 6G will further possess characteristics such as open architecture, energy efficiency, security, and intelligence to cope with the rapid development of integrated virtual and physical environments and the emergence of new applications like autonomous driving, remote healthcare, and the Meta Universe. The aforementioned applications of the 6G network will require the integration and collaboration of sensing and communication, which promotes signal processing in both areas and enables individual performance improvements through mutual assistance.

There is a notable distinction in signal processing between sensing and communication. Sensing technology aims to extract relevant information about the environment by observing, collecting, and analyzing data from noisy surroundings. On the other hand, communication focuses on transmitting and recovering messages through specialized signal design from noisy environments. The ultimate goal of sensor and communication fusion is to unify these two operations and pursue a balance between them while enhancing their respective performance. On one hand, sensor and communication fusion is expected to significantly improve spectrum and energy efficiency while reducing hardware and signal processing costs. This is achieved by attempting to merge sensing and communication into a single system, where they previously competed for various types of resources. On the other hand, sensor and communication fusion also seeks a deeper level of integration, where these two functionalities are no longer seen as separate end goals but are co-designed to achieve mutual benefits through communication-assisted sensing and sensing-assisted communication. However, the realization of such technologies relies on the collaboration of interference-free and mutually supportive sensor communication systems, leading to the rapid evolution of sensor fusion and integration technologies.

In the framework of ISAC, it is necessary to divide the entire architecture into different aspects. Based on different levels, it can be divided into system-level division and physical-level division. Based on these two divisions, further discussions can be conducted on various aspects. In light of this direction, recent research regarding sensing and communication fusion can be further divided into system-level collaboration, physical-layer collaboration, and physical-layer integration. The following chapters will introduce the research progress and the latest developments in these three areas.

System-level collaboration: In the system-level division, the discussion revolves around how to integrate the two systems of sensing and communication. Classification is based on the principle of who assists whom, thus resulting in the categorization of ISAC systems into communication-assisted sensing (CAS) systems and sensing-assisted communication (SAC) systems. In the field of sensing, cooperative communication between different devices is employed to communicate the signals obtained from each sensing device, breaking the limitations of a single hardware unit to achieve higher precision and lower cost in positioning or environmental sensing. This is known as communication-assisted sensing systems. On the other hand, sensing-assisted

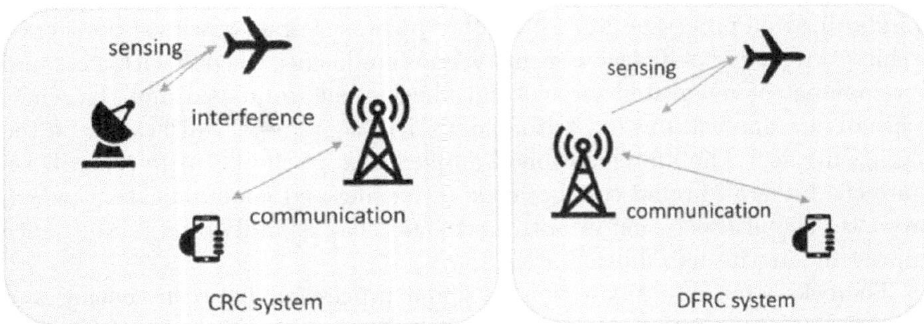

FIGURE 9.1
Illustration of CRC and DFRC systems.

communication systems utilize sensing-related technologies to obtain more accurate channel information, thereby increasing communication capacity and reducing error rates caused by environmental estimation errors. The challenges in system-level collaboration mainly lie in the communication between different systems, requiring the establishment of clear standards. Additionally, due to the diverse application scenarios of sensing and communication, most existing research focuses on specific environments and specialized model designs, making it challenging to extend to various environmental applications. A standardized framework needs to be defined.

Physical-layer collaboration/integration: In terms of physical-layer collaboration/integration, as shown in Figure 9.1, it can be broadly divided into two categories based on whether sensing and communication share hardware devices: collaborative systems with different hardware (also known as communication and radar coexistence, CRC) and integrated systems with shared hardware (also known as dual-functional radar and communication, DFRC) in the field of radar sensing. Unlike system-level collaboration, which focuses on improving application-layer performance, research in the physical layer aims to reduce mutual interference between the two functionalities and further enhance the performance of both systems through signal processing at the physical layer. Current research on physical-layer collaborative systems focuses on eliminating mutual interference between different hardware devices. This can be achieved by properly allocating natural resources such as time, frequency, space, and energy, or by adding encoding to increase additional degrees of freedom and reduce or avoid interference. The research direction in this regard can involve related issues in interference handling in traditional wireless communication. However, mathematical models need to consider certain signal processing characteristics specific to radar sensing, such as clutter, first-order reflection, and radar cross-section (RCS). Depending on the presence or absence of a central control unit, different processing approaches can be adopted, which will be discussed subsequently.

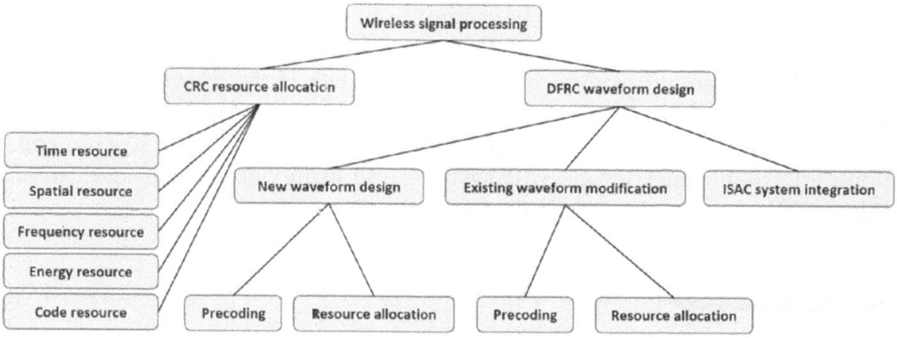

FIGURE 9.2
Illustration of the research directions in ISAC physical-layer collaboration/integration.

As for physical-layer integrated systems, the goal is to achieve both communication and sensing functionalities using a single hardware unit. The design focus is on enhancing the performance between sensing and communication. Current research primarily considers designs for specific environmental applications. The design challenges at the physical layer mainly revolve around the lack of clear performance evaluation metrics, the absence of appropriate benchmarks for comparison, and significant differences in design scenarios. Currently, there is no recognized testing environment. Moreover, there is no consensus on the direction of technical development, which can be categorized into three main approaches: design based on traditional communication waveforms, design based on traditional sensing waveforms, or the development of an entirely new waveform. Systems based on traditional communication waveform design need to overcome issues such as poor sensing performance or the need for more complex signal processing. Systems based on traditional sensing waveform design have good sensing performance (longer range, higher resolution, lower false alarm rates), but the challenge lies in how to add communication signals to the sensing signals while meeting the high transmission rate and low error rate requirements of existing communication networks. Designing a completely new waveform requires developing a suitable sensing fusion metric and designing a waveform that meets its standards, which is highly challenging and would require significant changes to the hardware design of existing devices. Therefore, its commercial strategic value is relatively low. We summarize the research directions of physical-layer collaboration and physical-layer integration in Figure 9.2 for easy reference. Although the three approaches mentioned above each have their advantages and disadvantages, the network applications of sensing fusion communication are highly diverse. Therefore, the design direction can be determined based on specific scenarios. For example, in the case of connected vehicles, higher sensing accuracy is required, making a system based on traditional sensing waveform design a suitable approach. On the other hand, applications such as drone surveillance and the Internet of Things (IoT) require lower sensing performance but have higher communication requirements. In such cases, a system based on traditional communication waveform design can be considered. To understand the unique requirements of different ISAC scenarios, we further discuss common performance metrics in ISAC systems and bring out the consideration of different ISAC scenarios in the next section.

9.3 Performance Requirements in Common ISAC Scenarios

When evaluating ISAC system performance, it is necessary to use mathematical performance metrics to assess the quality of ISAC systems. Owing to the fact that both sensing and communication have their own performance metrics, this section will explore the trade-offs between the metrics and performance used in communication and sensing systems to bring out the current requirements in terms of sensing and communication functionalities in common ISAC scenarios. We first introduce the commonly used performance metrics of current sensing systems.

Detection: Detection refers to making binary/multinomial decisions on the state of sensing targets under given noise and/or interference conditions [179]. The states of these sensing targets typically include the presence or absence of the targets. If system decisions are involved, they can be categorized as follows: the detection rate (or true positive rate) when the target truly exists, the miss detection rate (or false negative rate) when the target truly exists but is incorrectly determined as absent, the false alarm rate (or false positive rate) when the target truly does not exist but is incorrectly determined as present, and the correct rejection rate when the target truly does not exist and is correctly determined as absent, as shown in Table 9.2. When evaluating the performance of sensing systems, the detection rate and false alarm rate are often the main metrics used to assess the system's detection capability. The detection rate provides an indication of the system's ability to correctly identify the presence of targets, while the false alarm rate represents the likelihood of falsely detecting targets in the absence of actual

TABLE 9.2
The diagram of making a decision on the state of a sensing target.

	System decision: target exists	System decision: no target
Ground truth: target exists	Detection rate	Miss detection rate
Ground truth: no target	False alarm rate	Correct rejection rate

targets. These two metrics are commonly used because the miss detection rate can be derived from the detection rate, and the correct rejection rate can be derived from the false alarm rate. In order to determine the detection rate and false alarm rate, detectors often employ a constant false alarm rate (CFAR) mechanism [180] to determine a threshold for identifying peaks of targets in the Range–Doppler (RD) map.

Estimation: Estimation refers to the extraction of useful parameters of the sensed object from observations contaminated by noise and/or interference [179]. This may include estimating the target's distance, velocity, angle, or other relevant parameters. The performance of estimation is often evaluated using metrics such as mean square error (MSE) and Cramer Rao Bound (CRB) to assess the estimation capability. The mean square error is defined as the average of the squared difference between the true parameter value and its estimated value. It provides a measure of the accuracy of the estimation. On the other hand, the CRB is defined as the lower bound on the variance of any unbiased estimator for the parameter. It is calculated as the reciprocal of the Fisher information, which quantifies the accuracy of an estimator [181]. In the context of sensing and communication, the CRB, particularly for angles, can be used as a performance metric to optimize the waveform of radar sensing and communication systems.

Ambiguity function: Information about the parameters of the target is obtained by comparing the received echo signal with the transmitted signal. Interference or noise limits the ability to detect the presence of the sensing echo signal. Similarly, noise also limits the reliability of detecting features within the sensing echo signal. In order to achieve reliable detection, the echo signal should be larger than the noise. The influence of this noise can typically be minimized by using a matched filter. The matched filter maximizes the signal-to-noise ratio of the echo signal in the presence of additive noise. The matched filter can be applied to detect unknown signals by correlating the known signal with the unknown signal. This is equivalent to convolving the unknown signal with the delayed conjugate of the known signal. This concept is used for optimal detection of ambiguous functions, where the ambiguous function represents the response of the matched filter to both the matched signal and the Doppler-shifted signal. The ambiguous function can be written in the following form:

$$\mathcal{X}(\tau, f_d) = \int_{-\infty}^{\infty} s(t)s^*(t-\tau)e^{j2\pi f_d t}dt. \tag{9.1}$$

Here, $s(t)$ represents the transmitted signal, τ represents the time delay, and f_d represents the Doppler shift. While the ambiguous function is rarely used as a basis for practical system design, it serves as a representation of the limitations and utility of specific sensing waveforms. It serves as a design and evaluation tool [182, 183].

Conditional mutual information: In ref. [184], the author approaches the performance of sensing waveforms from the perspective of information

TABLE 9.3
The comparison between sensing mutual information and communication capacity.

System	Received signal	Interest	Condition MI (mono-static radar) MI (at user with CSIR)
Radar (waveform design)	$\Upsilon = \mathbf{S}^T\mathbf{G} + \mathbf{Z}$	G	$I(\mathbf{G};\Upsilon \mid \mathbf{S}) = N_R \log\left[\det\left(\sigma_n^{-2}\mathbf{S}^T\sum_G \mathbf{S}^* + \mathbf{I}_{N_S}\right)\right]$
Commun.	$\mathbf{Y} = \mathbf{HS} + \mathbf{N}$	S	$I(\mathbf{S};\mathbf{Y}) = N_S \log\left[\det\left(\sigma_n^{-2}\mathbf{H}\sum_S \mathbf{H}^H + \mathbf{I}_{N_R}\right)\right]$

theory. It is assumed that the parameters of the target have statistical characteristics, typically assumed to follow a Gaussian distribution. In the case of a known transmitted waveform, the degree of dependence between the observed echo and the target parameters indicates the ability to extract the target parameters from the received echo. When using conditional mutual information as a performance metric, the transmitter needs to assume the second-order statistical characteristics of the known target parameters [185, 186]. Unlike the channel capacity in communication, which focuses on the statistical characteristics of the transmitted signal and assumes a known channel, perception is interested in channel information, i.e., the target parameters, while assuming a known transmitted signal. The comparison between communication and perception from the perspective of information theory is shown in Table 9.3 [128, 131]. The advantage of using conditional mutual information as a performance metric is that it allows for waveform design similar to communication channel capacity. For example, in refs. [184, 185], the water-filling technique is used for subcarrier power allocation in OFDM. In ref. [131], it has been shown that maximizing conditional mutual information is equivalent to minimizing the mean square error of the target pulse response. Although there is currently no literature that rigorously proves the direct relationship between optimizing conditional mutual information and improving sensing performance, there is a strong correlation between optimizing conditional mutual information and optimizing signal-to-noise ratio. Therefore, many studies still use this performance metric for optimization design. We then introduce the common performance metrics of current communication systems.

Reliability: During wireless transmission, various factors such as interference, path loss, and fading, as well as thermal noise generated by electronic components at the receiver, can introduce errors during signal demodulation. Reliability, therefore, measures the ability of a communication system to reduce or correct these erroneous signals. Commonly used metrics for reliability include outage probability, bit error rate, symbol error rate, and frame error rate.

Efficiency: Successful transmission of information comes at the cost of wireless resources, such as spectrum, space, and energy. Efficiency, therefore, serves as a benchmark to evaluate how much information can be successfully transmitted from the transmitter to the receiver given the limited available resources [187]. Spectral efficiency and energy efficiency are widely adopted and defined as achievable rates per unit bandwidth/energy, measured in bits per second per hertz or bits per channel use, and bits per second per joule, respectively. Additionally, channel capacity, coverage range, and the maximum number of serviced users are also important efficiency indicators.

Signal-to-interference-plus-noise ratio (SINR): The SINR represents the ratio of the desired signal power to the combined interference and noise power. It is a performance metric in wireless communication that measures communication quality. SINR has a high correlation with channel capacity, hence improving SINR can enhance both communication reliability and channel capacity. Furthermore, improving SINR in sensing systems also contributes to enhanced detection rates. Therefore, SINR can be utilized in both sensing and communication systems.

In the previous section, we discussed an overview of common applications and scenarios in ISAC systems, along with their respective challenges. In this section, we further provide the required (or inherently existing) communication and sensing system parameters for these scenarios, as shown in Table 9.4, to conclude the discussions.

9.4 The Trade-off between Communication and Sensing Performance

Based on the above discussions, one can notice that wireless resources, such as spectrum and power resources, are shared between the sensing and communication functionalities. As a result, it is important to know those trade-offs so that ISAC system designers can provide advanced designs to meet the desired requirements of both sensing and communication functionalities simultaneously with the limited resources budget. In this section, we will provide several real examples of ISAC designs for this purpose.

Detection vs. communication: In the first example, we consider the design scenario in ref. [188] to balance the detection probability as a sensing performance metric and the achievable rate as a communication performance metric. In the considered scenario shown in Figure 9.3, an ISAC transceiver transmits useful information to a communication user and transmits a radar signal to provide surveillance service simultaneously. To do so, a communication waveform $s_{\mathrm{C}}(t)$ with transmit power P_{C} and a sensing waveform $s_{\mathrm{R}}(t)$ with transmit power P_{R} are transmitted from the transmitter side. To avoid mutual interference, it is assumed that orthogonal resource planning is applied in the time or frequency domain during the transmission process.

TABLE 9.4
The comparison between sensing mutual information and communication capacity.

Scene	Application	Maximum Detectable Distance (m)	Maximum Detectable Speed (m/s)	Distance Resolution (m)	Speed Resolution (m/s)	Data Rate per User	Angular Resolution (degree)
Smart Homes	Indoor positioning	20	2	0.5	0.1	9.5Gbps (peak)	4
	Fall detection	10	3	0.5	0.1	9.6Gbps (peak)	3
	Sleep monitoring	1	2	0.5	0.1	9.6Gbps (peak)	5
	Respiration and heart rate detection	1	2	0.4	0.1	9.6Gbps (peak)	5
	Intruder detection	20	5	0.5	0.1	9.6Gbps (peak)	5
Base Station Sensing	Drone monitoring	500	40	0.2	1	40Mbps (peak)	1
	Cellular network positioning and tracking	300	10	0.5	5	100Mbps (average) /10Gbps (peak)	3
	Crowd recognition and sensing	300	5	0.5	5	100Mbps (average) /10Gbps (peak)	3
	Area imaging	200	0.1	5	X	100Mbps (average) /10Gbps (peak)	3
	Channel information and sensing map construction	300	5	0.5	5	100Mbps (average) /10Gbps (peak)	3
	Passive sensing network	300	30	0.5	X	100Mbps (average) /10Gbps (peak)	10
Connected Vehicles	Raw data exchange and high-precision positioning	300	30	1	X	1Gbps (average) Between vehicles and between vehicles and base stations	4
	Hands-free communication security	300	0.1	X	5	100Mbps (average) /10Gbps (peak)	4
	Vehicle platooning	100	3	0.01	1	1Gbps (average) Between vehicles and between vehicles and base stations	2
	Simultaneous localization	300	30	0.1	5	100Mbps (average) /10Gbps (peak)	3
	Extended sensors	300	3	0.1	0.1	300Mbps (average)	2
Remote Sensing	Drone swarm synthetic aperture imaging	1000	40	0.1	X	20Mbps	X
	Satellite imaging and broadcasting	10000	X	1	X	100Mbps	X
Environmental Monitoring	Weather forecasting	500	X	0.1	X	100Mbps (average) /10Gbps (peak)	X
	Pollution monitoring	200	X	0.01	X	100Mbps (average) /10Gbps (peak)	X
	Rainfall measurement	200	X	0.5	X	100Mbps (average) /10Gbps (peak)	X
	Insect monitoring	200	X	0.1	X	100Mbps (average) /10Gbps (peak)	X

Moreover, a passive sensing receiver is also employed to finish the sensing process by comparing the received sensing signals from the direct channel and the

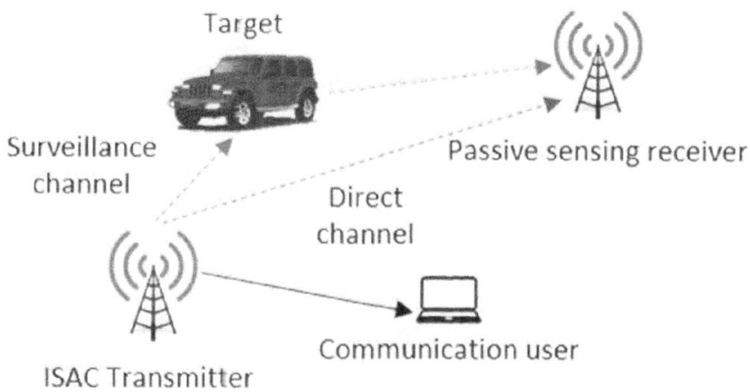

FIGURE 9.3
Illustration of the joint passive sensing and communication scenario.

surveillance channel to detect whether targets exist in the region of interest or not, as shown in Figure 9.3. To tackle the interested power resource allocation problem, an optimization problem is considered:

$$\max_{P_C, P_R} \mathcal{P}_D \quad \text{s.t.} \quad R \geq R_{th}, P_R + P_C = P_T. \tag{9.2}$$

In Eq. (9.2), $\mathcal{P}_D$ is the radar detection probability, $R = \log(1 + P_C\gamma_C)$ is the achievable rate, where γ_C is the communication channel gain over the noise power, R_{th} is the achievable rate threshold, and P_T is the total power constraint. Hence, the physical meaning of Eq. (9.2) is to maximize the radar detection probability while satisfying the achievable rate threshold by allocating the transmit power resource in the transmitter side. Furthermore, from the sensing perspective, the passive sensing receiver can detect the existence of targets in the surveillance channel by correlating the received signals from the surveillance channel and the direct channel. Hence, for the interested detection problem, by sampling the received signals as L time-domain samples, the hypothesis when targets do not exist in the surveillance channel can be expressed as:

$$\mathcal{H}_0 : \begin{cases} \mathbf{y}_d = \gamma_d \mathbf{G}_d \mathbf{s}_R + \mathbf{n}_d \\ \mathbf{y}_s = \mathbf{n}_s \end{cases}, \tag{9.3}$$

where $\mathbf{y}_d$ and $\mathbf{y}_s$ are the received signals from the direct and surveillance channels, $\mathbf{G}_d$ is a $L \times L$ unitary delay-Doppler operator metric for the direct channel, γ_d is a scalar coefficient of the direct channel, and $\mathbf{n}_d$ and $\mathbf{n}_s$ are the additive white Gaussian noise (AWGN) with noise power as σ^2. Similarly, the hypothesis when targets exist in the surveillance channel can be expressed as:

$$\mathcal{H}_0 : \begin{cases} \mathbf{y}_d = \gamma_d \mathbf{G}_d \mathbf{s}_R + \mathbf{n}_d \\ \mathbf{y}_s = \gamma_s \mathbf{G}_s \mathbf{s}_R + \mathbf{n}_s \end{cases}, \tag{9.4}$$

where $\mathbf{G}_s$ is a $L \times L$ unitary delay-Doppler operator metric for the surveillance channel and γ_s is a scalar coefficient of the surveillance channel. By

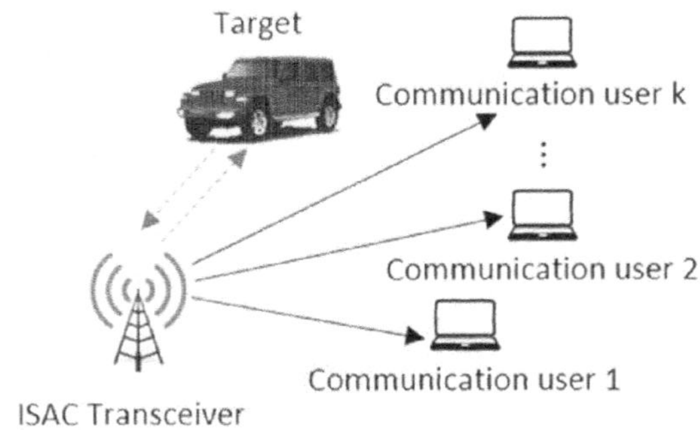

FIGURE 9.4
Illustration of the joint active sensing and communication scenario.

employing a generalized likelihood ratio test (GLRT) detector, the radar detection probability $\mathcal{P}_{\rm D}$ in the high direct-path signal-to-noise ratio region can be approximated as

$$\mathcal{P}_{\rm D} \approx Q_1(\sqrt{\frac{2P_{\rm R}|\gamma_d|^2}{\sigma^2}}, \sqrt{2\gamma}), \tag{9.5}$$

where $Q_1(a, b)$ is the first-order Marcum Q-function with parameters a and b, and γ is the detection threshold. Then, since the power resources constraint $P_{\rm R} + P_{\rm C} = P_{\rm T}$ holds, the detection probability can be further expressed[3] as

$$\mathcal{P}_{\rm D} \approx Q_1(\sqrt{2(P_{\rm T} - \frac{1}{\gamma_{\rm C}}(2^{R_{\rm th}} - 1))\frac{|\gamma_{\rm d}|^2}{\sigma^2}}, \sqrt{2\gamma}). \tag{9.6}$$

In Eq. (9.6), one can notice that the radar detection probability $\mathcal{P}_{\rm D}$ is related to the communication achievable rate $R_{\rm th}$. More precisely, higher $\mathcal{P}_{\rm D}$ will result in a sacrificed $R_{\rm th}$ and vise versa, being the design trade-off of the considered ISAC system.

Estimation vs. communication: Next, we further consider a more complex case, where time, frequency, and power resources are reused for sensing and communication purposes simultaneously to see the trade-off between sensing performance and communication performance. To do so, we consider a scenario shown in Figure 9.4, where an ISAC transmitter with N_t transmit antennas and $N_r \geq N_t$ receive antennas are communicating with K single-antenna users and monitoring an interested target simultaneously. By employing a joint ISAC waveform $\mathbf{X} \in \mathbb{C}^{N_t \times L}$ with L timeslots for both purposes, the echo signal received by the transmitter for sensing purposes can be expressed as

$$\mathbf{Y}_{\rm R} = \mathbf{G}\mathbf{X} + \mathbf{N}_{\rm R}, \tag{9.7}$$

[3]Please see ref. [188] for the derivation details.

where $\mathbf{G} \in \mathbb{C}^{N_r \times N_t}$ is the target response matrix and $\mathbf{N}_{\mathrm{R}}$ is the AWGN matrix with noise power as σ^2. On the other hand, the received signal on the user side for communication purposes can be expressed as

$$\mathbf{Y}_{\mathrm{C}} = \mathbf{H}\mathbf{X} + \mathbf{N}_{\mathrm{C}}, \tag{9.8}$$

where $\mathbf{H} = [\mathbf{h}_1, \mathbf{h}_2, ..., \mathbf{h}_K]^H \in \mathbb{C}^{K \times N_t}$ is the communication channel matrix and $\mathbf{N}_{\mathrm{C}}$ is the AWGN matrix with noise power as σ^2.

Assuming maximum likelihood estimation (MLE) of $\mathbf{G}$ is adopted as $\hat{\mathbf{G}} = \mathbf{Y}_{\mathrm{R}}\mathbf{X}^H(\mathbf{X}\mathbf{X}^H)^{-1}$ [4], the mean square error of the estimation can be expressed as

$$\mathbb{E}\{||\mathbf{G} - \hat{\mathbf{G}}||^2\} = \frac{\sigma_R^2 N_r}{L}\mathrm{tr}(\mathbf{R}_{\mathbf{X}}^{-1}), \tag{9.9}$$

where $\mathbf{R}_{\mathbf{X}} = \frac{1}{L}\mathbf{X}\mathbf{X}^H$ is the sample covariance matrix of $\mathbf{X}$. Hence, the ISAC waveform $\mathbf{X}$ design problem concerning both target estimation performance and communication performance can be expressed as

$$\min_{\mathbf{X}} \mathrm{tr}(\mathbf{R}_{\mathbf{X}}^{-1}) \quad \text{s.t.} \quad ||\mathbf{X}||_F^2 \leq LP_T, c_i(\mathbf{X}) \trianglerighteq C_i, \forall i. \tag{9.10}$$

In Eq. (9.10), $\trianglerighteq$ can be $\leq$, $\geq$, or $=$, and $c_i(\mathbf{X})$ represents a communication quality of service function constrained by C_i, such as per-user signal-to-interference ratio (SINR), sum-rate, and SNR. Furthermore, from the sensing perspective, $\mathbf{X}$ must be full-rank (i.e., with rank N_t) to guarantee the existence of $\mathbf{R}_{\mathbf{X}}^{-1}$. In other words, all available spatial degrees of freedom (DoF) will be utilized for sensing purposes. However, from the communication perspective, the number of DoFs is limited by $\min(N_t, K)$ and $K \leq N_t$ holds for almost all the real cases, especially for massive MIMO scenarios. Typically, for a rank-K data matrix $\mathbf{S}_{\mathrm{C}} \in \mathbb{C}^{K \times L}$ to support K streams for K users, a linear precoder can be employed as $\mathbf{X} = \mathbf{W}_{\mathrm{C}}\mathbf{S}_{\mathrm{C}}$, where $\mathbf{W}_{\mathrm{C}} = [\mathbf{w}_1, \mathbf{w}_2, \cdots, \mathbf{w}_K] \in \mathbb{C}^{N_t \times K}$ to ensure the rank of $\mathbf{X}$ as K. In light of this direction, in the normal case, where $K < N_t$, to ensure the sensing functionality, one way to do so is to augment the data matrix $\mathbf{S}_{\mathrm{C}}$ by adding at least $N_t - K$ sensing streams $\mathbf{S}_{\mathrm{C}}$, containing random information but ensure the orthogonality to the data streams $\mathbf{S}_{\mathrm{C}}$. Hence, the precoding matrix will also be augmented to let the precoding process be as

$$\mathbf{X} = [\mathbf{W}_{\mathrm{C}}, \mathbf{W}_{\mathrm{A}}] \begin{bmatrix} \mathbf{S}_{\mathrm{C}} \\ \mathbf{S}_{\mathrm{A}} \end{bmatrix} = \mathbf{W}_{\mathrm{C}}\mathbf{S}_{\mathrm{C}} + \mathbf{W}_{\mathrm{A}}\mathbf{S}_{\mathrm{A}}, \tag{9.11}$$

ensuring the full-rank of $\mathbf{X}$. However, adding those sensing streams is harmful to communication performance. To elaborate, the per-user SINR from the communication perspective can be expressed as

$$\gamma_k = \frac{|h_k^H w_k|^2}{\sum_{i=1,i\neq k}^{K} |h_k^H w_i|^2 + |h_k^H \mathbf{W}_{\mathrm{A}}|^2 + \sigma_C^2}, \forall k. \tag{9.12}$$

[4]Please see ref. [189] for the derivation details.

In Eq. (9.12), the first term in the denominator is the multi-user interference and the second term is the interference introduced by the sensing streams. In conclusion, one can notice that to ensure the normal functioning of the estimating procedure, communication performance is sacrificed by the reduced per-user SINR. The above two examples reveal the conflicts of sensing and communication performance due to the resource-sharing nature. As a result, how to design an ISAC system to satisfy both needs by optimally allocating limited and shared resources becomes the core design concept and the most important research topic in this area. For readers who want to further study the trade-off in ISAC systems from the theoretical perspective, we refer to refs. [190–192] for more information. In the next chapter, we will provide a systematic review of the current research progress in this direction to let readers become familiar with the ISAC transceiver design topics.

10

ISAC Transceiver Design Principles

10.1 Radar-Assisted Communications

In this section, we provide case studies to see how the ISAC framework can help current systems achieve the strict performance requirements of different applications. In terms of radar sensing-assisted communication, we present the following three case studies to demonstrate its functionalities.

Radar sensing-assisted design of highly directional communication signals [193]: Establishing high-speed vehicle-to-vehicle (V2V) communication links with narrow beamwidths presents challenges due to different network spectrum technologies. A beam that is too narrow may miss the intended receiver, while a beam that is too wide can result in signal-to-noise ratio (SNR) losses. Therefore, radar-assisted communication utilizes the high-precision sensing capabilities of radar to establish V2V communication links. The V2V communication link, as shown in Figure 10.1, involves multiple vehicles and objects distributed evenly in the instantaneous time domain. In this simulation scenario, the yellow autonomous vehicle (ego vehicle) seeks to communicate with the receiver of another vehicle based on its own position and driving angle information. The core idea is to leverage the objects detected by the radar sensors of the autonomous vehicle to identify a specific object in the environment that corresponds to the intended vehicle receiver. Finally, a highly directional narrow beam signal is designed to be received by that vehicle's receiver. What makes this approach special is that, in addition to utilizing the high-precision sensing results of radar, it can also combine distributed methods to enhance communication performance. This involves using the radar sensing results of each vehicle and the GPS results communicated by neighboring vehicles to calculate data correlations. Data correlation entails associating the coordinates of all objects detected by radar sensing with the GPS data of the vehicles. This method relies on global coordinate system transformations and utilizes the computation of Kullback–Leibler divergence (KLD) as shown in Eq. (10.1) to solve optimization allocation problems. Ultimately, it is demonstrated that in V2V communication, when high-reliability relative and absolute position information is available, radar sensing-assisted design of

DOI: 10.1201/9781003408239-10

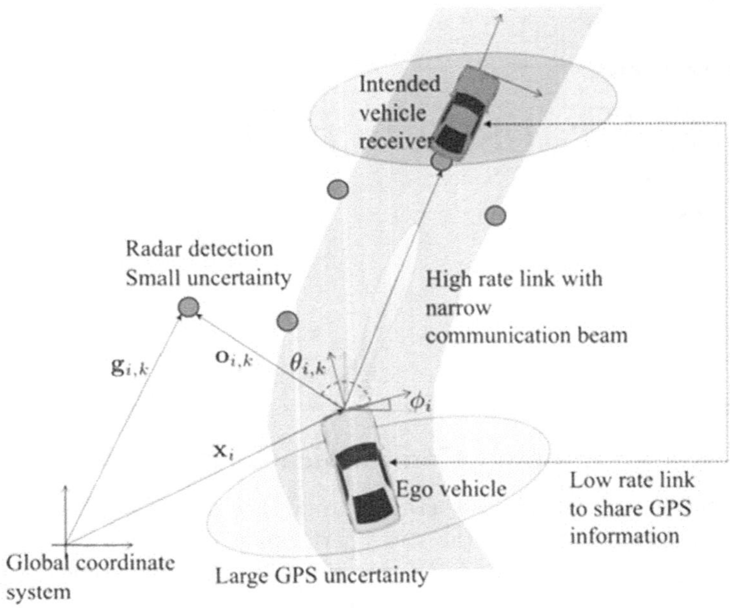

FIGURE 10.1
Illustration of the considered V2V communication scenario.

highly directional communication signals can avoid the use of time-consuming channel estimation methods and achieve higher SNR communication performance.

$$\begin{aligned} D(f\|g) &= \int f(\mathcal{X}) \log\left(\frac{f(\mathcal{X})}{g(\mathcal{X})}\right) \delta\mathcal{X} \\ &= f(\Theta)\log(\frac{f(\Theta)}{g(\Theta)}) + \int f(\mathcal{X})\log(\frac{f(\mathcal{X})}{g(\mathcal{X})})dx \\ &= D(\rho_f\|\rho_g) + \rho_f D(\rho_f\|\rho_g). \end{aligned} \tag{10.1}$$

Radar sensing-assisted secure communication [194]: In a dual functional radar and communication (DFRC) system, the integration of radar sensing and communication functions is achieved through the design of probing waveforms that carry communication signals and data. Clearly, such operations raise security concerns, which are often overlooked in existing DFRC literature. It is well-known that typical radar sensing requires concentrating the transmitted power in the direction of interest to obtain accurate estimations of targets. However, in the case of DFRC transmission, the embedded critical information in the probing waveform may be exposed to radar sensing targets, which could potentially be eavesdroppers. Therefore, information security must be considered in DFRC design. Given the dual functionality of DFRC systems, confidentiality issues can be addressed from ei-

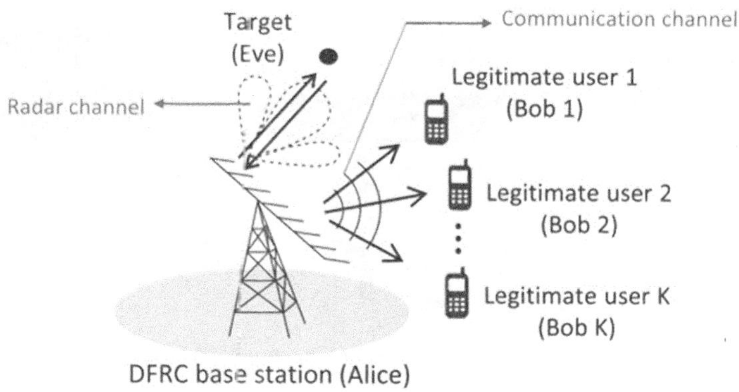

FIGURE 10.2
Illustration of the dual-function radar communication system with target objects as potential eavesdroppers.

ther the radar sensing or communication perspective. To the best of our knowledge, most of the existing works on secure transmission in DFRC systems rely on the assumption of accurate knowledge of channel state information (CSI) at the transmitter. This case study considers a dual-function multiple-input multiple-output (MIMO) DFRC system, consisting of a DFRC base station, legitimate users, and potential eavesdropper targets, as shown in Figure 10.2. The DFRC system is equipped with a uniform linear array (ULA) that serves multiple legitimate users while sensing target objects. Importantly, these target objects are assumed to be potential eavesdroppers who may intercept the communication messages transmitted from the base station to the legitimate users. To ensure transmission confidentiality, this case study introduces artificial noise (AN) at the transmit signal and formulates an optimization problem by minimizing the SINR received at the radar sensing targets while guaranteeing the desired SINR for legitimate users. Initially, the ideal scenario is that both the target angles and CSI are accurately known. The scenario is further extended to a more general case with target position uncertainties and CSI errors, leading to the proposal of robust optimization methods to ensure performance under worst-case scenarios.

Radar sensing-assisted beam alignment [195]: In future vehicle networks, sensing and communication functionalities will be intertwined. This case study investigates the use of DFRC technology for radar sensing-assisted predictive beamforming design in vehicle-to-infrastructure (V2I) communications. The case study considers an mmWave mMIMO Roadside Unit (RSU) serving multiple vehicles on the road. To communicate with the RSU, each vehicle is equipped with MIMO arrays on both sides of the vehicle body. It is assumed that the vehicles travel on a single-lane straight road parallel to the RSU antenna array, and the RSU communicates with each vehicle through line-of-sight (LoS) channels. Additionally, uniform linear arrays are deployed

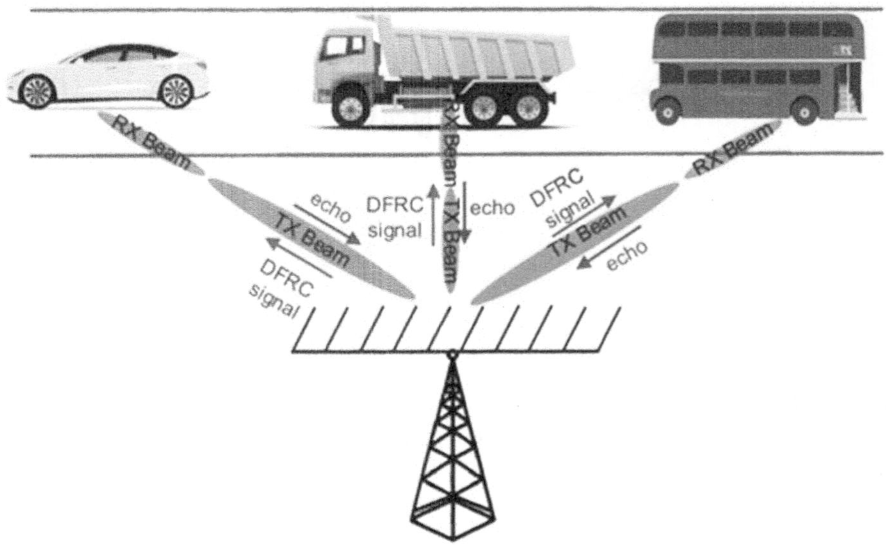

FIGURE 10.3
Illustration of the considered V2I scenario model.

on both the RSU and vehicles, as shown in Figure 10.3. The core idea of this case study is to achieve a reliable communication link through the design of a general framework. To fulfill this requirement, the RSU needs to acquire accurate information about the vehicle azimuth angles. On the other hand, the RSU also needs to know the angles of the RSU relative to each vehicle. Through this approach, both the RSU and vehicles can use their respective antenna arrays to form narrow beams precisely directed toward each other. Conventionally, beam alignment is achieved through beam training, where periodic pilot signals are transmitted and received over all possible beams to find the strongest beam pair. However, such schemes inevitably introduce significant latency and communication overhead due to the need for pilot signaling and feedback between the transmitter and receiver. To reduce the overhead generated during beam training, beam tracking can be performed by updating beam information based on the temporal correlation between consecutive transmission blocks, requiring only a small number of pilot signals. However, considering the high mobility of vehicles, it is preferable to predict the angle parameters of all vehicles rather than merely tracking them. To accomplish measurement and communication tasks in delay-critical vehicle networks, this case study proposes a low-overhead beam prediction framework based on DFRC technology, as shown in Figure 10.4. Experimental results demonstrate that this approach, utilizing the radar functionality of the RSU, significantly reduces the communication beam tracking overhead. Furthermore, to improve sensing accuracy while ensuring downlink communication

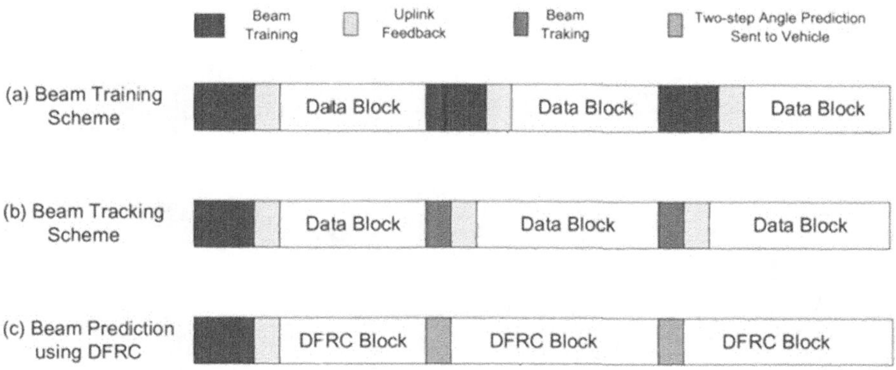

FIGURE 10.4
Illustration of the considered low-overhead beam prediction framework based on DFRC technology.

and rates, a multi-vehicle power allocation scheme is proposed. Simulation results indicate that the proposed DFRC-based beam tracking approach outperforms radar-unassisted techniques in terms of tracking performance. Additionally, the designed power allocation method achieves favorable performance trade-offs between sensing and communication.

10.2 Communication-Aided Sensings

In terms of communication-assisted sensing, we present the following two case studies to demonstrate its functionalities.

Communication-assisted sensing for localization [196]: Vehicle positioning is a critically important topic that has gained considerable attention in recent times. Navigation, vehicle tracking, and location-based services are emerging applications that require significant demand for accurate location information. The Global Positioning System (GPS) has become the practical standard solution for vehicle positioning issues. However, GPS still suffers from accuracy and reliability issues. For instance, GPS receivers as standalone positioning devices may experience positioning errors of up to 50m in multipath environments such as urban canyons, in addition to potential signal blockages, such as signal attenuation in tunnels. Despite the consideration of alternative positioning methods by many researchers, such as the integration of inertial navigation systems (INS) to enhance GPS positioning, studies have shown that this approach performs well in short-term GPS signal interruptions. However, long-term GPS signal interruptions and severe multipath conditions still pose significant challenges for achieving accurate and reliable vehicle positioning, which are defects that many applications cannot tolerate. Therefore, it is suggested to adopt a multi-measurement approach to enhance positioning

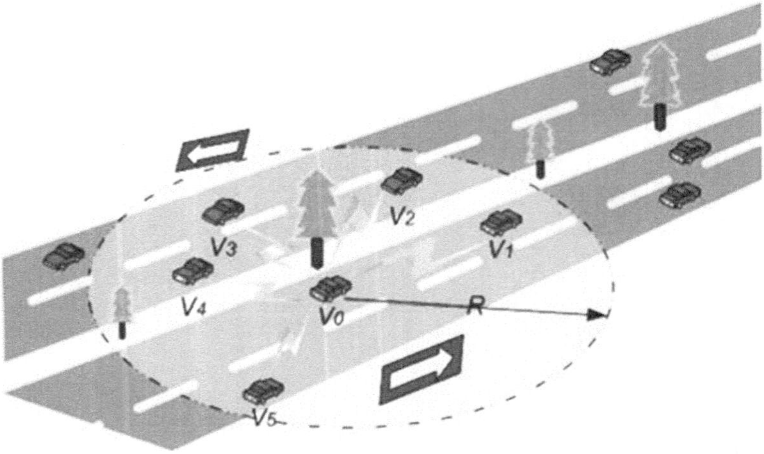

FIGURE 10.5
Illustration of the considered VANET environment.

in the vehicular ad hoc network (VANET) communication environment, as depicted in Figure 10.5. The core concept is to employ Intervehicle Communication-Assisted Localization (IVCAL) where vehicles with high uncertainty in their positioning accuracy transmit requests to nearby vehicles. Through information sharing with nearby vehicles that have higher positioning accuracy, the vehicle's movement information is combined with GPS position estimation to update its own positioning result, thus achieving highly accurate vehicle localization.

Next, the execution method of workshop communication-assisted sensing for localization is explained, in which the Kalman filter and workshop communication system collaboration are employed to improve the robustness and accuracy of each vehicle's localization in the workshop communication system network. The two main units that interconnect the workshop communication system and the Kalman filter are as follows: the Multipath Detection Unit (MDU) and the Localization-Enhancement Unit (LEU). The MDU is capable of detecting the presence of multipath effects in the output of the Kalman filter. Subsequently, the LEU retrieves information about neighboring vehicles from the workshop communication system and feeds back the optimized position estimation results to the Kalman filter. The schematic diagram of this method is shown in Figure 10.6. Assuming that each vehicle in the VANET environment is equipped with a GPS receiver, INS, and VANET transceiver, each vehicle utilizes a GPS/INS Fusion Unit (GPS/INS-FU) to integrate the measurements from INS and GPS with the Kalman filter. The workshop communication system of each vehicle extracts information related to the position estimation of nearby vehicles. This information includes the distance between vehicles, position estimations, and the uncertainty of

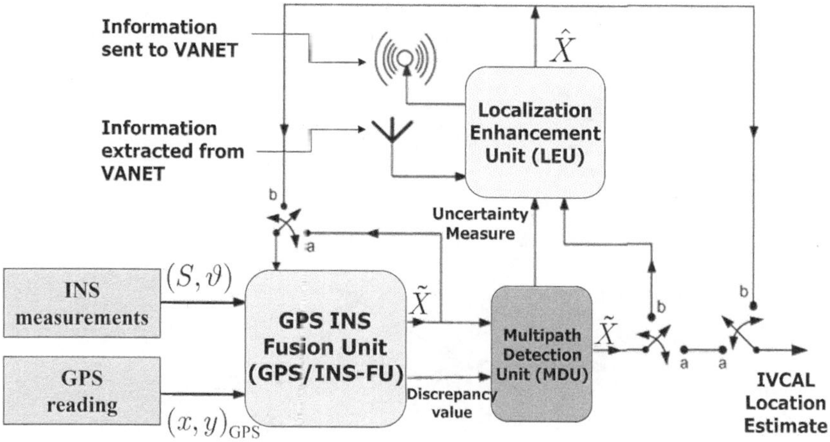

FIGURE 10.6
Architecture of the workshop communication-assisted localization technique.

position estimations of neighboring vehicles. This information can be used to compensate for the loss of satellite signals and correct errors caused by multipath effects (assuming not all vehicles in the network are simultaneously affected by multipath effects). Following this is the multipath detection unit, which enables vehicles to detect the presence of multipath effects in their position estimation, triggering the localization-enhancement unit to minimize positioning errors. The occurrence of multipath effects introduces noise into the position estimation. When multipath effects are present, the localization result of the Kalman filter is not optimal. Therefore, the difference between the estimation values of the Kalman filter and the measurements from GPS can be used to determine the presence of multipath effects. This case proposes the use of a neural network for classifying these patterns, designing a classifier capable of identifying measurements affected by multipath influences. The present study proposes a neural network for classifying these patterns and designs a classifier capable of recognizing multi-path interference measurements. The case study selects a feedforward backpropagation network (FFBN), which belongs to the category of artificial neural network (ANN) supervised learning, as the classifier for multi-path interference measurements. Furthermore, the study introduces enhanced units for localization. To mitigate the degradation of signals affected by multi-path interference, each vehicle communicates with its neighboring vehicles in the VANET environment. These neighboring vehicles are either unaffected or less affected by multi-path interference. Thus, it is crucial for the system to identify the measurement values of neighboring vehicles that can serve as indicators. The measurement values obtained through inter-vehicle communication assist vehicles in selecting the best three neighboring vehicles, enabling more accurate self-localization, as depicted in Figure 10.7.

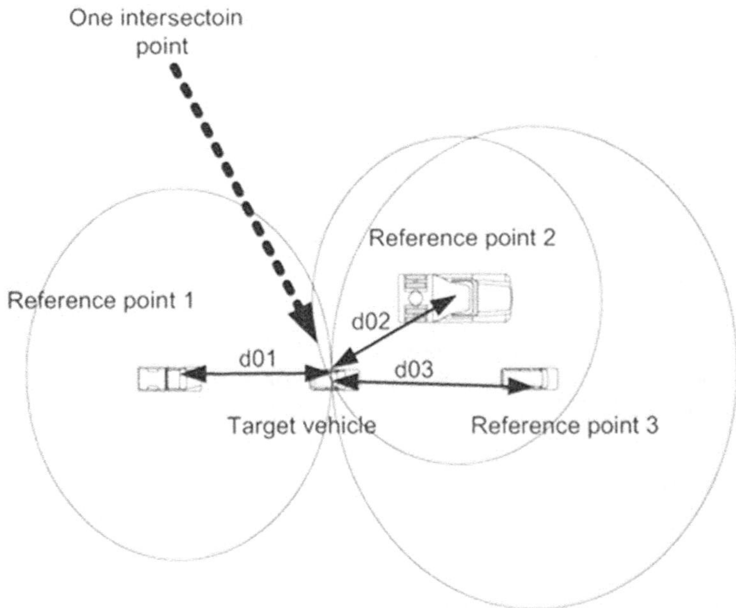

FIGURE 10.7
Example scenario of the optimization problem in an ideal situation.

In this case study, the accuracy of vehicle positioning is represented by the variation between the time-updated estimation values of the Kalman filter and the measurements from the GPS receiver. When a vehicle does not experience any multi-path effects, the accuracy of its GPS is high, and the variation is low, making it suitable as a neighboring vehicle that aids in vehicle localization under inter-vehicle communication conditions. In other words, the position of this vehicle has minimal uncertainty and can be used as a location anchor for its neighbors, and vice versa. Typically, after experiencing multi-path effects, the Kalman filter estimation requires periodic updates to converge to a low-error vehicle position estimation. Each vehicle retains a record of previous variation values; therefore, even after the vehicle leaves the multi-path region, the expected level of uncertainty in the position estimation is anticipated to persist for a considerable period. Vehicles encountering such situations will not serve as location anchors. Vehicles with high levels of uncertainty correct their position estimations with the assistance of communication with three neighboring vehicles, as shown in Eq. (10.2).

$$\hat{\mathbf{x}}^i_{k+1/k+1} = \underset{\mathbf{x}^i_{k+1/k+1}}{\arg\min} \sum_{j\in I} (d_{i,j} - ||\tilde{\mathbf{x}}^i_{k+1/k+1} - \mathbf{x}^j_{k+1/k+1}||)^2. \tag{10.2}$$

Minimization of interference in vehicle radar sensing assisted by vehicular communications [174]: With the development of autonomous vehicles, an increasing number of vehicles are equipped with multiple automotive radars, leading to significant mutual interference between radars. Vehicle-to-everything (V2X) communication is a common approach to coordinate automotive radars and reduce mutual interference between them. In this case study, the positional relationship between two mutually interfering radars is analyzed. The assumption is that each vehicle is equipped with one radar in the front and one radar on the side, with different sensing directions. The interference between the rear radar (primarily used for parking) and the other radars can be ignored. Each vehicle is equipped with a V2X system, operating in different frequency bands from the automotive radars. The interference from different types of radars is evaluated based on a Poisson point process (PPP). A novel framework is proposed to allocate spectrum resources to automotive radars using the V2X communication system to reduce interference. Similar to the traditional frequency-division multiplexing (FDM) approach, the frequency band is divided into several sub-bands. However, the number of sub-bands is limited, while the number of automotive radars is large, making it impossible for the mobile communication system to assign non-overlapping orthogonal frequency bands to each radar. Therefore, in the proposed framework of this case study, the core idea is as follows: the base station uses a location- and direction-based greedy algorithm to allocate reusable spectrum resources, aiming to minimize interference between radars. The minimum spectrum resources required for zero interference are analyzed for different scenarios.

Finally, we consider the following use case for joint communications and target localization

Unmanned aerial vehicle (UAV) trajectory design for ISAC and target localization [197]: The objective of this case study is to achieve an ISAC system in the sky, as depicted in Figure 10.8. The scenario involves a UAV, a communication user receiving ISAC signals, and a target being sensed by the UAV. The core idea is as follows: in the ISAC system in the sky, the UAV serves as both a base station for communication purposes and a single-site radar for sensing and localization, flying over a designated area. It transmits downlink communication signals to ground communication users. Simultaneously, the same transmitted signals are used for radar sensing and localization. Specifically, the UAV emits signals toward the target, and the target's position is estimated by receiving and processing the reflected echoes. By designing an optimized UAV flight trajectory, the performance of communication and sensing can be improved. The UAV's flight waypoints, hovering points, and flight speed are determined to simultaneously minimize the Cramér-Rao bound (CRB) for target localization and maximize the downlink communication rate for communication users. However, in previous studies on UAV applications for ISAC, the focus was mainly on the energy consumption related to signal transmission. In practical scenarios, the energy required for UAV flight and hovering is a major component of energy consumption. Therefore, considering the limited capacity of the UAV's battery,

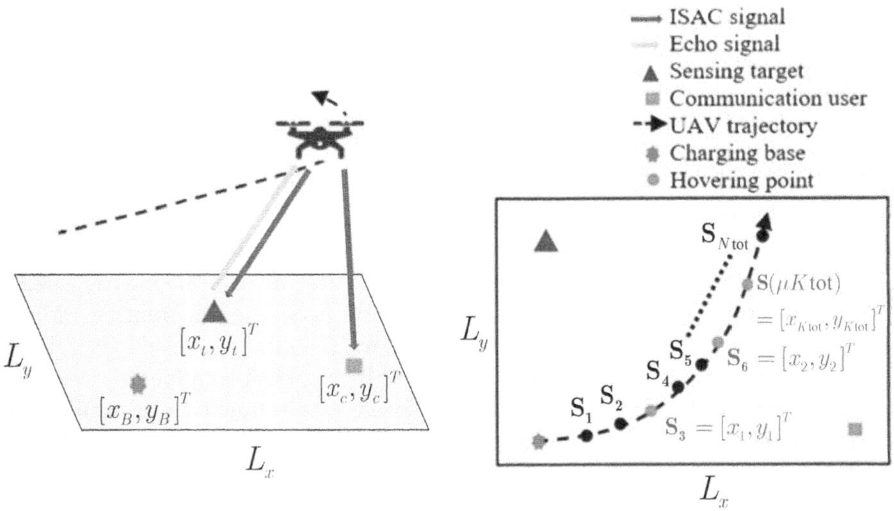

FIGURE 10.8
Illustration of the UAV-based ISAC system.

this case study incorporates energy constraints in the design of UAV flight trajectories. In addition, in reality, the UAV does not initially know the target's position. To design precise UAV trajectories for improving ISAC performance, this study proposes a multi-stage trajectory design method to develop effective target localization algorithms. Finally, the superior performance of the proposed trajectory design is validated.

10.3 Waveform Design Principles

Simultaneously meeting the requirements of sensing and communication tasks poses significant challenges to signal processing in the context of signal and information integrated receiver (SIIR). Generally, an SIIR receiver should be able to demodulate useful information from the communication signal while detecting/estimating targets from the echoes. If the two signals do not overlap, traditional signal processing methods can be employed since sensing and communication do not interfere with each other. However, if the two signals fully/partially overlap in the time and frequency domains, mutual interference occurs, which is the cost incurred to achieve integrated gain.

New waveform (without considering specific waveforms): In current SIIR research, no waveform with a fixed mathematical form specifically designed for sensing and communication has been devised. The concept of designing new waveforms disregards existing sensing and communication waveforms and focuses on achieving sensing and communication effects simultaneously through the design of second-order statistical characteristics of the

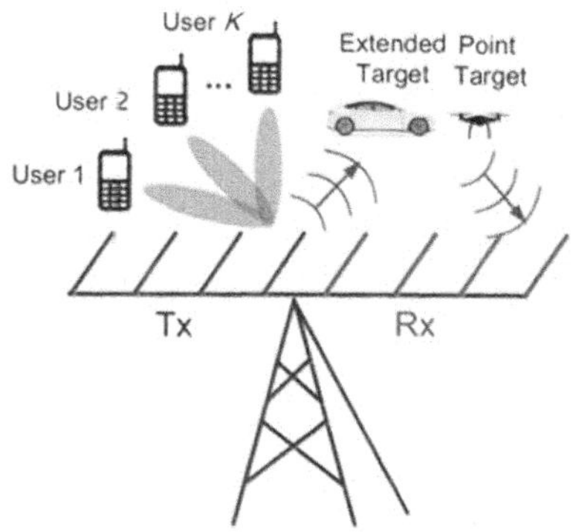

FIGURE 10.9
Illustration of the scenario with multiple users and a single target in the downlink signal transmission phase.

transmitted signals. In the design of new waveforms, waveform design for the transmitted signal is based on channel state information (CSI) to optimize the second-order statistical characteristics of the transmitted signal, aiming to achieve high achievable rates and reduce user interference. In the SIIR framework, as sensing functionality is incorporated, the design of precoding needs to allocate degrees of freedom to sensing. Therefore, the precoding design in SIIR optimizes sensing and communication in the spatio-temporal resource domain. However, for sensing, prior knowledge, specifically knowledge of the angles of the sensed targets, is required for precoding design. Thus, many papers skip the so-called search mode and directly enter the tracking mode, assuming perfect knowledge of the angles of the sensed targets before performing optimal precoding design. In the previous sections, individual performance metrics for sensing and communication were mentioned. In precoding design, the focus is on specific performance metrics, which can be classified as precoding designs centered around the performance metrics of radar sensing, centered around the performance metrics of communication, or designs that trade-off both performance metrics.

Radar sensing performance-centered precoding designs optimize the design based on the performance metrics of radar sensing. In ref. [181], a mono-static DFRC system with multiple users and a single target in the downlink signal transmission scenario is considered, with a focus on the estimation performance of sensing, as shown in Figure 10.9. The CRB is used as the objective function to minimize, aiming to minimize the CRB while imposing the users' SINR as a constraint in precoding design. This paper considers point targets and extended targets, which lead to different solutions in terms of the CRB. In the case of a single-user single-point target scenario, the solution is simple and intuitive, where the designed beamforming weights fall into the subspace of the steering vector and the communication channel's span. For the case of

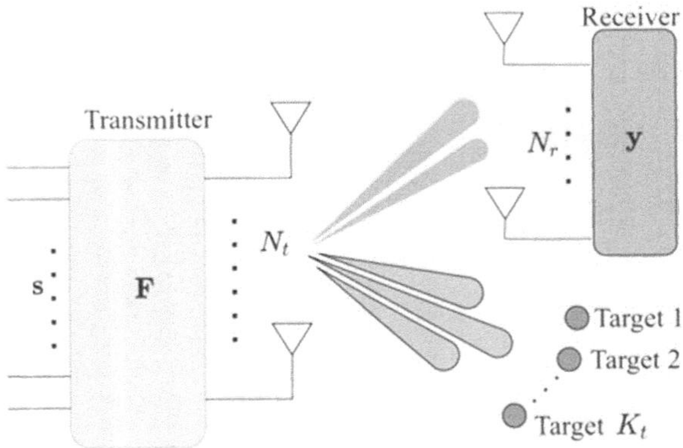

FIGURE 10.10
Illustration of the scenario with single multiple-antennas user and multiple targets in the downlink signal transmission phase.

multiple users and a single point target, semi-definite relaxation (SDR) is used to relax the problem to a convex problem, and for the case of extended targets, additional probing signals (orthogonal to the original transmitted signals) are added to the transmitted signal matrix since the Fisher matrix needs to be full rank. Similar to ref. [181], ref. [198] considers the downlink signal transmission scenario with multiple users and a single target but assumes imperfect CSI, accounting for channel errors with known probability distribution, mean, and standard deviation. The performance metric used for users is the outage probability, aiming to constrain the outage probability within a fixed threshold probability. For sensing, the objective is to concentrate the transmitted signal's energy as much as possible on the desired observed angle (assuming it is known). Therefore, the power in a specific direction is related to the covariance matrix of the transmitted power and the echo, which is associated with precoding. Thus, the problem aims to maximize the power at a specific angle while satisfying the communication's outage probability constraint.

Communication performance-centered precoding designs focus on optimizing communication performance while meeting the required sensing performance. In ref. [56], the precoding design for downlink signal transmission with multiple antennas and multiple targets is considered, as shown in Figure 10.10. The objective is to minimize the minimum Euclidean distance between noiseless received signal vectors while considering sensing constraints and given power constraints. The goal is to make the received signal vectors seen at the communication receiver as separable as possible, which is equivalent to reducing symbol error rates and increasing channel capacity. The sensing constraint aims to shape the transmit beampattern to approximate the desired sensing beampattern. Common communication metrics such as achievable rate, spectral efficiency, or energy efficiency assume Gaussian distribution of the transmitted signal. However, the amplitude of Gaussian signals is infinite, leading to distortion in signals transmitted by power-limited transmitters. Additionally, Gaussian signal detection at the receiver becomes more complex due to the continuous probability distribution function. Furthermore, waveform designs based on the assumption of Gaussian inputs may not perform well when applied to practical systems with finite alphabet inputs. This paper considers ISAC beamforming design with more practical finite alphabet inputs, reducing the problem to a quadratically constrained quadratic programming (QCQP) problem, which can be transformed into a semi-definite programming (SDP) problem. The global optimum solution is obtained using the SDR method. Due to the high complexity of the aforementioned methods, this paper also provides an approximate solution using deep learning methods.

Precoding designs balancing both performance metrics utilize weighted sums for design, allowing adjustments in weights to achieve a trade-off between sensing and communication. Consequently, this type of research has relatively more publications. In ref. [199], the precoding design for downlink signal transmission with multiple users and a single target is considered, combining the functionality of MIMO radar sensing with rate splitting multiple access (RSMA). RSMA is a powerful downlink communication scheme based on linear precoding rate splitting, which can partially decode multi-user interference (MUI) and treat it as noise, while having inherent robustness to partial channel state information (CSIT). With RSMA, DFRC precoding is optimized in the presence of partial CSIT, aiming to maximize the weighted sum rate under service quality constraints and minimize the mean squared error between the DFRC beamforming pattern and the MIMO radar sensing beamforming pattern under ideal conditions. In ref. [200], the precoding design for downlink signal transmission with a single antenna and multiple users and targets is considered, as shown in Figure 10.11. The objective is to simultaneously minimize the interference energy between users and have the minimum error between the transmit beamforming pattern and the pure radar sensing transmit beamforming pattern in the transmit beam direction. In this paper, the defined multi-user interference (MUI) refers to the difference between the transmitted signal after passing through the channel response and the desired demodulated symbols. By adjusting the weights, the overall performance of DFRC can be improved. Two cases are considered in this paper. The difference between the two design approaches lies in the assumption of omnidirectionality, which assumes that the waveforms emitted by each antenna are orthogonal to each other. This allows the formation of a virtual antenna array to improve angular resolution, making it well-suited for the search mode. In contrast, the directional approach concentrates the energy on a few specific angles to improve SINR and enhance the accuracy of detection. In ref. [201], downlink link scenarios with multiple antennas and multiple users with a single target and multiple targets are considered. The transmitted signal adopts a time-division multiplexing approach, where the transmission time is divided into segments for sensing and communication purposes. The design jointly optimizes the waveforms for the two segments, using communication channel capacity and

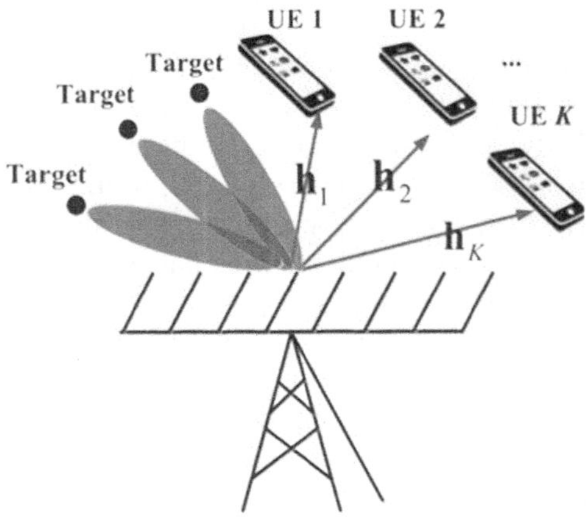

FIGURE 10.11
Illustration of the scenario with multiple users and multiple targets in the downlink signal transmission phase.

mutual information for sensing with weighted sums as the performance metrics. Due to the consideration of mutual information for sensing, the waveform design is relatively simple (similar to the sum rate in communication). The weights can be adjusted to balance sensing and communication proportions. This problem is solved using the water-filling algorithm. In the case of multiple antennas and multiple users with a single target, the considered problem will become non-convex, and the paper proposes using weighted minimum mean square error (WMMSE) to solve it.

The aforementioned pre-coding design exploits simultaneous time-space resources for waveform design. Apart from waveform design, preceding this process, power allocation for training and data symbols is performed [202], based on channel estimation errors to determine the distribution of power weights. Considering a typical data packet-based signal structure, comprising training and data symbols, this paper first derives the relevant conditionally mutual information between sensing and communication under the consideration of training overhead and communication channel estimation errors. Subsequently, this paper establishes a lower bound on channel estimation errors and optimizes the energy arrangement between training and data symbols to minimize channel errors. Both training and data symbols can be used for sensing purposes, whereas data symbols are employed for data transmission. Based on the optimal energy arrangement, this work provides optimal spatio-temporal power designs for three scenarios: maximizing mutual information exclusively for communication and sensing, and maximizing the weighted sum of mutual information between sensing and communication. In ref. [203], multiple distributed DFRC transmitters, radar sensing receivers, and communication receivers are considered, capable of simultaneously performing target

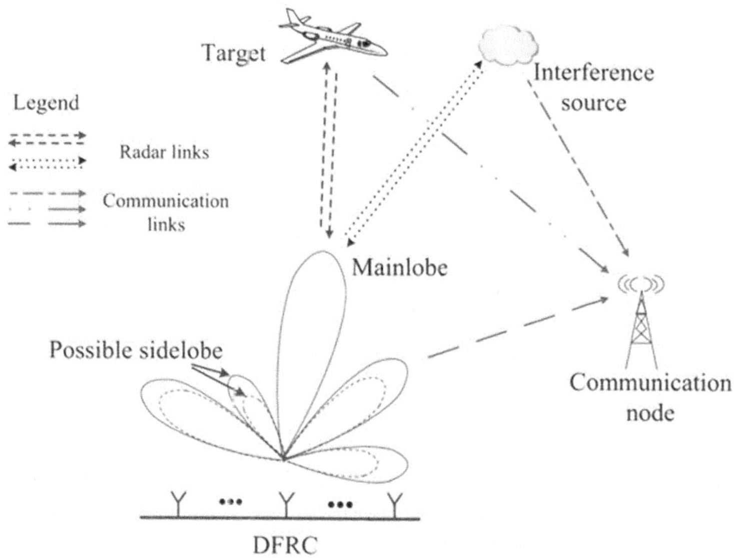

FIGURE 10.12
Illustration of the considered scenario.

state estimation and message transmission tasks. In this research, the authors propose a power allocation scheme based on low probability of intercept (LPI) for distributed MIMO-DFRC systems. The key mechanism of this scheme is to minimize the total system power by optimizing the transmission power allocation of different DFRC transmitters, subject to predefined target parameter estimation accuracy for radar sensing purposes and certain wireless communication performance. This approach enhances the LPI performance of the distributed MIMO-DFRC system for communication purposes, thereby improving overall system efficiency. Closed-form expressions for the Cramér-Rao Lower Bound (CRLB) are derived to assess target position and velocity estimation performance, with data transmission rate adopted as the performance metric for communication data transfer.

Traditional waveform variations: The widely used communication waveform is orthogonal frequency division multiplexing (OFDM), which exhibits high flexibility and has been proposed for radar sensing applications [204]. Integrating sensing and communication using OFDM holds significant promise and is in line with 3GPP's envisioned waveforms for 6G systems. Ref. [205] explores a downlink scenario with a single-user single-target, considering an interference source that affects the receivers of both DFRC and communication base stations, as depicted in Figure 10.12. The objective is to focus the main beam on the target of interest while using sidelobes for communication message transmission. This requires joint optimization of pre-coding for each subcarrier, DFRC receiver combiner, and user receiver combiner, with

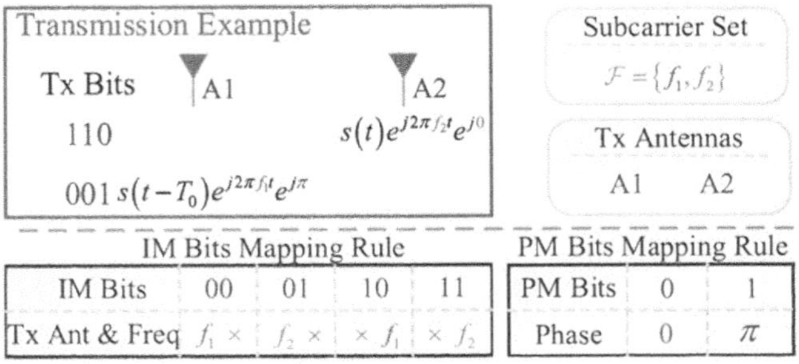

FIGURE 10.13
Illustration of the index modulation schemes.

the primary goal being sensing. Since optimizing based on detection rate is complex, this work employs the Kullback–Leibler divergence (KLD), which is related to the detection rate, as the performance metric. KLD measures the difference in probability distributions between received signals when the target is present and when it is absent. The communication part is constrained using the error rate. The proposed Alternating Direction Sequential Relaxation Programming (ADSRP) algorithm efficiently addresses the non-convex optimization problem. Ref. [206] considers a downlink scenario with single-antenna multiple-users and multiple-targets. The goal is to design transmit waveforms to minimize both communication mutual interference and the weighted sum of sensing transmit beam patterns, while adhering to peak-to-average power ratio (PAPR) constraints. Ref. [207] investigates a multi-antenna, multi-user, single-target downlink scenario with incomplete channel state information at the receiver. To make the transmit beam pattern independent of the selected data symbols, the work utilizes differential phase shift keying (DPSK) as the non-coherent modulation scheme. The optimization problem aims to maximize the radar objective function while achieving improved sensing for target localization, under average transmit power, specific direction power, and user error rate constraints. The objective function concerns sensing performance indicators for each subcarrier, related to SINR, defined in detail in ref. [207]. The work provides a solution based on a three-step framework involving alternating optimization, convex restriction, and minimax optimization.

In radar sensing waveforms, frequency modulated continuous wave (FMCW) waveforms are commonly used in automotive applications due to their superior estimation capability. However, their limited data rate presents a drawback when combined with communication. To address this, the concept of index modulation has been proposed [208–210], as illustrated in Figure 10.13. In this context, let us consider a simple example, as depicted in the aforementioned figure. Index modulation employs distinct resource indices to encode bits. Specifically, for antenna and frequency, we can distinguish different signal bit combinations. For instance, let us assign the first antenna with frequency as f_1, which represents bit sequence "00". Similarly, the first antenna with frequency can be assigned as f_2, representing "01". Furthermore, the second antenna with frequency could be assigned as f_1, representing "10", while the second antenna with frequency could be assigned as f_2, representing "11". Introducing additional factors such as phase and time would further augment the number of possible combinations, resulting in an increased total number of bits, as illustrated in Figure 10.14.

Among traditional waveforms, OFDM offers high flexibility and is therefore utilized in the ISAC framework for resource allocation [211]. In ref. [211], a typical V2X scenario is considered, with a frequency-sensitive target located in a frequency-selective channel. Hence, efficient allocation of transmission power on each subcarrier is crucial to ensure both sensing and communication performance. The work proposes a two-stage adaptive power allocation scheme, as shown in Figure 10.15. In the target search stage, joint optimization of subcarrier assignment and power allocation is performed to minimize the total transmit power of DFRC while considering conditional mutual information for target detection and specific data rates for communication. In the parameter estimation stage, a multi-objective optimization framework is designed to achieve a trade-off between communication channel capacity and radar sensing parameter estimation accuracy. Ref. [212] explores a single-user single-target downlink scenario and investigates the Power Minimization-based Joint Subcarrier Assignment and Power Allocation (PM-JSAPA) problem for DFRC, as shown in Figure 10.16. The PM-JSAPA strategy jointly optimizes subcarrier and power resource allocation to minimize the overall transmit power of DFRC, achieving improved energy efficiency while satisfying conditional mutual information for target parameter estimation and specific data rates for communication. The problem falls under mixed-integer nonlinear programming (MINLP), known to be NP-hard. To address this, an efficient three-step resource allocation framework is proposed to solve the resulting optimization problem. The first step determines subcarrier assignment for radar and communication purposes. The second and third steps perform power allocation for different applications based on the results obtained in the first step. Finally, numerical simulations demonstrate the superior energy-saving performance of the proposed PM-JSAPA strategy and the effectiveness of the solution approach, which adopts classic water-filling for power allocation on

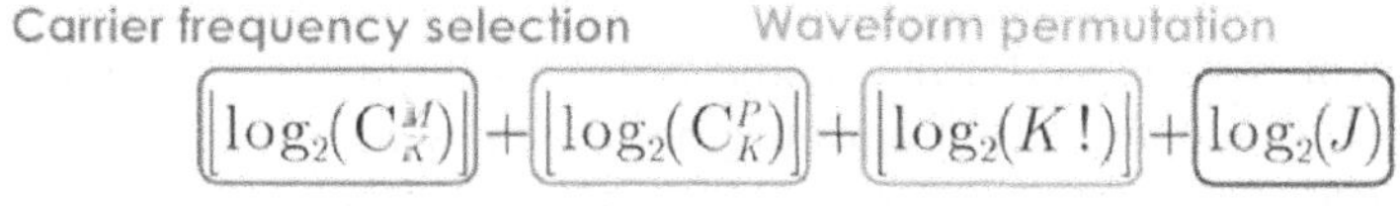

FIGURE 10.14
Illustration of the total number of data bits.

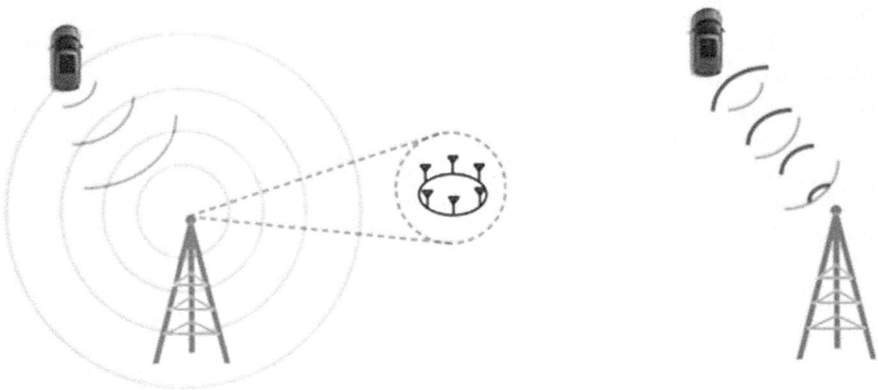

FIGURE 10.15
Illustration of the target search mode and parameter estimation mode.

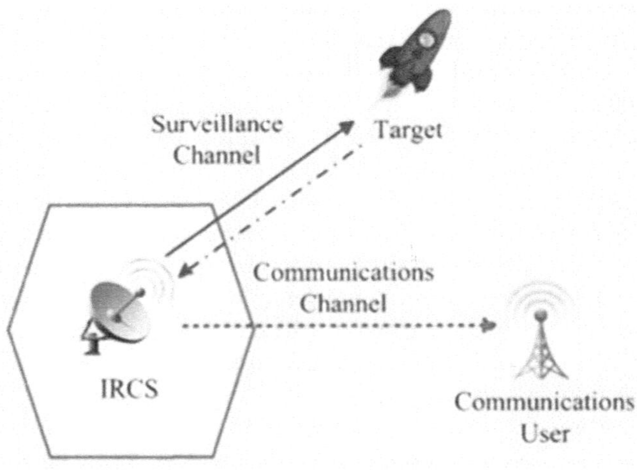

FIGURE 10.16
Illustration of the considered scenario.

OFDM subcarriers. Ref. [213] discusses power allocation in the ISAC system and layout design for pilot placement in time-frequency resources, as shown in Figure 10.17. The study analyzes the impact of power allocation and pilot placement on system performance and proposes channel estimation and pilot power allocation designs that optimize both radar sensing and communication performance.

ISAC special integrated architecture techniques: The ISAC system incorporates extended techniques based on traditional methods, such as hybrid precoding and symbol-level precoding. Conventional precoding designs typically involve digital precoders, which necessitate RF chains equal to the

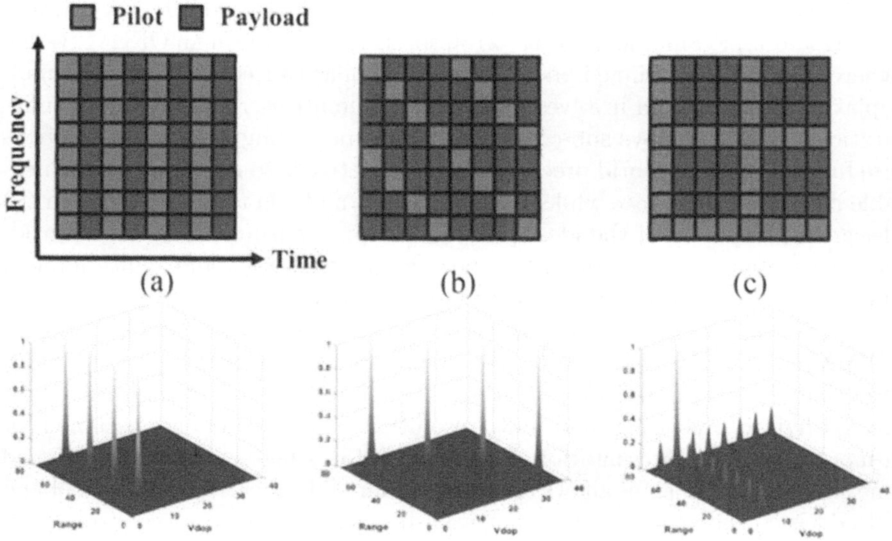

FIGURE 10.17
Illustration of the different pilot placement and power allocation.

number of transmit antennas. However, the cost of RF chains is prohibitively high, leading to the consideration of hybrid precoding that combines fewer RF chains with analog precoding. This approach aims to reduce costs while approaching the performance of pure digital precoding. Hybrid precoding is considered highly prospective for 6G technology, especially in scenarios requiring a large number of antennas to provide sufficient spatial degrees of freedom for user and target allocation. Two common types of hybrid precoding are illustrated in Figure 10.18 [214].

The classification in the figure is based on the number of phase shifters connected to the RF chain. A lower number of phase shifters result in cost reduction, though it may lead to performance degradation. In ref. [215], the concept of sub-connected hybrid precoding is introduced. In this approach, each transmit antenna requires two phase shifters to achieve beamforming flexibility and increase degrees of freedom. This research considers outdoor scenarios, where noise sources such as reflections from trees or buildings need to be taken into account. Two cases are studied: single-user multi-antenna single-target, and single-target multi-user single-antenna scenarios. Optimal designs of DFRC hybrid precoding, user receivers' hybrid precoding, and pure digital combiners (for both user receivers and DFRC receivers) are conducted using an alternating optimization approach. First, the design of DFRC's pure digital combiners is optimized while fixing DFRC hybrid precoding and user combiners. Then, the obtained optimal DFRC's pure digital combiners are substituted into the original problem, and consensus Alternating Direction

Method of Multipliers (consensus-ADMM) is used to design DFRC hybrid precoding and user combiners. In ref. [215], the study focuses on an OFDM system where each user is assigned one subcarrier (similar to frequency-division multiplexing). The system involves multi-antenna multi-user single-target configuration. DFRC employs sub-connected hybrid precoding, while user receivers use fully connected hybrid precoding. The objective is to maximize the achievable communication rate while satisfying constraints on the error between the designed precoder and the ideal precoder for pure sensing. This work considers two scenarios: multi-antenna single-user single-target and multi-antenna multi-user single-target. In the former case, an ADMM approach is used to solve the optimization problem, while in the latter case, weighted MMSE (WMMSE) is employed. In ref. [216], the focus shifts to multi-antenna multi-user single-target downlink scenarios. The DFRC system utilizes fully connected hybrid precoding, and finite-resolution phase shifters are assumed for a more practical representation. Unlike ref. [215], where each user is assigned one subcarrier, ref. [216] allows all users to share all subcarriers, and wideband OFDM is considered. As a result, spatial-frequency response needs to be considered in the presence of path fading, leading to time misalignment between receiving antennas (inability to separate steering vectors). In the context of sensing, the goal is to minimize the spatial spectrum matching error (SSME) between the spatial-frequency response of the DFRC system and the ideal spatial-frequency response. For communication, the spectral efficiency (SE) is considered. The optimization problem involves a weighted sum of these two performance metrics. Due to the complexity of the objective function, the combiners at the user end are initially assumed to be pure digital to re-

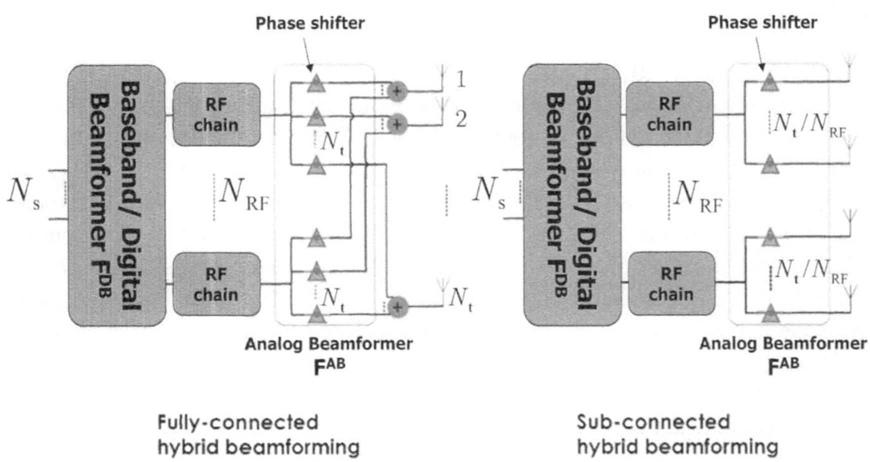

FIGURE 10.18
Illustration of two hybrid precoding architectures.

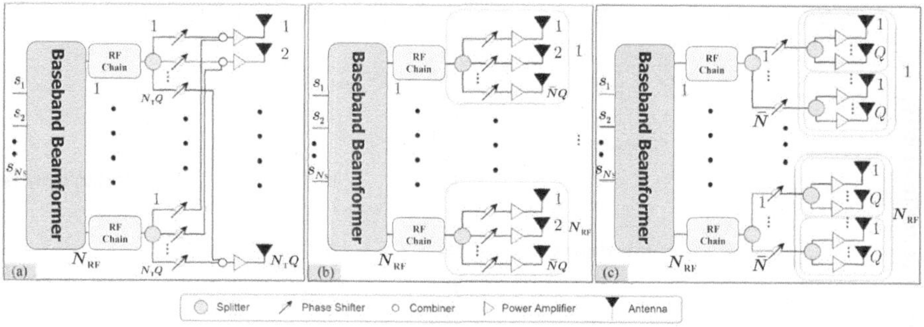

FIGURE 10.19
Illustration of the fully-connected, sub-connected, and GoSA architectures.

move the last constraint. The consensus-ADMM is then utilized to solve the optimization problem. In ref. [217], the consideration is for single-user multi-target downlink scenarios using OFDM in the terahertz frequency band. This scenario demands a large number of transmit antennas to counteract severe path attenuation. The paper considers a two-dimensional rectangular antenna array, applied to short-range vehicular communications. Due to the need for a large antenna array, this work employs a group-of-subarrays (GoSA) that is more cost-effective than sub-connected hybrid precoding, as depicted in the comparison in Figure 10.19. Unlike sub-connected, GoSA further divides subarrays into Q groups, each sharing the same phase shifters, effectively reducing the number of phase shifters by Q times.

Conventional block-level precoding designs (as mentioned in previous chapters) target the second-order statistics of transmitted signals, updating the precoding parameters periodically. Symbol-level precoding, on the other hand, customizes each symbol in every symbol time, involving nonlinear transformations (in contrast to the linear transformations in block-level precoding). Additionally, a key difference lies in handling user interference. Block-level precoding aims to reduce interference from other users by designing the precoder, whereas symbol-level precoding can focus interference energy into the correct decoding region, based on the modulation scheme (which determines the decoding region). The drawback is the high hardware cost, and the algorithms can be challenging to implement, as shown in Figure 10.20. In ref. [218], the focus is on single-antenna multi-user multi-target downlink scenarios. Symbol-level precoding is employed to design the ISAC system. This research considers QPSK modulation, where the green region denotes the correct decoding region. The vector represents the received signal vector. For communication, the goal is to ensure that the received signal vector falls within the correct decoding region. Any deviation beyond the decoding region would be considered an error due to noise interference. As a result, the communication performance metric is defined as the distance from the received signal vector to the boundary of the correct decoding region. For sensing, the instantaneous transmit

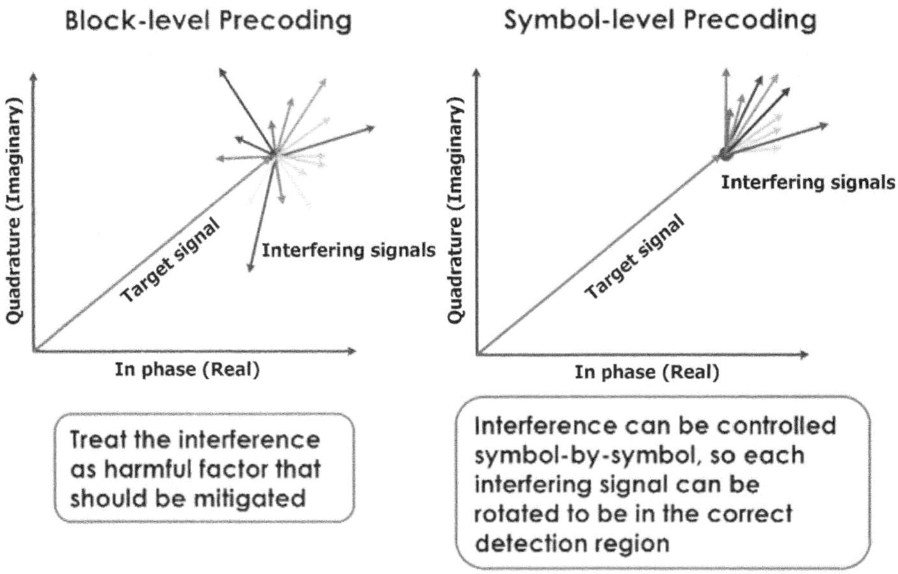

FIGURE 10.20
Illustration of the block-level precoding and symbol-level precoding.

beamforming pattern serves as the performance metric, with the aim to minimize the error between the designed and ideal transmit beamforming patterns for pure sensing.

10.4 Resource and Power Allocation

The primary application scenario of physical layer cooperation is the coexistence of sensing and communication using different hardware devices, also known as CRC in the radar sensing field. Due to the separate devices used for sensing and communication, cooperation between them is particularly important. By designing appropriate resource allocation to reduce the possibility of interference or control the interval of interference for signal processing, the functionality of sensing and communication can be achieved. Depending on the performance requirements, natural resource allocation intervals and components (such as time, frequency, space, and energy) need to be designed, or additional degrees of freedom can be achieved through encoding for artificial orthogonal resource allocation. In the design process, various interference scenarios of signal models for sensing and communication need to be considered in different application contexts. The resource allocation can be categorized into five types: time resource allocation, frequency resource allocation,

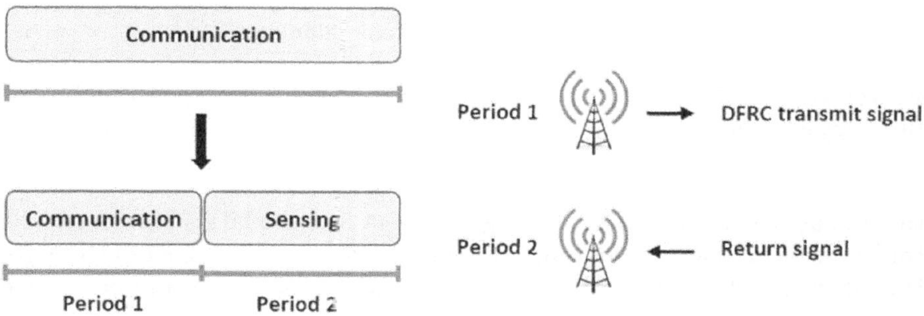

FIGURE 10.21
Illustration of the time resource allocation.

space resource allocation, encoding resource allocation, and energy resource allocation, depending on the specific resources to be allocated. In a broader sense, resource allocation problems are not necessarily limited to CRC systems where sensing and communication use different hardware. The main distinction is that the same resource does not possess both sensing and communication functionalities. However, in the context of this project, the description of physical layer cooperation particularly emphasizes resource allocation on different hardware. The following will introduce the current developments in research in this area.

Time resource allocation: For time resource allocation, the resources are divided into different time slots, each dedicated to either sensing or communication, as shown in Figure 10.21. Intervals between sensing and communication time slots are required to avoid interference. This approach has the advantage of being operationally straightforward and easily implementable in existing systems, leading to its wider commercial application. There have been various improvements designed for current standards, as well as related research on automotive environments. In time resource allocation, due to the lack of a unified performance evaluation metric for sensing and communication, the design can be based on the practical requirements of both systems to determine the duration of resource allocation for each. Alternatively, sensing-assisted communication or communication-assisted sensing can be used to unify the performance evaluation metrics of the two systems. Here is a simple example: Research concept: By allocating more time to sensing, the accuracy of channel estimation in communication can be improved, but it also reduces communication transmission time. Therefore, it is necessary to allocate sensing time appropriately. By obtaining multipath channels through sensing and utilizing the results with maximum ratio combining (MRC), the communication performance can be enhanced. The duration of sensing affects the accuracy of multipath sensing, while the duration of communication affects the transmission capacity. By combining both aspects, a new metric can be defined, such as the actual overall achievable transmission rate [219]. The indicator considers

FIGURE 10.22
Illustration of utilizing beamforming training period for sensing purposes.

the probability of correctly detecting an object and the false alarm rate. The transmission rate after adopting MRC in the two scenarios mentioned above is analyzed, and its expected value is taken as the actual overall achievable transmission rate to measure the time resource allocation. This is applicable to scenarios involving sensor-communication cooperation systems, such as automotive and unmanned vehicle communication. The simulated scenarios in the paper are relatively simple, focusing only on evaluating the channel environment of dual-site radar sensing and a single obstacle. However, it can be extended to more complex scenarios, albeit with increased computational complexity and time resource consumption. Additionally, this study does not consider the impact of interval spacing.

Time resource allocation is relatively straightforward in implementation. Another mainstream research direction focuses on studying existing communication standards and attempts to replace training periods [220], such as beamforming training periods in 802.11ad, with sensing periods, as shown in Figure 10.22. In 802.11ad, the beamforming training phase consists of five stages: Initiator Sector Sweep (ISS), Responder Sector Sweep (RSS), Sector Sweep Feedback (SSF), and Sector Sweep Acknowledgement (SSA). These stages involve transmitting training signals to achieve beam alignment between the base station and users. In research, one can explore the use of these five time periods' signals for radar sensing. In practice, time resource allocation is commonly used in automotive systems. In the application of vehicular networks, time periods can be divided into sensing periods and communication periods for communication with roadside units (RSUs) or other vehicles [221]. The current communication scenario requires the adoption of two different modes: communication with vehicles and communication with RSUs, based on the M/M/1N queuing theory model. In the above description, we can observe that research on time resource allocation mainly focuses on the design of existing standards or practical applications. This is because time resource allocation is relatively straightforward for hardware design.

Frequency resource allocation: Frequency resource allocation is similar to time allocation and is easy to implement. It requires frequency band spacing to reduce interference between sensing and communication bands. The mainstream research in frequency resource allocation is based on the OFDM waveform, which provides flexibility to adopt various waveforms compared to time resource allocation. Specifically, depending on different communication channel conditions, required sensing and communication performance, and transmit power budget, we can design the allocation of OFDM subcarriers to achieve frequency resource allocation. Frequency resource allocation for non-OFDM waveforms is less common in research due to the scarcity of spectrum resources. The methods that involve frequency allocation with band spacing are rarely mentioned, although a few studies have addressed it. In terms of design, it is necessary to establish a unified indicator. In the referenced paper [184], a transmission capacity specific to radar sensing is defined and integrated with communication transmission capacity. The definition of radar sensing transmission capacity considers the radar sensing resources in terms of angle, time, and frequency as communication bits, and transmission capacity is defined accordingly. In the design process, the required frequency band allocation is determined by optimizing the total transmission capacity of sensing and communication. Since the mathematical form is similar to communication transmission capacity, optimization methods for communication capacity can be applied. In terms of practicality, this approach is less practical due to the lack of consideration for band spacing. Additionally, the definition of sensing capacity has not been widely recognized in academia. Therefore, this research brings more of a conceptual inspiration. Regarding the use of OFDM waveforms, there is more research available. It involves allocating subcarrier resources to sensing or communication for frequency resource allocation. Adjusting the energy on subcarriers according to individual channel conditions is also possible. Due to the waveform characteristics of OFDM, there is no need to consider band spacing, and it provides more flexibility. As an example, the communication signal can be treated as either interference or useful information for sensing, and a mathematical model can be formulated using mutual information as the evaluation metric for radar sensing. Carrier allocation and energy distribution can be designed based on this. It can be observed that the above radar sensing mutual information expressions are similar to traditional communication capacity, and this mathematical form is valid only when the channel model follows a Gaussian distribution and the waveform is OFDM. Although the applicable model is relatively ideal, optimizing based on radar sensing mutual information provides a good direction.

Spatial resource allocation: Spatial resources refer to the angular diversity brought by multiple antennas. In communication, it can provide antenna diversity gain and offer more degrees of freedom to achieve multiplexing or enhance signal stability. In terms of sensing, multiple antenna technology can be used to perform beamforming to enhance sensing performance. With the assistance of multiple antennas, in addition to estimating distance and velocity, radar sensing can also detect angle parameters, thereby obtaining three-dimensional spatial information. For the fusion of sensing and communication technologies, spatial resource allocation refers to the allocation of antenna resources. The antennas are divided into sensing antennas and communication antennas, and the quantity and distribution of antennas can be determined according to the respective requirements of sensing and communication. Common indicators include angular resolution and beamforming gain. In terms of technology, to avoid interference between sensing and communication, it is necessary to design separate antennas for beamforming, so that

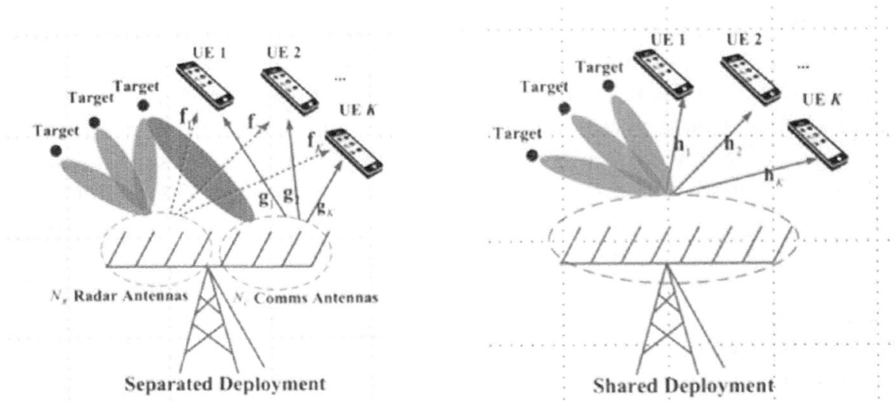

FIGURE 10.23
Illustration of spatial resource allocation.

the signals of the two can point in different directions. Alternatively, in signal space, they can be projected onto each other's null space to avoid interference generation [222]. Of course, in practice, antenna resources can also be left unallocated, and the functionality of sensing and communication can be achieved by designing beam patterns to point in multiple directions, as shown in Figure 10.23. However, this belongs to an example of integration at the physical layer and is not a simple cooperation problem.

Coding resource allocation: In coding resource allocation, unlike the natural resources of time, frequency, and space, coding resources increase the additional degrees of freedom for signal transmission through the design of artificial orthogonal codes. The orthogonal properties of coding are used to ensure that sensing and communication signals do not interfere with each other. However, the drawback is that additional computational complexity or hardware architecture is required to implement coding, resulting in extra costs. Therefore, coding is more commonly used in extreme scenarios where the SNR is excessively high and the allocation of natural resources cannot meet the performance requirements of sensing and communication. In such cases, coding can be used to enhance performance. Adding a switch control to adopt coding to solve the performance degradation caused by excessive noise in low SNR scenarios. In terms of hardware design, the introduction of a switch control in ref. [223] increases the cost of hardware, and the adoption of coding also increases computational complexity. Therefore, there is relatively less research related to coding resource allocation.

Energy resource allocation: In energy resource allocation, due to the lack of orthogonality in energy, although it is possible to achieve non-orthogonal multiple access (NOMA) multiplexing through the allocation of energy levels, there are significant limitations in its practical application. Complex scenarios result in strong fading, leading to poor performance of

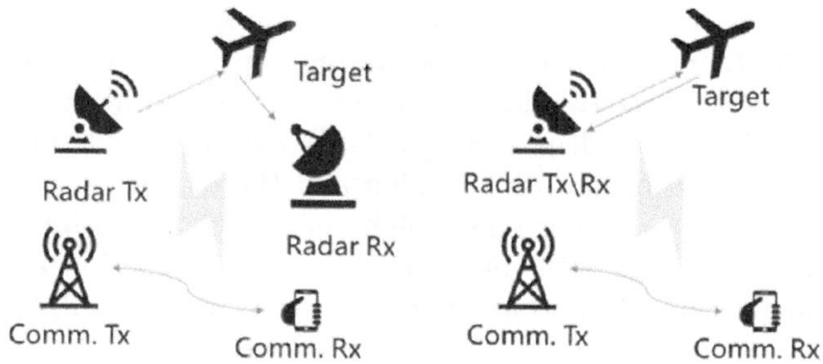

FIGURE 10.24
Illustration of multi-location radar sensing and single-location radar.

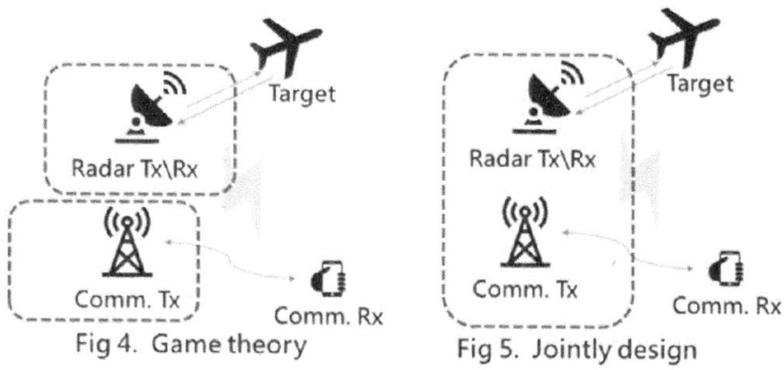

FIGURE 10.25
Illustration of game theory-based solution and joint design.

NOMA. Since this chapter focuses on the analysis of physical-layer collaboration, NOMA technology is mostly used for physical-layer integration. Therefore, energy allocation here refers to the allocation of resources among different hardware and the handling of mutual interference between different hardware. Under different Cooperative Radar Communication (CRC) systems, the positioning of radar sensing can be categorized as multi-location radar sensing or single-location radar sensing. These two have distinct mathematical differences in the signal model, as their received steering vectors are not the same, as shown in Figure 10.24.

In addition, due to different environments and the distance of the detected objects, the system can be classified into complex surrounding environments (such as urban or indoor) that require clutter consideration, models with secondary reflection interference when the target object is close to the communication user, or models with only line-of-sight interference when the distance between the two is far apart. Regarding the solution for resource allocation, it can be divided into game theory-based solutions for resource competition and centralized joint design for optimization, as shown in Figure 10.25. In the joint design approach, it is only necessary to define the mathematical model based on the environment and use optimization algorithms to compute the solution. In the game theory approach, since the systems cannot communicate with each other, each participant can only optimize their own utility to achieve Nash equilibrium. In related research, energy allocation under orthogonal frequency division multiplexing (OFDM) waveform is determined based on comparing the performance of sensing and the channel capacity of communication [184]. Considering line-of-sight interference, clutter interference, and secondary reflection path interference, energy allocation is determined based on channel capacity and sensing signal-to-noise ratio [224–226]. Game theory is employed to solve energy allocation problems in non-cooperative systems [227, 228]. Energy allocation is designed based on energy utilization efficiency and low interpretability [229].

Part IV

Learning-Based Enhancements

11

Machine Learning Techniques

11.1 Brief History of Deep Learning

The idea of deep learning (DL), especially neural networks, was first proposed in 1943 by McCulloch and Pitts [230], trying to utilize mathematical models to mimic the nervous activity of human beings. In 1958, the first neural network with automatic learning capability (i.e., perceptrons) was proposed by Rosenblatt, adjusting trainable parameters via gradient descent algorithm in the neural network to perform binary classification tasks [231]. This inspires the revolution in research of the first artificial intelligence wave for years to come, till the first AI winter. In 1969, Marvin Minsky and Seymour Papert proved in their book [232] that a single-layer perceptron model can only solve linear classification problems and cannot solve simple XOR and XNOR problems, triggering the first winter of neural network research. In this period, research activities were still performed, developing the ideas of some famous and fundamental neural network architectures, such as the convolutional neural network architecture (CNN) and the recurrent neural network (RNN). Complex nonlinear problems were first addressed by Geoffrey Hinton, Rumelhart, and Williams's work [233] in 1986 and thus rebuilt the confidence of the research community toward this direction. In this work, the famous backpropagation algorithm (BP) is first utilized in the training process of multi-layer perceptrons (MLPs), which constantly updates the network parameters to minimize network loss based on the chain rule method. This work triggers the second wave of artificial intelligence research to focus on the development of shallow learning. In this period, the architectures of CNN and RNN were completed and are still employed nowadays. LeCun proposed the biologically inspired CNN model based on the BP algorithm in 1989, establishing the foundation of DL for modern computer vision. Moreover, Cybenko proved that a single hidden layer perceptron containing a finite number of neurons can approximate any continuous function [234] in 1989, greatly stimulating the imagination of the power of artificial intelligence (AI). Specifically, with the increasing number of hidden layers, the fitting capability of neural networks is also increasing significantly. As a result, complex and nonlinear problems can be addressed. However, with the compelling number of hidden layers, the error propagation process triggered by the BP algorithm becomes

DOI: 10.1201/9781003408239-11

malfunctioning and consequently leads to extremely slow convergence of the training of neural networks. This so-called gradient vanishing problem was pointed out by Hochreiter's work in 1991 [235]. Constrained by the limited computational capacity back in 1991, it was impossible to support the training of deep neural networks efficiently. Moreover, other machine learning models with statistical backgrounds, such as support vector machine (SVM) [236], were also developed in the middle of the 1990s. From then on, the research of neural networks suffered a second major roadblock for almost two decades.

Finally, in the last two decades, the previous challenges of the training of neural networks have started to be solved along with the fast advances in terms of computation hardware, big data, and theoretical backgrounds. Thanks to those advances, the power of DL is revealed and becomes non-negligible by providing state-of-the-art solutions to various information and communication technologies (ICT) domains. Specifically, In 2006, the gradient vanishing problem was first overcome by Hinton's work [237]. In this work, a layer-by-layer training procedure (i.e., deep belief networks) is employed, training and fixing a target layer to train each layer of a deep neural network sequentially to avoid the gradient vanishing problem. In 2008, Andrew NG's group at Stanford started advocating for the use of GPUs for training DNNs to speed up the training time by many folds [238]. Starting from here, training neural networks on huge volumes of data becomes possible to pave the way for efficient and practical DL. In 2009, Fei-Fei Li, a professor at Stanford, launched ImageNet, which is a database of 14 million labeled images [239]. It would serve as a benchmark for the DL researchers who would participate in ImageNet competitions (ILSVRC) every year and greatly advocate the dominant position of neural networks in the computer vision research domain. In 2011, Bengio developed a novel but simple way to prevent the vanishing gradient problem by adjusting the nonlinear function inside the neural network (i.e., activation function) [240]. This means that now, apart from heavy GPU uses when employing deep belief networks to tackle gradient vanishing problems, DL community has another tool to avoid issues of longer and impractical training times of deep neural networks and facilitate the end-to-end training of neural networks instead of layer-by-layer training. In 2012, AlexNet, a GPU-implemented CNN model designed by Alex Krizhevsky [241], won Imagenet's image classification contest with an accuracy of 84%. It is a huge jump over 75% accuracy that earlier models had achieved. This win triggers a new DL boom globally. Being the earliest ancestor of generative AI, generative adversarial network, also known as GAN was created by Ian Goodfellow in 2014 [68]. GANs open a whole new door to the applications of DL, such as medical imaging and highly convincing image/video generation thanks to their ability to synthesize real-like data. In 2016, DeepMind's deep reinforcement learning (RL) model beat a human champion in the complex game of Go [242]. The game is much more complex than chess, so this feat captures the imagination of everyone and take the promise of DL to a whole new level. In 2019, Yoshua Bengio, Geoffrey Hinton, and Yann LeCun won the Turing

Award 2018 for their immense contribution to advancements in the area of DL and AI. This is a defining moment for those who had worked relentlessly on neural networks when the entire machine learning community had moved away from it in the 1970s. In 2022, OpenAI released ChatGPT, which is an AI chatbot with human-like dialogue capability and showed the power of neural networks in the natural language processing domain using both supervised and RL techniques. Most recently, John Hopfield and Geoffrey Hinton won the Nobel Prize in physics for their contributions to the development of artificial neural networks.[1] Given its success in computer vision (CV), natural language processing (NLP), and increasingly more emerging research areas, the ability of DL to extract features automatically has motivated researchers to employ DL technology on different research problems.

In light of this direction, DL technology has been introduced into signal processing and communication research areas, showing high performance and low complexity when compared to conventional optimization-based algorithms. In fact, in the recent specification Release 18 defined by 3rd Generation Partnership Project (3GPP), three critical use cases, including channel state information (CSI) feedback, beam management, and positioning, are selected to investigate the improvements brought by DL-aided designs. Besides that, one can also notice that DL-based solutions have started to play important roles in the automotive sensing research area due to its provided robustness against imperfections. In conclusion, we anticipate that increasingly more communication and sensing system designs can be further enhanced with the power of DL, being an important motivation for introducing this exciting and novel research area to readers in this book. In the remaining sections of this chapter, we aim to introduce the basics of current neural networks. Then we will focus on the benefits and design concepts of introducing DL into wireless communication, sensing, and eventually, integrated sensing and communication research areas in the following chapters in this book.

11.2 Training Methods of Deep Learning Models

In the previous section, one can notice that the development of DL is inspired by the biological activity of human beings, trying to mimic the amazing capability of human brains to process information mathematically. Specifically, given input from the environment, human brains can process the input and provide appropriate output to respond to the environment, being the main target to be mimicked by DL models. For example, when a driver sees a stop sign (i.e., input image) on the road, the driver will hit the brake (i.e., output)

[1]For a more detailed description of the history and development of deep learning, readers can refer to the chronological review article [243].

to slow down the vehicle gradually, thanks to the brain's capability to process the input and produce appropriate output in a very short time. Moreover, even though there are some variations when seeing different stop signs in different conditions, such as the lightness or the stains on the signs, we are still able to recognize those stop signs without any hesitation. Specifically, this is the major advantage provided by DL-based solutions since it is almost impossible to overcome this issue if using traditional rule-based solutions. Thus, in the research of DL models, our main goal is to replace the above procedures using mathematical models so that different applications can be executed intelligently and automatically without the assistance or supervision of human beings. Although the final goal is to build general-purpose DL models (i.e., artificial general intelligence (AGI)) to fully duplicate human brains' capability, current AGI research is still in its infant stage. As a result, we will mainly focus on single-propose DL models to introduce how to build a DL model to perform a specific task after appropriate training in this section.

Given a predefined input $\mathbf{x}^{(i)}$ as ith sample with the dimension of $N_{\mathrm{I}} \times 1$ in the training set[2], our goal is to build a mathematical function f_{DL} to process the input and generate an appropriate output $\hat{\mathbf{y}}^{(i)}$, which is close to $\mathbf{y}^{(i)}$ and both as $N_{\mathrm{O}} \times 1$ vectors. Hence, the inferring procedure can be expressed as:

$$\hat{\mathbf{y}}^{(i)} = f_{\mathrm{DL}}(\mathbf{x}^{(i)}; \boldsymbol{\Theta}), \tag{11.1}$$

where $\boldsymbol{\Theta}$ is the trainable weight of the DL model f_{DL}. Using Eq. (11.1), various tasks can be defined and learned through the appropriate training of DL models. For example, in the computer vision domain, image classification tasks can be performed by setting input $\mathbf{x}$ as an image and output $\mathbf{y}$ as the desired image class; in the natural language processing domain, language translation can also be executed by setting input $\mathbf{x}$ as a Chinese sentence and output $\mathbf{y}$ as the corresponding English sentence. Moreover, in the wireless communication and sensing domain, beamforming vector design can also be aided by DL models for instance, by setting input $\mathbf{x}$ as a channel state information and output $\mathbf{y}$ as the desired beamforming weight to maximize the quality of service from the user perspective. After defining the input/output relationship for a desired task, the next step to train a DL model is to employ an optimization framework to obtain the optimal trainable weight $\boldsymbol{\Theta}^*$ in an iterative manner (i.e., offline training phase). After that, the optimal weight will be fixed and the DL model will be ready to be used (i.e., online inferring phase) for generating output based on unseen new samples. Based on the availability of the desired output $\mathbf{y}$, different learning methods, including supervised learning, unsupervised learning, reinforcement learning, and generative learning, can be utilized to construct the optimization framework for training purposes. We provide a brief introduction to each learning method in this section.

Supervised learning: Being the most naive training method, supervised learning can be adopted when the corresponding output $\mathbf{y}$ to each input $\mathbf{x}$

[2]Two-dimensional matrices or multidimensional tensors can be vectorized in advance to follow the form in this section.

is available in the training phase. In light of this direction, assuming that there are N_{sample} samples in the training phase,[3] the following optimization framework can be employed to train a DL model in a supervised learning manner:

$$\mathbf{\Theta}^* = \arg\min_{\mathbf{\Theta}} \sum_{i=1}^{N_{\text{sample}}} \text{Loss}\{\hat{\mathbf{y}}^{(i)} = f_{\text{DL}}(\mathbf{x}^{(i)}; \mathbf{\Theta}), \mathbf{y}^{(i)}\}, \tag{11.2}$$

where $\text{Loss}\{\cdot\}$ is a pre-defined loss function used to measure the similarity inversely between $\hat{\mathbf{y}}^{(i)}$ and $\mathbf{y}^{(i)}$ so that a similar pair of $\{\hat{\mathbf{y}}^{(i)}, \mathbf{y}^{(i)}\}$ will result in lower value and vice versa. As a result, by minimizing Eq. (11.2), we are actually searching for an optimal weight $\mathbf{\Theta}^*$, which can maximize the similarity of the produced output $\hat{\mathbf{y}}$ and ground truth $\mathbf{y}$. Based on different types of applications, different loss functions can be employed or designed to better evaluate the behavior of DL models and provide better supervision in the training process. For example, mean square error (MSE) is a commonly used loss function to supervise regression tasks of DL models, defined as:

$$\text{MSE}\{\hat{\mathbf{y}}^{(i)} = f_{\text{DL}}(\mathbf{x}^{(i)}; \mathbf{\Theta}), \mathbf{y}^{(i)}\} = ||\mathbf{y}^{(i)} - \hat{\mathbf{y}}^{(i)}||^2, \tag{11.3}$$

where i is the index of samples in the training set $i \in \{1, \cdots, N_{\text{sample}}\}$. On the other hand, multi-class cross entropy loss is often employed to supervise multi-class classification tasks of DL models, defined as:

$$\text{CE}\{\hat{\mathbf{y}}^{(i)} = f_{\text{DL}}(\mathbf{x}^{(i)}; \mathbf{\Theta}), \mathbf{y}^{(i)}\} = -\sum_{j=1}^{N_O} \mathbf{y}_j^{(i)} \log \hat{\mathbf{y}}_j^{(i)}, \tag{11.4}$$

where $\mathbf{y}_j^{(i)}$ represents the jth element of ith sample in the training set. In this case, $\mathbf{y}^{(i)}$ is defined as $\mathbf{y}^{(i)} \in \{1, 0\}^{N_O}$ using 1 to represent the corresponding input is belonging to a specific class and vice versa. When $\mathbf{y}_j^{(i)} = 1$ exists, the prediction $\mathbf{y}_j^{(i)} = 1$ will result in the loss as 0. On the other hand, the prediction $\mathbf{y}_j^{(i)} = 0$ will lead to the loss as ∞. By doing so, the multi-class cross entropy loss function can be used to supervise the training of the DL models to provide a desired output when facing multi-class classification problems. Moreover, the output generated in this way has the physical meaning of probability, which further enables more downstream applications in communication or sensing research areas, such as soft decision using log-likelihood ratio (LLR) [245,246]. After finishing the loss function design, gradient-based optimization algorithms can be used to solve Eq. (11.2) to obtain the optimal trainable weight, that is,

$$\mathbf{\Theta} \leftarrow \mathbf{\Theta} - \epsilon \cdot \sum_{i=1}^{N_{\text{sample}}} \frac{\partial \text{Loss}\{\hat{\mathbf{y}}^{(i)} = f_{\text{DL}}(\mathbf{x}^{(i)}; \mathbf{\Theta}), \mathbf{y}^{(i)}\}}{\partial \mathbf{\Theta}}, \tag{11.5}$$

[3] In a standard DL training procedure, all available data in the training phase should be partitioned into three mutually exclusive sets: training set, validation set, and testing set. For the details regarding this training setup, readers can refer to the book [244]. Here, we assume N_{sample} as the training set size to conduct our discussion.

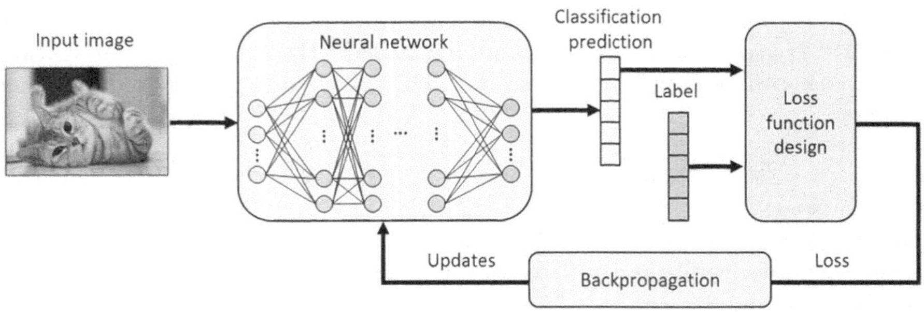

FIGURE 11.1
Illustration of the supervised learning flow diagram.

where ϵ is the predefined learning rate. Considering the complex architecture of the current DL models, a mature training algorithm, the backpropagation algorithm, utilizes the concept of the chain rule to compute gradient information of all trainable weights efficiently, working with gradient-based optimization algorithms to finish the training procedure in an iterative manner. The whole procedure is illustrated in Figure 11.1. In the figure, using image classification as an example, a neural network is trained to perform decision-making to tell which object is mainly presented in the input figure. Thus, the classification prediction result is a vector describing the decision probability of each predefined category. This prediction will then be compared with the corresponding label to generate a loss value. Finally, the backpropagation method is utilized to tell how each trainable weight should be adjusted to yield to improved loss value by means of gradient descent optimizations to finish the training on a single sample.

In literature, supervised learning has a lot of applications due to its easy implementation. Moreover, it is often considered a useful benchmark to evaluate the potential and practicality of adopting DL solutions to tackle a specific problem. Specifically, as long as the ground truth can be provided, various applications can be considered to build a forecasting model to perform regression or classification tasks. For example, a regression model can be developed to forecast house prices given property information, such as location and historical price. Also, a classification model can also be built to perform automatic image classification given the original image. In the direction of wireless communication and sensing, supervised learning also can be employed to aid system designs for better performance or lower complexity. For instance, using a well-known weighted minimum mean square error (WMMSE) beamforming algorithm to generate beamforming weight corresponding to channel matrixes in the training set, a DL model can be built as a beamformer to mimic the behavior of WMMSE beamforming algorithm in a low-complexity manner [247]. Moreover, given the log information of communication or radar systems, a DL classification model can also be trained to perform anomaly detection by extracting error patterns in log information automatically [248].

Unsupervised learning: Although supervised learning can already be employed to support several exciting applications in various research domains, the availability of the required huge amount of labels limits the broader usage of supervised learning. In fact, labeling is a time-consuming and expensive procedure since high-quality data with precise labels takes effort to collect. Moreover, human power is still required heavily in nowadays labelling process. Alternatively, unsupervised learning is developed to train DL models without the need for labels.[4] As a result, instead of mimicking the behavior of the model algorithm or human behaviors, unsupervised learning can be understood as an automatic optimizer, minimizing the predefined loss function by finding the optimal trainable weights automatically. However, to avoid the labeling effort and to enjoy the benefit of unsupervised learning, the loss function design of unsupervised learning is not trivial. Specifically, unlike supervised learning can utilize the similarity between DL model predictions and ground truth to train the DL models. Without labels, the goal of unsupervised learning is to design an objective function to evaluate a DL model's output directly. Moreover, to allow the normal function of the backpropagation algorithm, the design of the above objective function should be differentiable with the perspective of trainable weights so that all trainable weights can be adjusted based on gradient information to minimize the objective function gradually, which is the major difficulty in designing such functions.

Given the challenging nature of the unsupervised learning objective function designs, the volume of unsupervised learning research is much less than supervised learning research. However, there are still some ingenious designs to enable unsupervised learning to be employed to perform desired tasks and to enjoy the provided benefits. For example, in the computer vision research domain, super-resolution is a classic task, where DL models are expected to generate high-resolution images based on given low-resolution images. While supervised learning is straightforward to be adopted to do so by training a DL model to learn the input/output mapping of low-resolution/high-resolution pairs, ref. [251] develops an unsupervised learning framework to allow super-resolution training utilizing unpaired low-resolution and high-resolution images. Moreover, in the wireless communication and sensing research domain, utilizing the beamforming design problem as an example, ref. [252] designs an unsupervised learning framework to allow direct beamforming designs without labels generated by other reference algorithms, as shown in Figure 11.2. As a result, the proposed DL model can learn to directly maximize sum-rate (i.e., the customized loss function in this case) by generating a beamforming

[4] There are also some variants of unsupervised learning, such as semi-supervised learning and self-supervised learning. Semi-supervised learning stands between supervised learning and unsupervised learning, where only a subset of labels can be utilized in the training process. Self-supervised learning designs tasks that can process unlabeled data to obtain useful representations that can help with downstream learning tasks. For a more detailed description of those novel learning categories, readers can refer to refs. [249, 250].

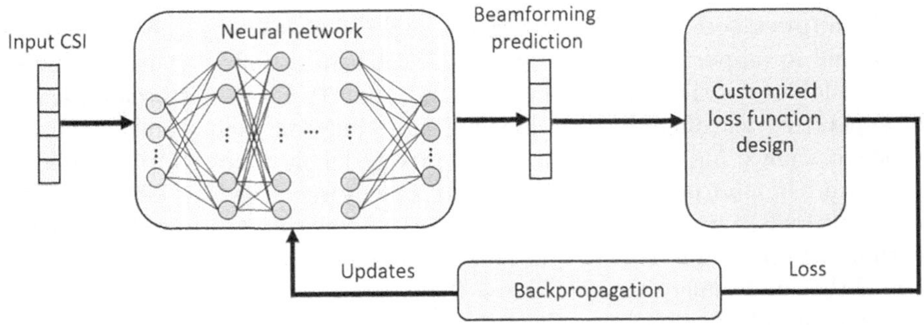

FIGURE 11.2
Illustration of the unsupervised learning flow diagram.

matrix for a given channel matrix. In this case, the derivative of the sum-rate with the perspective of trainable weights ensures the normal function of the backpropagation algorithm, so that the whole learning framework, similar to supervised learning, can still be performed in the training process. Note that while the performance of supervised learning-based beamforming algorithms is naturally bounded by the reference algorithm, unsupervised learning-based beamforming algorithms can often provide superior performance by directly optimizing the sum-rate instead of mimicking other reference algorithms.

Reinforcement learning: Following the logic of unsupervised learning, in the case that we aim to skip the labeling effort, we will be motivated to design an objective function satisfying the differentiable property to adopt the unsupervised learning framework. However, designing such objective functions can be even very challenging. For example, in Go games, given the current status of the Go board as an image input, a classification problem can be formed to output the next action, but how to evaluate the goodness of each action using mathematical methods is still unknown. Due to this difficulty, it is impossible to train mature DL models to play Go games intelligently instead of just memorizing Go records from master players using either supervised learning or unsupervised learning. Moreover, in the Go game, evaluating the goodness of every single action is not making sense since some strategic moves might take a series of actions to see the consequences instead of making efforts in every single action, also being the pain point to utilize supervised learning or unsupervised learning since model input and output should be defined and fixed in advance and thus cannot handle the design of the aforementioned strategic moves with arbitrary length. To solve the above two issues, RL is developed to train DL models to perform complex and strategic tasks, such as Go games, human-like chatbots, and video games.

Similar to the idea of other DL models, RL also tries to find a function to act as the brain of the RL algorithms so that the model can take appropriate action (i.e., output) to respond to environment input (i.e., input). We use a fundamental but classic reinforcement algorithm, Q-learning, as an example to continue our discussion here. To introduce RL mathematically, we assume $\mathbf{s} \in \mathcal{S}$ is the environment input vector and $\mathbf{a} \in \mathcal{A}$ is an action vector containing all possible N_{action} actions as a $N_{\text{action}} \times 1$ vector where $\mathcal{S}$ and $\mathcal{A}$ represents the state space (i.e., the set of all possible state) and action space (i.e., the set of all possible action). Assuming that we are interested in utilizing Q-learning to solve a 2D maze game, then $\mathbf{s}$ will be the current position of the character, which will be fed into the RL algorithm as an input, and $\mathbf{a}$ will be a 4×1 action vector representing the probability to guide the character in the maze games to move right, left, forward, or backward. Then our goal is to train an action value function, which is defined as:

$$Q(\mathbf{s}, \mathbf{a}) : \mathcal{S} \times \mathcal{A} \to \mathcal{R}, \tag{11.6}$$

where $\mathcal{R}$ is the reward space. In this example, the reward $R \in \mathcal{R}$ can be the product of -1 and the number of expected steps to finish the maze games. The physical meaning of the above action value function is to evaluate the quality (i.e., reward R) of each state–action combination by returning the corresponding reward. By doing so, if a well-trained action value function is available in the online inferring phase, for any given state $\mathbf{s}$, we can always select the optimal action $\mathbf{a}$ to maximize the reward function and thus minimize the expected steps to finish the maze games. One can notice that the reward function is similar to the objective function of supervised learning and the unsupervised learning framework. However, to better evaluate the goodness of a series of actions for the learning of strategic moves, the long-term effects should be considered in the design of the reward function, being the main difference between RL and previous learning frameworks. To explain, for example, suppose that in the 2D maze game, one can design a reward function so that if we successfully arrive at the exit of the maze, we will gain 1 point as an immediate reward, but we lose 1 point to take each step. In this process, not only the final step but also a series of steps before that should be considered as desired behavior and be rewarded. By doing so, the RL algorithm can not also learn to finish the maze games but also be motivated to find the most efficient way to finish the maze games. To advocate this important long-term effect, the updating rule of action value function $Q(\mathbf{s}, \mathbf{a})$ in Q-learning is defined as:

$$Q(\mathbf{s}, \mathbf{a}) \leftarrow Q(\mathbf{s}, \mathbf{a}) + \epsilon[r + \gamma \max_{\mathbf{a}'} Q(\mathbf{s}', \mathbf{a}') - Q(\mathbf{s}, \mathbf{a})], \tag{11.7}$$

where $\gamma < 1$ is a pre-defined discount factor, $\mathbf{s}'$ is the next state after we take action $\mathbf{a}$, and $\mathbf{a}'$ is the next action we take when facing next state $\mathbf{s}'$ using current action value function. In Eq. (11.7), ϵ is the learning rate and the whole term $[r + \gamma \max_{\mathbf{a}'} Q(\mathbf{s}', \mathbf{a}') - Q(\mathbf{s}, \mathbf{a})]$ is the error term of the previous prediction, which will be utilized to update the current action value function. In this error term construction, besides the immediate reward r, the future expected reward $\gamma \max_{\mathbf{a}'} Q(\mathbf{s}', \mathbf{a}')$ is also considered to realize the evaluation of a series of moves. To explain, after we take action $\mathbf{a}$ as the response of current state $\mathbf{s}$, we will enter the next frame and face $\mathbf{s}'$ as the new state.

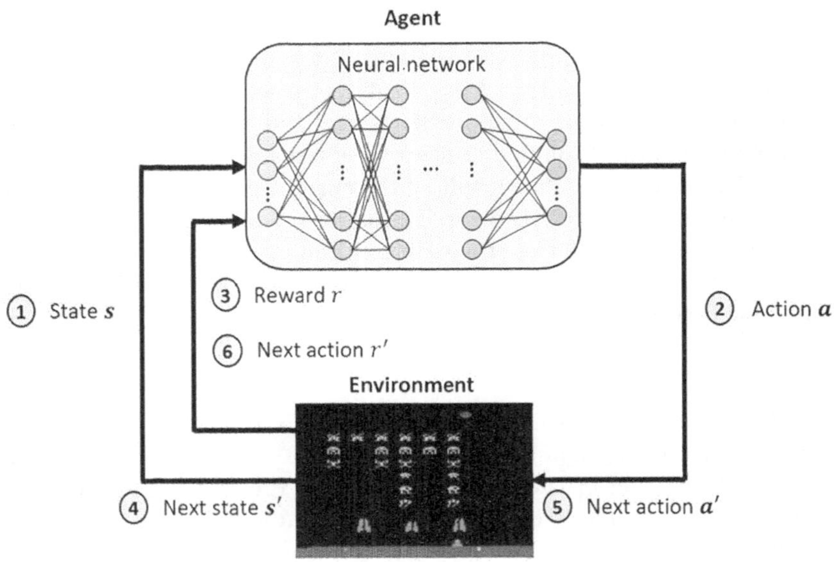

FIGURE 11.3
Illustration of the reinforcement learning flow diagram.

The term $\max_{\mathbf{a}'} Q(\mathbf{s}', \mathbf{a}')$ represents the best outcome we can obtain after entering $\mathbf{s}'$ by choosing optimal action $\mathbf{a}'$ according to the action value function recommendation. One can also notice that the structure of this updating rule is quite similar to the gradient-based updating rule in Eq. (11.5), the only difference is that Eq. (11.5) utilizes the similarity between prediction and ground truth to obtain error term and Eq. (11.7) models the error term as the difference between the maximum future reward and current predicted reward. The whole training procedure and how the agent interacts with the underlying environment are also illustrated in Figure 11.3. In the figure, in the first timeslot, a state $\mathbf{s}$ is given by the environment so that the agent can output the decision $\mathbf{a}$ based on the given state. After the action $\mathbf{a}$ is executed and fed into the environment, the immediate reward r is generated so that the actual reward R can also be obtained for RL training purposes. Then the next scene is provided to the agent and the above procedure starts over again until the termination of the game.

As shown in Figure 11.4, if we consider the naive reward function design (i.e., only arriving at the final destination yields 1 point), we assume that we are in a position so that the final two steps to solve the 2D maze game from the current position are to move forward and then turn left. In this case, although walking forward to move a step closer to the destination will not get positive reward immediately, the term $\max_{\mathbf{a}'} Q(\mathbf{s}', \mathbf{a}') = 1$ still indicates this decision is the most efficient decision for now since the maze game can be solved in the next step. In other words, in the considered maze game, as

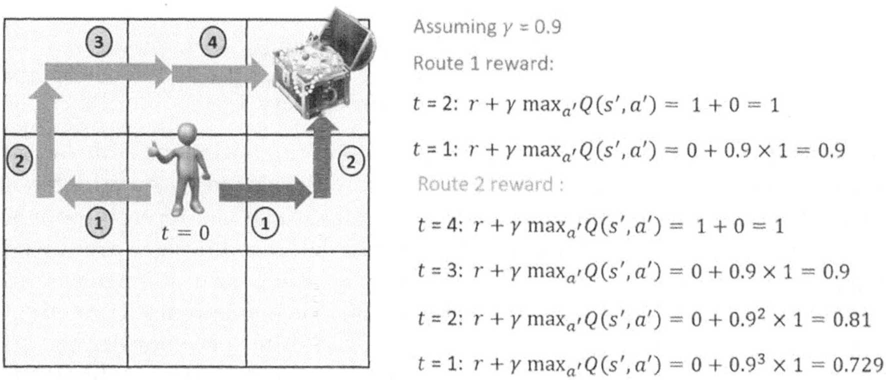

FIGURE 11.4
Illustration of the reward mechanism in reinforcement learning training procedure.

shown in Figure 11.4, where we present a numerical example, although only arriving at the final destination will result in a positive reward, spending more steps turning around is not encouraged owing to the design of the discount factor. Using this way, the agent will be guided to find the most efficient way to the final destination to maximize the total expected reward instead of only considering the immediate reward. To further accelerate the convergence, one can further consider adding the penalty term when taking each step as aforementioned, but solely considering the long-term reward can also lead to convergence thanks to the discount factor design. In the above discussion, one can notice that the only way we can update the reward when the agent chooses a specific action when facing a specific scene is through the multiple times of trial and error procedures. This is actually the main idea of the RL algorithm: the exploration and exploitation mechanism. In fact, in the training procedure of current RL algorithms, the so-called adaptive $\epsilon-$greedy algorithm is also employed to advocate the exploration behavior in the early training stage and emphasize the exploitation behavior in the mature training stage. To explain with the aforementioned Q-learning algorithm, at the early training stage, the Q-table records almost random values since almost no iterations are conducted to update the action value table. As shown in Eq. (11.8), instead of following the recommendation of the Q-table, there is ϵ probability that the agent will perform a random action, exploring different strategies in the early training stage. As the training epoch goes on, ϵ will be set to decrease accordingly. As a result, in the later training stage, the agent will have a higher probability of following the Q-table recommendation since the Q-table has been updated and is close to its convergence to offer mature recommendations in this stage. Following this logic, RL can also be understood as an efficient search method to test different strategies and record the optimal one with the highest expected reward without labeling.

$$\mathbf{a} = \begin{cases} \arg\max_{\mathbf{a}} Q(\mathbf{s}, \mathbf{a}), & \text{with probability } \varepsilon \\ \text{random action}, & \text{with probability } 1 - \varepsilon. \end{cases} \tag{11.8}$$

Depending on the problem setting, such as discrete/continuous state/action space, a number of different RL algorithms are developed to serve different scenarios. Moreover, although the fundamental idea of the recently popular large language model (LLM) is still RL, several advanced designs are further employed to supervise the RL agent to tackle complex natural language processing tasks. For a more detailed survey of current mainstream RL methodologies, interested readers can refer to refs. [253–258] for more details. Although different algorithms introduce special features to further enhance the learning capability of RL algorithms in different scenarios, the basic ideas, such as the exploration and exploitation concept and updating rule, are still utilized as the core of different RL algorithms.

Generative adversarial learning: While most regression or classification tasks can be tackled by the above training methods, a special type of task, structure learning, cannot be handled by the previously introduced learning framework. Hence, generative adversarial learning is developed to serve those special needs. Specifically, in terms of structure learning, we wish DL models to generate various outputs following a special structure instead of having a standard answer. For example, given a prompt sentence, we aim to ask a DL model to generate a highly convincing photo containing the theme in the prompt sentence instead of other themes. Another example is the algorithmic composition, which is anticipated to generate a specific type of music, such as Jazz or Classical music, according to the given input. In the above two examples, it is hard to either prepare the "ground truth" answer to each input sample or design an appropriate function to evaluate the model output, while human beings can easily tell the quality of the generated photos or music. In fact, given a photo or music, a classification model trained by the aforementioned supervised learning can be used to tell if the theme (e.g., cats) exists in the photo or what type of music (e.g., Jazz) it is. The above phenomenon reveals DL model actually can extract and detect subtle differences, that cannot be formulated mathematically, and utilize those features to finish the classification task. In light of this direction, the development of generative learning comes from an interesting but naive idea: Can we employ another neural network (i.e., discriminator) to supervise the generation procedure of the original network (i.e., generator)? By doing so, even without an elegant way to directly evaluate the quality of the generated product through loss functions (i.e., supervised learning or unsupervised learning), indirect supervision can still be utilized via the feedback of the discriminator since the discriminator has the ability to perform classification and provide the probability to the desire question. The flow diagram of generative adversarial learning using the human face image generation problem as an example is illustrated in Figure 11.5. To give generative adversarial learning the ability to address the desired

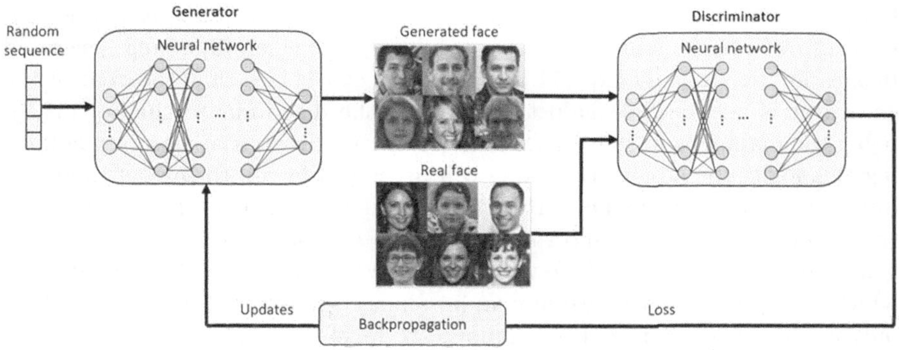

FIGURE 11.5
Illustration of the generative adversarial learning flow diagram.

structure learning problem, in the figure, two neural networks, a generator and a discriminator, are built in the proposed framework. By feeding a random sequence into the generator, the generator will work as an image generator in this example, producing various human face images. Furthermore, to empower the generator to produce highly convincing human face images, a discriminator is trained as a binary classifier to distinguish between generated human face images and real human face images. By doing so, the discriminator actually offers a special loss metric, the generative adversarial loss, to supervise the generator by comparing and evaluating the pixel distribution (i.e., pattern) in the generated and natural human face images (e.g., a common human face should contain eyes, nose, and mouth in the "right" position).

Mathematically speaking, the loss function of the whole generative adversarial learning framework can be expressed as:

$$\min_{\mathbf{G}} \max_{\mathbf{D}} L_{\text{adv}}(\mathbf{G}, \mathbf{D}), \tag{11.9}$$

where $\mathbf{G}$ and $\mathbf{D}$ represent the generator neural network and the discriminator neural network, respectively. The generative adversarial loss L_{adv} can be further expressed as:

$$\mathbb{E}_{\mathbf{x} \in S_{\text{real}}}[\mathbf{D}(\mathbf{x})] - \mathbb{E}_{\tilde{\mathbf{x}} \in S_{\text{fake}}}[\mathbf{D}(\tilde{\mathbf{x}})], \tag{11.10}$$

where $\mathbf{x}$ is the sample from a real human face dataset S_{real} and $\tilde{\mathbf{x}}$ is the generated human face from the fake human face dataset S_{fake}. It is noteworthy that the fake human face dataset S_{fake} can be generated by simply inputting random noise into the generator and collecting the generated products as the fake human face dataset. In Eqs. (11.9) and (11.10), from the generator's perspective, the only way that can achieve its goal of minimizing Eq. (11.9) is to generate highly convincing human face images to fool the discriminator by using generated human face images to obtain higher scores so that the second term in Eq. (11.10) can be maximized. On the other hand, the discriminator attempts to discriminate between real human face images and generated human face images so that a comparatively high score will be given when inputting real images while a low score will be given when inputting generated images for maximizing Eq. (11.9). Here, one can notice that the goal of the generator is adversarial against the goal of the discriminator in Eq. (11.9), being the main reason to be named as generative adversarial learning. Following this logic, the adversarial game will be played for up to several thousand iterations until convergence. After the iterations until convergence, the distinguishing result of the discriminator is close to a random guess since the generator can already produce various human face images with ample details, which is very close to the real human face image dataset. It is noteworthy that only a single generator will be employed to finish the human face generation problem in the testing phase. In other words, the role of the discriminator is just to provide a special loss function based on the distributional similarity of generated human face images and the real face images in the training phase to supervise the behavior of the generator.

Similar to the RL case, several different generative adversarial learning algorithms have been developed to serve different scenarios and generate increasingly convincing results. For example, several good papers work on the convergence analysis of the above training procedure and provide improved mathematical functions as the loss functions of the generator and discriminator to accelerate the convergence. Moreover, one can notice that in the aforementioned generative adversarial training, the products of the trained generator will still show huge variations. For example, it is hard to ask the trained generator to produce a woman's face or a man's face since all inputs are random noise vectors. In light of this direction, conditional generative adversarial networks are also developed, so that one can specify constraints to ask the trained generator to produce desired products. This is especially useful when we try to employ generative adversarial networks to address engineering problems since we often require a design under constraints instead of random designs. For a more detailed survey of current mainstream generative adversarial learning methodologies, interested readers can refer to refs. [259–261] for more details. In those variations of the original version of generative adversarial learning algorithms, the setting of multiple neural networks and the concept of the adversarial game can still be found as the main concept.

11.3 Network Structures of Deep Learning Models

Similar to the case of different training methods of neural networks, neural network architectures are also further specialized and developed to better adopt data with different features. In this section, we classify mainstream neural networks based on their structure, then provide a brief introduction and certain examples of each category for our continued discussions

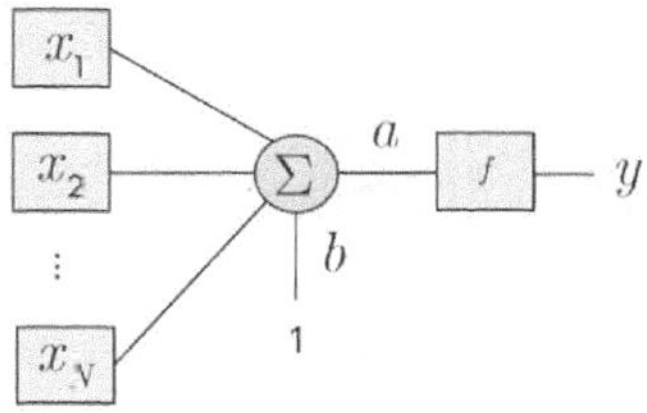

FIGURE 11.6
A mathematical model of an elemental component of fully connected neural network (FCNN), where N input features $x_0, x_1, \ldots, x_N$ with different weights and a bias term b are summed together to give a, which is then fed into a sigmoid activation function to generate an output y.

Fully connected neural network: Fully connected neural network (FCNN) is the most basic neural network of the DL model. Figure 11.6 shows the elementary component of a FCNN, which is also known as the perceptron. From the previous section, we know that the training phase of the DL model can be comprehended as an optimization process. That is, by designing an interested key performance indicator (i.e., loss function), the model will adjust the trainable weights to achieve a better performance using the backpropagation algorithm iteratively. As shown in the figure, all the input features are fed into a model simultaneously and the summation of weighted inputs and bias will be sent into a nonlinear function to generate the output of this neuron. The aforementioned procedure actually mimics the activation process in the neural activity of human beings. Imaging a neuron is responsible for detecting the pains of human body, a larger weight will be given to the sensory receptor, which can detect extremely high temperatures so that scald can be detected in a timely manner. As a result, once extremely high temperature appears, we are more sensitive to that and can react accordingly in a short period of time. Similarly, the aforementioned weighting sum architecture can adjust those weights and biases to let the FCNN have a more severe reaction when the interested features are detected. Mathematically speaking, the above operations can be expressed as:

$$y = f(\sum_{i=1}^{N} x_i w_i + b), \tag{11.11}$$

where x_i is the input element, w_i is the corresponding weight, and b is the bias term. If we adopt this one-layer FCNN as the neural network architecture, the aforementioned trainable weights actually refer to those weights and bias corresponding to each input (i.e., $\boldsymbol{\Theta} \triangleq \{w_i, b\}$ for $i = \{1, \ldots, N\}$). With the above different training methods, our goal is to systematize this adjusting procedure to automatically search for optimal results based on any given loss function. In other words, the most effective features can be extracted automatically instead of handcrafted features.

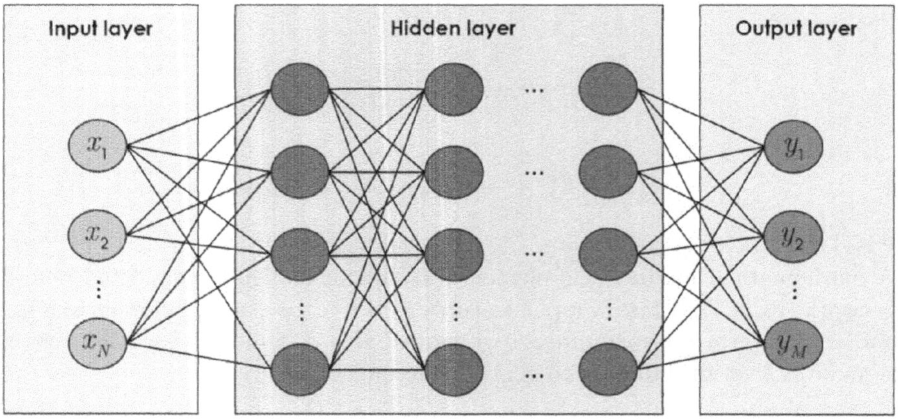

FIGURE 11.7
A deep FCNN includes input, hidden, and output layers. The multiple hidden layers are stacked together to capture high-level features.

According to ref. [262], any relationship between input features and output results can be approximated by a one-layer neural network with a sufficient number of neurons. However, it is well-conformed that a deeper model is more efficient than a one-layer neural network in terms of the required complexity to achieve a certain performance requirement. The reason is that a deeper neural network has a stronger ability to capture high-level features to further improve desired tasks. To explain, each layer of the neural network with the summation and nonlinear operation can be understood as a transformation function. If the number of layers increases, a more complex transformation can be made through the sequential combination of a lot of layers. For example, through those transformations, nonlinear separable data can be transferred into linear separable data in the higher-dimensional space so that just simple operations are needed to separate those data in this higher-dimensional space. Following this logic, the aforementioned elementary component can be stacked to form a deep FCNN, as shown in Figure 11.7, for enhanced capability through a number of nonlinear transformations. Actually, the capability of deep FCNNs to design weighting parameters automatically has encouraged researchers to apply this model in several signal processing works, especially for heterogeneous data fusion problems.

Convolutional neural network: Convolutional neural network (CNN) is a special case of FCNN to enhance performance on image-based tasks (i.e., image classification tasks) by considering image portieres in the architecture designs. To elaborate, FCNN can only support a one-dimension (1D) input. Therefore, FCNN needs to vectorize a 2D image to satisfy the input limitation. Yet, images have a strong 2D local structure and a lot of features only persist in 2D representations (e.g., we can easily recognize human faces in 2D representations, but it is hard to tell after 1D vectorization since the features we utilize to recognize human faces are destroyed.). Consequently, the 2D local structure is destroyed after performing 1D vectorization, bounding the performance of FCNN in image-based tasks. On the other hand, CNN contains two important properties that deal with the above drawbacks. First, CNN can support 2D structure input to preserve those image features. Second, in most image-based tasks (e.g., image classification and super-resolution image-based tasks.), we are caring about if interested features appear in the image or not, regardless of the actual location of those features in the image. Thus, after designing a set of weights to detect a specific feature, CNN also uses the concept of convolution to go through the whole image to sense the desired feature efficiently, irrespective of where the features appear in an image. On the other hand, FCNN can only utilize redundant parameters (i.e., almost the same weights) to sense the same feature in different locations of the image due to the lacking of convolutional operations. Furthermore, the property of the actual location of features in the image in image-based tasks can also be used to further promote CNN's efficiency. Specifically, the convolution operation is presented on the automatically learned feature filters and input features to compute the values of correlation and record them on the output feature maps. However, in a CNN layer, several tens or even hundreds of filters have to be employed to detect certain features, resulting in an increasing amount of output feature maps and consequently a computational overhead when tacking those feature maps with larger sizes in the next layer. To counter this challenge, most of the CNNs position a pooling layer after a convolution layer. That is, a down-sampling process is operated in the pooling layer, resulting in a smaller size of input features for the subsequent convolution layer. In this process, although the location information will be discarded, the existence of each features can still be preserved and passed to the subsequent convolution layer, which can already satisfy most of image-based tasks. Thus, in location-sensitive image-based tasks (e.g., object localization), the pooling layer will not be employed to maintain the precise location information in those tasks. In summary, a typical working flow of the CNN is shown in Figure 11.8. In fact, CNN is usually utilized as a feature extractor and concatenated with a FCNN as the final layer to enhance performance of desired tasks based on the extracted features. Deep CNN is also employed in many engineering applications, such as certain communication works [263–265], by treating the channel or other resources as an image and providing promising performance.

Recurrent neural network: In the aforementioned FCNN, all input features are fed into a model simultaneously. Thus, when it comes to time-series data, such as stock prices and Electrocardiography data, the information associated with the timeline cannot be emphasized and extracted easily when using FCNN to tackle such type of data. Conversely, as shown in Figure 11.9, recurrent NN (RNN) permits its input to enter in different time slots. Consequently, RNN can grasp features of data according to context. Specifically, a special design of RNN is that a state mechanism is created to pass messages along with the timeline; further, the output in every time slot is decided by

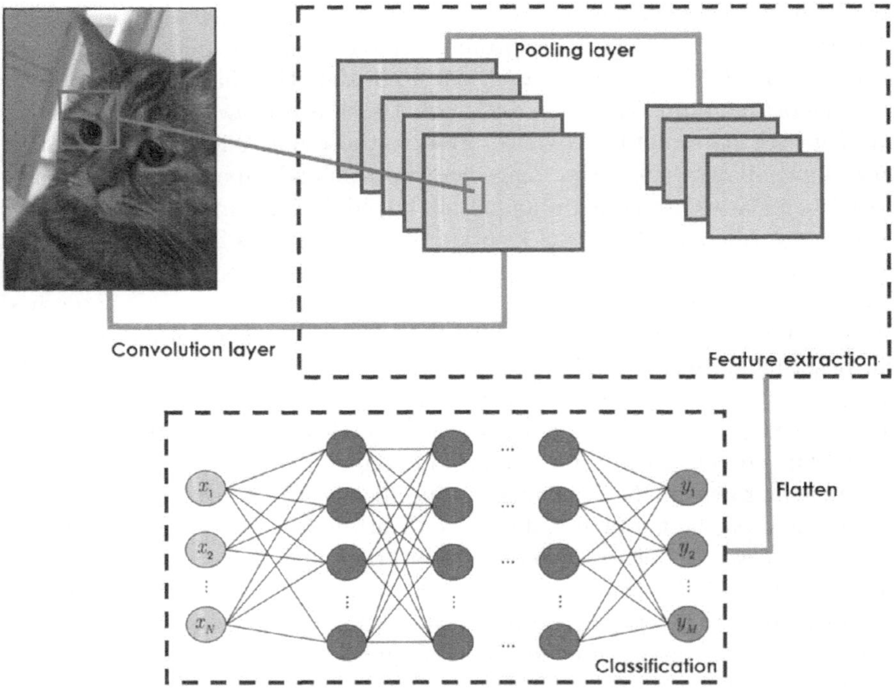

FIGURE 11.8
A convolutional neural network (CNN) architecture with a convolution and pooling layer before flattening. By inputting a partial image at one time, the convolution layer can learn and determine the correlation between the input and the feature (e.g., an eye of a cat) filters to detect the existence of features. The pooling layer can then down-sample the output from the convolution layer.

the current state (i.e., information extracted from the previous input features) and the new input feature. By doing so, RNN can extract important messages automatically and store them in each state for the subsequent decision-making process to better understand the events that happened in different time slots. Although providing superior capability when tackling time-series data, training a deep RNN will suffer from several numerical problems (i.e., gradient vanish and exploding problems). To overcome this drawback, several modified models, such as long short-term memory (LSTM) and gated recurrent unit (GRU) [266,267], were proposed to mitigate the difficulty of training deep RNN models. Owing to its outstanding performance, the LSTM is widely used in areas such as communication systems [268–270], providing a more advanced structure than RNN.

Graph neural network: Finally, the idea of graph NN (GNN) is quite similar to that of CNN. In graph theory, a graph is used to describe a situation

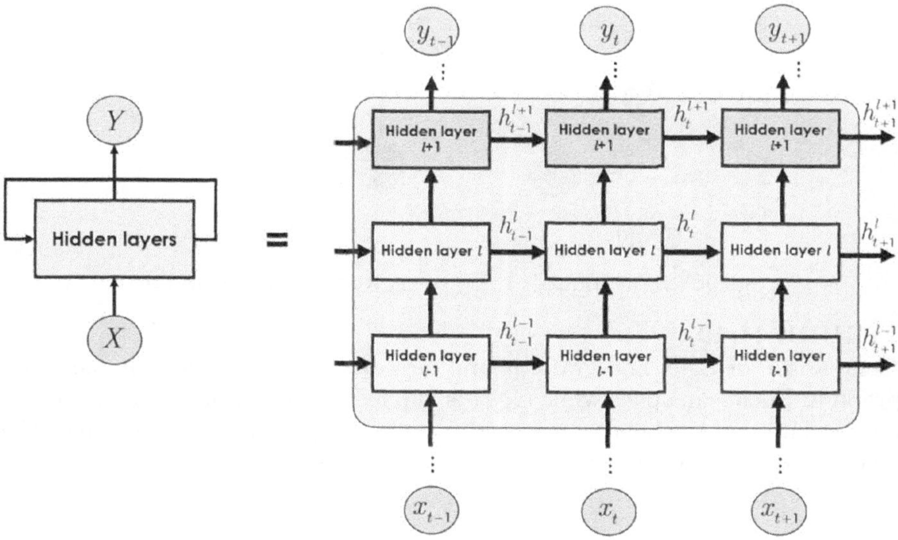

FIGURE 11.9
A recurrent neural network (RNN) architecture is designed by inputting features in different time slots (x_{t-1}, x_t, x_{t+1} in the figure represent the inputs in different timeslots). The current outputs in each time slot are dependent on the current inputs and stored features h, which are detected by RNN based on the former inputs automatically, to enhance prediction. Moreover, the RNN can be stacked (hidden layer l in the figure stands for l-th hidden layer of the RNN) as a deep NN to further improve the prediction result.

of several nodes and the connections between those nodes (edges). An image can be treated as a special case of the graph with regular and well-organized connections (i.e., edges) in Euclidean space, where pixels in an image are regarded as the nodes of the graph. The aforementioned CNN is designed to extract the spatial features of a graph only in Euclidean space and cannot directly apply to general graph cases without performance degeneration. However, increasingly more data in real-time applications (e.g., traffic networks, social networks) can also be presented precisely as a graph in non-Euclidean space. Thus, to promote the use of DL in those applications, GNN was developed to extract spatial features of a graph with arbitrary connection in a non-Euclidean space. The comparison of GNN and CNN is shown in Figure 11.10. By mapping to the corresponding neural network structure in a conventional DL model, the GNN can be divided into a recurrent GNN, a convolutional GNN, graph autoencoders, and spatial-temporal GNNs to satisfy different requirements by inheriting the advantages of different underlying architecture. In the wireless communication research areas, the radio link through wireless

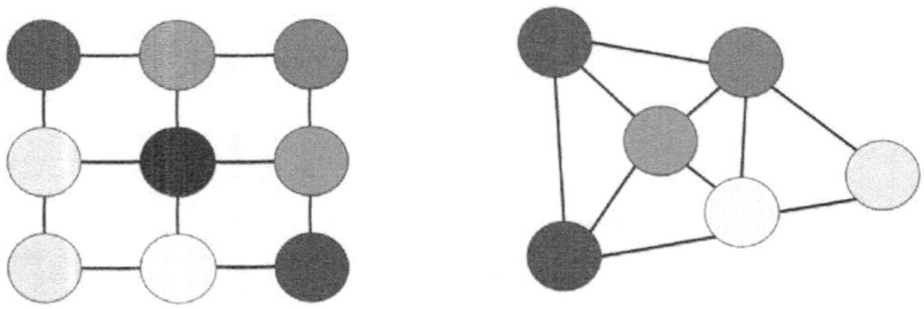

FIGURE 11.10
Comparison of CNN and graph neural network (GNN). CNN is applied to a graph in Euclidean space while GNN is applied to a graph in a non-Euclidean space. The non-Euclidean space indicates a more arbitrary space than the Euclidean space owing to its arbitrary connections between nodes.

connectivity can also be treated as a graph in non-Euclidean space since the distance and link strength varies a lot in different scenarios. Thus, there are also some great works utilizing GNN to better tackle wireless communication engineering problems [271–273]. In the above all examples, one can conclude that different training methods and neural network structures should be selected carefully to better fit the considered problem, thus leading to improved system performance or efficiency compared to traditional optimization-based designs.

11.4 When to Use Deep Learning Models

After discussing the common training methods and the classic neural network architecture of current mainstream DL models, we aim to discuss the motivation to introduce DL solutions to any interested problem, especially in wireless communication and sensing areas, to conclude this section. From the previous discussions, one can notice that there are two features of the current DL models. First, when applying DL-based algorithms, a clear and realistic system model is often not required. Given ample training data or a well-built environment, learning-based solutions can be utilized to automatically extract the input/output relationship of the interested problem and adjust the trainable weights inside the neural network for accurate predictions. Second, compared to optimization-based solutions, the computational complexity of neural networks in the online testing phase is very simple since only standard matrix operations are involved. Although the computational complexity of neural networks in the offline training phase can be really high, those procedures can often be performed in powerful servers or data centers in advance. In other words, compared to optimization-based solutions, DL-based solutions can offload the computation overheads from the online testing phase to the offline training phase, facilitating the development of real-time operations in real communication systems. Following this logic, one can expect to benefit from employing DL-based algorithms when the following deficits exist.

Model deficit: Conventional optimization-based algorithms are developed according to a predefined and accurate mathematical model. If the mathematical model is imprecise or if it cannot even be formulated, optimization-based algorithms cannot be designed appropriately, resulting in performance loss or functional disability. Conversely, instead of using a mathematical model, by feeding abundant data, DL-based algorithms can learn input output relationship without any prior knowledge. Therefore, DL-based algorithms are preferable in the model deficit condition. An example of a model deficit condition in the communication research area is a DL-based decoding algorithm for molecule communication [274–276]. Specifically, unlike the electromagnetic wave channels, molecule communication channel has no authoritative mathematical expression to characterize the transmission process. Consequently, traditional optimization-based decoding algorithms cannot be employed in this scenario without performance degeneration. Conversely, DL-based algorithms can infer the input-output relationship in the molecule transmission process automatically, resulting in its application in molecule communication with acceptable performance.

Algorithm deficit: If a precise mathematical model is available, optimization-based algorithms can be developed. However, the usage of optimization-based algorithms does not always guarantee their efficiency. Consequently, certain problems are still solved by inefficient methods (e.g., brute force algorithms and exhaustive search) or iterative solutions. As a non-iterative and low complexity algorithm, DL-based algorithms are considered an efficient alternative in an algorithm deficit condition. A case of algorithm deficit in the communication research area is the CSI feedback problem [277–280]. Instead of compromising the overhead incurred while transmitting a full version CSI, a compression mechanism can be designed to generate a compressed version of the CSI. The mechanism is aimed at maintaining information invariance after the compression. Conventional compressed sensing (CS) methodology can be exploited as a solution; however, it suffers from high complexity owing to the iterative processes. Conversely, the DL-based algorithm offers an alternative without iterations. Moreover, because the DL-based algorithm can properly utilize spatial correlations while CS algorithms cannot, the DL-based algorithms provide a better performance on the CSI feedback problem.

12

DL-Based Signal Processing in Communication Systems

12.1 DL-Based Channel Estimation

Being a fundamental signal processing functionality, researchers also investigate how to utilize the power of deep learning (DL)-based algorithms to assist channel estimation procedures. Specifically, if we treat the single-input single-output-orthogonal frequency division multiplexing (SISO-OFDM) channel as an image with two dimensions (i.e., time and frequency), it starts to be possible to employ mature research outcomes from the computer vision (CV) domain into the wireless communication domain by treating the channel matrix as a special image. Specifically, in the typical channel estimation procedure of SISO-OFDM systems, some time-frequency resources will be placed with pilots (i.e., known signals) so that the experienced channel can be easily estimated. Then, interpolation methods can be further applied to obtain channel estimates in the time-frequency resources with data-symbols (i.e., unknown signals) for subsequent detection and decoding processes. Mathematically speaking, the input-output relationship in the k-th timeslot and i-th subcarrier of a SISO-OFDM system can be expressed as:

$$y_{i,k} = h_{i,k}x_{i,k} + z_{i,k}, \tag{12.1}$$

where $y_{i,k}$, $h_{i,k}$, $x_{i,k}$, and $z_{i,k}$ represent the received signal, channel, transmitted OFDM symbol, and white Gaussian noise, respectively. To estimate the experienced channel in the pilot positions with pilot length N_{P}, a commonly used way is to consider the least square (LS) channel estimates as a diagonal matrix $\mathbf{H}_{\mathrm{p}}^{\mathrm{LS}} \in \mathbb{C}^{N_{\mathrm{P}} \times N_{\mathrm{P}}}$ and solving $\mathbf{H}_{\mathrm{p}}^{\mathrm{LS}}$ through the optimization below:

$$\hat{\mathbf{H}}_{\mathrm{p}}^{\mathrm{LS}} = \arg\min_{\mathbf{H}_{\mathrm{p}}} ||\mathbf{y}_{\mathrm{p}} - \mathbf{H}_{\mathrm{p}}\mathbf{x}_{\mathrm{p}}||_2^2, \tag{12.2}$$

where $\mathbf{x}_{\mathrm{p}} \in \mathbb{C}^{N_{\mathrm{P}} \times 1}$ and $\mathbf{y}_{\mathrm{p}} \in \mathbb{C}^{N_{\mathrm{P}} \times 1}$ is the known pilot values and corresponding estimates.

The optimal solution of Eq. (12.2) can be obtained as $\hat{\mathbf{h}}_{\mathrm{p}}^{\mathrm{LS}} = \mathrm{diag}(\hat{\mathbf{H}}_{\mathrm{p}}^{\mathrm{LS}}) = \mathbf{y}_{\mathrm{p}} \oslash \mathbf{x}_{\mathrm{p}}$. Furthermore, the minimum mean square error (MMSE) estimator

DOI: 10.1201/9781003408239-12

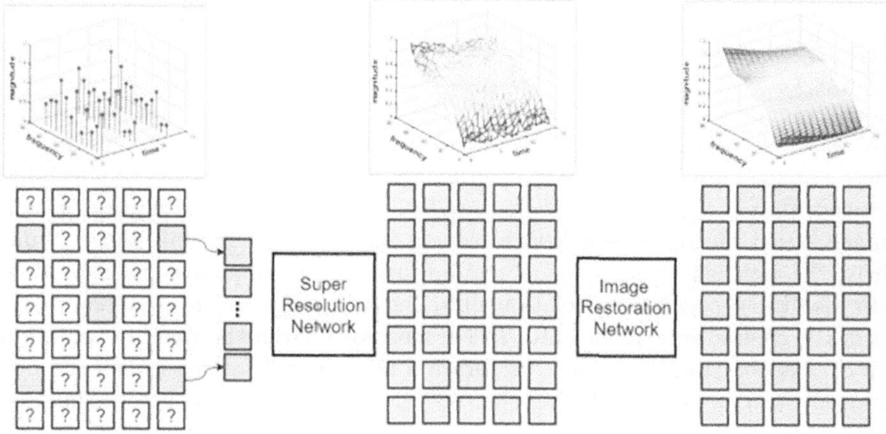

FIGURE 12.1
The proposed pipeline for DL-based channel estimation (copyright from ref. [281]).

can be used to extend the channel estimates to non-pilot positions via the filtering matrix designs as below:

$$\hat{\mathbf{h}}^{\mathrm{MMSE}} = \mathbf{A}_{\mathrm{MMSE}}\hat{\mathbf{h}}_{\mathrm{p}}^{\mathrm{LS}}, \tag{12.3}$$

where $\hat{\mathbf{h}}^{\mathrm{MMSE}} \in \mathbb{C}^{N_{\mathrm{L}}\times 1}$ is the channel estimates with total length N_{L} and $\mathbf{A}_{\mathrm{MMSE}} \in \mathbb{C}^{N_{\mathrm{L}}\times N_{\mathrm{P}}}$ is the filtering matrix. The optimal filtering matrix designs can be obtained by minimizing the MSE of channel estimates to obtain the designed filtering matrix as:

$$\mathbf{A}_{\mathrm{MMSE}} = \mathbf{R}_{\mathbf{hh}_{\mathrm{p}}}(\mathbf{R}_{\mathbf{h}_{\mathrm{p}}\mathbf{h}_{\mathrm{p}}} + \sigma_n^2(\mathbf{x}_{\mathrm{p}}\mathbf{x}_{\mathrm{p}}^H)^{-1})^{-1}, \tag{12.4}$$

where the matrix $\mathbf{R}_{\mathbf{hh}_{\mathrm{p}}} = \mathbb{E}[\mathbf{h}\mathbf{h}_{\mathrm{p}}^H]$ denotes the channel correlation matrix between the desired sub-frame and pilot symbols and the matrix $\mathbf{R}_{\mathbf{h}_{\mathrm{p}}\mathbf{h}_{\mathrm{p}}} = \mathbb{E}[\mathbf{h}_{\mathrm{p}}\mathbf{h}_{\mathrm{p}}^H]$ is the channel correlation matrix of the pilot symbols. Thus, it is clear that error-free correlation matrices should be obtained as prior knowledge to perform channel estimation successfully, or an error propagation effect will appear and affect the channel estimation performance. However, obtaining high-precision correlation matrices is not trivial and often increases system overheads. To solve this issue, ref. [281] develops a learning-based solution to extract that information in a data-driven manner, thus avoiding system overheads and providing high-quality channel estimates simultaneously.

In ref. [281], a two-step channel estimation method is proposed, utilizing the power of recent learning-based achievements to perform intelligent interpolation for the above purposes. Specifically, the proposed solution contains a super-resolution module and an image restoration module to do so, as shown in Figure 12.1. Given the LS channel estimates $\hat{\mathbf{h}}_p^{\mathrm{LS}}$ as the model input, the super-resolution module aims to perform interpolation by extending LS channel estimates from pilot position to non-pilot position as the functionality of the MMSE estimator but without the correlation matrices information. This procedure is very similar to the typical super-resolution task in the CV research domain. That is, given the abundant pairs of low-resolution and high-resolution images, the employed neural network can learn the underlying transformation and can be used to predict the corresponding high-resolution image based on an unseen low-resolution image in the testing phase. Similarly, through observing a lot of training pairs, the super-resolution module can also learn possible correlation matrices in a data-driven manner to perform the desired interpolation task. Furthermore, given the raw interpolation result, the concatenated image restoration module will further polish the raw interpolation result by reducing the reconstruction noise to generate the final channel estimates. Mathematically speaking, the whole procedure to generate final channel estimates can be presented as:

$$\hat{\mathbf{H}} = f(\boldsymbol{\Theta}; \hat{\mathbf{h}}_p^{\mathrm{LS}}) = f_{\mathrm{R}}(f_{\mathrm{S}}(\boldsymbol{\Theta}_{\mathrm{S}}; \hat{\mathbf{h}}_p^{\mathrm{LS}}); \boldsymbol{\Theta}_{\mathrm{R}}), \tag{12.5}$$

where f presents the whole neural network operations, including super-resolution module f_{R} and image restoration module f_{S} with all trainable parameters $\boldsymbol{\Theta}$ consisting of the trainable parameters from the super-resolution module $\boldsymbol{\Theta}_{\mathrm{S}}$ and the trainable parameters from the image restoration module $\boldsymbol{\Theta}_{\mathrm{R}}$. Furthermore, to train all trainable parameters in Eq. (12.5), Mean square error (MSE) is selected as the cost function to evaluate the similarity between channel estimates and ground truth, that is,

$$C = \frac{1}{||T||} \sum_{\mathbf{h}_{\mathrm{p}} \in T} ||f(\boldsymbol{\Theta}; \hat{\mathbf{h}}_p^{\mathrm{LS}}) - \mathbf{H}||_2^2, \tag{12.6}$$

where T represents the total number of training samples. The results show that the performance of the proposed method is highly competitive with the MMSE algorithm with full correlation matrices information.

Similar to the idea of ref. [281], ref. [282] extends the prior work [281] to MIMO-OFDM scenario and further enables channel prediction in the temporal domain. Specifically, the channel matrix of the interested MIMO-OFDM system can be expressed as a 4D tensor $\mathbf{H} \in \mathbb{C}^{N_{\mathrm{t}} \times N_{\mathrm{r}} \times T \times N}$, where N_{t} is the number of transmitter antennas, N_{r} is the number of receiver antennas, N is the number of subcarriers and T is the number of OFDM symbols. In ref. [282], the LS channel estimation method will be performed first to obtain the channel estimates in the pilot position. As illustrated in Figure 12.2, to obtain non-pilot position channel estimates, the MIMO-OFDM channel matrix of t-th OFDM symbol $\mathbf{H}_t \in \mathbb{C}^{N_{\mathrm{t}} \times N_{\mathrm{r}} \times N}$ (note that non-pilot positions remain as zeros for this input) will be fed into two sequential convolutional neural networks (CNNs) for frequency-domain interpolation to fill the zero parts with the inferred channel estimates. Then, to perform time-domain channel prediction, two parallel long-short-term memory (LSTM) neural networks (i.e., bi-directional LSTM (BiLSTM) architecture) are employed to effectively predict

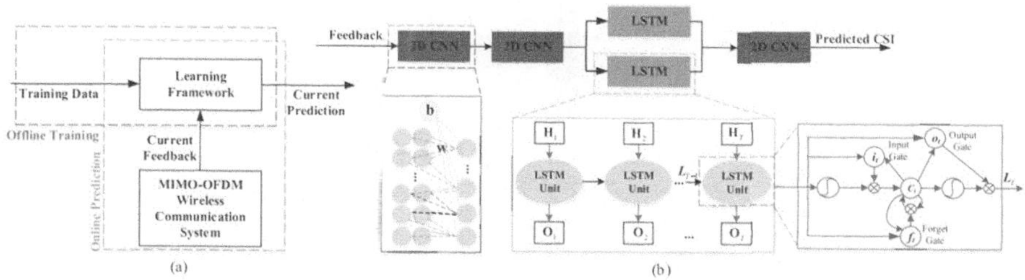

FIGURE 12.2
(a) Framework of proposed DL-based channel estimation algorithm. (b) The structure of proposed learning network. Since the input of the learning network is a 4D tensor, we use time distributed 2D CNN, which is an independent convolution for each time step signal. Similarly, the LSTM network is a recurrent convolutional neural network; both its input transformations and recurrent transformations are 2D convolutional (copyright from ref. [282]).

$(t+1)$-th OFDM symbol based on previous channel estimates. In particular, one LSTM network is designed to perform forward prediction while another LSTM network is responsible for performing backward prediction. This design aims to tackle possible error propagation issue for normal LSTM, especially for scenarios with long previous channel estimates from multiple-timeslots OFDM symbols. Finally, another CNN neural network will be utilized to fuse the prediction from the forward LSTM and backward LSTM to generate the final channel estimates of $t+1$-th OFDM symbol $\mathbf{H}_{t+1} \in \mathbb{C}^{N_t \times N_r \times N}$. Simulation results confirm that the proposed method is suitable for channel estimation of fast-moving targets by delivering superior performance compared to conventional LS and LMMSE channel estimation algorithms. Also concentrating on utilizing temporal correlation to aid the channel estimation process, ref. [283] further provides a systematic channel estimation method with reduced pilot overheads. Specifically, D successive coherence intervals of channels will be grouped as a channel estimation unit (CEU), where only the first coherence interval will utilize full pilots and the rest coherence intervals will only employ partial pilots to perform channel estimation. To recover the reduced prior information from the partial pilots utilization, all previous pilots in the CEU will also be provided to the channel estimation neural network to fuse the information from current and previous coherence intervals. With this design, the temporal correlation can be extracted by the proposed neural networks to compensate for the information loss from the reduced pilots to benefit the channel estimation process. Finally, there are also several works utilizing generative adversarial networks (GANs) to perform channel estimation for different wireless communication scenarios [284–286]. Different from the above works, which utilize conventional cost function (i.e., Mean square error or Mean absolute error) to evaluate the similarity between channel estimates and ground truth, the basic idea of introducing GANs into the channel esti-

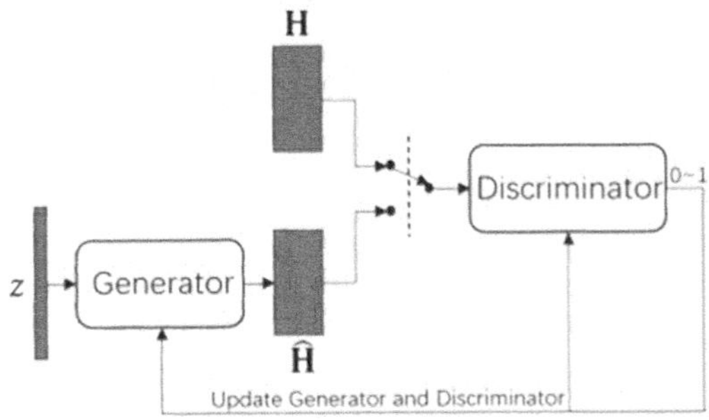

FIGURE 12.3
The architecture of the GAN (copyright from ref. [287]).

mation pictures is to employ discriminator to provide a special cost function to evaluate the similarity between channel estimates distribution and ground truth distribution, as shown in Figure 12.3. By doing so, the generated channel estimates can be even more convincing thanks to the distribution supervision, also resulting in improved channel estimation performance.

12.2 DL-based Codebook-Based Precoding/Beam Selection

In order to pursue faster and larger data transmissions to provide a better user experience, beamforming is one of the key methods in current wireless communication systems. In Chapter 4, we discuss codebook-based precoding/beamforming and conclude that it is a promising way to realize fast beamforming designs. Specifically, there are two ways to perform codebook-based beamforming, that is, with and without channel information. While with channel information, the optimal beam can always be selected for data transmission, it also brings considerable system overhead from the required channel estimation and channel feedback. Especially, the complexity and system overhead even increase as the array size increases, becoming a challenge to work with a large array in high-throughput communication systems. On the other hand, exhaustive beam switching is often required to obtain near-optimal beams without channel information, prohibiting its use when a large codebook with narrow beams is adopted. To present highly efficient beam selection without the need of channel information, the authors propose a novel DL-based image reconstruction approach to aid beam selection in ref. [288]. Inspired by the research on DL applications in the fields of medical imaging [289–291], in the proposed method, only a small number of beams (i.e., eigen-beams) will be switched to obtain the received power of those beams.

Then, beam domain image reconstruction (BDIR) is performed via the proposed DL model, utilizing the received power of eigen-beams to infer the received power of the rest of the possible beams. By considering the received power of all beams as an image, the proposed method mimics the medical image reconstruction workflow in computer vision research areas to perform efficient received power prediction of all beams without actual transmission, thus reducing system overhead significantly without channel information.

Specifically, this work considers a downlink mmWave multiple-input single-output (MISO) transmission scenario without channel knowledge at the transmitter end. It is assumed that the base station (BS) and mobile station (MS) are equipped with a uniform planar array (UPA) antenna and an omnidirectional antenna, respectively. For downlink transmission, the BS, performing beamforming techniques by a codebook-based beam selection with N_{BS} array elements, is assumed to communicate with the MS with an omnidirectional antenna. Note that the MS does not estimate the channel response and conduct channel feedback before the beamforming operation, which reduces the overhead and time consumption on beamforming. In the considered three dimension (3D) beamforming case, it is assumed that there are $N_{\mathrm{beam,TX}} = N^{\mathrm{v}}_{\mathrm{beam,TX}} \times N^{\mathrm{h}}_{\mathrm{beam,TX}}$, where $N^{\mathrm{v}}_{\mathrm{beam,TX}}$ represents the number of beams along the vertical dimension and $N^{\mathrm{h}}_{\mathrm{beam,TX}}$ represent the number of beams along the horizontal dimension, and the goal is to determine the optimal beam from $N_{\mathrm{beam,TX}}$ choices to serve any specific MS by best utilizing the underlying channel characteristics. In the general beam selection approach, by switching the transmitting beam at the BS, a power matrix of dimension $N^{\mathrm{v}}_{\mathrm{beam,TX}} \times N^{\mathrm{h}}_{\mathrm{beam,TX}}$ can be obtained accordingly. Then the straightforward solution searches the global optimum value of the receive power map. It can either be an exhaustive or a numerical search method for achieving the best beamforming performance. However, the search overhead of exhaustive search is not affordable and the performance of numerical search methods highly depends on the search overhead. Figure 12.4 illustrates the transformation of the receive power matrix to the receive power map, named beam-domain received power map (BDRPM). From this point of view, the beam selection problem can also be considered as a peak searching issue in an image (receive power map). Figure 12.5 further shows the illustration of the selected indoor scenario with several users demanding data transmission. Both line-of-sight (LoS) and non-line-of-sight (NLoS) scenarios exist for the communication links between access points (AP) and user equipment (UE).

To address the drawbacks of existing beam selection methods, the authors propose a DL framework addressing beam selection as a BDIR problem. Thus, the search overhead of online beam selection can be significantly reduced. The first research question that needs to be addressed is: how to determine a representative subset of beams (i.e., eigen-beams) to switch as a good seed to facilitate following BDIR? To do so, the proposed framework first employs a data-driven approach to obtain eigen-beam set as the seeds for BDIR and exploits the collected data to train the deep neural network (DNN), which

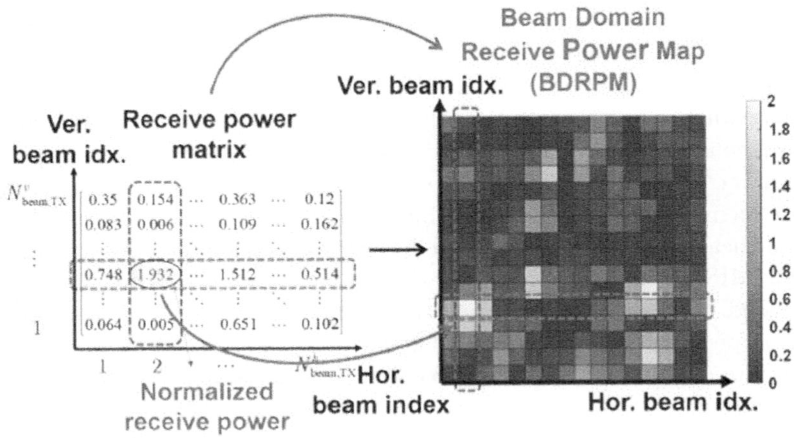

FIGURE 12.4
Illustration of BDRPM in 3D beamforming case (copyright from ref. [288]).

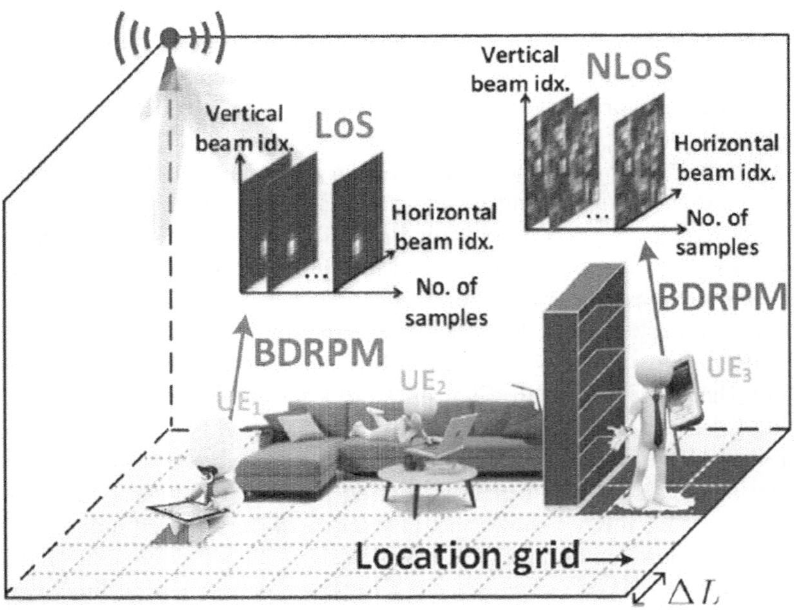

FIGURE 12.5
Illustration of the considered indoor scenario (copyright from ref. [288]).

performs BDIR. Thus, in the online stage, eigen-beams can be utilized to collect the received power as the seeds for BDIR by switching on the eigen-beams and recording their corresponding received power to obtain a BDRPM with only a few colored pixels, known as low-resolution BDRPM. DNN is then

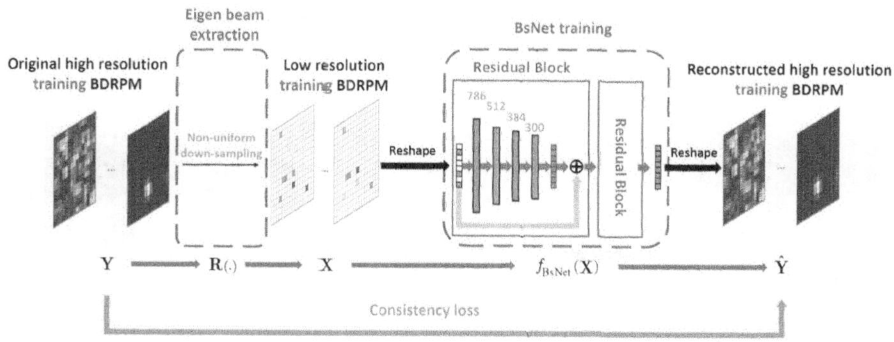

FIGURE 12.6
Overview of the off-line training stage: the off-line training stage consists of eigen-beam extraction and BsNet training (in the red circles). The eigen-beam set offers good seeds for BsNet. BsNet is expected to reconstruct a BDRPM that is close to the original BDRPM (copyright from ref. [288]).

employed to reconstruct the complete BDRPM (recovered high-resolution BDRPM) from the low-resolution BDRPM to address the optimal beam index by switching one more beam. The operation procedure is illustrated in Figure 12.6. Specifically, the following two operations need to be done in the offline stage: (1) Eigen-beam Extraction: To address the pixels assigned to collect the receive power in the online stage for generating a low-resolution BDRPM, a learning-based approach is employed to extract eigen-beams that are suitable for the given scenario. Let **Y** and **X** denote the high-resolution and low-resolution BDRPM, respectively. The relationship between them can be expressed as:

$$\mathbf{X} = \mathbf{R}(\mathbf{Y}), \tag{12.7}$$

where **R** is the mask function designed according to the eigen-beams.
(2) BsNet Training: DNN is also trained as a predictor for the online prediction (reconstructing high-resolution BDRPM from low-resolution BDRPM). Our goal is to reconstruct **X** to an image $\hat{\mathbf{Y}}$, that is

$$\hat{\mathbf{Y}} = f_{\text{BsNet}}(\mathbf{X}). \tag{12.8}$$

The above procedure is quite similar to the super-resolution image reconstruction task in the computer vision research area and the input/output relationships of the proposed method are shown in Figure 12.7 for reference. Note the Eigen-beam set can be obtained in many ways, including statistical, learning or data-driven methods, depending on its purpose and whether it matches the subsequent processing scheme. In fact, in follow-up works, different Eigen-beam set designs are proposed, but the main idea remains the same: to select a representative subset of beams, then perform image reconstruction to infer out-of-subset beam behavior. Also note that, different from the conventional

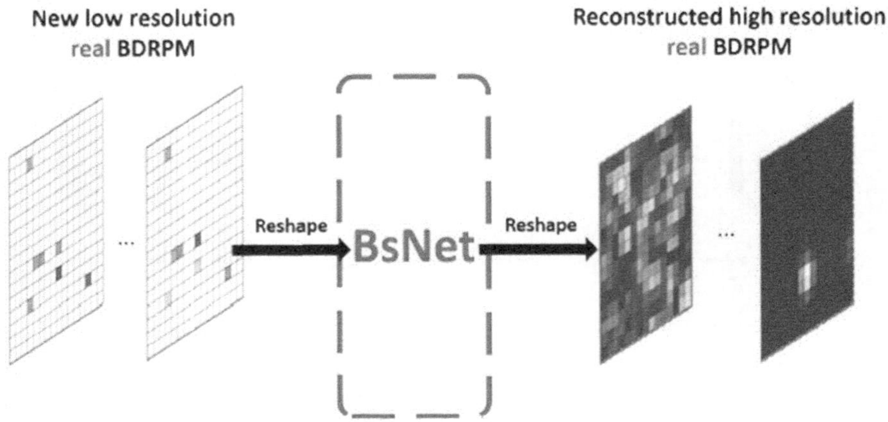

FIGURE 12.7
Illustration of the online prediction stage: a degenerated BDRPM with a few colored pixels is used as the input of BsNet. Then, BsNet can reconstruct a high-resolution BDRPM to solve the beam selection problem without the ground truth BDRPM (copyright from ref. [288]).

IR problems, the seeds for reconstruction algorithms are the pixels that are uniformly sampled from the desired image. The BDIR problem shows a noticeable difference where only the pixel with the highest power (i.e., optimal beam) is important, thus suggesting nonuniform sampling over uniform sampling for improved efficiency. Note that the same idea is also utilized in MRI reconstruction to present higher reconstruction quality in the region of interest. By doing so, the search overhead of the proposed method can be further reduced because only the peak value of BDRPM for optimal beam selection is emphasized, which is different from the general IR problems.

In the original paper, a learning-based approach is employed to train the Eigen-beam set as the image seeds for BDIR in this work. Specifically, k-nearest neighbors (k-NN) [292] is employed to classify the BDRPM collected from M-point locations in a given scenario (e.g., a classroom) into $N_{\text{beam,TX}}$ categories and rank them in order, denoted by $\mathbf{z}$. Assuming that the size of Eigen-beam set is N_{EB}, the entries of Eigen-beam set can be chosen by selecting the N_{EB} beams corresponding to the first N_{EB} in terms of $\mathbf{z}$. Thus, the Eigen-beams of the given scenario can be determined to work as the initial seeds for the following BDIR procedure. N_{EB} is a parameter that significantly impacts the search overhead of the online prediction that changes with the application scenario and requirements since the persentage of the searching overhead compared to exhaustive search can be approximated as $N_{\text{EB}}/N_{\text{beam,TX}}$. By increasing N_{EB}, the performance of beam selection (i.e., average spectral efficiency (SE) and accuracy of beam selection) can be greatly improved. However, the time consumption of online prediction also increases with N_{EB}.

Thus it is a trade-off between accuracy and time consumption. In the simulation results of different scenarios, this work demonstrates that it requires only 10% of the total beams and the average SE can achieve up to 99% by exhaustive search in a given scenario by the proposed method. It significantly reduces the search overhead while maintaining a good performance.

In Figure 12.7, assuming that the size of the considered codebook is $N_{\text{beam,TX}}$, the input feature of BsNet is an $N_{\text{beam,TX}} \times 1$ normalized real-value vector, which is a low-resolution BDRPM after vectorization and contains N_{EB} non-zero values and $N_{\text{beam,TX}} - N_{\text{EB}}$ zeros. The input vector enters several FCNN-based residual blocks, which are inspired by the design of ResNet [293], to stepwise reconstruct the high-resolution image from the low-resolution image. A residual block contains four hidden fully connected layers that are made up of 786, 512, 384, and 300 neurons, respectively. Behind each layer, a parametric rectified linear unit [294] is introduced as the activation function to offer nonlinearity. It is noteworthy that the dimension of the output layer is designed as $N_{\text{beam,TX}} \times 1$, which is the same as the input feature in the BsNet design to mimic the image reconstruction process. The number of residual blocks is set as two to further refine the input low-resolution image. The simulation result shows that a good performance can be obtained by using only two residual blocks. Adding more residual blocks into BsNet does not lead to significant improvement in terms of performance but causes additional computational complexity.

As for the BsNet training, end to end learning procedure is employed to train all trainable weights and the bias $\boldsymbol{\Theta}$ in BsNet. By treating BsNet as a function of low-resolution BDRPM input and trainable parameters $\boldsymbol{\Theta}$, the mean square error (MSE)-based loss function is employed to supervise BsNet training, which can be expressed as follows:

$$L(\boldsymbol{\Theta}) = \sum_{i=1}^{D} (\hat{\mathbf{Y}}_i - f_{\text{BsNet}}(\mathbf{X})_i; \boldsymbol{\Theta})^2. \tag{12.9}$$

Here, $\hat{\mathbf{Y}}_i$ is a high-resolution BDRPM from the training set, $\mathbf{X}_i$ is a low-resolution BDRPM input, which is the downsampled result from $\hat{\mathbf{Y}}_i$, and D is the number of total data in the training set. Finally, the gradient descent-based optimizer ADAM is utilized to iteratively reduce the loss of each epoch with the initial learning rate of 0.00005. The batch size is set to 512. After 1500 epochs, the values of $\boldsymbol{\Theta}$ are recorded, and BsNet training is completed. To evaluate the practicality of the proposed method, ray-tracing simulator, a site-specific radio propagation software that can generate highly convincing channel behavior via high-performance computing, is employed to generate the training and testing data in this work. By doing so, the realistic radio propagation behaviors, including diffused scattering effect, multipath effect, and penetration loss, are considered in the simulator. It can output the received power, channel matrix, and channel impulse response (CIR) of the given communication system and environment setting for training and evaluation purposes in this work. The test scenarios include three environments: a conference room (CR), a living room (LR), and an enterprise cubicle (EC), which are specified by the IEEE 802.11ad task group. The CR is the smallest environment and contains more complicated cases owing to multi-path effect. The EC is the largest environment and contains many NLoS cases owing to the presence of more furniture. To test the performance in each scenario, several UEs are deployed in the room according to the height specifications. The spacing of UEs in each room is set to a value no more than 0.65(m) and multi-sample data from each location is collected for training and evaluation purposes. The total number of training data and testing data is 120000, 180000, and 320000 samples with different size of codebooks for CR, LR, and EC scenarios, respectively. In the training data set, 30% data is randomly picked for validation to decide the hyperparameters in the proposed framework. All results are obtained as the average of the testing data set, excluding the training and validation data set. With the above settings, simulation results confirm that the near-optimal performance (i.e., above 99% of the optimal spectral efficiency obtained via exhaustive search) can be achieved with 90% reduced beam selection overhead. Encouraged by the success of this work, this research direction catches a lot of attention recently, interested readers can refer to follow-up works [295–298] for more information.

12.3 DL-Based Spectrum Sensing

In Chapter 4, various solutions are discussed to improve the data rate of communication systems using a given bandwidth, and we also discuss DL-based channel estimation and beamforming/precoding designs in the previous sections. Besides those research directions, another straightforward idea to improve the capability of communication systems is to increase the frequency resource utilization rate, and DL-based solutions already dominate this research direction, owing to their capability to automatically extract radio utilization status from a contested and noisy environment. Thus, in this section, we aim to discuss an important and popular research direction, DL-based spectrum sensing, to see how DL-based solutions can aid in the considered scenario. Specifically, since most wireless communication scenarios will only occupy frequency resources intermittently, those wasted frequency resources can be collected and re-used to improve overall system performance. In this direction, several research papers [299, 300] reveal that the spectrum under-utilization problem, which is caused by the existence of idle channels, occurs in current communication systems, reducing the overall spectral efficiency. To address this issue, spectrum management [299, 300] has been developed by detecting idle spectrum and temporarily assigning that spectrum to the demanding user, thus improving the overall spectral efficiency. Among all the research topics in spectrum management, spectrum sensing is the most widely discussed in literature [299–302] since it is a prerequisite for mitigating error propagation. For example, V2X data-coordination scenario [303] is a practical usage to employ spectrum sensing algorithms in wireless communications environment. Typically, vehicle-to-infrastructure (V2I) connections will be assigned specific bands for the high bandwidth entertainment applications transmission (e.g., video streaming). On the other hand, vehicle-to-vehicle (V2V) connections may wish to occasionally perform safety message (e.g., vehicle position, speed and heading) transmission using idle bands assigned to the V2I connections. By employing spectrum sensing (SS) algorithms in V2X communications, the underutilized spectrum can be reused and consequently leads to a better overall spectrum efficiency [303]. When it comes to the wide band scenario, sub-Nyquist sampling [299, 300] must be introduced to avoid costly hardware requirements. As a result, compressed sensing (CS) algorithms have been introduced to support spectrum reconstruction from measurements of sub-Nyquist sampling [301] in the past decade. However, they also suffer from high computational complexity to reconstruct the under-sampled signal. Recent research suggests that learning-based compression and reconstruction outperforms traditional CS algorithms since the local correlation is considered to reconstruct the desired output, being the main topic we discuss in this section.

As shown in Figure 12.8, consider a small cell including a base station (BS) with N_r receiver antennas and several UEs, each with N_t transmitter antennas for vehicular communications. In the downlink phase of the considered, the BS may occupy frequency bands from f_a to f_b by using some of the N_s subcarriers to perform V2I connections. Considering a transmission pair between the BS and i th UE, the complex baseband signal can be expressed as:

$$\mathbf{y}_i = \mathbf{H}_i\mathbf{x}_i + \mathbf{n}_i, \tag{12.10}$$

where $\mathbf{y}_i \in \mathbb{C}^{N_\mathrm{t}}$ is the received signal, $\mathbf{H}_i \in \mathbb{C}^{N_\mathrm{t}\times N_\mathrm{r}}$ is the channel matrix, and $\mathbf{x}_i \in \mathbb{C}^{N_\mathrm{r}}$ is the transmitted signal, respectively. Assuming a perfect sampling process with a Nyquist sampling rate of $T = 1/2f_b$, discrete-time signals can be obtained from Eq (12.10). When a UE aims to create V2V connections with surrounding UEs to share safety messages, a SS should be performed to detect the idle bands from existing V2V and V2I connections. Yet, in a wideband scenario, to emulate the needed hardware burden, CS must be introduced to aid the reconstruction of the compressed measurements obtained from sub-Nyquist sampling. Considering the aforementioned wideband communication scenario, a combining operation can be conducted at the UE to get the time domain measurements r_i, shown as:

$$\mathbf{r}_i = (\mathbf{w}_i^*)^T\mathbf{y}_i = (\mathbf{w}_i^*)^T\mathbf{H}_i\mathbf{x}_i + (\mathbf{w}_i^*)^T\mathbf{n}_i, \tag{12.11}$$

where $\mathbf{w}_i^* \in \mathbb{C}^{N_\mathrm{r}}$ is the combining weighting. Then the time domain signals of the considered wideband system can be expressed as $\mathbf{r} = \mathbf{s} + \xi =$

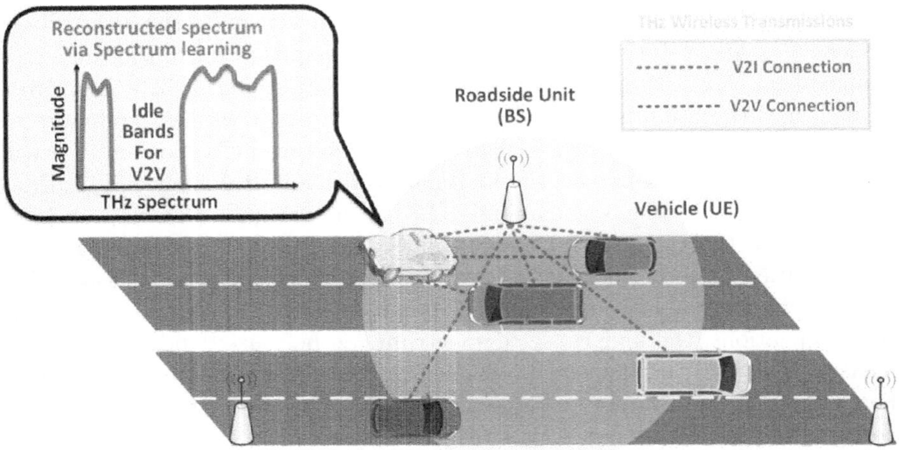

FIGURE 12.8
Considered THz communications for vehicular environment: In downlink phase, the base station (BS) will create several transmission links to different user equirements (UEs) for V2I connections. However, a UE aims to create V2V connections with surrounding UEs to share safety messages. As a result, the UE should perform SS first to obtain the information of idle spectrum, then use that spectrum to perform V2V connections (copyright from ref. [304]).

$[r_1, ..., r_i, ..., r_{N_s}] \in \mathbb{C}^{N_s}$, where $\mathbf{s} = [(\mathbf{w}_1^*)^T\mathbf{H}_1\mathbf{x}_1, ..., (\mathbf{w}_i^*)^T\mathbf{H}_i\mathbf{x}_i, ...(\mathbf{w}_{N_s}^*)^T \mathbf{H}_{N_s}\mathbf{x}_{N_s}]$ and $\xi = [\mathbf{w}_1^*\mathbf{n}_1, ..., \mathbf{w}_i^*\mathbf{n}_i, ...\mathbf{w}_{N_s}^*\mathbf{n}_{N_s}]$. Let $\boldsymbol{F}$ denote a N_s-point discrete Fourier transform (DFT). If the signal is sampled at a sub-Nyquist rate, then the relationship between the clean spectrum $\mathbf{s} \in \mathbb{C}^{N_s}$ and under-sampled measurements $\mathbf{z} \in \mathbb{C}^{N_m}$ can be expressed as:

$$\mathbf{z} = \boldsymbol{\Phi}\boldsymbol{F}\mathbf{r} = \boldsymbol{\Phi}\boldsymbol{F}(\mathbf{s} + \xi), \tag{12.12}$$

where $\boldsymbol{\Phi}_{N_m \times N_s}$ is the complex-valued sensing matrix. From Eq. (12.12), the goal is to design the sensing matrix and the corresponding reconstruction algorithm so that the clean spectrum $\boldsymbol{F}\mathbf{s}$ can be recovered from the under-sampled measurements $\mathbf{z}$ by the reconstructed spectrum $\hat{\boldsymbol{F}}\mathbf{s}$. It is noteworthy that once a high quality reconstructed spectrum is available, a simple energy detector can be employed to trivially identify the unused frequency bands. Moreover, the reconstructed spectrum with high quality can enable more complex spectrum sharing design in different coexistence models of heterogeneous communication systems [299, 300]. Hence, the motivation is to develop spectrum reconstruction methods. In this direction, while classic CS solutions, such as Lasso and its various variants, the unacceptable computation burden prohibits their usage in practical scenarios, such as vehicular communications and delay-sensitive scenarios. Thus, given raw spectrum data with under-sampling noise, the goal is to reconstruct the originl spectrum from it efficiently and without a heavy computational burden. In light of this direction, one can notice that this working flow is quite similar to the image denoising task in the computer vision research area. Thus, the introduction of learning-based solutions is reasonable and expected to provide a better trade-off between performance and complexity compared to CS solutions, aiding SS task in wireless communication with efficiency. Let $\boldsymbol{\Theta}$ stand for the trainable weight in the neural network and $f(\mathbf{x}; \boldsymbol{\Theta})$ is the nonlinear transformation with $\boldsymbol{\Theta}$, loss function can be written as below for the model updating, that is

$$Loss = ||\boldsymbol{F}\mathbf{s} - f(\boldsymbol{\Phi}\boldsymbol{F}\mathbf{s}; \boldsymbol{\Theta})||^2. \tag{12.13}$$

In Eq. (12.13), the goal is to develop a DL-based function, which takes the noisy and under-sampled measurement $\boldsymbol{\Phi}\boldsymbol{F}\mathbf{s}$ as input, to output the reconstructed clean spectrum close enough to $\boldsymbol{F}\mathbf{s}$. In the literature, there are already some prior attempts to tackle the similar problem with the aid of DL-based solutions. For example, in ref. [305], a popular GAN framework is further introduced to provide a more highly convincing spectrum reconstruction results. However, it still employs a randomly generated sensing matrix to conduct the compression, failing to perform joint optimization of the compression and reconstruction to obtain the best reconstruction results. Also, the consequently heavy overhead of the training process of GAN-based SS algorithm creates an implementation challenge to be employed in real scenario. In conclusion, the development of a practical SS algorithm, which can be employed in real ultra-wideband communications, is an unsolved problem. In this paper, the authors develop a DL-based spectrum reconstruction algorithm, named compression and reconstruction network (CRNet), to offer an efficient SS solution for THz communications. The authors propose a compression and reconstruction network (CRNet) for efficient spectrum sensing applications in THz communications. There are two features of the CRNet. First, conventional SS algorithms, including existing DL-based SS solutions, essentially employ randomly selected (i.e., *unstructured*) sensing matrix to perform compression to get under-sampled measurements, implying there is no special design of the sensing matrix. As an alternative, CRNet firstly introduces the joint design of compression and reconstruction by developing a *structured* sensing matrix and corresponding reconstruction algorithm in an end-to-end learning manner, offering a superior performance compared to existing SS algorithms. Secondly, compared to GAN-based SS algorithms, the training overhead of CRNet is reduced significantly. To be more specific, the under-sampled measurements obtained from the structured sensing matrix are more informative compared to those from an unstructured sensing matrix, and the reconstruction can be finished by a low-complexity CNN based-model to get a promising reconstruction result.

As shown in Figure 12.9, there are three modules in CRNet, compression, coarse reconstruction, and fine reconstruction modules. The compression module is a specially-designed one-layer CNN to produce under-sampled

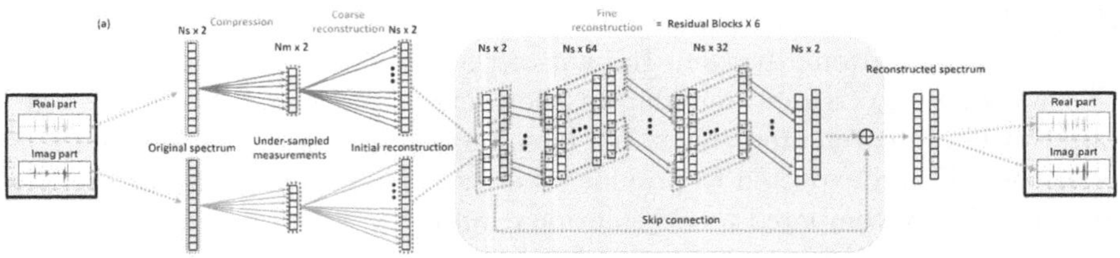

FIGURE 12.9
The model architecture of the CRNet. In CRNet, there are three modules: compression, coarse reconstruction, and fine reconstruction. In the compression and coarse reconstruction modules, the real and imaginary parts of the original spectrum are compressed and reconstructed, separately. Then, the fine reconstruction module with ResNet-structure is employed to perform fine-scale reconstruction to obtain high-fidelity reconstructed spectrum. Note that the compression and reconstruction process are performed in a end-to-end training model. As a result, the joint optimization can be performed to obtain optimal weights (copyright from ref. [304]).

measurements by making the trainable weights in the compression module act as the content of the sensing matrix, being expressed as:

$$\mathbf{z}_{\mathrm{DL}} = \mathbf{\Phi}_{\mathrm{DL}}\boldsymbol{F}\mathbf{r}, \tag{12.14}$$

where $\boldsymbol{F}\mathbf{r} \in \mathbb{C}^{N_r}$ is the original spectrum, $\Phi_{\mathrm{DL}} \in \mathbb{C}^{N_m \times N_r}$ is the sensing matrix designed by the compression module, and $\mathbf{z}_{\mathrm{DL}} \in \mathbb{C}^{N_m}$ is the under-sampled measurements from the designed sensing matrix. Given an original spectrum $\boldsymbol{F}\mathbf{r}$, in order to feed it into DL-based model, the input of the compression module $\boldsymbol{F}\mathbf{r}$ is presented as a real matrix with the size of $N_s \times 2$, containing the real part and imaginary part of the original complex vector. After the operation of the *compression module*, the output $\mathbf{z}_{\mathrm{DL}}$ is a real matrix with the size of $N_m \times 2$, standing for the real part and the imaginary part of the under-sampled measurements. To be more specific, a 1-dimension (1D) CNN layer is constructed with N_m filters in the compression module. In each filter, the trainable weight will be created as a matrix with the size of $N_s \times 2$, and then inner product operation between the input and the trainable weight will be conducted on the real part and imaginary part separately to obtain the computed result with the size of 1×2 representing the real part and imaginary part. This operation reflects the matrix operation between each row of the sensing matrix $\mathbf{\Phi}_{\mathrm{DL}}$ and the input $\boldsymbol{F}\mathbf{r}$. As there are N_m filters in this CNN layer, where the size of the output matches Eq. (12.14) to get the compressed measurements $\mathbf{z}_{\mathrm{DL}}$ for the following reconstructions. It is noteworthy that although in the compression module, the computation is performed on the real part and imaginary part separately, the operation is exactly equivalent to the inner product on a complex vector, as shown in Eq. (12.14). Moreover, note that there is no activation function in this 1D CNN layer to ensure the whole compression module as a linear operation. Finally, once the training of the compression module is finished, the trainable weights in each filter (i.e., each row of the sensing matrix) can be represented as a pseudo-random (PN) sequence as shown in ref. [305]. By mixing the received signal with N_m PN sequences (as there are N_m rows in the sensing matrix $\mathbf{\Phi}_{\text{DL}}$) and passing through a low-pass filter, the compressed measurements $\mathbf{z}_{\text{DL}}$ can be obtained. For a real scenario, there are no implementation issues to employ the practical CRNet DL-based spectrum reconstruction algorithm.

After CRNet obtains the compressed measurements $\mathbf{z}_{\text{DL}}$, the *coarse reconstruction module* aims to provide an initial reconstruction for the following refinements. To do so, another 1D CNN layer is employed, which has N_s filters with the size of $N_m \times 2$. After the CNN layer, batch normalization (BN) and parameter-Relu (PRelu) are employed to accelerate convergence and provide nonlinearity, respectively. By doing so, an initial reconstruction with the size of $N_s \times 2$, which is the same as the original spectrum, is obtained for the following fine reconstruction module. As for the architecture of the fine reconstruction module, the ResNet-structure [306] gradually refines the initial reconstruction result. To be more specific, the spectrum reconstruction problem is treated as a special image reconstruction problem and employs computer vision techniques to perform meticulous reconstruction. There are two main advantages to introducing ResNet-structure into the design of the proposed spectrum reconstruction algorithm. To explain, a typical DL model with ResNet-structure usually contains several residual blocks, which is built by several CNN layers. Instead of asking a DL model to provide a reconstruction result with high quality from scratch, the ResNet-structure lets a residual block refine the current reconstruction result based on the knowledge from previous residual blocks. As a result, all the residual blocks can coordinate with each other to synergistically produce a final reconstruction result. Another advantage of ResNet-structure is that a DL model with ResNet-structure is more unlikely to suffer from the over-fitting as the special skip-connection mechanism lets the DL model control the number of efficient weightings. For the exact architecture of the fine reconstruction module, in each residual block, three 1D CNN layers with a number of filters 64, 32, and 2, respectively, are built to refine the initial spectrum reconstruction. Behind each layer, BN and PRelu are also employed as the setting in the initial reconstruction module. In this paper, the CRNet comprises six residual blocks to perform fine-scale reconstruction as experiments with higher numbers of residual blocks do not improve overall performance but increase computational complexity. From Figure 12.9, a fully CNN architecture is used in the CRNet DL-based algorithm design. There are two main reasons to use the CNN architecture. First, the number of trainable weights can be decreased significantly as a result of the weight-sharing mechanism of the CNN. Simulation results confirm that the CRNet model outperforms existing DL models with significantly lower trainable parameters. Secondly, the CRNet fine reconstruction module aims to capture the occupied spectrum to perform fine-scale reconstruction. As the domain knowledge suggests that the occupied spectrum may appear anywhere in the whole spectrum, the convolution operation introduced by CNN can be used to capture the pattern of occupied spectrum regardless the location and number of the occupied spectrum to perform fine-scale reconstruction. An end-to-end learning is employed to jointly update all the trainable parameters in CRNet. As a result, the whole compression and reconstruction process can be designed simultaneously to achieve better performance. Let $\boldsymbol{\Theta}_{\text{CR}}$ stand for the trainable weight in the coarse estimator and Θ_{FR} represent the trainable weight in the fine estimator and $f(\mathbf{x}; \boldsymbol{\Theta}_{\text{CR}}.\boldsymbol{\Theta}_{\text{FR}})$ is the nonlinear transformation with $\boldsymbol{\Theta}_{\text{CR}}$ and $\boldsymbol{\Theta}_{\text{FR}}$, The mean square error (MSE) loss function is set for the model updating, that is

$$Loss = ||\boldsymbol{F}\mathbf{s} - f(\boldsymbol{\Phi}_{\text{DL}}\boldsymbol{F}\mathbf{s}; \boldsymbol{\Theta}_{\text{CR}}.\boldsymbol{\Theta}_{\text{FR}})||^2. \tag{12.15}$$

Note that during each training cycle, $\boldsymbol{\Phi}_{\text{DL}}$, $\boldsymbol{\Theta}_{\text{CR}}$, and $\boldsymbol{\Theta}_{\text{FR}}$ will be updated jointly via the back propagation process to gradually minimize the training loss until convergence, generating optimal structured sensing matrix and trainable weights. Finally, as for the training specifics of the scenario in this paper, the Adam optimizer minimizes the aforementioned loss function. The initial learning rate is set as 0.0005, and the number of epochs is set as 20. The mini-batch mechanism is employed with a batch size of 128 to facilitate fast convergence. Simulation results reveal that CRNet outperforms existing algorithms and can provide a realistic reconstruction result as the structured sensing matrix design and the corresponding reconstruction module design. To be more specific, CRNet can offer superior performance with only 44% training overhead as compared to existing DL-based solutions. It is noteworthy that CRNet assumes that blind spectrum reconstruction is performed in this paper, which means the proposed method can only obtain information from the under-sampled measurements. While the above assumption is practical, extra information can be provided to enhance the achieved performance of the proposed SS solution. Thus, as for future work in this direction, one can consider the case when some of the additional information (e.g., user locations, channel statistics) is provided to further improve the spectrum reconstruction for superior SS performance.

12.4 DL-Based Signal Detection

Based on the above discussions, it is clear that learning-based solutions are especially useful when a noisy or unclear channel model is involved in wireless communication procedures, thus the knowledge from ample training data can provide extra information in that scenario to overcome the noisy or unclear channel model deficits. In fact, a lot of wireless communication procedures are actually operated in such scenarios for the needs or constraints of low-overhead

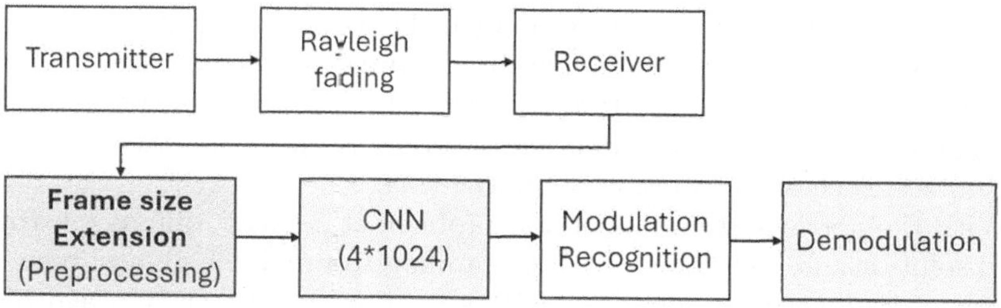

FIGURE 12.10
Proposed system model in ref. [310].

over-the-air data exchanges. For example, besides the previous discussed spectrum sensing, another downstream wireless communication procedure after spectrum sensing also shows impressive progress thanks to the development of learning-based achievements, that is automatic modulation classification task. In spectrum sensing scenarios, the control signal exchange between transmitter and receiver is minimized, being the motivation to develop solutions for automatic band utilization recognition. Similarly, in the same scenario, the receiver is very likely to also lack modulation and coding selection information from control signal, being the main motivation for the development of automatic modulation classification algorithms. Without any control message exchanges, automatic modulation classification algorithms can take raw IQ data as inputs to perform modulation classification, providing probability to different modulation schemes for final decision, as shown in Figure 12.10. Its effectiveness and straightforward adaption from image classification tasks soon attract researchers to contribute the development of learning-based automatic modulation classification methods.

To facilitate our discussion regarding the recent development of learning-based automatic modulation classification solutions, we first present a system model for the considered scenario. Specifically, the general system model of automatic modulation classification problem can be expressed as:

$$y[n] = x[n, \mathbf{H}_k] + w[n], \tag{12.16}$$

where $y[n]$ is the received radio signal, $x[n, \mathbf{H}_k]$ is the transmitted signal with channel effects, H_k is the underlying channel and $w[n]$ is the additive white Gaussian noise (AWGN) effect. Furthermore, the transmitted signal with channel effects term can be further expressed as:

$$x[n, \mathbf{H}_k] = A_n e^{2\pi f_0 nT + \theta_n} \sum_{k=-\infty}^{\infty} x[k]h(nT - kT + \epsilon_T), \tag{12.17}$$

where A_n is the signal amplitude of symbol n, f_0 is the carrier frequency offset, θ_n is the varying phase offset, T is the symbol duration interval, $h()$ is the baseband channel, $x[k]$ is the symbol sequence modulated with a specific modulation scheme, and ϵ_T is the timing offset between the transmitter and the receiver. Under this general system model, the goal of automatic modulation classification is to determine the specific modulation scheme used to modulate the transmitted symbol without any prior knowledge from the transmitter and channel (since channel estimation also requires prior knowledge from the transmitter, such as pilot resource information.)

With the above system model and problem statement, different learning-based solutions are developed, with the aiming to provide high accuracy classification results even under low SNR scenarios. In the early stage of the development of learning-based automatic modulation classifications. Several AMC methods have been proposed by leveraging the fundamental neural network architecture, fully connected neural networks, to learn the handcrafted features. In ref. [307], authors utilized various features in both the time and frequency-domain as the inputs to the automatic modulation classifier. The classifier is built using a simple three-layer fully connected neural network with ReLU activation layers. The network is trained by the loss function, which employs negative-log-likelihood to compute the distance between model output and ground truths. Inspired by this work, refs. [308] and [309] further improve the feature selection mechanism by utilizing mutual information or correlation coefficient to select the best subset of features to reduce the complexity of model learning. In this stage, the neural network design of those works is still basic and not specialized, thus manual feature selection still plays an important role in aiding the learning of neural networks. In response to the specially designed network architecture, RNN and CNN are soon introduced into this topic by utilizing their strengths to provide better modulation classification results. On the one hand, in ref. [311], authors proposed a data-driven model for AMC by exploiting a LSTM network to process the time domain amplitude and phase (AP) samples of modulated signals at the receiver. This deep network was designed with two LSTM layers (each layer has 128 LSTM units) for feature extraction and one fully connected layer for classification, outperforming the aforementioned solutions. In ref. [312], similar to the idea we introduced in the previous section, a LSTM-based classifier was deployed for sub-Nyquist rate wideband spectrum sensing, where the regular network was able to extract the temporal dependencies between input samples. The performance of this approach was investigated in terms of correct classification accuracy under different channel models (e.g., frequency-selective Rayleigh fading channel and Rician fading channel model with time-varying and Doppler). Moreover, over-the-air validation is also provided for demonstration purposes. Also in this direction, an accurate LSTM-based method was introduced in ref. [313] to effectively deal with various channel impairments (Rayleigh fading and mixture AWGN) by additionally exploiting a temporal attention layer to enrich features. Based on the simulation results of four digital formats (e.g., BPSK, QPSK, 8-PSK, and 16QAM), the network proved the superiority over traditional classifiers and other regular LSTM networks. On the other hand, authors in ref. [314] deployed two CNNs by adapting the architectures of VGG and ResNet with one-dimensional asymmetric convolution filters in layers. Based on the simulation results obtained on the dataset, two CNNs significantly outperformed traditional AMC methods thanks to the automatic feature extraction capability offered by convolutional layers. Furthermore, two CNNs were investigated with different signal lengths and different numbers of training samples under a synthetic channel impairment, involving multipath fading, carrier frequency offset, symbol rate offset, and AWGN.

To facilitate the varying input dimensions, a novel CNN with two training stages [315] was deployed for classifying the modulations of long symbol-rate observation sequences. The proposed CNN, namely CNN-AMC, incorporated the raw signal data with the estimation of symbol SNR as the supplemental information to enhance pattern learning. The first stage (pre-training) trained CNN-AMC on a basic dataset in the presence of AWGN and the second stage (fine-tuning) tuned the trained model with a new dataset by replacing some top layers with random parameter initialization. As a result, this training strategy can deal with different channel conditions and adapt various modulation patterns (i.e., being easy to update model with a new dataset). For performance evaluation, several simulation results were provided to show the effectiveness of two-training strategy and also the robustness under channel deteriorations. Besides, the proposed CNN-AMC revealed the preeminence in terms of accuracy and complexity over some existing FB approaches. Numerous CNNs have specified sophisticated convolutional blocks and advanced processing modules by cleverly incorporating multiple convolutional layers with others operation layers to improve diversified features. From the above literature, one can notice that handcrafted features are abandoned, and deep neural network architectures are employed instead to utilize the automatic feature extraction capability from those deep models, being the main reason of the provided improvement in terms of classification accuracy. Besides that, also thanks to another benefit of employing neural network, the provided flexibility makes it easy to offer additional functionality more than pure AMC. For example, to deal with crashed signals due to additive noise, a CNN-based framework [316] was introduced as a multi-function neural network with three modules: SNR prediction, classification, and signal processing, in which the input signals estimated with low SNR were first re-constructed by U-Net for signal re-construction and enhancement and then provided to a CNN for classification. Ref. [317] introduced a novel frequency selection layer to first detect the frequency band of signals and then filter out the out-of-band noise.

Apart from raw signal data with IQ and AP samples, the constellation image and spectrum image of digital modulations have been utilized to automatically identify modulations by CNNs [318], where the modulation classification is regarded as an image classification task. For example, ref. [319] evaluated the classification accuracy of AlexNet [320] and GoogleNet [321] on different datasets of gray-scale and color constellation images. Moreover, the performance of CNNs was investigated with various image resolutions and network configurations. Although this approach performs modulation classification more accurately than traditional AMC methods, it is more complex and requires a longer processing time. To reduce the network complexity, ref. [322] proposed a lightweight CNN with few convolutional layers to capture the representational features from regular constellation images and contrast enhanced grid constellation images. Moreover, this network exploited intra-class compactness and inter-class separability using a compressive loss constraint to improve the accuracy of higher-order digital modulation. Ref. [323] proposed a hierarchical CNN-based modulation classifier, in which one CNN was designed to classify low-order digital modulation formats using IQ samples and another one was specified to discriminate high-order digital modulation formats by learning visual features from constellation images. In the abovementioned CNNs, the activations resulted by the last convolutional layer were flattened to directly connect with neurons in the first fully connected layer, which dramatically increases the number of trainable parameters. This poor design strategy can be addressed by arranging a global average pooling layer before the first fully connected layer to reduce the number of parameters and also prevent overfitting.

13

DL-Based Interference Mitigation in Communication Systems

13.1 DL-Based Resource Allocation Techniques

Multiple-input multiple-output (MIMO) technology is an indispensable component in current systems as it has contributed to successfully improve system capability without requiring more time-frequency resources; Thus, it has also encouraged the development of accompanying MIMO system technologies, such as beamforming to fully utilize the provided spatial diversity and to better exploit channel characteristics. While optimization-based algorithms perform beamforming designs by solving an optimization problem, which is usually non-convex, and is computationally prohibitive in a real-time manner. Some researchers recently focused on exploiting deep learning (DL) models in communication system designs, as DL models use simple linear and nonlinear operations to address complicated problems efficiently. In light of this direction, ref. [324] is a pioneering work to explore DL-based beamforming designs. Considering a multi-user MIMO system, where a transmitter with P antennas is serving K users each with Q antennas at the same time, the received signal y_k at user k can be expressed as:

$$\mathbf{y}_k = \mathbf{H}_k \mathbf{s} + \mathbf{n}_k, \tag{13.1}$$

where $\mathbf{H}_k \in \mathbb{C}^{Q \times P}$ is the channel matrix, $\mathbf{s} \in \mathbb{C}^{P \times 1}$ is the transmitted signal, and n_k is the noise. The transmitted signal is obtained via beamforming operation as:

$$\mathbf{s} = \sum_{k=1}^{K} \mathbf{W}_k \mathbf{x}_k, \tag{13.2}$$

where $\mathbf{x}_k$ is the data vector for user k and $\mathbf{W}_k$ is the linear beamformer designed for user k. Then the beamforming design problem can be formulated as:

$$[\mathbf{W}_1, ..., \mathbf{W}_k] = \arg\max \sum_k u_k R_k, \tag{13.3}$$

DOI: 10.1201/9781003408239-13

satisfying power constraint $\sum_{k=1}^{K} \text{tr}(\mathbf{W}_k\mathbf{W}_k^H) \leq p_{\max}$. In the above equation, $R_k = \log\det(\mathbf{I}_k + \mathbf{W}_k^H\mathbf{H}_k^H\mathbf{J}^{-1}\mathbf{H}_k\mathbf{W}_k)$ is the achievable rate for user k and u_k is the weight for user k. With the above setting, the loss function of the proposed unsupervised learning beamforming design can be expressed as:

$$l(\mathbf{\Theta}; \mathbf{H}; \hat{\mathbf{W}}) = -\sum_{k=1}^{K} \log\det(\mathbf{I}_k + \mathbf{W}_k^{\mathbf{H}}\mathbf{H}_k^H\mathbf{J}^{-1}\mathbf{H}_k\mathbf{W}_k), \tag{13.4}$$

where $\mathbf{\Theta}$ is the trainable weight in the proposed neural network and $\hat{\mathbf{W}}$ is the designed beamforming weights for all users based on the channel condition $\mathbf{H} = [\mathbf{H}_1, ..., \mathbf{H}_k]$. In the training process, $\mathbf{\Theta}$ will be optimized until convergence, then the neural network is ready to use by producing $\hat{\mathbf{W}}$ for a new given H in the testing process. Simulation results confirm that the computation time can be reduced by at least 80% compared to the well-known WMMSE beamforming algorithm, although the performance cannot outperform the result of the WMMSE beamforming algorithm.

To further improve the beamforming performance of DL-based solutions but still maintain the real-time implementation advantage, ref. [252] proposes ResNet-Inspired beamforming (RI-BF), which inherits the advantages of both optimization-based and DL-based beamforming by employing deep unfolding techniques for efficient beamforming. Specifically, two DL-based modules, including a coarse estimator module and a deep unfolding module, are built in RI-BF. The coarse estimator module alone can already generate good beamforming designs. Then the deep unfolding module, which is developed based on the concept of the gradient ascent beamforming algorithm, can further enhance the performance of the final beamforming result. That is, traditional gradient ascent beamforming suffers from slow convergence and consequently, remarkable computational complexity. Although a well-designed initial beamforming, which can be provided by our coarse estimator module, can already reduce the number of iterations significantly, several iterations are still needed to reach convergence. Instead, by employing the deep unfolding technique to reference the optimization process of gradient ascent beamforming for the neural network architecture design, the beamforming efficiency can be improved significantly. Simulation results demonstrate that the performance and computational complexity of the proposed RI-BF improve significantly compared to the existing DL-based and optimization-based algorithms, respectively.

In this work, a typical interference channel system containing K pairs of a transmitter with N_{T} antennas and a receiver with N_{R} antennas is considered. Each transmitter communicates with its corresponding receiver, causing interference to other users simultaneously due to the propagation nature of the wireless channels. The received signal at the i th receiver $\mathbf{y}_i \in \mathbb{C}^{N_{\text{R}}\times 1}$ is described as follows:

$$\mathbf{y}_i = \mathbf{H}_{ii}\mathbf{S}_i + \sum_{j=1, j\neq i}^{K} \mathbf{H}_{ij}\mathbf{S}_j + \mathbf{n}_i, \tag{13.5}$$

where $\mathbf{H}_{ij} \in \mathbb{C}^{N_{\mathrm{R}} \times N_{\mathrm{T}}}$ is the channel between the i th transmitter and the j th receiver, $\mathbf{S}_i$ represents the transmitted vector and $\mathbf{n}_i$ is the noise vector. The first term is the desired signal sent from the i th transmitter and the second term represents the interference from other transmitters.

It is assumed that the transmitted vector can be further expressed as $\mathbf{S}_i = \mathbf{P}_i \mathbf{x}_i$ by means of beamforming, where $\mathbf{P}_i \in \mathbb{C}^{N_{\mathrm{T}} \times N_{\mathrm{T}}}$ and $\mathbf{x}_i \in \mathbb{C}^{N_{\mathrm{T}} \times 1}$ represents the beamformer and data symbol of the i th transmitter. Similarly, a combiner $\mathbf{M}_i$ can be applied at the i th receiver and the decoded output vector can be written as:

$$\hat{\mathbf{x}}_i = \mathbf{M}_i \mathbf{H}_{ii} \mathbf{S}_i + \mathbf{M}_i \sum_{j=1, j \neq i}^{K} \mathbf{H}_{ij} \mathbf{S}_j + \mathbf{M}_i \mathbf{n}_i. \tag{13.6}$$

$\mathbf{H} \in \mathbb{C}^{K^2 \times N_{\mathrm{R}} \times N_{\mathrm{T}}} = [\mathbf{H}_{11}, .., \mathbf{H}_{KK}]$ as block matrices containing all the channel information is denoted for later formulation uses. The goal of this work is to maximize the sum-rate of all users $\tilde{R}_{\text{sum}}$ to offer a better communication quantity. Focusing on designing the beamformer $\mathbf{P}_1, ..., \mathbf{P}_K$ subject to a given transmit power constraint E_{tx}, the optimization problem can be formulated as:

$$\begin{aligned} [\mathbf{P}_1, ..., \mathbf{P}_K] &= \arg\max_{\{\mathbf{P}_i\}} \tilde{R}_{\text{sum}} \\ &= \arg\max_{\{\mathbf{P}_i\}} \sum_{i=1}^{K} \log_2 \left| \mathbf{I} + \mathbf{H}_{ii} \mathbf{P}_i \mathbf{P}_i^H \mathbf{H}_{ii}^H (\mathbf{\Pi}_i^{-1}) \right| \\ &\text{subject to} \quad \sum_{i=1}^{K} \mathrm{tr}(\mathbf{P}_i^H \mathbf{P}_i) \leq E_{\text{tx}} \end{aligned} \tag{13.7}$$

where $\mathbf{\Pi}_i \in \mathbb{C}^{N_{\mathrm{R}} \times N_{\mathrm{R}}}$ represent the effective noise and interference covariance matrix at the i th receiver, being further expressed as $\mathbf{\Pi}_i = \sum_{j=1, j \neq i}^{K} \mathbf{H}_{ij} \mathbf{P}_j \mathbf{P}_j^H \mathbf{H}_{ij}^H + \sigma_n^2 \mathbf{I}$. To address the above problem efficiently, the proposed RI-BF algorithm is shown in Figure 13.1. Specifically, two neural network modules, a coarse estimator module and a deep unfolding module, are built in RI-BF. A given channel sample $\mathbf{H}$, will be fed into the coarse estimator module first, outputting initial beamforming. Then the following deep unfolding module will produce the final beamforming based on the initial beamforming, further improving the achieved performance.

As for the design of the proposed coarse estimator, two parallel neural networks, namely, the beamforming designer and power allocation actor, are constructed to focus on a specific part of complete beamforming designs, that is, beamforming weighting design and power design, respectively. The entire operation of the beamforming designer can be expressed as: $[\hat{\mathbf{P}}_1^{\text{BD}}, \hat{\mathbf{P}}_2^{\text{BD}}, ..., \hat{\mathbf{P}}_K^{\text{BD}}] = F_{\text{BD}}(\mathbf{H}; \mathbf{\Theta}_{\text{BD}})$, where $\mathbf{\Theta}_{\text{BD}}$ stands for all trainable weightings in the beamforming designer. Similarly, the operation of the power allocation actor can be described as $[\hat{\mathbf{P}}_1^{\text{PAC}}, \hat{\mathbf{P}}_2^{\text{PAC}}, ..., \hat{\mathbf{P}}_K^{\text{PAC}}] = F_{\text{PAC}}(\mathbf{H}; \mathbf{\Theta}_{\text{PAC}})$ where $\mathbf{\Theta}_{\text{PAC}}$ is the trainable weights in the power allocation actor. As a whole, the output of the

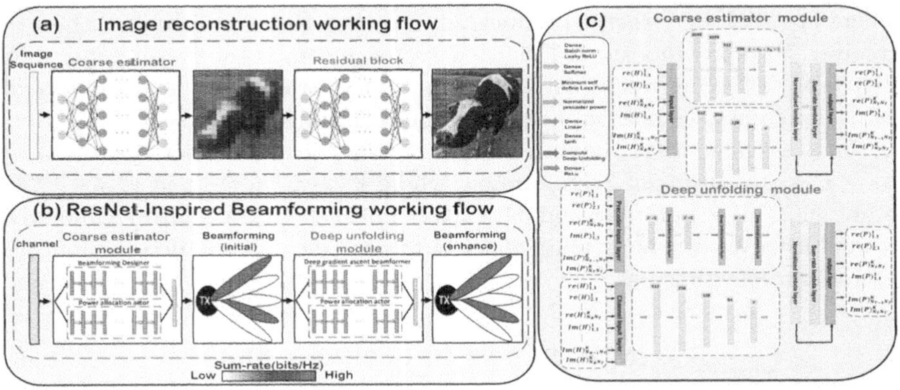

FIGURE 13.1
(a) Working flow of a typical ResNet-based DL model (b) RI-BF based on the concept of ResNet (c) NN architecture of RI-BF (copyright from ref. [252]).

coarse estimator, which is the combination of the outputs of the beamforming designer and the power allocation actor, can be represented as follows:

$$\hat{\mathbf{P}}_i^{\mathrm{CE}} = \sqrt{\frac{E_{\mathrm{tx}}\hat{\mathbf{P}}_i^{\mathrm{PAC}}}{\mathrm{tr}(\hat{\mathbf{P}}_i^{\mathrm{BD}}(\hat{\mathbf{P}}_i^{\mathrm{BD}})^H)}}\hat{\mathbf{P}}_i^{\mathrm{BD}}. \tag{13.8}$$

As for the design of the deep unfolding module, two neural networks, a deep gradient ascent beamformer, and a power allocator were constructed in the deep unfolding module, similar to the coarse estimator. Recalling that the goal is to maximize the sum-rate, which can be further expressed as:

$$\begin{aligned}(\tilde{\mathbf{P}}_{\mathbf{1}}^{\star}, ..., \tilde{\mathbf{P}}_{\mathbf{K}}^{\star}) &= \arg\max_{\{\tilde{\mathbf{P}}_i\}} \tilde{R}_{\mathrm{sum}} \\ &= \arg\max_{\{\tilde{\mathbf{P}}_i\}} \sum_{i=1}^{K} \log_2 \left|\tilde{\mathbf{\Phi}}_i\right| - \left|\tilde{\mathbf{\Pi}}_i\right|\end{aligned} \tag{13.9}$$

$\tilde{\mathbf{\Phi}}_i \in \mathbb{C}^{[\mathbf{N}_{\mathrm{R}} \times \mathbf{N}_{\mathrm{R}}]}$ and $\tilde{\mathbf{\Pi}}_i \in \mathbb{C}^{[\mathbf{N}_{\mathrm{R}} \times \mathbf{N}_{\mathrm{R}}]}$ are further defined as follows:

$$\tilde{\mathbf{\Phi}}_i = \sum_{j=1}^{K} \beta_j \mathbf{H}_{ij}\tilde{\mathbf{P}}_j\tilde{\mathbf{P}}_j^H\mathbf{H}_{ij}^H + \sigma_n^2\mathbf{I}. \tag{13.10}$$

$$\tilde{\mathbf{\Pi}}_i = \sum_{j=1, j\neq i}^{K} \beta_j \mathbf{H}_{ij}\tilde{\mathbf{P}}_j\tilde{\mathbf{P}}_j^H\mathbf{H}_{ij}^H + \sigma_n^2\mathbf{I}. \tag{13.11}$$

In Eq. (13.8) and Eq. (13.9), a beamforming result $\mathbf{P}_j^{\mathrm{ori}}$ of j th transmitter can be expressed as $\mathbf{P}_j^{\mathrm{ori}} = \sqrt{\beta_j}\tilde{\mathbf{P}}_j$, where $\beta_j = \frac{\hat{E}_{\mathrm{tx}}^{(j)}}{\mathrm{tr}(\tilde{\mathbf{P}}_j\tilde{\mathbf{P}}_j^H)}$ and $\hat{E}_{\mathrm{tx}}^{(j)}$ is the real power of j th transmitter. Then, the gradient of Eq. (3.7) can be derived as:

$$\begin{aligned}\nabla_{\tilde{\mathbf{P}}_k}\tilde{R}_{\text{sum}} &= \frac{2\beta_k}{\ln 2}\sum_{i=1}^{K}\mathbf{H}_{ik}^H\tilde{\mathbf{\Phi}}_i^{-1}\mathbf{H}_{ik}\tilde{\mathbf{P}}_k \\ &- \frac{2\beta_k^2}{\hat{E}_{\text{tx}}^{(k)}\cdot\ln 2}\sum_{i=1}^{K}\text{tr}(\tilde{\mathbf{P}}_k^H\mathbf{H}_{ik}^H\tilde{\mathbf{\Phi}}_i^{-1}\mathbf{H}_{ik}\tilde{\mathbf{P}}_k)\tilde{\mathbf{P}}_k \\ &- \frac{2\beta_k}{\ln 2}\sum_{i=1,i\neq k}^{K}\mathbf{H}_{ik}^H\tilde{\mathbf{\Pi}}_i^{-1}\mathbf{H}_{ik}\tilde{\mathbf{P}}_k \\ &+ \frac{2\beta_k^2}{\hat{E}_{\text{tx}}^{(k)}\cdot\ln 2}\sum_{i=1,i\neq k}^{K}\text{tr}(\tilde{\mathbf{P}}_k\mathbf{H}_{ik}^H\tilde{\mathbf{\Pi}}_i^{-1}\mathbf{H}_{ik}\tilde{\mathbf{P}}_k)\tilde{\mathbf{P}}_k.\end{aligned} \tag{13.12}$$

By substituting $\hat{\boldsymbol{P}}^{\boldsymbol{CE}}{}_i$ in Eq. (3.5) for $\mathbf{P^{ori}}_\mathbf{j}$, the result of the coarse estimator can be further improved by adopting the updated equation as follows:

$$\tilde{\mathbf{P}}_k(l) = \tilde{\mathbf{P}}_k(l-1) + \delta\nabla_{\tilde{\mathbf{P}}_k}\tilde{R}_{\text{sum}}, \tag{13.13}$$

where δ and l denote the step size and number of iterations, respectively of the gradient ascent algorithm.

Here, we define $\mathbf{G}_{j,l}^k = F_{\text{Lea}}(\mathbf{W}_{j,l}^k \cdot F_{\text{Bat}}(\text{Vec}(\tilde{\mathbf{P}}_k)))$, where $F_{\text{Lea}}(.)$ denotes "LeakyRelu" activation function, F_{Bat} stands for batch normalization, $\mathbf{W}_{j,l}^k \in \mathbb{R}^{1\times(N_{\text{T}}\times N_{\text{T}}\times 2)}$ are the trainable weights, and Vec(.) represents column-wise vectorization operation. In particular, the l-th layer of deep gradient ascent beamformer can be characterized as:

$$\begin{aligned}\nabla_{\tilde{\mathbf{P}}_k}\tilde{R}_{\text{sum}} &= \mathbf{G}_{1,l}^k\cdot\frac{2\beta_k}{\ln 2}\sum_{i=1}^{K}\mathbf{H}_{ik}^H\tilde{\mathbf{\Phi}}_i^{-1}\mathbf{H}_{ik}\tilde{\mathbf{P}}_k \\ &- \mathbf{G}_{2,l}^k\cdot\frac{2\beta_k^2}{\hat{E}_{\text{tx}}^{(k)}\cdot\ln 2}\sum_{i=1}^{K}\text{tr}(\tilde{\mathbf{P}}_k^H\mathbf{H}_{ik}^H\tilde{\mathbf{\Phi}}_i^{-1}\mathbf{H}_{ik}\tilde{\mathbf{P}}_k)\tilde{\mathbf{P}}_k \\ &- \mathbf{G}_{3,l}^k\cdot\frac{2\beta_k}{\ln 2}\sum_{\substack{i=1\\ i\neq k}}^{K}\mathbf{H}_{ik}^H\tilde{\mathbf{\Pi}}_i^{-1}\mathbf{H}_{ik}\tilde{\mathbf{P}}_k \\ &+ \mathbf{G}_{4,l}^k\cdot\frac{2\beta_k^2}{\hat{E}_{\text{tx}}^{(k)}\cdot\ln 2}\sum_{\substack{i=1\\ i\neq k}}^{K}\text{tr}(\tilde{\mathbf{P}}_k\mathbf{H}_{ik}^H\tilde{\mathbf{\Pi}}_i^{-1}\mathbf{H}_{ik}\tilde{\mathbf{P}}_k)\tilde{\mathbf{P}}_k\end{aligned} \tag{13.14}$$

and Eq. (3.11) can be transformed as follows:

$$\tilde{\mathbf{P}}_k(l) = \tilde{\mathbf{P}}_k(l-1) + \mathbf{G}_{5,l}^k\cdot\nabla_{\tilde{\mathbf{P}}_k}\tilde{R}_{\text{sum}}. \tag{13.15}$$

All trainable weightings in the deep gradient ascent beamformer can be expressed as:

$$\mathbf{\Theta}_{\text{DGA}} = \mathbf{W}_{1,l}^k, \mathbf{W}_{2,l}^k, \mathbf{W}_{3,l}^k, \mathbf{W}_{4,l}^k, \mathbf{W}_{5,l}^k \ \textit{for}\ l = 1, ..., 10. \tag{13.16}$$

Using Eq. (3.13), the beamforming result of the coarse estimator can be updated efficiently. $[\hat{\mathbf{P}}_1^{\mathrm{DGA}}, \hat{\mathbf{P}}_2^{\mathrm{DGA}}, ..., \hat{\mathbf{P}}_K^{\mathrm{DGA}}] = F_{\mathrm{DGA}}(\mathbf{H}; \boldsymbol{\Theta}_{\mathrm{DGA}}; \hat{\mathbf{P}}^{\mathrm{CE}})$ denotes the entire operation of coarse estimator mathematically. Then, to meet the power constraint, another power allocation actor, which has the same architecture as the former one, is built in the deep unfolding module. The operation is denoted as $[\hat{\mathbf{P}}_1^{\mathrm{PAC2}}, \hat{\mathbf{P}}_2^{\mathrm{PAC2}}, ..., \hat{\mathbf{P}}_K^{\mathrm{PAC2}}] = F_{\mathrm{PAC2}}(\mathbf{H}; \boldsymbol{\Theta}_{\mathrm{PAC2}}; \hat{\mathbf{P}}^{\mathrm{CE}})$ where $\boldsymbol{\Theta}_{\mathrm{PAC2}}$ are trainable weightings in the power allocation actor. Overall, the output of the deep unfolding module, which is the combination of the output of the deep gradient ascent beamformer and the power allocation actor, can be represented as follows:

$$\hat{\mathbf{P}}_i^{\mathrm{RI}} = \sqrt{\frac{E_{\mathrm{tx}} \hat{\mathbf{P}}_i^{\mathrm{PAC2}}}{\mathrm{tr}(\hat{\mathbf{P}}_i^{\mathrm{DGA}} (\hat{\mathbf{P}}_i^{\mathrm{DGA}})^H)}} \hat{\mathbf{P}}_i^{\mathrm{DGA}}. \tag{13.17}$$

Note that sum-rate is introduced as a loss function to train all the trainable weightings in the deep unfolding module, expressed as:

$$L(\boldsymbol{\Theta}_{\mathrm{DGA}}; \boldsymbol{\Theta}_{\mathrm{PAC2}}; \mathbf{H}; \hat{\mathbf{P}}) = -\tilde{R}_{\mathrm{sum}}. \tag{13.18}$$

With the above designs, and thus no labeling effort is needed to conduct the above training process, simulation results confirm that RI-BF outperforms the existing pure optimization-based and DL-based algorithms in terms of sum-rate while the complexity is significantly lower than WMMSE in various system configurations.

13.2 DL-Based Scheduling Techniques

In the above section, we discussed how DL-based solutions can be used to aid beamforming designs to solve the desired NR-hard optimization problems with efficiency. Similarly, in multi-user scheduling problem, DL-based solutions are introduced again as an efficient optimization solver, being the main topic of this section. Let us consider an orthogonal frequency division multiplexing (OFDM) system with time-division duplex mode in ref. [325]. In the considered system, a base station with M antennas aims to serve L single-antenna users in the coverage. Due to the limited spatial degree of freedom, the base station can only serve N users subject to $N \leq L$ and $N \leq M$. To do so, L users periodically send orthogonal pilot sequences to the base station for channel estimation purposes, and then the base station selects a set of N users for data transmission through beamforming based on their estimated channel and guides those selected users to utilize appropriate modulation schemes for uplink transmission using control channel message. In the uplink procedure, at each transmission time interval (TTI), the uplink signal model can be expressed as:

$$\mathbf{y} = \mathbf{H}\mathbf{u} + \mathbf{n}, \tag{13.19}$$

where $\mathbf{y} \in \mathbb{C}^{M\times 1}$ is the received signal at the base station, $\mathbf{H} \in \mathbb{C}^{M\times N}$ is the channel matrix, $\mathbf{u} \in \mathbb{C}^{N\times 1}$ is the transmitted information vector from N users, and $\mathbf{n} \in \mathbb{C}^{M\times 1}$ is the noise vector. Note that in such a system, the value of N can vary in each TTI based on channel condition and base station load, and we assume $N \leq N_{\max}$ holds as its upper bound. In a 5G OFDM system, considering resource block (RB) as the smallest scheduling unit, the interested scheduling optimization problem can be formulated as:

$$\begin{aligned} &\arg\max_{x^t_{l,b}} \sum_b^B \sum_l^L w^t_{l,b} x^t_{l,b}, \\ &\text{s.t.} \sum_l^L x^t_{l,b} \leq N_{\max} \end{aligned} \tag{13.20}$$

where $x^t_{l,b}$ is a binary selection indicator of user l at TTI t and RB b.

$$w^t_{l,b} = \frac{r^t_{l,b}}{\sum_b^B p^t_{l,b}} \tag{13.21}$$

is the weighted rate based on the instantaneous rate $r^t_{l,b}$ and received rate $p^t_{l,b}$, which is further defined as:

$$p^t_{l,b} = \begin{cases} p^{t-1}_{l,b} + r^{t-1}_{l,b}, \text{if } x^{t-1}_{l,b} = 1 \\ p^{t-1}_{l,b}, \text{otherwise} \end{cases} \tag{13.22}$$

Note that by introducing the received rate to normalize the instantaneous rate in Eq. (13.22), the user fairness can be guaranteed even when some users are with poor channel conditions. Otherwise, those users might never be selected due to the poor channel conditions and consequently limited contribution to the total rate. Furthermore, it is also noteworthy that Eq. (13.20) is an NP-hard problem since it can be reformulated as an integer linear programming (ILP) problem. Thus, there are no efficient solutions to solve the aforementioned optimization problem, and exhaustive search remains the mainstream method to obtain the optimal solution to this problem. To present an efficient solver to this problem with the aid of DL-based solutions, the authors in this work propose a novel scheduler based on the combination of soft actor-critic (SAC) reinforcement learning (RL) method and k-nearest neighbor (KNN) method. Specifically, although providing good performance in several real-time control problems, SAC cannot be directly employed to solve the problem owing to the fact that SAC is designed to handle continuous action spaces instead of the considered discrete action spaces. To adapt it to the considered problem, KNN algorithm is utilized to discretize SAC results to a discrete action space. As a result, the decision of the proposed method can be employed as a scheduling decision in real communication systems.

Specifically, as for the design of the proposed SAC method, the state space of user l at TTI t is defined as $\mathbf{s}_t^l = [\gamma_t^l, f_t^l, g_t^l] \in \mathbf{S} = [\mathbf{\Gamma}, \mathbf{F}, \mathbf{G}]$, where γ_t^l is the maximum achievable spectral efficiency of user l at TTI t, f_t^l is the total amount of transmitted data by user l until TTI t, and g_t^l is the user group label of user l at TTI t. The action space of TTI t can be expressed as $a_t \in \mathcal{A}$, where $\mathcal{A}$ is the total set of all possible user scheduling combinations so the space size grows exponentially as $\sum_{i=1}^{N_{\max}} \binom{L}{i}$. Finally, the instantaneous reward is defined as:

$$R_t = \beta \gamma_t^{\text{total}} + (1 - \beta) JFI_t, \tag{13.23}$$

where γ_t^{total} is the sum rate achieved by all scheduled users at TTI t, JFI_t is Jain's fairness index (JFI), and β is an important factor to control the ratio between the first term and the second term. In Eq. (13.23), the JFI reward can be further expressed as:

$$JFI_t = \frac{(\sum_{l=1}^{L} f_l^t)^2}{L \sum_{l=1}^{L} (f_l^t)^2}. \tag{13.24}$$

To elaborate, the JFI reward can be used to promote fairness between users. Thus, the reward design in Eq. (13.23) aims to consider the system performance and user fairness jointly in the scheduling process, where pre-defined parameter β can be used to adjust the reward emphasis to adopt different requirements in real communication systems. After SAC RL method generates an appropriate action in the defined action space, this action cannot be used in the MIMO scheduling decision directly until discretization. Moreover, as aforementioned, the action space size grows exponentially, further prohibiting its use in real communication systems. To solve this problem, the authors in this work propose to use KNN for the discretization. Specifically, after SAC method generates a so-called proto continuous action (e.g., a real number in $[-1, 1]$) from the defined action space, KNN is employed to calculate the l^2-norm between the proto action and pre-defined discrete actions to generate K discrete actions. Those discrete actions will be fed into the critic network in SAC structure to compute the expected rewards, so that the one with the highest expected reward will be selected as the final action. The whole procedure is illustrated in Figure 13.2 for reference. In both simulations and real-world evaluations, the proposed method shows improvements compared to conventional scheduling designs in terms of spectral efficiency and JFI score. In other words, by generating better scheduling results (i.e., better user grouping, for example, grouping users with lower channel correlation together to avoid mutual interference), both performance matrices in the reward function design is improved to demonstrate the benefits provided by the proposed RL-based solutions.

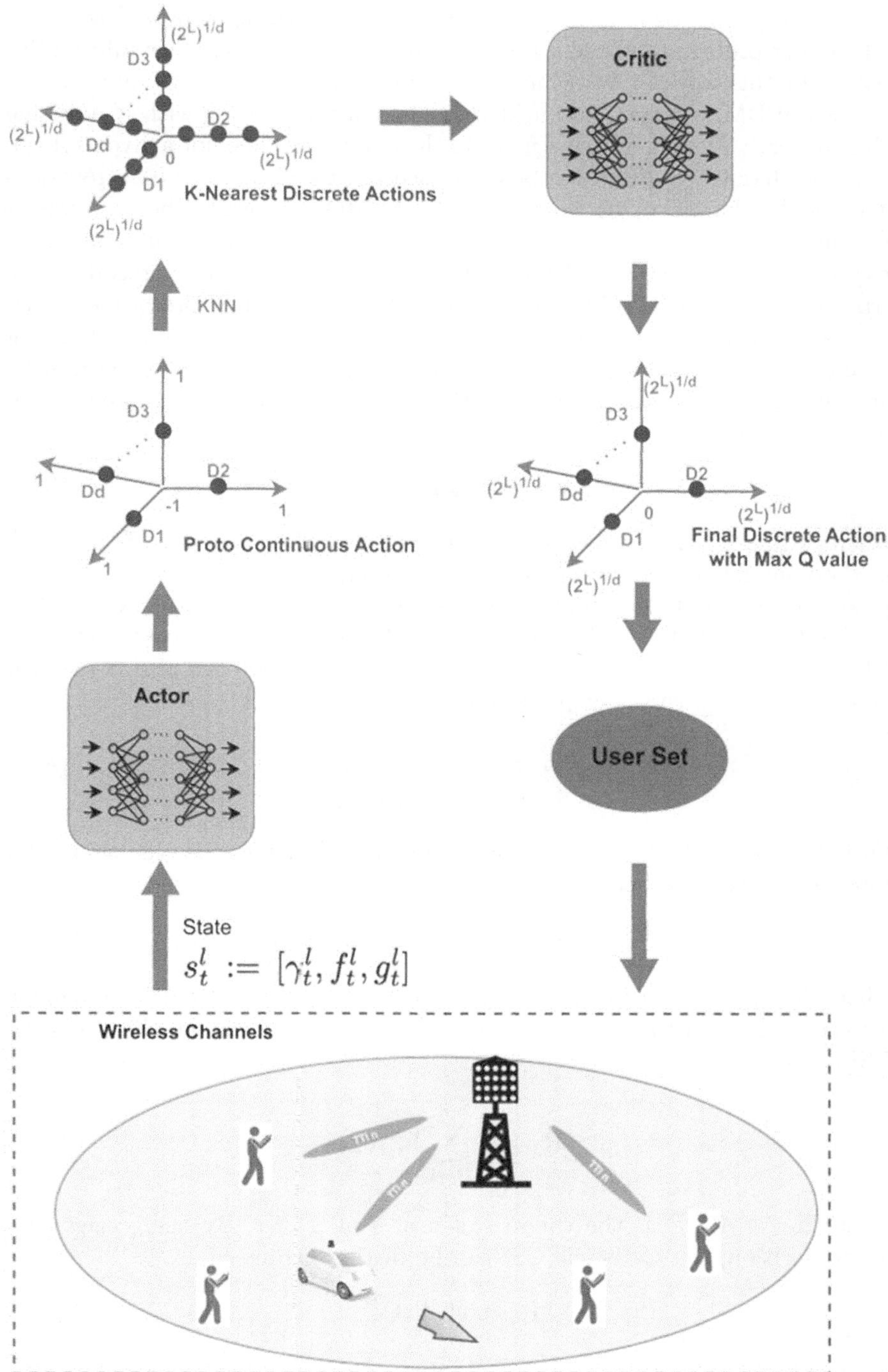

FIGURE 13.2
SMART architecture proposed in ref. [325] (reproduced from ref. [325]).

Similarly, ref. [326] also develops an advanced RL algorithm to solve the scheduling problem and relax the need for precise channel state information (CSI). In this work, a downlink orthogonal frequency division multiplexing access (OFDMA) system is considered, where a base station with N antennas aims to serve J single-antenna users in the coverage. Thus, for a given subcarrier, the channel response can be expressed as $\mathbf{H} \in \mathbb{C}^{J\times N}$ and each row of $\mathbf{H}$ denotes the channel of jth user. With the MIMO advantage, the base station is able to spatially multiplex K different users using the same time-frequency resources (i.e., RB in 5G OFDM system) by employing a linear precoding matrix $\mathbf{W} \in \mathbb{C}^{N\times K}$. If hybrid precoding is adopted, the linear precoding matrix can be further expressed as $\mathbf{W} = \mathbf{W}_{\text{RF}}\mathbf{W}_{\text{BB}}$, where $\mathbf{W}_{\text{RF}} \in \mathbb{C}^{N\times K}$ is the analog precoder and $\mathbf{W}_{\text{BB}} \in \mathbb{C}^{K\times K}$ is the digital precoder. Assuming K users are selected out of J users to be served in a specific TTI, the system model of such a system can be expressed as:

$$\mathbf{y} = \tilde{\mathbf{H}}\mathbf{W}\sqrt{\mathbf{P}} \in \mathbf{x} + \mathbf{z}, \tag{13.25}$$

where $\mathbf{y} \in \mathbb{C}^{K\times 1}$ is the received signal vector of K served user, $\tilde{\mathbf{H}} \in \mathbb{C}^{K\times N}$ is the channel matrix of the K selected users, $\mathbf{P} \in \mathbb{R}^{K\times K}$ is a diagonal power matrix with the power allocated to each selected user, $\mathbf{x} \in \mathbb{C}^{K\times 1}$ is the transmitted signal, and $\mathbf{z} \in \mathbb{C}^{K\times 1}$ is the noise vector. Defining $\mathbf{M} = [m]_{i,j} = \tilde{\mathbf{H}}\mathbf{W}\sqrt{\mathbf{P}} \in \mathbb{C}^{K\times K}$, the signal to interference-plus-noise ratio of selected user k can be expressed as:

$$\gamma_k = \frac{|m_{k,k}|^2}{\sigma^2 + \sum_{j\neq k}^{K} |m_{k,j}|^2}. \tag{13.26}$$

Furthermore, if 256-QAM is employed for data transmission, the achievable rate can be expressed as:

$$r_k = N_{\text{sc}}N_{\text{symb}} \min\{\log_2(1+\gamma_k), 8\}(\text{bits/TTI}), \tag{13.27}$$

where N_{sc} is the number of subcarriers and N_{symb} is the number of symbols for the considered time-frequency resources. In such a system, the statistical CSI can be expressed as:

$$\mathbf{R}_j = \frac{1}{T}\sum_{t=1}^{T} \mathbf{h}_{t,j}^H \mathbf{h}_{t,j}, \tag{13.28}$$

where $\mathbf{h}_{t,j} \in \mathbb{C}^{1\times N}$ is the channel of user j at TTI t. By using eigenvalue decomposition, the statistical CSI can be further expressed as:

$$\mathbf{R}_j = \mathbf{U}_j\mathbf{\Lambda}_j\mathbf{U}_j^H, \tag{13.29}$$

where $\mathbf{U}_j \in \mathbb{C}^{N\times N}$ and $\mathbf{\Lambda}_j \in \mathbb{C}^{N\times N}$. With the above setting, user grouping can be finished via K-means clustering algorithm to split J users into C clusters. In this work, the dominant column eigenvector $\mathbf{u}_{j,1}$ is input to K-means algorithm to cluster the users. The centroid of each cluster is updated using the mean of the dominant eigenvectors of the users currently belonging to the cluster. Then, the association of users to the clusters is updated by assigning each user to the cluster whose centroid is closest to its dominant eigenvector $\mathbf{u}_{j,1}$ until convergence. Regarding the hybrid precoding design, a simple analog precoder for the k selected users can be designed as:

$$\mathbf{W}_{\mathrm{RF}} = \frac{1}{\sqrt{N}}[e^{j\angle \mathbf{u}_{1,1}} \ \dots \ e^{j\angle \mathbf{u}_{k,1}}], \tag{13.30}$$

by using the phase of the dominant eigenvector of each user. Then, the zero-forcing (ZF) digital precoder can be designed as:

$$\mathbf{W}_{\mathrm{BB}} = \frac{\mathbf{H}_{\mathrm{eq}}^{H}(\mathbf{H}_{\mathrm{eq}}\mathbf{H}_{\mathrm{eq}}^{H})^{-1}}{||\mathbf{H}_{\mathrm{eq}}^{H}(\mathbf{H}_{\mathrm{eq}}\mathbf{H}_{\mathrm{eq}}^{H})^{-1}||_{\mathrm{F}}}, \tag{13.31}$$

where $\mathbf{H}_{\mathrm{eq}} = \tilde{\mathbf{H}}\mathbf{W}_{\mathrm{RF}} \in \mathbb{C}^{K\times K}$. Assuming users belonging to different clusters are low-correlated with each user, the interference among users from different clusters becomes negligible by using ZF digital precoding, which can be used to reduce the number of possible compositions of groups of users. That is, assuming the reward d is set as the system achievable rate of scheduled K users, the original action space is:

$$A = \tbinom{J}{K}. \tag{13.32}$$

Using the previous negligible interference assumption and assuming there is a virtual agent of each group to select users in the group to be scheduled for services, the reduced action space is given as:

$$A_c = \tbinom{J_c}{K_c}, \tag{13.33}$$

where J_c is the total number of scheduled users and K_c is the total number of users of cluster c. Assuming that each virtual cluster has its action values stored in $\mathbf{d}_c = \mathbb{R}^{A_c\times 1}$, the real action value of an action is defined as the mean received reward when the action is selected. Thus, by the law of large numbers, the incremental average updating method can be used to update $\mathbf{d}_c$, further expressed as:

$$\mathbf{d}_c(a_c) = \mathbf{d}_c(a_c) + \frac{1}{\mathbf{n}_c(a_c)}(d - \mathbf{d}_c(a_c)), \tag{13.34}$$

where a_c is the given action and $\mathbf{n}_c$ is the vector containing the number of times that each action was selected. With the above setting, the proposed RL algorithm is able to perform user scheduling without precise CSI. After user scheduling, only the selected users will be required to provide feedback on CSI for precoder design purposes. That is, the base station will take the following steps in each TTI: (1) schedule users using the proposed RL method; (2) require selected users to perform CSI feedback to design an analog precoder;

(3) compute a digital precoder based on the designed analog precoder. Note that this procedure is different from most existing solutions, which often require CSI knowledge to perform user scheduling, instead of using statistical CSI in this work. In other words, RL is employed to perform user scheduling based on statistical CSI. Simulation results confirm that the proposed method outperforms the conventional scheduling method by offering improved system throughput and reducing needed CSI feedback simultaneously.

13.3 DL-Based NOMA Transceiver Designs

In the direction of learning-based NOMA transceiver designs, there are also several impressive works, utilizing the power of DL solutions to offer a better trade-off between achieved performance and computational complexity. In ref. [327], a DL model is proposed to perform power allocation to improve both sum data rate and energy efficiency of MIMO-NOMA systems. In the considered scenario, a base station with M antennas is severing D multi-antenna users, each with $N_\mathrm{r} \geq M$ antennas. To adopt MIMO-NOMA system, all the users are grouped into M clusters and each cluster is with K users (i.e., $D = KM$). Letting $\boldsymbol{s} \in \mathbb{C}^{M\times 1}$ presents the information vector in the base station, which can be further expressed as:

$$\mathbf{s} = \begin{bmatrix} \bar{s}_1 \\ \vdots \\ \bar{s}_M \end{bmatrix} = \begin{bmatrix} \beta_{1,1}s_{1,1} + \cdots + \beta_{1,K}s_{1,K} \\ \vdots \\ \beta_{M,1}s_{M,1} + \cdots + \beta_{M,K}s_{M,K} \end{bmatrix}. \tag{13.35}$$

In Eq. (13.35), $s_{m,k} \sim \mathcal{CN}(0, \epsilon)$ is the information signal transmitted to the user k in the cluster m. The transmit power per symbol is denoted as ϵ and $\beta_{i,j}$ is the NOMA power allocation coefficient. The information vector will be processed in the base station before transmission, showing as:

$$\mathbf{x} = \mathbf{P}\mathbf{s}, \tag{13.36}$$

where $\mathbf{P} \in \mathbb{C}^{M\times M}$ is the precoding matrix. Thus, if we denote $\mathbf{H}_{m,k} \in \mathbb{C}^{N_\mathrm{r}\times M}$ as the channel matrix for the user k in the cluster m, the received signal at the user k in the first cluster can be expressed as:

$$\mathbf{y}_{1,k} = \mathbf{H}_{1,k}\mathbf{P}\mathbf{s} + \mathbf{z}_{1,k}, \tag{13.37}$$

where $\mathbf{z}_{1,k}$ is the noise vector. Then, by the NOMA general assumption that $\beta_{1,1} \leq \beta_{1,2} \leq \cdots \leq \beta_{1,K}$, the channel gain can also be ordered as:

$$|\mathbf{v}_{1,K}^H\mathbf{H}_{1,K-1}\mathbf{p}_1|^2 \leq |\mathbf{v}_{1,K-1}^H\mathbf{H}_{1,K-1}\mathbf{p}_1|^2 \leq \cdots \leq |\mathbf{v}_{1,1}^H\mathbf{H}_{1,1}\mathbf{p}_1|^2, \tag{13.38}$$

where $\mathbf{v}_{1,k}$ represents the detection vector at user k and $\mathbf{p}_i$ is the i-th column of $\mathbf{P}$. Assuming classical maximum, ratio combining (MRC) is utilized for detection purposes, given a fixed $\mathbf{P}$, the detection vector satisfies

$$\mathbf{v}_{1,k}^H \mathbf{h}_{m,1k} = 0, \tag{13.39}$$

where $\mathbf{h}_{m,1k}$ is the $m$column of $\mathbf{H}_{1,k}$. Under this constraint, $\mathbf{v}_{1,k}$ can be obtained as:

$$\mathbf{v}_{1,k} = \mathbf{U}_{1,k}\mathbf{n}_{1,k}. \tag{13.40}$$

In Eq. (13.40) $\mathbf{U}_{1,k}$ can be obtained by including all left singular vectors of matrix $[\mathbf{h}_{2,1k}, \mathbf{h}_{3,1k} \cdots \mathbf{h}_{M,1k}]$. Also, the normalized vector $\mathbf{n}_{1,k}$ can be obtained as:

$$\mathbf{n}_{1,k} = \frac{\mathbf{U}_{1,k}^H \mathbf{h}_{1,1k}}{|\mathbf{U}_{1,k}^H \mathbf{h}_{1,1k}|}. \tag{13.41}$$

By utilizing this detection vector, the received signal in the user k of the first cluster can be expressed as:

$$\begin{aligned}
\mathbf{v}_{1,k}^H \mathbf{y}_{1,k} &= \mathbf{v}_{1,k}^H \mathbf{H}_{1,k}\mathbf{P}\mathbf{s} + \mathbf{v}_{1,k}^H \mathbf{z}_{1,k} \\
&= \mathbf{v}_{1,k}^H \mathbf{H}_{1,k}\mathbf{p}_1(\beta_{1,1}s_{1,1} + \cdots + \beta_{1,K}s_{1,K}) \\
&+ \sum_{m=2}^{M} \mathbf{v}_{1,k}^H \mathbf{H}_{1,k}\mathbf{p}_m\bar{s}_m + \mathbf{v}_{1,k}^H \mathbf{z}_{1,k}.
\end{aligned} \tag{13.42}$$

Furthermore, assuming perfect SIC is performed to let each user in the same cluster decode the desired information successfully, the SINR of user k in the first cluster can be expressed as:

$$\gamma_{1,k}^k = \frac{|\mathbf{v}_{1,k}^H \mathbf{H}_{1,k}\mathbf{p}_1|^2 \beta_{1,k}^2}{\sum_{n=1}^{k-1} |\mathbf{v}_{1,k}^H \mathbf{H}_{1,k}\mathbf{p}_1|^2 \beta_{1,n}^2 + \sum_{m=2}^{M} |\mathbf{v}_{1,k}^H \mathbf{H}_{1,k}\mathbf{p}_m|^2 + |\mathbf{v}_{1,k}|^2 \frac{1}{\eta}}, \tag{13.43}$$

where η is the transmitted SNR. The data rate of user k in the first cluster can be further expressed as:

$$R_{1,k} = \log_2(1 + \gamma_{1,k}^k), \tag{13.44}$$

and the sum rate is

$$R_{\text{sum}} = \sum_{m=1}^{M} \sum_{k=1}^{K} R_{m,k}. \tag{13.45}$$

With the above setting, the interested optimization problem in this work can be formulated as:

$$\begin{aligned}
&\max_{\mathbf{p}_m, \beta_{i,j}} R_{\text{sum}} = \max_{\mathbf{p}_m, \beta_{i,j}} \sum_{m=1}^{M} \sum_{k=1}^{K} R_{m,k} \\
&\text{s.t. } 0 \le \beta_{m,k} \le 1, \forall m, k \text{ and } 0 \le ||\mathbf{p}_m|| \le P_{\text{tr}}, \forall m.
\end{aligned} \tag{13.46}$$

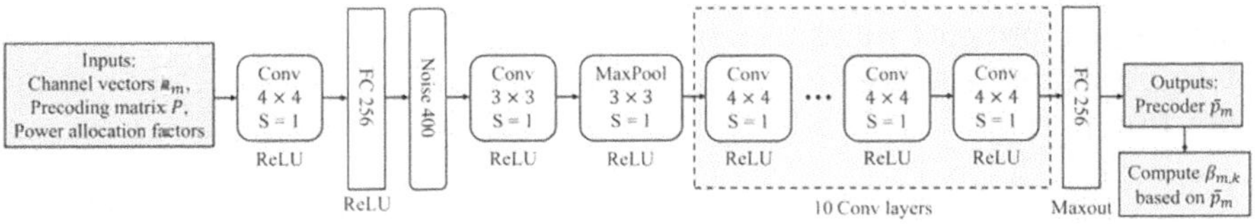

FIGURE 13.3
A CDNN-based MIMO-NOMA framework (Conv represents the convolutional layer, FC represents the fully connected layer, MaxPool means the maxpoooling operation, and S represents the strides). This framework consists of 11 Conv layers and 2 FC layers, as well as one Maxpooling layer. Also, precoding $\mathbf{P}$ contains precoders for each antenna, and the output precoder p_m and its corresponding power allocation factors are the optimal power allocation results (copyright from ref. [327]).

The total power constraint and minimum data rate requirement can be further considered in Eq. (13.46) by satisfying $\sum_{m=1}^{M} ||\mathbf{p}_m|| \leq P_{\text{tr}}, \forall m$, $\sum_{m=1}^{M}\sum_{k=1}^{K}\beta_{m,k} = 1$, and $R_{m,k} \geq R_{\min}$, where P_{tr} is the pre-defined power constraint and $R_{\min}$ is the pre-defined minimum data rate. Owing to the non-convex nature of the optimization problem Eq. (13.46), there is no efficient solution to address this problem so the motivation for introducing DL-based solutions is to solve the underlying algorithm deficit. In this work, a DL model is proposed to perform an efficient exhaustive search to address the interested power allocation problem without heavy computational burden. Specifically, as shown in Figure 13.3, the model inputs include the channel matrix $\mathbf{h}_m$, the precoding matrix $\mathbf{P}$, and the power allocation factors $\{\beta_{1,1}, \cdots \beta_{i,j}, \cdots \beta_{M,K}\}$ and the output of the model is normalized power allocation results $\{||\bar{\mathbf{p}}_1||, \cdots ||\bar{\mathbf{p}}_2||, \cdots ||\bar{\mathbf{p}}_M||\}$. The loss of the proposed model is formulated as:

$$\begin{aligned} l_{\text{CDNN}} &= \mathbb{E}\left[-\sum_{k=1}^{K}\sum_{m=1}^{M} R_{m,k} + \tau\sum_{k=1}^{K}\sum_{m=1}^{M}(R_{\min} - R_{m,k}) - \rho\sum_{m=1}^{M} p_m\right] + l_2 \\ &= -\frac{1}{NMK}\sum_{j=1}^{N} R_{\text{sum}}(\beta_{i,j}, p_m) \\ &\quad + \tau\sum_{k=1}^{K}\sum_{m=1}^{M}(R_{\min} - R_{m,k}) - \rho\sum_{m=1}^{M} p_m + l_2, \end{aligned} \tag{13.47}$$

where N is the number of training samples and l_2 norm of trainable weights are considered for training robustness. While Eq. (13.37) can be considered as an unsupervised optimizer to maximize R_{sum} by adjusting power allocation variables $\beta_{i,j}$ and p_m, the aforementioned power constraints in the original optimization problem are also formulated as the penalty terms in Eq. (13.37), causing loss decreasing if violated. Thus, by using the above loss function to train the proposed model, the model will be trained to find the optimal power allocation results according to the input to maximize the loss function through the backpropagation algorithm. Once the optimal mapping from a given input to the outputted power allocation results is obtained, the optimal trainable weight will be saved so that real-time power allocation can be performed in the testing phase by inputting an input from a new scenario. Simulation results confirm that the proposed method outperforms existing solutions in terms of both data rate and energy efficiency. Moreover, the computational complexity is also improved since only simple matrix operations are needed in the testing phase to output the power allocation results for a given scenario.

Besides this work, there are also several great survey papers discussing the recent achievements of learning-based NOMA solutions. interested readers can refer to refs. [328–330] for more information.

13.4 DL-Based FHSS

After discussing how to utilize the available spatial, time, and power resources to provide interference mitigation capability to a wireless communication system with the aid of DL-based solutions, we conclude this chapter by introducing the development of learning-based FHSS solutions for anti-interference purposes. Specifically, besides the interference caused by other users in the same time-frequency resources, interference introduced by malicious users (i.e., jammers) should also be considered, especially for military communication purposes. Since the goal of the jammer is to try its best to destroy legitimate communications and the goal of the legitimate transceiver is to maintain the communication quality, the problem is also formulated as a game with two players. In this direction, RL has already shown impressive capability to address such problems, such as video games or Go games against other players. Thus, several great works utilize reinforcement learning for the FHSS transceiver design, aiming to obtain the optimal channel selection policy to improve the system capacity during malicious attacks, being the main topic we will discuss in this section.

In this direction, ref. [331] is a pioneering work for the development of RL anti-jamming FHSS system design. In this work, a wireless communication scenario, where a transmitter-receiver pair is operating under a jamming attack, is considered as shown in Figure 13.4. In this work, it is assumed that there is a reinforcement agent, that can make anti-jamming decisions based on receiver observations, and then send the decision to the transmitter for decision execution. It is also assumed that the wideband sensing mechanism is performed on the receiver side so that the transmitter can adjust the transmitting frequency band without notifying the receiver without interrupting communication. At the same time, the jammer can also switch its jamming

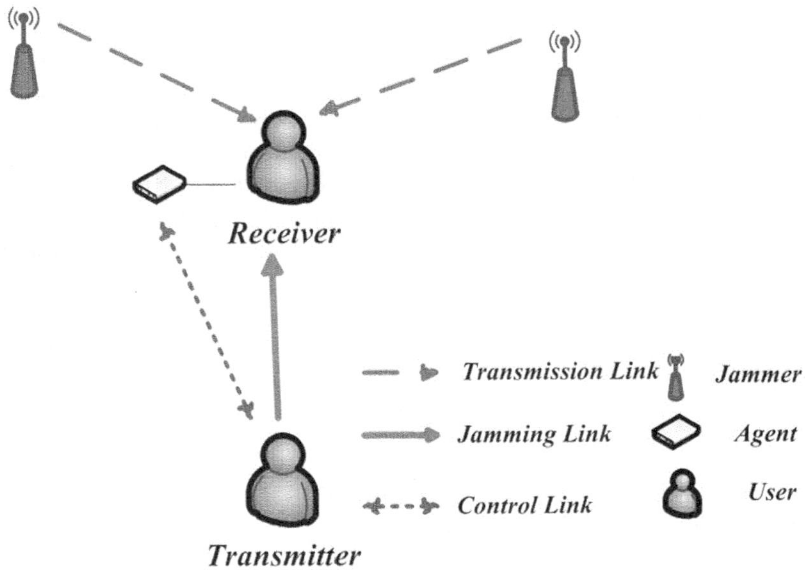

FIGURE 13.4
System model in ref. [331].

band based on different jamming patterns. Thus, through wideband sensing, the historical jamming information can be detected and stored to learn the jamming pattern and adjust the transmitter decision accordingly. To do so, the spectrum vector of the communication band at time t is denoted as $\mathbf{P}_t = (p_{t,1}, p_{t,2}, \cdots, p_{t,i}, \cdots, p_{t,n})$, where $p_{t,i} = 10\log[\int_{i\Delta f}^{(i+1)\Delta f} S(f + f_L)df]$ is the spectral energy of frequency i at time t, $S(f)$ is the power spectral density (PSD) function, and Δf is the spectrum resolution. Then the time-frequency characteristic $\mathbf{S}_t$ (i.e., spectral waterfall diagram) can be expressed as:

$$\mathbf{S}_t = \begin{bmatrix} \mathbf{P}_{t-1} \\ \vdots \\ \mathbf{P}_{t-M} \end{bmatrix} = \begin{bmatrix} p_{t-1,1} & p_{t-1,2} & \cdots & p_{t-1,N} \\ \vdots & \vdots & \ddots & \vdots \\ p_{t-M,1} & p_{t-M,2} & \cdots & p_{t-M,N} \end{bmatrix}. \tag{13.48}$$

Note through the introduction of the time-frequency characteristic, the jamming behavior can be tackled as a pattern recognition task in this time-frequency image, thus powerful DL-based solutions can be employed from the computer vision research area without heavy modifications. Assuming the central frequency of the legitimate link is f_t, the central frequency of the jamming link is f_j, the user's transmission bandwidth is b, the PSD function of noise is $n(f)$, the PSD function of the jamming signal is J_t, the transmission power is p, the channel gain is g_t and the channel gain of the jamming link is

g_j, the SINR of the received signal can be expressed as:

$$\eta(f_t, f_j) = \frac{g_t p}{\int_{f_t - b/2}^{f_t + b/2} \{n(f) + \sum_{j=1}^{J} g_j J_t(f - f_j)\} df}. \tag{13.49}$$

Denote $u(f_t, f_j)$ as the indicator function for successful transmission, which can be expressed as:

$$u(f_t, f_j) = \begin{cases} r_m, \ \eta(f_t, f_j) \geq \eta_{\text{th}} \\ 0, \ \eta(f_t, f_j) < \eta_{\text{th}} \end{cases}, \tag{13.50}$$

η_{th} in the equation is the pre-defined SINR threshold to tell if the legitimate communication is successful or not and r_m is a positive reward when conducting successful transmissions. Assuming that the communication frequency range equals the jamming frequency range as $B_{\text{u}} = B_{\text{j}}$ and the user's transmission band is represented as b_{u}, the number of frequency selection choices the transmitter can make is $n = \frac{B_{\text{u}}}{b_{\text{u}}}$. Thus, the action space can be expressed as $A = \{a_1, a_2, \cdots, a_n\}$, and $a(t) \in A$ represents the channel selection result at time t. However, switching channels also comes with cost, which can be expressed as:

$$W(f_t, f_j) = \begin{cases} 0, \ a(t) = a(t-1) \\ c, \ a(t) \neq a(t-1), \end{cases} \tag{13.51}$$

where c is the channel switching cost. With the above prior, the interested optimization problem can be formulated as:

$$\max_{f_t' \in A} U = \sum_{t=0}^{\infty} \gamma^t (u(f_t, f_j) - W(f_t, f_j)), \tag{13.52}$$

where $\gamma \in (0, 1)$ is the discount factor. In this work, the goal is to make anti-jamming decisions by adjusting the transmitting frequency band to maximize the cumulative future reward shown in Eq. (13.52) to maximize the transmission success rate while minimizing switching costs. From the above discussion, one can notice that the jamming pattern is the key to anti-jamming decisions. If the jamming pattern can be learned from the historical data, the transmitter can use this information to adjust the future selected frequency band to increase the transmission success rate. To do so, a DL model is proposed to perform jamming pattern classification. As shown in Figure 13.5, the proposed model is based on CNN to perform efficient classification based on given time-frequency characteristics. After the jamming mode classification, the result will be passed to the RL agent to make the FHSS decision, as shown in Figure 13.5. Specifically, classical Q-learning is adopted for the development of the proposed RL agent. Utilizing the current channel status as the state, the agent is trained to perform frequency band selection under the previously discussed action space. Then based on whether the channel is jammed or not in the next timeslot, the aforementioned reward can be computed for RL

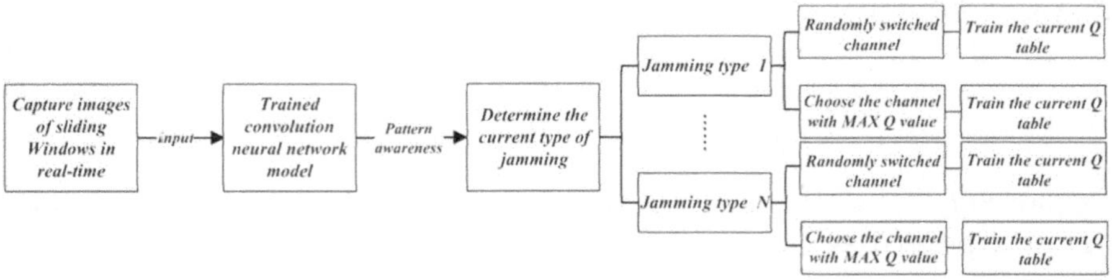

FIGURE 13.5
Algorithm flowchart in ref. [331].

training purposes. Note that to simplify the training process and relax the need for huge data samples, several RL agents will be trained to learn the best frequency switching policy under different jamming patterns. However, in the testing phase, only one agent will be triggered at one time based on the jamming mode classification result of the underlying environment. Simulation results confirm the effectiveness and practicability of the proposed anti-jamming communication method by showing improved throughput and lower switching times compared to existing works.

Following similar ideas, RL is widely used in the FHSS anti-jamming designs. Interested readers are suggested to refer to refs. [332–335] for the recent development in this direction.

14

DL-Based Signal Processing in Radar Systems

14.1 DL-Based Radar Waveform Designs

Similar to the concept of deep learning (DL)-based communication waveform designs, DL-based algorithms have also been introduced to aid waveform-related research activities in recent years. Specifically, current related literature can be categorized into several directions. First, we notice that several works utilize learning-based schemes to recognize radar waveforms for jamming purposes [336–339]. To elaborate, besides commercial usage, radar systems also play a critical role in military usage to offer accurate information about detected objects, such as airplanes or battleships. In contrast, jamming can be used as an effective way to avoid information collection from the enemy. Thus, if we can recognize the radar waveform that the enemy currently uses, it starts to be possible to perform more precise jamming to interrupt radar sensing with minimal resources. For this reason, learning-based schemes are widely used to recognize different radar waveforms in realistic environments. Compared to traditional statistic-based recognition methods, the capability of learning-based algorithms to extract features helps to increase the detection rate while maintaining the false alarm rate in such scenarios. Second, to tackle the above precise jamming to the transmitted radar waveforms, anti-jamming methods are also developed utilizing the capability of DL-based algorithms by introducing learning-based schemes to select radar waveforms according to the current environment [340–343]. Specifically, if the employed radar transceiver is capable of operating several different radar waveforms, then reinforcement learning (RL)-based algorithms can be employed straightforwardly to improve the radar transceiver by offering adaptive radar transition capability. To elaborate, after sending out a specific radar waveform, the selected waveform, and the corresponding key performance metrics can be used as the state to the RL agent, and then the RL agent is trained to select the next waveform to be transmitted as the action to improve the key performance metrics (i.e., reward). With sufficient offline training, the RL agent can learn how to perform anti-jamming radar sensing under a jamming attack. Moreover, the above framework is flexible and can be easily extended by adding more states or

DOI: 10.1201/9781003408239-14

actions. For example, channel estimation results can also be added as additional inputs to let RL agents have more understanding of the underlying radio environment, thus improving the action selection capability. Besides the model selection, which is mainly modeling the waveform design problem as a classification problem, there is good work in introducing generative adversarial learning to design radar waveforms from scratch [344]. The considered scenario can be described by the following signal model:

$$x(t) = \alpha s(t) + n(t) + b(t), \tag{14.1}$$

where $x(t)$ is the detector's received signal, $s(t)$ is the emitted radar signal, and α is the energy attenuation coefficient. Furthermore, $n(t)$ is the noise vector, and $b(t)$ is the RF background signal(s) present in the spectrum. The aim of this work is to design a radar signal $s(t)$ that has good radar performance and is difficult to detect. For this purpose, to quantize radar performance, the ambiguity function $\hat{A}(\tau, F_D)$ is defined as:

$$\hat{A}(\tau, F_D) = \int_{-\infty}^{\infty} s(t)e^{(j2\pi F_D t)} s^*(t-\tau)dt, \tag{14.2}$$

where τ is the time shift and F_D is the Doppler shift. With this definition, the ideal behavior can be captured by the ambiguity function as:

$$\hat{A}_T(\tau, F_D) = \begin{cases} \xi, & \text{if}(\tau, F_D) = (0,0) \\ 0, & \text{otherwise} \end{cases} \tag{14.3}$$

which is the ideal ambiguity function with only a clean variable target peak ξ. Thus, we can evaluate the radar detecting performance by comparing the generated ambiguity function with the above target ambiguity function. For the mainlobe peak, the similarity can be captured as:

$$L_{\text{main}} = \frac{1}{M}||\text{ReLU}(\hat{A}_T(0,0) - \hat{A}_G(0,0))||_2^2, \tag{14.4}$$

where M is the batch size of the generated waveforms and $\hat{A}_G$ is a batch of ambiguity functions computed from the generated waveforms. For the sidelobe parts, the similarity can be captured as:

$$L_{\text{side}} = \frac{1}{M}||(\hat{A}_T(\tau, F_D) - \Gamma \circ \hat{A}_G(\tau, F_D))||_2^2, \tag{14.5}$$

to evaluate all sidelobe parts (i.e., $(\tau, F_D) \neq (0,0)$). In Eq. (14.5), Γ is a weight matrix that increases the loss for the zero-Doppler slice, which is provided in the original paper [344]. This design aims to prevent optimizing to minima with low Doppler sidelobes and high zero-Doppler sidelobes. The above two equations are then combined as the final ambiguity loss function as below to supervise the detection performance of the generated radar waveform.

$$L_{\text{ambig}} = L_{\text{main}} + L_{\text{side}}(\tau, F_D)|_{(\tau, F_D) \neq (0,0)}. \tag{14.6}$$

Besides the radar detection performance guarantee, the aim of this work is to also develop radar waveforms satisfying a low probability of being detected. In this direction, traditional methods struggle since it is hard to model the optimization problem to directly optimize waveforms, allowing a low probability of being detected. Thanks to the power of generative adversarial learning, this low probability of being detected can be optimized by optimizing the similarity of the generated waveforms and the background radio environment. Specifically, a discriminator will be developed as a binary classifier, utilizing the feature extraction capability to distinguish the generated waveform and background radio environment, and then a generator will be trained to develop the generated waveform, which can fool the discriminator. In this setting, the only way to fool the discriminator is to design a radar waveform that blends into the radio environment, thus achieving a low probability of being detected purpose. The challenge here is that the radio environment is time-dependent. To elaborate, although the long-term statistics of the radio environment might be consistent, the radio environment shows a nonstationary and highly fluctuating nature in short-term statistics. To tackle this challenge, an advanced conditional generative adversarial network (cGAN) is employed to enable the generator to produce waveforms from the conditional distribution $\mathbb{P}_{g|y}$, where y is the instantaneous RF background measurement taken at or near the time of inference and g is the generated waveform distribution from the generator. The final adversarial loss to supervise the generator can be expressed as:

$$L_W = \min_G \max_D \mathbb{E}_{x \sim \mathbb{P}_b}[D(x|y)] - \mathbb{E}_{\bar{x} \sim \mathbb{P}_{g|y}}[D(\bar{x}|y)] + \lambda L_{\text{regularization}}, \tag{14.7}$$

where G is the generator, D is the discriminator, $\mathbb{P}_b$ is the background radio environment distribution and x is a sample set sampling from this distribution, $\bar{x}$ is a sample set sampling from the generated waveform dataset, λ is a preset constant to control the ratio of regularization loss, and $L_{\text{regularization}}$ is the regularization loss developed by 1-Lipschitz constraint to improve the training stability. Finally, the final loss function can be expressed as below by combining the ambiguity loss and cGAN loss:

$$L_{\text{total}} = L_W + \eta L_{\text{ambig}}, \tag{14.8}$$

where η is a scaling parameter to set the strength of the ambiguity loss. The above training procedure is also expressed in Figure 14.1 for reference. The evaluation results demonstrate that the generated waveforms achieve up to a 90% reduction in detectability when compared to traditional waveform design methods and retain desirable ambiguity characteristics. Furthermore, the proposed method is able to tune the generated waveforms for a desired trade-off between detectability and sensing accuracy.

Besides the above two applications, there are several great works directly employing end-to-end learning to aid radar transceiver designs [345,346]. The idea is similar to using an autoencoder to perform communication system design, thus avoiding sophisticated joint modeling and optimization when

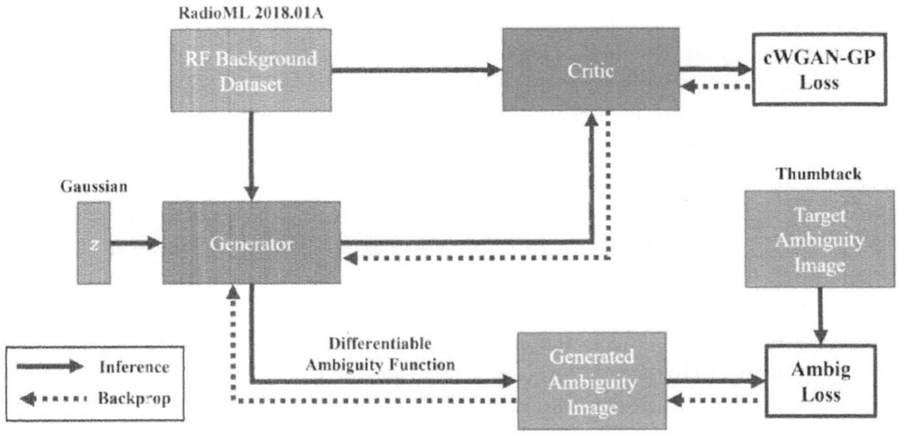

FIGURE 14.1
Adversarial Training Framework: Training process for our cWGAN- GP, where our generator learns to produce waveforms with desirable ambiguity functions that mimic the RF background. Training is initially performed without the ambiguity loss and alternates between training the critic and the generator. Once converged, the generator is fine-tuned for a small number of epochs with both the cWGAN-GP loss and ambiguity loss (copyright from ref. [344]).

designing radar transceivers. Specifically, this representative work [345] addresses the problem of data-driven joint design of transmitted waveforms and detectors in radar systems by introducing two novel learning-based approaches. Unlike traditional radar design methods that rely on rigid mathematical models, the proposed frameworks in this work directly leverage end-to-end learning to perform radar waveform designs for enhanced adaptability to environmental uncertainties. The first approach employs an alternating training strategy, where the detector undergoes supervised learning for a fixed waveform while the transmitter is optimized using RL for a fixed detector. The second approach simultaneously trains both components, potentially accelerating convergence by reducing the number of required radar transmissions. Furthermore, the proposed framework can be extended to incorporate further operational waveform constraints, such as peak-to-average power ratio (PAPR) and spectral compatibility. Theoretical analyses establish the effectiveness of these methods in dynamically adapting transmitted waveforms to environmental conditions while maintaining design constraints. In conclusion, the contributions of this work include the development of a radar system architecture based on feedforward multilayer neural networks, the formulation of two end-to-end learning algorithms, and a theoretical comparison of alternating and simultaneous training paradigms. Additionally, the authors also justify RL-based transmitter training by analyzing its gradient properties in relation to an ideal model-based likelihood function. Through these advancements, this work demonstrates the potential of learning-based radar design to improve detection performance and robustness in complex environments.

14.2 DL-Based Range and Doppler Estimation

Based on the discussion in Chapter 6, the constant false alarm rate (CFAR) algorithm is an important step in radar range and Doppler estimation. It tells which peak in the range-Doppler (RD) map is associated with the real targets instead of noise or clusters. Specifically, the core concept of target detection schemes is to evaluate the noise level in a RD map according to the neighbor cells surrounding the cell under test (CUT) and set an appropriate threshold for detecting targets. The cell-averaging CFAR (CA-CFAR) [347] utilizes the arithmetic mean of the power of reference cells as noise level estimates. One of its variants, namely the greatest of cell-averaging CFAR (GOCA-CFAR) [348], can improve the false alarm rate of the original scheme. Although both schemes perform well in homogeneous scenarios, their performance degrades in multi-target scenarios based on erroneous noise level estimation. The smallest of cell-averaging CFAR (SOCA-CFAR) [349] has been proposed to improve performance in multi-target scenarios. However, it does not improve performance significantly in dense multi-target scenarios. Order-statistic CFAR (OS-CFAR) [350] can handle such problems, but it introduces significant computational complexity. Therefore, it is necessary to develop a robust scheme for multi-target scenarios with lower computational complexity. Recently, the authors in ref. [351] proposed exploiting DL techniques to improve the detection performance with CA-CFAR with a lower false alarm rate. In ref. [352], a support vector machine model for selecting the best CFAR scheme for a given reference window is proposed. However, the performance of these schemes is only marginally better than that of conventional schemes because they simply train models to emulate the conventional ones or choose an optimal conventional scheme. Moreover, it is also observed that the degeneration of well-known CFAR schemes occurs at high signal-to-noise ratios (SNR) and in multi-target scenarios. In such cases, strong side-lobes associated with target returns cause overestimated noise levels, resulting in lower detection rates. Hereafter, this phenomenon is referred to as the side-lobe issue for the sake of brevity. If side-lobes can be eliminated, superior noise estimation results can be achieved. In refs. [353, 354], the authors employ a DL model to tackle the denoising problem. Their works suggest that DL models can recognize the structured pattern and remove noise patterns in an image. Inspired by these works, ref. [355] proposes a novel CFAR detection scheme aided by a DL model called DL-CFAR. The proposed model learns the structures of targets in an RD map and eliminates those structures to acquire an RD map with pure noise, which can be used to estimate noise levels more accurately and facilitate CFAR detection that is robust in various scenarios. This paper is the first attempt to enhance the noise estimation process of CFAR. As a result, the proposed scheme outperforms conventional CFAR schemes significantly. Furthermore, the performance of conventional CFAR schemes is influenced by parameter settings (guard cell size, order selection). Poor parameter settings in conventional CFAR schemes may lead to significant performance decay. In contrast, no parameter settings are required for DL-CFAR. As a result, the proposed method is a practical CFAR target detection scheme for real-world implementation.

Target information (i.e., range and Doppler velocity of each target relative to the observer) is acquired by performing target detection based on self-radiated returns from targets. In frequency modulated continuous wave (FMCW) radar and orthogonal frequency division modulation radar (OFDM), time-frequency channel coefficients are exploited to derive target information. In a general form, the channel coefficient of the kth frequency and lth time sample of a target return can be represented as:

$$(\mathbf{H})_{k,j} = \sum_{h=0}^{H-1} b_h e^{-2\pi f_{\mathrm{D},h} l T_{\mathrm{s}}} e^{-2\pi \tau_h k \Delta f} e^{j\phi_h}, \tag{14.9}$$

where $\mathbf{H}$ is a channel coefficient matrix (CCM) consisting of H target returns, and T_{s} and Δf are the sample period and frequency spacing, respectively. b_h, $f_{\mathrm{D},h}$, τ_h, and ϕ_h are the complex amplitude, Doppler shift, round-trip time, and random phase rotation, respectively, associated with the hth target return. Specifically, considering FMCW radar as an example, we transmit a frame composed of M chirps (a chirp is a sinusoid whose frequency increases linearly with time), then mix the transmitted and received chirps into M intermediate frequency (IF) signals in chirp-by-chirp fashion. Next, we extract N samples of the IF signal in each chirp using a predetermined sampling period. The CCM is constructed from cascading columns of these chirp-by-chirp samples, as illustrated in Figure 14.2. In this case, the sampling period is T_{s}, and the frequency spacing of the adjacent samples in each chirp is Δf. The 2D fast Fourier transform (FFT) operation is a well-known solution for identifying sinusoids in a discrete-time signal. An RD map is generated by performing a 2D FFT on a CCM $\mathbf{H}$, which can be represented as:

$$\begin{aligned} \mathrm{RDM}(n,m) &= |\text{2D FFT}(\mathbf{H})(n,m)|^2 \\ &= \left| \sum_{k=0}^{N-1} \sum_{l=0}^{M-1} (\mathbf{H})_{k,l} e^{j2\pi lm/M} e^{j2\pi kn/N} \right|^2, \end{aligned} \tag{14.10}$$

where N and M are the FFT lengths in the frequency and time domains, respectively. An RD map is composed of $N \times M$ RD bins. Maximum ratio combining (MRC) is achieved at RD bins for which the range and Doppler velocity of targets match both discrete sinusoidal components in both domains. Such RD bins are referred to as MRC RD bins. Note that a processing gain of

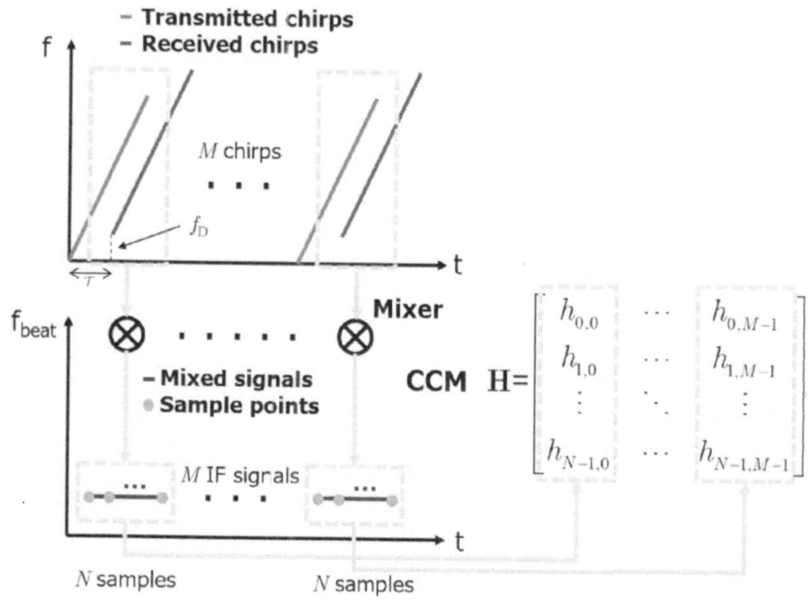

FIGURE 14.2
Illustration of CCM generation for FMCW radar (copyright from ref. [355]).

$10\log_{10}(NM)$ dB can be achieved after performing the 2D FFT operation to enhance the power of target returns. Therefore, to derive target information, a CFAR detection scheme can be applied to identify peaks.

The proposed scheme deals with an RD map with the size of $N_{\mathrm{W}} \times M_{\mathrm{W}}$. For an RD map of a large size, one can further divide it into several RD maps with the size of $N_{\mathrm{W}} \times M_{\mathrm{W}}$ such that they are processed independently. Our scheme focuses on the development of a neural network that can recognize and remove target patterns from an original RD map to obtain more accurate noise level estimations. DL-CFAR is summarized in Figure 14.3. In Figure 14.3 (a), an input RD map is first fed into a truncating operator as a truncated RD map. The resulting RD map is then treated as an input for the proposed neural network. The truncated RD map can be considered as a combination of a truncated target RD map and a noise RD map, as shown in Figure 14.3(b). The proposed neural network can remove target patterns in the truncated RD map, allowing us to approximate a pure noise RD map and estimate noise levels at each RD bin in the RD map accurately. The proposed neural network must be able to handle input RD maps with various magnitudes. A conventional solution to this problem is to employ a normalization operator prior to processing by the neural network. However, the noise would be normalized to nearly zero for every RD bin in the RD map at high SNR, making it fail to update the weights of the neural network properly. Therefore, a truncating operator is utilized to handle different RD map magnitudes.

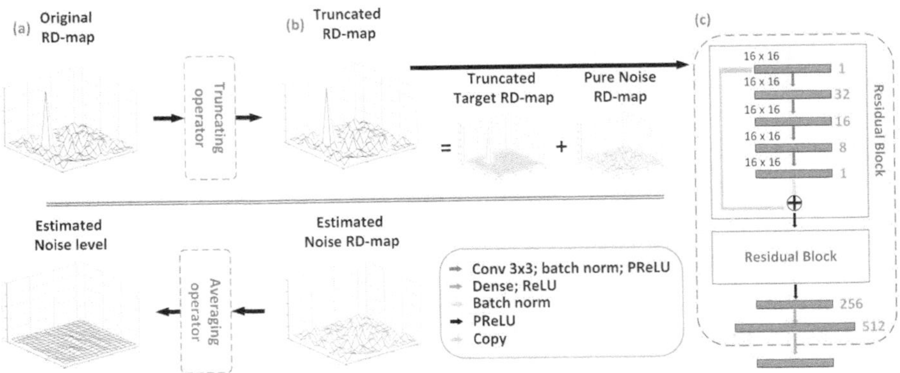

FIGURE 14.3
(a) Overview of the proposed DL-CFAR. (b) Truncated RD map combining a truncated target RD map and pure noise RD map. The proposed neural network can remove the truncated target RD map. As a result, the output of the proposed neural network is similar to the pure noise RD map. (c) The architecture of the proposed neural network (copyright from ref. [355]).

Although strong target patterns may be truncated, the proposed neural network still has the ability to recognize and remove such patterns from the RD map, as demonstrated by the simulation results discussed later in this paper. Figure 14.3(c) presents the architecture of the proposed neural network. The inputs for the proposed neural network are fed into several residual blocks first. In each residual block, convolutional neural networks (CNNs) are employed, and each CNN layer further utilizes kernels with dimensions of 3×3 to exploit the spatial local correlations in input RD maps. The number of feature maps in each layer are 32, 16, 8, and 1, respectively, and the output size for each layer in the residual block is the same as the size of the input RD map, which is $N_{\mathrm{W}} \times M_{\mathrm{W}}$. Batch normalization [356] is performed after each CNN to speed up convergence and avoid over-fitting problems. Finally, parametric rectified linear unit (PReLU) [357] is used as an activation function to introduce nonlinearity. The output of the residual block chain will be fed into two fully-connected layers. The number of neurons in the two layers is 512 and $N_{\mathrm{W}}M_{\mathrm{W}}$, respectively. Following these layers, a rectified linear unit (ReLU) is employed as an activation function. It is noteworthy that the output of a residual block has the same size as the input of a residual block. This design was inspired by ref. [358] and was adopted to remove target patterns step by step. Another feature of the proposed neural network is that it does not include any pooling layers in the design of the neural network because the authors want to keep all information from an RD map to estimate noise levels precisely. The same design concept can be found in refs. [359,360]. The number of residual blocks is set to two. Additional residual blocks and layers do not improve performance, but they do increase computational complexity.

Finally, end-to-end learning is utilized to train all trainable kernels and bias Θ in the neural network. An input RD map is denoted as $\text{RDM} = \mathbf{T} + \mathbf{N}$, where $\mathbf{T}$ represents a pure target RD map and $\mathbf{N}$ represents a pure noise RD map. The resulting truncated RD map can be expressed as $\text{RDM}_\text{T} = \tilde{\mathbf{T}} + \mathbf{N}$, where $\tilde{\mathbf{T}}$ represents the truncated target RD map. The proposed neural network is a supervised learning algorithm, and the mean square error is employed as a loss function as follows:

$$L(\Theta) = \sum_{i=1}^{D} (\text{N}_i - f(\text{RDM}_{\text{T},i}; \Theta))^2, \tag{14.11}$$

where D is the total number of samples in the training dataset. During the training process, Adam [361], a gradient-decent-based optimizer, is used to update all the trainable parameters iteratively. The initial learning rate is set as 0.00005, and the batch size is set to 128. After 500 epochs, the training process is completed, and the weights are recorded.

In terms of performance comparison with other existing CFAR methods, DL-CFAR outperforms all the other CFAR schemes significantly. It is noteworthy that, in the regime of low false alarm rates, the performance of DL-CFAR is clearly superior to that of other schemes. This phenomenon reveals that DL-CFAR can attain a significant improvement in detection rate even in an extremely low false alarm rate. In conclusion, although other CFAR schemes employ guard cells to handle the side-lobe issue, their performance drops because the fixed-size guard cells cannot handle various types of side-lobes. By employing neural networks, DL-CFAR can recognize and remove target patterns in different scenarios. As a result, the performance of DL-CFAR is superior to that of all the other CFAR schemes with different SNRs. Moreover, although it has been revealed that some conventional CFAR schemes perform well in specific scenarios, such as CA-CFAR outperforming other conventional CFAR schemes in homogeneous scenarios, these schemes all provide poor performance in multi-target scenarios. In contrast, by employing neural networks, DL-CFAR always achieves the best performance in both single-target and multi-target scenarios. These results suggest that neural networks enable CFAR detection to adapt to different scenarios. Encouraged by the success of this work, several follow-up studies [362, 363] are proposed to enhance the DL-CFAR's design; interested readers can refer to those papers for more details.

14.3 DL-Based Target Tracking, Detection, and Recognition

The primary benefit of automotive radars compared to other sensors is their reliability in challenging conditions, such as heavy rain and dense fog. However, the effectiveness of conventional radar is largely influenced by radar processing techniques—including detection, clustering, tracking, and data association—which require manual fine-tuning to adapt to different environments. With the advent of machine learning, learning-based solutions have been widely implemented across various applications. In particular, advancements in computer vision have significantly enhanced object detection and tracking performance using cameras. Therefore, integrating learning techniques into radar systems to improve radar object detection and tracking has drawn attention in recent years.

To this end, one common approach is to replace the conventional signal processing-based extended Kalman filter (EKF) method for trajectory tracking and prediction in the radar system with a deep neural network (DNN)-based method, in which a sequence of range-Doppler-angle (RDA) maps is used as the input for generating the target tracking trajectories. For example, in ref. [364], a learning-based architecture for tracking indoor humans was proposed by replacing the overall EKF with the learning-based architecture. Specifically, as shown in Figure 14.4, by using a sequence of RDA maps, the learning-based architecture first conducts the conventional denoising to remove undesired clutter or static reflections. Then, the clustering based on the common density-based spatial clustering for applications with noise (DBSCAN) algorithm is adopted to generate targets from the RDA maps. These targets are then sent to the DNN-based tracker to generate tracking trajectories of targets, where the tracker proposed in Figure 14.4 could be either a denoising autoencoder or a sequence-to-sequence autoencoder, as illustrated in Figure 14.5. Note that when considering K radar observations of K consecutive time-slots, the input $\mathbf{Z}_{1:K}$ consists of the observed ranges, Doppler velocities, and angles of these K time-slots from the radar. Then, the output $\hat{\mathbf{X}}_K$ would be the K consecutive predicted locations of the targets. It should be noted that the approach described in Figure 14.4 is only used to generate tracking trajectories of targets. However, the association of targets to the trajectories need to be additionally handled by usingthe Hungarian algorithm with Euclidean distance or some other approaches. More implementation details can be found in ref. [364].

While the aforementioned approach replaces the prediction of tracking trajectories with a learning-based approach, it target detection still relies on the conventional clustering, i.e., the DBSCAN algorithm. Then, to further replace the target detection with learning-based approaches, the introduction of the famous You Only Look Once (YOLO) for object detection is commonly adopted, leading to a system in which both target detection and tracking are performed with learning-based architectures. Serving as a great example, ref. [365] proposed a novel DL-based multi-target detection and tracking architecture, in which YOLO is adopted for target detection and Deep Simple Online and Realtime Tracking (D-SORT) is used for tracking. It should be noted that while YOLO is known for providing good performance for target detection, it also provides object recognition functionality at the same time.

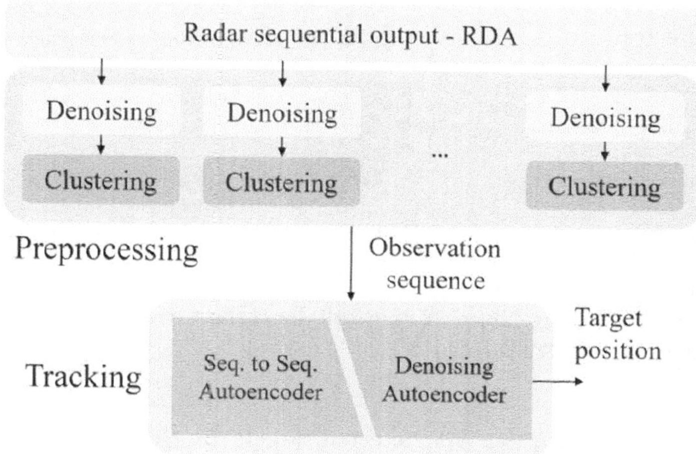

FIGURE 14.4
Illustration of learning-based architecture for tracking (copyright from ref. [364]).

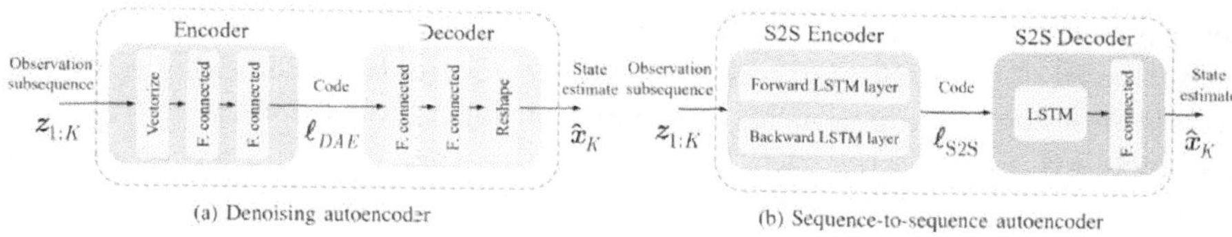

FIGURE 14.5
Illustration of the architectures of the trackers (copyright from ref. [364]).

As a result, when adopting YOLO for detection, the types of targets could also be predicted as a byproduct.

Specifically, as shown in Figure 14.6, a learning-based detection and tracking framework was developed by using YOLO and D-SORT, where the input of the framework is RDA maps of N_{frame} time-slots. Then, as a full RDA map is overly complex for subsequent processing and the complete Doppler domain information has less impact on target tracking and detection, the Doppler domain is compressed by selecting RDA bins with higher Doppler velocities to form the enhanced RA (ERA) map, defined as:

$$\left(\mathbf{M}_{\mathrm{P}}^{\mathrm{ERA}}\right)_{n,q} = \max_{m} \left(\mathbf{M}_{\mathrm{P}}^{\mathrm{RDA}}\right)_{n,m,q}, n = 1, ..., N, q = 1, ..., Q, \tag{14.12}$$

where $\left(\mathbf{M}_{\mathrm{P}}^{\mathrm{RDA}}\right)_{n,m,q}$ is a RDA bin, with n, m, and q being the indices of range, Doppler velocity, and angle. After the preprocessing, since D-SORT is designed to associate detected targets across different time-slots and assign them unique identities (IDs) for tracking, the ERA map is first processed by

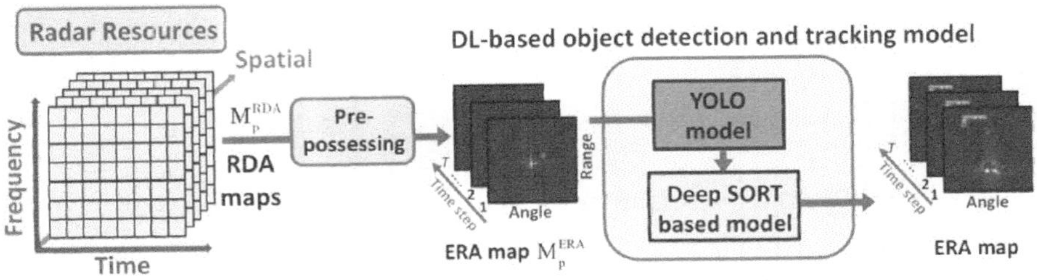

FIGURE 14.6
Illustration of learning-based framework for target tracking and detection (copyright from ref. [365]).

YOLO for target detection. The detected targets are then passed to D-SORT for tracking.

The details of the learning-based architecture for target tracking and detection are illustrated in Figure 14.7. Specifically, as YOLO is a machine learning model composed of convolutional and residual layers, its primary function is to detect objects and classify them by treating the input as an image and leveraging image-like features. In our case, the ERA map provides a 2D representation of objects through angle and range information. This allows YOLO to interpret the ERA map as an image and perform object detection. The output includes object categories, bounding boxes that define object sizes, and confidence scores that reflect YOLO's certainty in its detections. Since multiple bounding boxes can be present in an ERA map, the number of detected objects is inherently determined by the number of bounding boxes.

After the target detection, the detected targets by YOLO is fed to the D-SORT for trajectory tracking. Specifically, D-SORT is a tracking algorithm that utilizes Kalman filtering on images and performs frame-by-frame object association using the Hungarian method. The association metric considers both motion and appearance information. However, the performance of the conventional D-SORT is constrained by false positives caused by unreliable objects. Thus, the D-SORT in Figure 14.6 is modified by incorporating a two-stage low-confidence filtering (LCF) mechanism [366]. Precisely, as shown in Figure 14.6, the low confident filtering includes a two-stage filtering which is with a detection confidence threshold t_{d} and an average detection confidence threshold t_{avg}. In the first stage of LCF, the detected targets from YOLO are evaluated based on their confidence levels. If the confidence level exceeds t_{d}, the target is passed to D-SORT; otherwise, it is discarded. In the second stage of the LCF, a tracked target is evaluated by its average confidence level of a sequence of timeframes $(t), ..., (t-\tau)$, where t is the current timeframe for deciding whether a tracked object should be removed and τ is a set parameter. Then, if the average confidence level is larger than t_{avg}, this tracked target is kept; otherwise, the tracked target is discarded. As a consequence, this LCF mechanism allows for a lower t_{d} to reduce missed detections while increasing

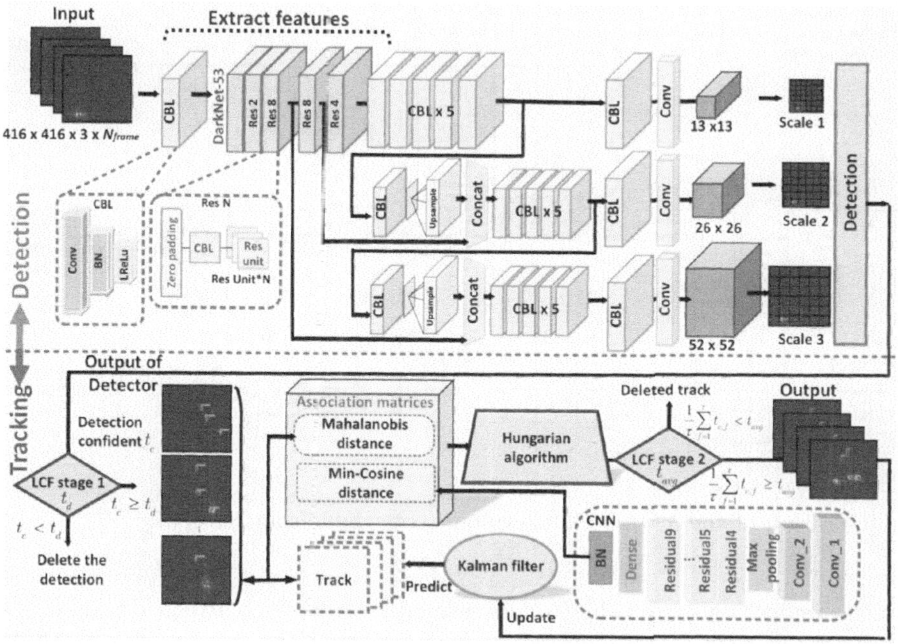

FIGURE 14.7
Illustration of the architecture for target tracking and detection (copyright from ref. [365]).

t_{avg} to filter out false-positive tracks by increasing t_{d}.

As mentioned above, the radar target detection based on radar images and YOLO not only can detect targets, but also can recognize the types of targets. Such target recognition is significantly helpful for many applications, including autonomous vehicles and intelligent transportation. However, as YOLO focuses mainly on the detection part, it target recognition commonly has some drawbacks. To remedy this, a common approach is to append YOLO with some additional learning-based modules to fix the recognition results of the standard YOLO. For example, in ref. [367], a YOLO-based object recognition approach was proposed by using radar range-angle (RA) maps as the input, where YOLO is concatenated with a recheck module. The overall object recognition system is then called YOLO-ORE (YOLO-Object REcheck) which consists of three main modules: radar image generation module, YOLO-based object recognition module, and object recheck module.

The overall YOLO-ORE system is shown in Figure 14.8. For radar image generation, a method called the target highlight technique (THT) is introduced, which converts the RA map produced by the radar system into a radar image (RR map) while enhancing potential targets. This radar image is then processed by a YOLO model to generate an initial object recognition result. However, YOLO's output may include overlapping detections and

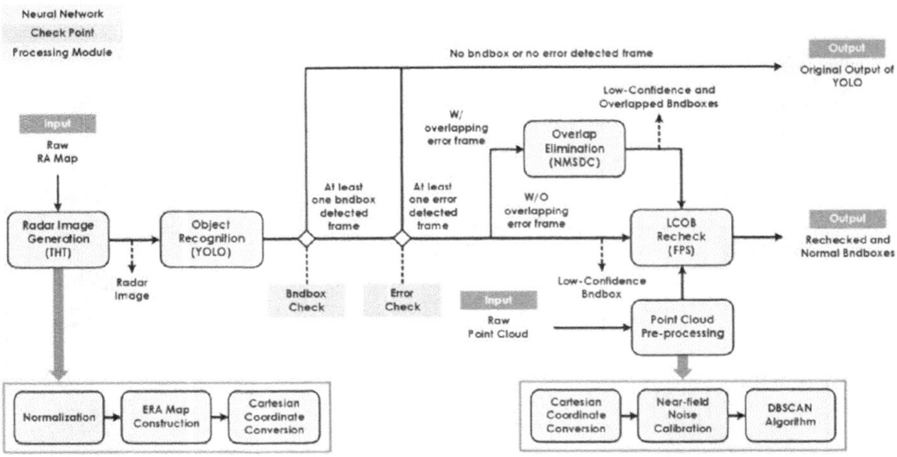

FIGURE 14.8
Illustration of the YOLO-ORE system (copyright from ref. [367]).

misclassification errors due to unreasonable overlaps between detected objects or incorrect target identification. To address these issues, the recognition results from YOLO are further processed by an object recheck system to eliminate overlapping and misclassification errors. More details of the design principle of the recheck module can be found in ref. [367]. Finally, it can be observed that the aforementioned target recognition approaches are basically using radar RA maps, which resemble the camera images. Thus, the recognition is conducted based on computer vision techniques. However, a radar system could have its own special features that help conduct object recognition from different domains. For example, ref. [368] adopts a 4D image radar that can conduct target recognition by using the spatial multi-representation offered by the 4D image radar. In addition, ref. [369] conducts the target recognition by using special micro-Doppler features of targets. These micro-Doppler features come from the small vibrations of targets, e.g., motor engines and turbines, and they can be captured by radar systems with high Doppler resolution.

14.4 DL-Based Vital Sign Monitoring

Besides the above radar-related applications enabled by learning-based solutions, we aim to introduce another radar research area, vital sign monitoring, in this section due to the increasing attention in this area recently. Specifically, vital sign monitoring plays a crucial role in assessing a patient's health status by providing real-time physiological information. While traditional methods such as electrocardiography (ECG) and respiratory inductance plethysmography (RIP) rely on contact-based devices, there has been growing interest in non-contact alternatives due to their advantages in patient comfort and continuous monitoring feasibility. Among various non-contact approaches, radar-based techniques—particularly those utilizing FMCW radars—have emerged as a practical solution due to their robustness against thermal noise and independence from external lighting conditions. However, existing FMCW radar-based methods often suffer from unstable and inaccurate vital sign estimation, especially in low SNR environments. To address this limitation, a pioneering work [370] introduced a novel DL-aided Newtonized orthogonal matching pursuit (NOMP) framework, which integrates neural networks as denoisers to enhance signal quality and improve estimation accuracy in low-SNR conditions. This approach leverages DL's capability to recognize structured signal patterns even in noisy environments, thereby enabling more reliable heart rate and respiratory rate detection. In this work, an FMCW radar waveform employs a chirp as the transmitted signal, which is a sinusoid whose frequency increases linearly with time, is considered to perform non-contact vital sign monitoring. To be more specific, a chirp can be characterized by a start frequency f_c, bandwidth B and duration T, expressed as follows:

$$s(t) = e^{j(2\pi f_c t - \pi \frac{B}{T} t^2)}. \tag{14.13}$$

Through comprehensive over-the-air (OTA) experiments, we validated the effectiveness of the DL-aided NOMP scheme and demonstrated its superiority over conventional methods. This work represents a significant advancement in radar-based vital sign monitoring, offering an energy-efficient and accurate solution for continuous health assessment. After the reflection by an object, the received chirp is delayed by a range-dependent value t_d of the transmitted signal, being expressed as follows:

$$r(t) = e^{j(2\pi f_c (t-t_d) - \pi \frac{B}{T} (t-t_d)^2)}. \tag{14.14}$$

Suppose that there is an object situated at a distance of R_0 to the radar but there are slight range variations around R_0 due to the chest wall displacement and the heart wall displacement. Then t_d can be further expressed as $\frac{2(R_0+x_t)}{c}$, where x_t is a time-varying value representing the distance variation around R_0 and c is the speed of light. To extract the range information of the object and infer vital signs from the obtained range information, the beat signal $b(t)$ can be obtained by mixing and filtering $r(t)$ and $s(t)$, which can be expressed as:

$$\begin{aligned} b(t) &= e^{j(2\pi f_c t - \pi \frac{B}{T} t^2) - j(2\pi f_c (t-t_d) - \pi \frac{B}{T} (t-t_d)^2)} \\ &= e^{j(2\pi f_c t_d + 2\pi \frac{B}{T} t t_d - \pi \frac{B}{T} t_d^2)} \\ &\approx e^{j(4\pi \frac{B(R_0+x_t)}{cT} t + \frac{4\pi}{\lambda}(R_0+x_t))} = e^{j(2\pi f_b t + \phi_b)}. \end{aligned} \tag{14.15}$$

In Eq. (14.16), it is obvious that the range information can be extracted by analyzing the phase term of the beat signal. Suppose that the imaginary and real parts of a complex beat signal are $Q(t)$ and $I(t)$, respectively. Then,

$$\begin{aligned} \phi_b &= \text{unwrap}\left[\tan^{-1}(\frac{Q(t)}{I(t)})\right] \\ &= \frac{4\pi}{\lambda}(R_0 + x_t). \end{aligned} \tag{14.16}$$

The output phase obtained from $\tan^{-1}$ is wrapped in the range $[-\pi, \pi]$. However, the phase can change beyond the range because of x_t, which is a physical displacement that can be greater than $\frac{\lambda}{4}$. Therefore, an unwrap function can be employed to obtain the actual displacement. In vital sign estimation tasks, by applying several chirps in continuous time, which is similar to the sampling of ϕ_b, we can obtain a sequence of distances $\{R_0 + x_t\}_{t=1}^{n}$ and as a result, also obtain the distance variation $\{x_t\}_{t=1}^{n}$ by the phase difference $\Delta\phi_b$. Suppose f_r and f_h are breathing rate and heart rate of the participant, we can consider the displacement sequence $\{x_t\}_{t=1}^{n}$ as a vibration due to heartbeat and respiration; this displacement sequence varies according to the two frequency components $-f_r$ and f_h. Hence, we can further express the equation as follows:

$$\begin{aligned} \mathbf{X}_b &= \{x_t\}_{t=1}^{n} = \mathbf{S}_t + \mathbf{Z}_t \\ &= A_r \text{sin}(2\pi f_r t + \phi_r) + A_h \text{sin}(2\pi f_h t + \phi_h) + z_t. \end{aligned} \tag{14.17}$$

where A_r and A_h are signal amplitudes for breathing and heartbeat, ϕ_r and ϕ_h are starting phases of sinusoids for breathing and heartbeat, $\mathbf{Z}_t \sim \mathcal{N}(\mathbf{0}, \sigma^2)$, respectively. In this work, SNR is defined as $\frac{|A_r|^2}{\sigma^2}$. The SNR difference between heartbeat and breathing is set as 5 dB, referencing the guide provided by TI and further confirmed through over the air experiment. Overall, the vital sign estimating problem can be transformed into a frequency estimation problem to infer breathing rate and heart rate in Eq. (14.17).

To estimate respiratory and heart rate frequencies from an observation sequence, Texas Instruments (TI) employs simple bandpass filters. However, this method is highly sensitive to noise, misalignment, and interference, leading to imprecise and unstable results. Moreover, fixed cutoff frequencies (6–30 BPM for respiration, 48–120 BPM for heart rate) limit applicability to individuals within this range, restricting practical deployment. To overcome these limitations, the authors in this work adopt NOMP for improved estimation. While NOMP enhances accuracy, it struggles in low SNR conditions, impacting power efficiency. Observing that target sinusoidal patterns remain discernible even at low SNR, the authors further integrate a neural network (NN) as a denoiser to enhance observation quality and improve NOMP performance. NOMP iteratively refines frequency estimates by first searching a discrete basis set and then applying Newtonized coordinate descent to mitigate basis mismatch errors. Iterations continue until the difference between the estimated and observed signals falls below a predefined threshold. To improve performance in low-SNR conditions, the authors propose a DL-aided NOMP framework, addressing two key objectives: (i) identifying the optimal domain for denoising and (ii) determining the best NN denoiser architecture. To do so, six combinational schemes are developed, including TD-CNN-NOMP, TD-DNN-NOMP, TD-LSTM-NOMP, FD-CNN-NOMP, FD-DNN-NOMP, and FD-LSTM-NOMP, where "TD" and "FD" indicate time or frequency-domain denoising. Supervised learning is employed in this work to learn the mapping from low-SNR input to high-SNR samples to achieve the aim of the developed model. Simulation results demonstrated that TD-CNN-NOMP can outperform other NN-based schemes significantly, even in a low SNR region, achieving the goal of enhancing power efficiency. Moreover, through an actual implementation, the superiority of TDCNN- NOMP in a low SNR region is also verified, achieving the aim to develop DL-aided NOMP schemes.

15

DL-Based Interference Mitigation in Radar Systems

15.1 Overview of DL-Based Interference Mitigation in Radar Systems

Based on our discussion in Chapter 8, one can notice that there are still some limitations when using traditional signal processing methods for interference mitigation in radar systems. This is especially true in the case that we need real-time processing capabilities or in the case that the observed data in a new environment is quite limited to obtain robust statistics, which will be further used to identify and suppress interferences. In both cases, deep learning (DL)-based solutions might be able to help for more efficient and effective interference mitigation in radar systems. More specifically, the following three limitations are observed as the deficits of traditional interference suppression techniques: (1) reliance on domain knowledge and the interference characteristics; (2) their inability to generalize; and (3) complexity resulting from detecting the precise location of the interference, and thus suggesting the usages of DL-based solutions alternatively. There is already ample literature discussing the usage of DL-based solutions from the following two perspectives:

Intelligent processing: By integrating artificial intelligence technologies, intelligent processing is applied to signals that contain electronic interference, environmental clutter, and radar target echoes. This approach extracts high-dimensional features of interference and clutter, thereby mitigating their impact and enabling intelligent detection, tracking, and identification of radar targets. Intelligent processing is key to cognitive intelligent radar, fundamentally leveraging artificial intelligence (AI) technologies and using deep neural networks to replace traditional detection, tracking, and identification algorithms. It can even achieve integrated detection-tracking-identification, uncovering deep features hidden in data that are difficult for humans to intuitively understand.

Intelligent scheduling: Using prior knowledge and real-time cognitive information, the results of intelligent processing are evaluated for efficiency, and predictions and reasoning are performed for intelligent decision-making. This enables multi-level intelligent scheduling of radar tasks, countermeasure

DOI: 10.1201/9781003408239-15

strategies, system resources, and processing algorithms, ultimately achieving closed-loop self-evolution of radar detection capabilities. Intelligent scheduling plays a central role in cognitive intelligent radar, functioning similarly to the human brain by analyzing and judging the processing results based on memory and experience, and making intelligent predictions about future changes in the electromagnetic environment. This determines the next set of tasks, waveforms, receiver filters, and processing algorithms that the radar will use, maximizing overall detection performance under resource constraints.

Following the above two directions, DL-based radar interference mitigation methods are widely discussed and already provide several state-of-the-art solutions to aid traditional methods. To let readers catch up with this tendency efficiently, we will further address the above two directions in the rest of this chapter.

15.2 DL-Based Intelligent Signal Processing

To employ DL-based solutions to aid interference mitigation in modern radar systems, a straightforward idea is to employ a trained DL module in the radar receiver as an "intelligent filter" to filter out interference signals before normal radar receiver processing. On the one hand, one can see a clear difference between the interfered signal and the normal signal in a certain domain. On the other hand, finding an appropriate domain that can best distinguish the interfered signal and the normal signal still remains a difficult challenge to traditional radar signal processing methods, especially since the aforementioned appropriate domain is often not a common time, frequency, or spatial domain, but some other high-dimensional domains. Thus, estimation of deep features of interference primarily leverages the feature extraction capability of deep neural networks, transforming low-dimensional features of interference signals into high-dimensional feature spaces. By combining classifiers and detectors, the deep features of interference are decoded, enabling recognition of interference types and estimation of parameters.

For example, one pioneering work to investigate DL-based radar interference mitigation is ref. [371]. In ref. [371], authors consider a common frequency modulated continuous wave (FMCW) radar system to transmit a radar signal consisting of k linear frequency chirps as:

$$f(t) = f_C + \alpha(t - kT_{\text{chirp}}), \tag{15.1}$$

where the time-domain received signal can be further presented as $\phi(t) = 2\pi \int_0^t f(t)dt = 2\pi(f_C t + \frac{1}{2}\alpha t^2 - \alpha k T_{\text{chirp}} t)$. In Eq. (15.1), T_{chirp} is the chirp duration, $\alpha = \frac{B_{\text{SW}}}{T_{\text{chirp}}}$ represents the slope of the FMCW waveform since B_{SW} is sweep bandwidth, and f_c is the carrier frequency of the transmit signal.

To remove high-frequency signals, an anti-aliasing filter is often used in the receiver as:

$$\phi_B(t) = \phi(t) - \phi(t-\tau) = 2\pi f_C \tau - \pi\alpha(\tau^2 - 2\tau t_k), \text{if } \tau < t < T_{\text{chirp}}. \quad (15.2)$$

However, in typical road usage, several vehicles with FMCW radar might interfere each other and consequently degrade each radar's performance in terms of range and velocity estimation. Moreover, the above anti-aliasing filter cannot completely remove this irregular interference. To tackle this situation, this paper proposes a DL module to filter interference before normal range and velocity estimation, thus maintaining the estimation performance in such congested scenarios. Specifically, the proposed DL architecture is built on the combination of bi-directional gated recurrent units (GRUs) and attention blocks, as shown in Figure 15.1. By doing so, the employed Bi-directional GRU can be used to extract the time-domain dependency of interfered signals and also get rid of the gradient vanishing problem, and then the attention block can further locate and then isolate the interference signal precisely to restore the desired signal without interference to achieve the goal. MSE is used to construct the loss function to allow alternate direction method of multipliers (ADMM) optimizer to obtain optimal trainable weight in the training stage with ample training samples. Then the model is ready to use in the receiver for real-time interference mitigation purposes. Simulation results suggest that 14 dB singal-to-interference-plus-noise ratio (SINR) gain can be provided with the proposed method compared to the original interfered signals, and the performance outperforms commonly used radar interference mitigation methods by at least 8 dB. Moreover, with minor modifications, the proposed method can also be used to serve orthogonal frequency division multiplexing (OFDM) radar, showing the flexibility of the DL-based solutions.

Another representative work in this direction is ref. [372], which shows the incorporation of DL-based and traditional solutions to yield better operational efficiency. Specifically, this work considers the radar signals after dechirping, which is a widely used signal processing mechanism in FMCW radar systems to reduce the sampling requirement for analog-to-digital converters. The radar signal after dechirping can be expressed as:

$$y(t) = s(t) + f(t) + n(t), \text{for } 0 < t < T_{\text{SW}} \quad (15.3)$$

where $n(t)$ is the thermal noise, $s(t) = \sum_{k=1}^{N} \sigma_k e^{j2\pi(-f_c\tau_k - K\tau_k t + \frac{1}{2}K\tau_k^2)}$ is the desired signal, and $f(t) = \mathcal{F}_{lp}[p^*(t)\sum_{m=1}^{M} f_m(t)]$ represents interferences. In Eq. (15.3), T_{SW} is the sweep duration, f_c is the center frequency, K is the chirp rate of FMCW waveform, and τ_k is the time-delay of the signal from the k th target. $p^*(t)$ is the reference signal used for dechirping, $f_m(t)$ is the m-th interference and $\mathcal{F}_{lp}$ represents low-pass filtering operation. In this work, the aim is to remove interference signals with different chirp rates, CW interference, and transient interference by capturing and removing interference with specific patterns to maintain the normal estimation performance of the

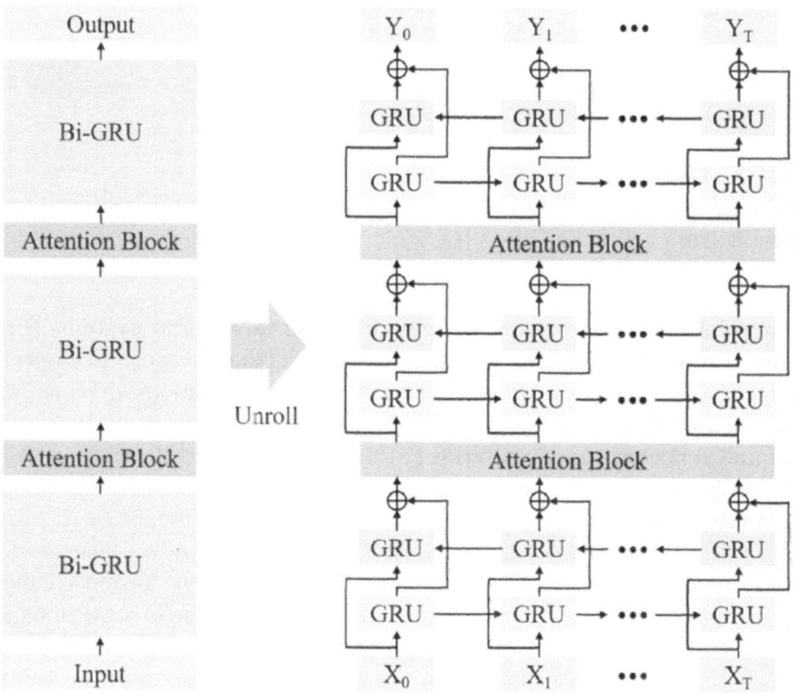

FIGURE 15.1
Deep learning architecture (copyright from ref. [371]).

radar system. Specifically, the authors in this paper notice that the difference between desired signals and interfered signals becomes compelling in the time-frequency domain after a short-time Fourier transformation (STFT). As a result, the workflow developed in this paper is shown in Figure 15.2, where the STFT is taken as a data preprocessing step before inputting into a neural network, and then the inverse STFT is employed to transform the output of the neural network back to the time-domain for normal radar signal processing without interference. Compared to the method in ref. [371], one can notice that domain knowledge really helps to simplify the neural network architecture since the pattern is more compelling in the time-frequency domain compared to the original time-domain. Consequently, a relatively simple neural network architecture (i.e., convolutional neural network (CNN) only) is good enough to finish the task by identifying and removing interference signals. Note that this is especially useful when it comes to limited data sample scenarios, since complex neural network architecture often needs more data samples to train to avoid over-fitting drawback.

Similar to the above works, a lot of follow-up works have been developed to investigate the usage of DL-based solutions in the radar receiver against different interference sources and in different scenarios. Ref. [373] introduces a

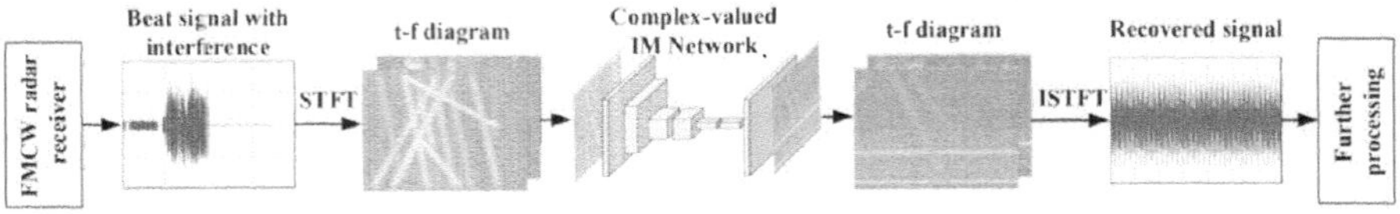

FIGURE 15.2
Signal processing flow of our propose approach for interference mitigation (IM) (copyright from ref. [372]).

novel approach to mitigate interference in automotive radar systems through the application of DL techniques, specifically utilizing a recurrent neural network (RNN) model with GRU architecture. In autonomous driving, radars play a crucial role in detecting target ranges and velocities by analyzing transmitted and reflected signals. However, the presence of interference signals can significantly elevate the noise floor, impairing the detectability of target objects. While previous studies have aimed to cancel interference or reconstruct original signals, conventional signal processing methods often face challenges and limitations in effectively addressing these issues. The proposed method demonstrates high performance across various interference conditions while maintaining low processing times. Notably, it is positioned as the first DL approach to mitigate interference in the time domain for radar systems. By leveraging advancements in DL, which have shown remarkable success in image and language processing, this technique effectively reconstructs transmitted signals even amidst diverse interference scenarios. The reconstructed signals can then be utilized for object detection via fast Fourier transform (FFT). Additionally, the algorithm operates through straightforward matrix calculations without requiring iterative structures or adaptive thresholds, enhancing its efficiency. Experimental results indicate that this DL-based method outperforms existing algorithms in environments where noise and interference coexist, thereby addressing a critical challenge in enhancing the reliability and performance of automotive radar systems essential for safe autonomous driving. Ref. [374] explores quantization techniques for CNN-based denoising and interference mitigation in automotive radar systems, specifically addressing the challenges posed by the increasing number of radar sensors and the unregulated frequency band, which leads to inevitable mutual interference. Given that specialized radar sensor hardware typically has strict resource constraints, including limited memory capacity and computational power, the authors focus on optimizing CNN models to fit these limitations. They analyze the quantization of both weights and activations across different CNN architectures to reduce memory requirements for model storage and inference. The study compares models with fixed and learned bit-widths, employing two methodologies for training quantized CNNs: the straight-through gradient estimator and training distributions over discrete weights. The findings highlight the significance of using small real-valued base models for effective quantization, demonstrating that learned bit-widths can yield significantly smaller models, achieving a memory reduction of approximately 80% compared to real-valued baselines. For practical implementation, the authors recommend an 8-bit representation for weights and activations, resulting in models that occupy only 0.2 megabytes of memory. Utilizing real-world FMCW and chirp sequence (FMCW/CS) radar measurements with simulated interference, the research illustrates that exceptionally small models can be quantized without substantial performance degradation. Furthermore, it shows how distributions over discrete weights can not only aid in denoising and interference mitigation but also provide uncertainty estimates for range-Doppler maps. This work contributes to advancing resource-efficient DL applications in automotive radar systems, addressing the critical issue of mutual interference in increasingly crowded electromagnetic environments.

Ref. [375] introduces a novel DL-based approach for interference mitigation in FMCW radars, which are essential sensors for environmental sensing in modern vehicles used for driving assistance and autonomous driving. The increasing number of radar sensors and the limited frequency bandwidth inevitably lead to mutual interference, which can degrade target detection capabilities and pose safety hazards. To address this challenge, the authors propose a method that employs dilated convolution to enhance the network's receptive field, allowing it to capture more contextual information from the radar signals. Additionally, they implement a designated contrastive learning strategy during training, which improves the network's ability to distinguish between interference and desired signals. The results from both numerical simulations and experimental data processing reveal that the dilated convolution-based IM network significantly outperforms traditional convolution-based networks, achieving higher signal-to-interference-plus-noise ratio (SINR) and improved target detection rates. Furthermore, the designated contrastive learning strategy contributes to more stable interference mitigation performance without adding complexity to the network, thereby facilitating faster signal processing. This research contributes valuable insights into DL applications for FMCW radars, addressing the critical need for effective interference mitigation in increasingly crowded automotive environments. Ref. [371] presents a novel DL-based approach for mitigating interference in automotive radar systems, which is essential for enhancing the safety and reliability of autonomous driving. The authors address the critical challenge of increased noise floors caused by interference signals, which significantly impair the detectability of target objects. While many previous studies have focused on traditional methods for canceling interference or reconstructing original signals, there has been limited exploration of DL techniques in this area. In this work, the authors enhance existing DL algorithms by incorporating an attention mechanism, which improves the model's ability to focus on relevant features in the radar signals. The proposed method is applied to both OFDM and FMCW radar environments, demonstrating its versatility and effectiveness across different radar technologies. Experimental results indicate that the DL-based method outperforms existing interference mitigation approaches, showcasing its potential for improving target detection in complex automotive scenarios. By addressing the pressing need for robust interference mitigation solutions in increasingly crowded electromagnetic environments, this research contributes significantly to the field of autonomous driving. Additionally, the implementation of this work is made available on GitHub, facilitating further exploration and validation by the research community. Ref. [376] presents a novel approach to mitigate automotive radar interference using a generative adversarial network (GAN). Unlike previous methods that address interference in the time domain, this approach focuses on recovering the complex signal in the frequency domain, specifically on the complex range profile obtained after the first fast Fourier transform of fast-time samples (RFFT spectrum). The proposed GAN employs gated convolution and an attention mechanism, enabling the generator network to learn both amplitude and phase information for missing data from the remaining signal. This approach is designed to handle more complex and diverse interference scenarios, including cases where multiple chirps are completely disturbed. Experimental results demonstrate that the proposed method significantly improves the SINR and maintains robustness in severely disturbed scenarios, even those more complex than the training dataset. This GAN-based approach addresses limitations of previous methods, such as the potential lack of robustness in preserving target peak values in CNN-based approaches, and the insufficiency of linear predictive coding (LPC) in scenarios with widespread interference across multiple chirps. The authors' work contributes to the growing body of research on DL applications for radar interference mitigation, offering a promising solution for handling complex and random interference patterns in FMCW/CS radars. While providing increasingly better performance, the aforementioned methods are applicable when there is sufficient training data. However, when training samples are limited, neural networks face overfitting issues, leading to a sharp decline in recognition performance. Refs. [377] and [378] propose solutions for these challenges: the former suggests an interference classification method based on a Siamese network, and the latter introduces a recognition method combining weighted ensemble CNN (WECNN) and transfer learning, enabling effective interference recognition in small-sample conditions.

If we further consider another common type of interference, jamming, there has also been recent progress in using DL-based solutions for anti-jamming purposes. Specifically, anti-jamming processing methods based on feature mining primarily focus on exploiting differences between interference and targets in low-dimensional feature spaces (e.g., time, spatial, and Doppler domains). However, when the target and interference become similar in these low-dimensional feature spaces, the aforementioned methods may fail. Deep neural networks have the capability to uncover differences between signals in high-dimensional feature spaces. Therefore, developing methods that leverage deep neural networks to distinguish between interference and targets in high-dimensional feature spaces is a critical and promising direction for future research. For example, based on CNNs, ref. [379] proposes an end-to-end target detection network in interference scenarios, enhancing target detection performance under intermittent sampling and forwarding jamming. Similarly, ref. [378] also focuses on the jamming signal classification problem and further discusses the impact of limited training data samples. To tackle the performance degeneration when employing normal CNN architecture, a specialized CNN is proposed to allow training convergence even with limited training data sample conditions for practical usages. Besides CNN architecture, a famous object detection network (i.e., YOLO 5 [380]) is also employed to capture interference signal in the time-frequency domain after STFT operation in ref. [381]. Furthermore, when it comes to multi-dimension classification, ref. [382] focuses on the pseudo noise code-integrated linear frequency modulation chirp jamming detection to protect normal radar functionality. In this work, besides in-phase and quadrature representations of the received signals, the time-frequency distribution, which can be obtained by performing the Choi–Williams distribution transformation, is employed to provide extra information for unknown jammer detection. The workflow proposed in this work is shown in Figure 15.3. One can notice that using time-frequency distribution, an outlier detector is performed first to catch the combinations of the reflected and disguised signal. If passing this outlier detection, an automatic modulation classifier is then further employed to distinguish whether the pure signal belongs to the pure reflected signal or the pure disguised signal transmitted by the unknown jammer. Simulation results confirm that the proposed method can provide a higher detection rate with a lower false alarm rate in different signal-to-jammer-noise-ratio conditions, outperforming the existing jammer detection method. Ref. [383] also proposes a specialized neural network architecture, utilizing four parallel 1-D CNNs to extract features from signal modulus, phase, real part, and imagnary part respectively and a 2-D CNN to extract features from the time-frequency domain after STFT, then a fusion network is used to combine the aforementioned extracted features and perform classification based on those features jointly. Simulation results confirm that the proposed solution can classify 12 different types of jamming signals effectively.

Besides the above works, ref. [384] presents a great work by introducing a learning-based autoencoder into a traditional interference detection filter to provide improved interference mitigation performance. Specifically, in a typical chirp sequence radar system, the transmit waveform containing L consecutive chirps can be expressed as:

$$T_x(t) = \sum_{l=0}^{L-1} x(t - lT) \tag{15.4}$$

and an individual transmitting chirp signal is given by:

$$x(t) = e^{j2\pi(f_c t + 0.5\alpha t^2)} \mathrm{rect}_T(t). \tag{15.5}$$

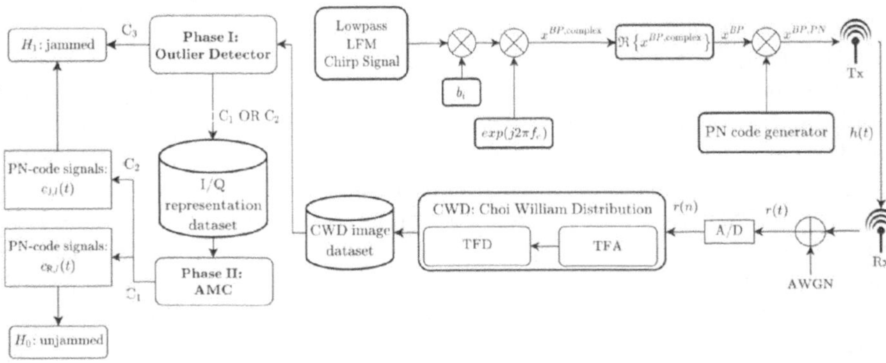

FIGURE 15.3
Illustration of radar communications in LPI networks with the proposed deceptive jammer detection algorithm. (copyright from ref. [382]).

In Eq. (15.5), f_c is the carrier frequency, $\alpha = B/T$ denotes the slope of the transmit signal with B and T denoting the sweep bandwidth and chirp duration, and $\text{rect}_T(t)$ is the square pulse of duration T. For l th chirp, the echo signal will be delayed by τ from the transmit signal with a normalized relative Doppler shift d if facing a target, as shown below:

$$r_l(t) = A_l x(t + (t + lT)d - \tau) + \text{v}_l(t), \tag{15.6}$$

where A_l is the received amplitude, $\tau = 2R/c$, $d = 2v/c$ and $+\text{v}_l$ denotes the complex Gaussian noise. In the above equation, R is the distance, v is the relative radial velocity between the radar and target and c is the speed of light. Then, in the radar receiver end, by mixing r_l with the complex conjugate of the transmitted signal as $\hat{y}_l(t) = r_l(t)x^*(t)$, the ADC samples after filtering and sampling (assuming the sampling period is T_s and N samples per chirp) can be obtained as:

$$\hat{y}_{l,n} \approx A_l e^{j2\pi(-\alpha\tau + f_c d)nT_s} e^{j2\pi(f_c dlT)} + \text{v}_{l,n}, \; n \in [0, N-1]. \tag{15.7}$$

Then, by denoting $\bar{y}_{l,n}$ as the additive in-band interference, the discrete ADC samples with and without interference can be further expressed as:

$$y_{l,n} = \begin{cases} \hat{y}_{l,n} + \text{v}_{l,n}, \; n \subseteq H \\ \hat{y}_{l,n} + \bar{y}_{l,n} + +\text{v}_{l,n}, \; n \subseteq K \end{cases} \tag{15.8}$$

where H is the set of object reflected samples without interference and K denotes the set of samples containing interference. Existing filters can be used to detect abnormal signal periods and obtain interference-pruned samples. Then, those samples distributed by interference will be masked before the proposed neural network is input to perform intelligent interpolation. With the proposed designs, simulation results confirm that the SINR can be improved by at least 17 dB after the intelligent interpolation in severe interference scenarios, demonstrating the benefits of the proposed learning-based interference mitigations.

15.3 DL-Based Interference-Aware Cognitive Radar

In addition to previously explored DL methods for radar signal processing, which primarily focus on suppressing interference within radar receivers, DL also introduces proactive strategies to address radar interference. One such strategy is cognitive intelligent radar interference mitigation scheduling, which integrates cognitive insights about interference with advanced scheduling algorithms. This approach facilitates interference management and resource allocation by forming a "perception-learning-decision-action" feedback loop, enabling radar systems to adapt dynamically to complex electromagnetic conditions. Evaluating the performance of anti-jamming strategies is equally essential for measuring the efficacy of these countermeasures. The sections below detail the essential elements of intelligent resource scheduling and performance evaluation for anti-jamming applications. Feature-mining-based anti-jamming methods typically exploit differences in amplitude, spatial distribution, and other attributes between target echoes and jamming signals to enhance target detection and identification under electronic interference. Deep neural networks, with their superior capability to extract features, enable the exploration of distinctions between jamming signals and target echoes in high-dimensional spaces. For instance, one approach utilized a signal direction map under mainlobe jamming as input and trained the network using the expected signal direction map. By employing error backpropagation, this method created deeper nulls at interference locations while preserving the main beam, resulting in improved angular measurement accuracy compared to conventional techniques. Another study introduced an end-to-end target detection network leveraging CNNs to enhance performance in scenarios involving intermittent sampling and relay jamming. Additional research has focused on the quantization characteristics of Digital Radio Frequency Memory (DRFM) electronic jamming. Neural network classifiers have been employed to identify DRFM deception jamming signals with minimal quantization bits. Further investigation into the impact of DRFM phase quantization on the jamming signal spectrum led to the development of an adaptive correlation and generalized likelihood ratio detection algorithm. This approach utilized error angles of jamming signals to differentiate between interference and target echoes, extending its applicability to frequency diverse array (FDA)-MIMO radar deception jamming. However, its effectiveness diminishes with higher quantization bit counts, as jamming signals increasingly resemble transmitted signals, thereby degrading target detection performance. While feature-mining-based anti-jamming methods excel in low-dimensional feature spaces, their performance is limited when target and jamming signals are closely aligned. Deep neural networks, capable of analyzing high-dimensional feature spaces, present new opportunities for developing sophisticated interference and target discrimination techniques. There is a pressing need for innovative methods that exploit these high-dimensional spaces to improve jamming and target signal differentiation through deep neural network architectures.

Intelligent resource scheduling for anti-jamming focuses on the dynamic and adaptive allocation of radar resources—including frequency, power, and waveform selection—based on real-time interference and operational demands. This ensures optimal radar performance, even when encountering unknown interference or advanced jamming techniques. Machine learning and optimization algorithms enhance scheduling by learning from past experiences and adapting to new interference patterns, boosting radar resilience. The primary challenges in this field include balancing detection precision, response speed, and resource limitations across diverse electromagnetic environments. To assess the effectiveness of these strategies, especially within reinforcement learning (RL) contexts, performance evaluation plays a critical role. Key metrics such as radar detection accuracy, signal-to-noise ratios, and target tracking performance under various interference scenarios provide insight into the effectiveness of anti-jamming measures. By analyzing scheduling outcomes and countermeasure results, systems can iteratively improve responses to jamming through adaptive learning. Robust performance metrics enable radars to maintain reliable operations in dynamic electromagnetic environments, enhancing anti-jamming capabilities. Collectively, these advancements contribute to a robust framework for intelligent anti-interference in cognitive radar systems. This framework supports efficient decision-making and resource management in response to evolving threats. As advanced jamming techniques such as agile, swarm, and intelligent jamming become more prevalent, traditional reactive radar technologies are increasingly insufficient. Proactive cognitive intelligent anti-jamming methods are now central to addressing complex electronic warfare challenges. These methods emphasize agile waveform transmission and the coordinated management of multi-domain resources across space, time, frequency, and energy domains. The "cognitive intelligence" aspect involves the radar's ability to intelligently manage and optimize resources and parameters in response to environmental changes, forming the aforementioned perception-learning-decision-action loop in adversarial scenarios. Intelligent resource scheduling remains a cornerstone technology for cognitive intelligent radar systems. Recent research has prioritized three main areas: intelligent scheduling of waveform parameters, dynamic allocation of power resources, and frequency domain resource optimization [132].

Specifically, intelligent waveform scheduling refers to the radar's ability to dynamically and adaptively adjust transmission signals, including waveform type, spatial, temporal, and frequency-energy parameters, based on real-time perception of the external environment. This makes it difficult for adversaries to predict radar transmission strategies and intentions. For example, ref. [345] addresses the problem of data-driven joint design of transmitted waveform and detector in radar systems by proposing two innovative learning-based approaches that utilize end-to-end training. The first approach involves alternating between supervised training of the detector for a fixed waveform and RL of the transmitter for a fixed detector, while the second approach trains both the transmitter and detector simultaneously, which can potentially accelerate the training process. The authors formulate a radar system architecture where both components are implemented as feedforward multi-layer neural networks, incorporating various operational waveform constraints such as peak-to-average-power ratio (PAPR) and spectral compatibility into the design. Unlike traditional radar design methods that often rely on rigid mathematical models with limited applicability, this work enhances radar learning by training the detector using synthetic data generated from multiple statistical models of the environment, thereby increasing robustness. Theoretical results presented in the paper establish connections between alternating and simultaneous training by computing the gradients of their respective loss functions, while also justifying the use of RL for transmitter training by comparing its gradient to that of an ideal model-based likelihood function. Overall, the proposed methods demonstrate a significant capability to adapt transmitted waveforms to varying environmental conditions while satisfying essential design constraints, marking a substantial advancement in radar system design methodologies. On the other hand, power resource scheduling involves dynamically adjusting radar transmission power based on real-time interference information to balance detection and stealth performance. Increasing transmission power enhances anti-jamming performance and detection range but also increases the risk of detection by adversaries. Conversely, reducing transmission power improves electromagnetic stealth but may compromise detection capabilities. The challenge is to optimize the trade-off between detection and survivability, particularly in the presence of advanced jamming techniques. Finally, frequency agility in radar operations—rapid shifts in working frequency between pulses or pulse groups—can effectively prevent jamming signals from locking onto radar frequencies. In this direction, ref. [385] introduces a cooperative game theoretic power allocation (CGTPA) strategy aimed at enhancing air defense efficiency within a distributed multiple-input multiple-output (D-MIMO) radar sensor network (RSN) for detecting low-altitude targets. The CGTPA mechanism focuses on precise power management to effectively mitigate variations and fading in the received signals, which are often caused by multipath effects. By leveraging the abundant scattering characteristics inherent in multipath environments, this approach enhances target detectability. The authors introduce the concept of multipath distance difference (MDD), which incorporates errors from space-based external information, and derive the Neyman-Pearson based detection probability alongside the SINR. SINR is subsequently utilized as a key metric for evaluating RSN target detection performance, leading to the establishment of an optimization model based on the max-min criterion. Given the non-convex nature of this optimization problem, cooperative game theory offers a flexible framework for facilitating collaborative detection among intelligent antennas within the RSN. The article analyzes the existence and uniqueness of the core in this cooperative game setting and employs Shapley values to address the power allocation problem while adhering to fairness principles. Additionally, weighted graph techniques are utilized to simplify computational complexity, resulting in a power allocation scheme that respects both Pareto optimality and fairness. The numerical results presented in the study validate the theoretical analysis and demonstrate the effective performance of the proposed algorithm, which approaches optimal resource allocation methods with notable immediacy. Frequency domain intelligent scheduling refers to the real-time sensing of interference frequencies and patterns, followed by adaptive adjustments of working frequency, bandwidth, and agility speed. This is crucial in countering increasingly sophisticated, intelligent, and agile jamming strategies. Recent research has focused on using deep RL and game theory to solve frequency resource scheduling problems. Ref. [386] introduces a decentralized spectrum allocation approach aimed at addressing the significant issue of mutual interference among automotive radars, which has become a growing concern in recent years. The proposed method employs RL to dynamically allocate frequency subbands to individual radars without necessitating communication between them, which is crucial given that each radar operates with limited information. The dynamic nature of the allocation is essential due to the mobility of the vehicles on which the radars are mounted. To enhance decision-making, a long short-term memory (LSTM) recurrent network is utilized to aggregate observations over time, enabling each radar to learn to choose a frequency subband by integrating both current and past observations. The authors conducted simulation experiments to evaluate the performance of their approach against other common spectrum allocation methods, such as random and myopic policies, demonstrating that their RL-based method significantly outperforms these alternatives. This research contributes to the broader field of spectrum sharing for automotive radars, offering an adaptive solution that effectively mitigates mutual interference in increasingly congested electromagnetic environments, thereby enhancing the reliability and safety of radar operations in autonomous driving scenarios. In the above three perspectives, we can see that game theory and deep RL have been widely used to optimize radar anti-interference/anti-jamming strategies due to their capability to deliver complex actions based on multiple environmental information (i.e., status). In conclusion, proactive cognitive intelligent anti-jamming techniques have become essential for countering complex jamming threats. Intelligent resource scheduling, leveraging game theory, RL, and optimization algorithms, plays a critical role in enhancing radar resilience in increasingly challenging electromagnetic environments.

Besides the above perspectives, some advanced works are working on radar transmitter optimization as well besides radar receiver designs by means of cognitive radar (i.e., cognitive radar transceiver co-design) for interference mitigation purposes. Note that in radar systems, the transmitter and receiver are often co-located in the same device and thus joint optimization or cooperation is much easier and practical compared to communication systems. For example, refs. [387, 388] are representative works, that propose a closed-loop perception-action cycle to allow the transmitter to perform waveform adaptation based on the spectrum sensing and spectrum interpretation/prediction results. By employing off-the-shelf spectrum sensing methods, such as the SBSS method presented in ref. [389], a CNN is built to perform spectrum prediction for future spectrum usages. Based on this result, different transmitter adaptation methods, such as carrier frequency and start time, can be adjusted by means of interference avoidance, thus yielding better performance. Specifically, given the predicted time-frequency interference status of the upcoming cycle, the below cumulated interference cost function is evaluated for optimal transmitter adaption selection:

$$I_{MII} = \frac{1}{\sum_{h=1}^{H}\sum_{w=1}^{W} m_{h,w}} \sum_{h=1}^{H}\sum_{w=1}^{W} m_{h,w} p_{h,w}, \tag{15.9}$$

where $h \in [1, H]$ and $w \in [1, W]$ represent the time-domain pixel and frequency-domain pixel of the considered time-frequency domain image, $m_{h,w}$ is the binary mask of occupied resources for the waveform adaptation, and $p_{h,w}$ is the normalized interference intensity of each pixel. With this cost function, grid-search optimization can be performed to evaluate all possible transmitter adaptation combinations to provide optimal transmitter design with minimized interference. Similarly, in refs. [390] and [391], RL is introduced to control the frequency-hopping decision of the radar transmitter based on the received environment status. Specifically, in the proposed solution, the radar observes the environment through the received pulses, including the pulse width of the received pulse, how long has the radar stayed on the same frequency, and whether it is jammed, then an RL agent, such as Q-table or DQN, is employed to perform decision-making based on the given status to maximize the expected long-term reward. The action defined in this work is to choose to hop to a different random frequency or not, and the reward considered in this work is the weighted sum of the integration efficiency factor and the probability of interception factor. By adjusting the weightings, the trade-off between integration efficiency factor and the probability of interception can be fine-tuned to fit different design requirements and provide anti-jamming capability. Ref. [392] investigates intelligent strategies for interference suppression in radar systems operating within complex electromagnetic environments, utilizing RL to enhance performance. The authors establish an interactive loop that models the dynamic interaction between the radar and its environment, effectively mapping interference suppression tasks to various RL components such as interference state sets, method sets, evaluation criteria across different domains, and principles of interference substate transformation. To implement this framework, two novel algorithms are developed: Retroactive-Q (R-Q) learning and Retroactive-Deep Q Network (R-DQN). These algorithms introduce a backtracking Q-value mechanism that links evaluations at each time step during a training round, thereby improving the learning process. The study further analyzes the selection probabilities of optimal implementation sequences for interference suppression and conducts comparative evaluations among the proposed R-Q learning and R-DQN algorithms against conventional Q learning and DQN in terms of output Q-values. The numerical results validate the effectiveness and robustness of the proposed suppression strategies across diverse scenarios, highlighting their potential to significantly enhance radar performance in the face of challenging electromagnetic interferences. This research contributes to advancing intelligent radar systems by leveraging machine learning techniques to adaptively mitigate interference, thereby improving operational reliability in complex environments. Although the above works can already lead to improved spectrum efficiency by the proposed spectrum allocation or frequency hopping schemes, the flexibility is limited since the non-overlapping band partition is needed in advance, and thus the whole learning system will need to be re-trained when aiming to deploy to a different band setting. Alternatively, ref. [393] provides enhanced flexibility to such RL-based radar transmitter designs. Specifically, this work considers the scenario when running a pulse-agile radar in a heavily congested environment. Unlike the above works, the waveform parameter selection problem is modeled as a continuous control task in this work, allowing the proposed solution to scale to any wideband and high-resolution setting without modifications. Using spectrum sensing energy in the time-frequency domain as the state, the proposed RL agent is trained to suggest the start and stop frequency for pulse-agile radar usage. Note that via this setting, besides band selection, the agent can also control bandwidth to adjust radar behavior based on environmental status and achieved performance. The reward function consists of two components, bandwidth utilization, and collisions with other users, to guide the agent to make the decision that maximizes bandwidth utilization but minimizes collisions with other users. Moreover, to allow real-time implementation when using USRP to demonstrate the proposed solution, the neural network architecture is also specially designed to employ recurrent attention-based structure for its parallel processing capability. The over-the-air tests confirm that the proposed solution is flexible and can adapt its behavior to a range of user objectives and priorities for practical radar deployments.

In ref. [394], a GAN-based neural network is used to generate realistic waveforms from a training set of already existing waveforms. The synthesized waveforms show a high similarity with the training dataset waveforms with negligible cross-correlations to the training waveform dataset, showing that the proposed model actually learns to generate realistic and distinct radar waveforms. Thus, those synthesized waveforms can be used to transmit simultaneously with the original waveforms for multi-target tracking with low interference. Moreover, the synthesized waveforms were also constrained to possess a constant modulus for practical usages. With a similar idea, refs. [395, 396] propose the idea to generate an adaptive radar waveform that blends into the environment and maintains radar functionality simultaneously utilizing deep generative learning. Specifically, the considered signal model can be expressed as:

$$x(t) = \alpha s(t) + n(t) + b(t), \tag{15.10}$$

where $x(t)$ is the received signal, $s(t)$ is the transmitted signal, α is the attenuation factor, $n(t)$ is the noise, and $b(t)$ is other signal presented in the radio environment, which can be observed in the receiver. To evaluate radar detection performance of transmitted signal $s(t)$, the ambiguity function can be employed and expressed as:

$$\hat{A}(\tau, F_D) = \int_{-\infty}^{\infty} s(t) \exp(j2\pi F_D t) s^*(t-\tau) dt, \tag{15.11}$$

where τ is the time shift and F_D is the Doppler shift. On the other hand, to evaluate the low probability of detection performance, one can utilize deep generative learning to generate radar waveform, which is distributively close to other waveforms presented in the radio environment, thus it is hard to detect. To do so, conditional GAN is employed to build the loss function as:

$$\min_G \max_D E_{x \sim P_b}[D(x|y)] - E_{\tilde{x} \sim P_{g|y}}[D(G(z|y)|y)], \tag{15.12}$$

where G represents the generator neural network, D stands for the discriminator neural network, P_b is the distribution of $b(t)$, $P_{g|y}$ is the distribution of the generated signal conditional on instantaneous radio environment measurement y, z is a random seed to trigger the generator. To stabilize the training process, different regularization terms can be added to the above equation to smooth the loss function. Note that in the above equation, when a signal with a similar distribution to the $b(t)$ is fed into the discriminator, the discriminator is designed to give a high score to maximize the loss function (see the first term). On the other hand, when a generated signal is fed, a low score will be given to maximize the loss function (see the second term). However, the goal of the generator is to confuse the discriminator to get a high score using the generated signals. Thus, until convergence, the generator is well-trained to provide highly convincing signal, which is similar to other signals in the radio environment for low probability of detection purposes. Simultaneously, to guarantee the detection performance when using generated signals, the ambiguity function will also be optimized during training. Thus, the final loss function is the weighted sum of Eq. (15.11) and Eq. (15.12), and the weighting can be used to control the generated signal behavior by emphasizing low probability of detection or detection performance.

16

DL-Based Signal Processing in ISAC Systems

16.1 DL-Based Waveform Design

While the waveform design for integrated sensing and communication (ISAC) systems is commonly conducted based on traditional optimization-based and model-based approaches, e.g., approaches discussed in Chapter 10, deep learning (DL)-based approaches have shown some advantages, especially from the low-complexity and online adaptivity aspects.

The fundamental idea of using DL to obtain low-complexity design approach is to replace the sophisticated optimization procedure with some neural networks (NNs) that can be efficiently implemented in real time. As a concrete example, ref. [397] considered an uplink ISAC system where the BS conducts the mono-static sensing and users at the same time transmit the uplink communication signals of K users to the BS. Then, the design criterion is obtained by first formulating the sensing rate (i.e., the mutual information between the environment and sensing signal) and the uplink sum-rate of users, and then maximizing their weighted-sum function give as:

$$\max_{\mathbf{S},\mathbf{w}_k,\forall k} \alpha R_s(\mathbf{S}) + (1-\alpha) R_c(\mathbf{S}, \mathbf{w}_k, \forall k), \tag{16.1}$$

where $R_s(\mathbf{S})$ is the sensing rate, $R_c(\mathbf{S}, \mathbf{w}_k, \forall k)$ is the uplink sum-rate, $\mathbf{S}$ is the sensing signal broadcast by the BS, $\mathbf{w}_k$ is the uplink precoding of user k, and α is the tradeoff factor between sensing and communication performance. Since Eq. (16.1) is in general non-convex, the convectional approach for solving Eq. (16.1) is to first convert the problem into a sequence of convex optimization problems, and then solve them. This inherently leads to very high complexity. Alternatively, when the DL is adopted, the approach is to train a NN that can directly output the design of $\mathbf{S}$ and $\mathbf{w}_k, \forall k$ along with some pre-processing and post-processing module using unsupervised learning, where the loss function is defined as:

$$L = -\frac{1}{N_s} \sum_{q=1}^{N_s} \alpha R_s(\tilde{\mathbf{S}}_q) + (1-\alpha) R_c(\tilde{\mathbf{S}}_q, \tilde{\mathbf{w}}_{k,q}, \forall k), \tag{16.2}$$

DOI: 10.1201/9781003408239-16

where N_s is the number of training samples, $\tilde{\mathbf{S}}_q$ is the output of the overall DL-based design module for the sensing signal of training sample q, and $\tilde{\mathbf{w}}_{k,q}$ is the output of the DL-based design module for the precoder of user k of training sample q. The detailed descriptions of the module can be found in ref. [397]. With the DL-based module, the sensing signal and precoding of users can be obtained in real time by the relatively simple computation of the module, rather than solving many convex optimization problems as with the conventional optimization-based approaches.

Ref. [398] shows another example where using the DL-based approach can lead to a low-complexity waveform design for ISAC systems. Specifically, ref. [398] considers a mono-static ISAC system where a BS simultaneously conducts sensing of Q targets and communication with K users using N_t transmit antennas. Then, the design goal is to design the preocder so that the sensing beampattern can be optimized while the communication quality of users can be guaranteed. To this end, the following design problem is considered:

$$\begin{aligned} \min_{\mathbf{w}} \quad & f(\mathbf{w}) \\ \text{s.t.} \quad & g(\mathbf{w}) \preceq \mathbf{0} \\ & h(\mathbf{w}) = P, \end{aligned} \tag{16.3}$$

where $\mathbf{w}$ is the precoder, $f(\mathbf{w})$ is the beampattern matching error, $g(\mathbf{w})$ is to force the received signals at users can fall within the correct signal regime for demodulation, and $h(\mathbf{w}) = P$ is the total power constraint. To use the DL to solve Eq. (16.3), the basic approach is to define the loss function give as:

$$L = f(\mathbf{w}) - \lambda_g \sum_{k=1}^{K_g} \|\mathrm{ReLU}(g_k(\mathbf{w}))\|^2 + \lambda_h \|h(\mathbf{w})\|^2, \tag{16.4}$$

where λ_g and λ_h are some penalty weights associated to the inequalities and equalities in Eq. (16.3). It follows that one can construct a NN model, and then use the above loss function with unsupervised learning to train the NN to output a suitable precoding vector $\mathbf{w}$ for the system. However, the direct use of NN without resorting to some specific structure and training strategy could easily lead to inefficient training and/or ineffective design of NN for providing suitable $\mathbf{w}$.

To resolve this issue, ref. [398] provides a specific NN structure based on CNN for solving Eq. (16.3), which is illustrated in Figure 16.1. It should be noted that the use of CNN is because a layer of CNN is more computationally efficient than that of the regular NN, leading to lower complexity when the overall NN module with multiple layers could maintain low. In addition to the special structure in Figure 16.1, a specifically designed training strategy is also mentioned in ref. [398]. The idea is to include an additional supervised learning procedure for training before using the regular unsupervised learning for training. As the precoding design problem can be effectively solved by using the block coordinate descent (BCD)-based approach in ref. [399] with

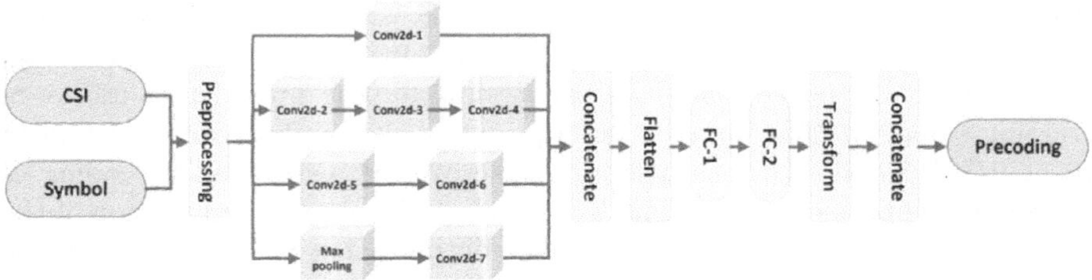

FIGURE 16.1
Illustration of the DL-based design module for symbol-level precoding (copyright from ref. [398]).

high complexity, one can use the approach to generate I labeled samples to construct a training dataset for supervised learning. Then, at the beginning of the overall training of the NN module, these samples are used to train the NN in a supervised manner with the loss function being

$$L_{\text{sup}} = \frac{1}{I}\sum_{i=1}^{I} \|\hat{\mathbf{w}}_i - \mathbf{w}_i\|^2 + \lambda_g \sum_{i=1}^{I}\sum_{k=1}^{K_g} \|\text{ReLU}(e^{g(\hat{\mathbf{w}}_i)} - 1)\|^2, \tag{16.5}$$

where $\hat{\mathbf{w}}_i$ is the output of NN with sample i and $\mathbf{w}_i$ is the precoder obtained by using the BCD-based approach for sample i. Hence, it is clear that the supervised learning with the above loss function is to let the NN module to learn from some ground truth samples generated by using the conventional optimization-based approach, and then proceed with the more general unsupervised learning. This can let the NN be trained with higher efficiency.

The above approaches mainly replace the optimization-based approaches by using some NN modules that are specifically designed for solving the problem. However, these approaches directly rely on the insights and domain knowledge of the original design problems and systems, while the structures of the original optimization-based approaches are not effectively leveraged. To better exploit the original optimization-based approaches, by the observations that the high-complexity optimization-based approaches commonly have iteration structures which help updating the solutions until some stopping criteria are satisfied, the unfolding learning-based approaches have been adopted. The idea of the unfolding learning is to embed AI models into the original optimization-based approaches by replacing multiple iterations conducted by the optimization solver with some NN layers that can be easily accelerated by parallel computing and other acceleration approaches. The idea of the unfolding learning is illustrated in Figure 16.2. From the figure, it can be observed that multiple iterations are replaced by a single NN module, where the NN modules are well-trained to be able to predict what should be done by the original iterative updates of the optimization solvers. By doing so, the optimization can be significantly accelerated by using NNs. However, the structure and training approaches of NNs are not coming from nowhere. They indeed follow

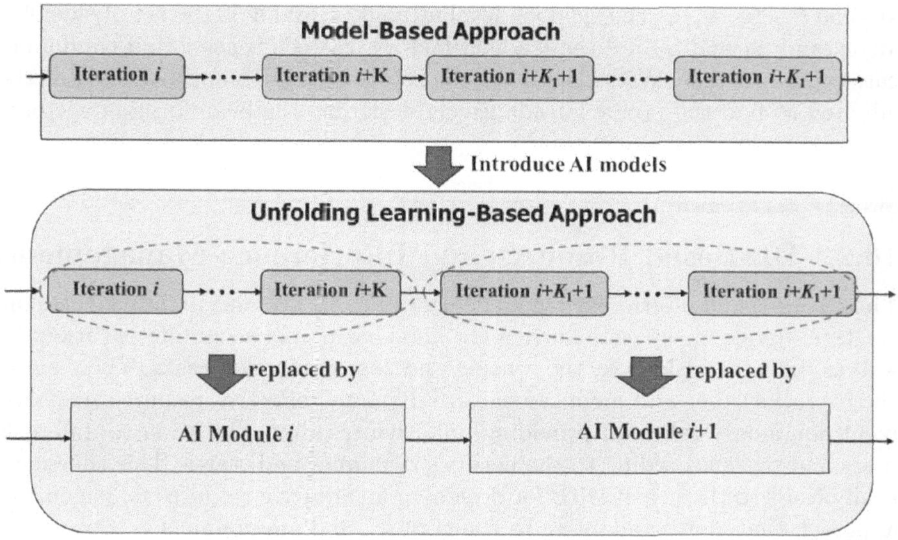

FIGURE 16.2
Illustration of the structure of unfolding learning.

the original structure of the optimization-based approaches. In addition, the loss functions for training are also derived based on the original optimization-based approaches. The main reason of maintaining the original structures and deriving the loss functions from the optimization-based approaches when applying the unfolding learning is to enjoy both the stability of solutions the optimization-based approaches as well as the flexibility of learning-based approaches, making the overall approach to learn from the structures of conventional optimization methods, while outperforming the conventional methods by involving certain data-driven features. One example of the use of unfolding learning-based design approach for ISAC systems is discussed in ref. [400].

Finally, when considering to design the waveform with environmental adaptivity, the use of deep reinforcement learning (DRL) might be a desired direction. For example, ref. [401] proposes a DRL approach that can adaptively change the beamforming weights according to the movement of the user and target. Specifically, it considers designing the beamforming vectors $\mathbf{w} = \frac{1}{N_t}[e^{j\theta_1}, ..., e^{j\theta_{N_t}}]$ for tracking a target while communicating to a user, where N_t is the number of transmit antennas and θ_k is the phase shift of antenna k, quantized by a set $\mathcal{D}_\theta$ within 0 and 2π. Then, it defines the state space as $S = [\theta_1, ..., \theta_{N_t}, \mathbf{H}_t]$, where θ_k should be at one of the quantized values of $\mathcal{D}_\theta$ and $\mathbf{H}_t$ is the channel matrix at time t. The action space is thus the selection of all possible quantized values of $\theta_k, \forall k$. Finally, the reward function at time t is defined as the weighted-sum of the sidelobe level of the beampattern and the received communication channel gain, given as $r_t = -\alpha_1 \max_{\phi\in\Phi}(P(\phi, \mathbf{w}_t)) + \alpha_2|\mathbf{H}_t\mathbf{w}_t|^2$, where $\mathbf{w}_t$ is the beamforming vector at time t; $P(\phi, \mathbf{w}_t)$ is the sidelobe level at angle ϕ and Φ is the set of sidelobe angle; and α_1 and α_2 are the tradeoff factors between sensing and communication performance. With the above, the actor-critic DRL approach is directly adopted to find the policy for adaptively designing the beamforming vectors.

16.2 DL-Based Resource and Interference Management

The resource and interference management of ISAC systems mainly focuses on the time-frequency-space management and sharing between different users as well as different tasks, i.e., the sensing and communication tasks. Then, since the overall system and environment are dynamic, effective resource management commonly relies on providing adaptive resource allocation to different users and tasks according to the network dynamics and states. This thus naturally leads to the use of DRL for designing an effective resource management approach that can be adaptive to the network and environment.

Serving as a concrete example, ref. [402] considers a ISAC system where the tracking tasks and communication tasks compete for time resources. Thus, the resource management is to determine how much amount of time should be allocated to conduct tracking for each time interval. Its goal is to develop a dynamic strategy that can balance between tracking and communication phases via using DRL. Note that in this ISAC system, the sensing and communication are individually but coordinated implemented, instead of being integrated. To this end, it considers n users to track and communication, and consider that the duration of a time interval is T_0. Then, by denoting τ_t^n as the amount of time for tracking user n in time interval t, the remaining amount of time for communications is $\tau_t^{\text{Com}} = T_0 - \sum_{n=1}^{N} \tau_t^n$, where N is the total number of targets to track. With this, the communication rate can be obtained as $R_t = \frac{\tau_t^{\text{Com}} B}{N} \log_2(1 + SNR_t^n(\theta_n, \hat{\theta}_n))$, where B is the system bandwidth shared by N users, $SNR_t^n(\theta_n, \hat{\theta}_n)$ is the communication signal-to-noise ratio (SNR) for user n at time t, with θ_n and $\hat{\theta}_n$ being the correct and estimated angles for beamforming the communication signal to the user, respectively. Note that if $\hat{\theta}_n$ deviates more from θ_n, the SNR in general would be lower. However, if we want to increase the angle estimation accuracy, we might need to spend more time on tracking phase, which leads to a tradeoff between the time allocation to tracking and communication phases. With the above, the design problem becomes:

$$\max_{\pi} \sum_{t=0}^{\infty} \gamma^t R_t \quad \text{s.t.} \quad \sum_{t=0}^{\infty} \gamma^t \left(T_0 - \sum_{n=1}^{N} \tau_t^n \right) \geq 0, \tag{16.6}$$

where the objective function is to maximize the long-term communication rate reward, subject to that the overall communication time should be non-zero, and π is the time allocation strategy we want to develop.

To obtain an effective strategy π, the DRL based on the double Q-learning network (DQN) has been developed in ref. [402]. First of all, to resolve the constraint in Eq. (16.6), one can convert the problem as for each time m, we want to solve:

$$\min_{\lambda_m \leq 0} \max_{\pi} \sum_{t=0}^{\infty} \gamma^{m+t} \left[R_{m+t} + \lambda_m \left(T_0 - \sum_{n=1}^{N} \tau_{m+t}^{n} \right) \right], \tag{16.7}$$

where λ_m is the Lagrange multiplier for time interval m, which is time-varying. Then, to solve this, one can define the state of the system at time t as $S = [s_t^1, ..., s_t^N]$, where s_t^n is the state of user n at time t, given as $s_t^n = \left[\tau_n^{t-1}, \sigma_{t-1,n}^2, \lambda_{t-1}\right]$, where τ_n^{t-1} is the allocated amount of time to track user n at the previous time interval and $\sigma_{t-1,n}^2$ is the angle estimation variance of user n at the previous time interval. Then, the action of the system can be defined as choosing between different amount of time for $\tau_n^t, \forall n$, with the action space of a τ_n^t being the quantized set within 0 and T_0. Finally, the reward at time t is defined as:

$$r_t = R_t + \lambda_t \left(T_0 - \sum_{n=1}^{N} \tau_t^n \right). \tag{16.8}$$

With the above, the DQN of DRL is directly introduced to establish the policy π, where the ϵ greedy exploration method is adopted.

In ref. [403], the joint power and bandwidth allocation problem for the multi-user ISAC systems where a BS needs to sensing the positions of M users, and then communication with the users. In the system, each time period within the coherence time is split into sensing and communication phases, where $R_{m,s}(t)$ is the amount of time for sensing user m and $R_{m,c}(t)$ is the amount of time for communication to user m at period t. Then, denoting $V_{m,s}(t)$ and $V_{m,c}(t)$ as the value of services of sensing and communication of user m at time period t, the reward of the system at time period t is given as $V(t) = \sum_{m=1}^{M} \omega_{m,s} V_{m,s}(t) + \omega_{m,c} V_{m,c}(t)$, where $\omega_{m,s}$ and $\omega_{m,c}$ are the weighting of priority of sensing and communication services of user m. We denote $N_{m,s}(t)$ and $N_{m,c}(t)$ as the number of subcarriers allocated to sensing and communication services of user m at period t, respectively, and denote $P_{m,s}(t)$ and $P_{m,c}(t)$ as the power allocated to sensing and communication services of user m at period t, respectively. Then, the goal is to optimize $N_{m,s}(t)$, $N_{m,c}$, $P_{m,s}(t)$, and $P_{m,c}(t)$ so that the overall system reward can be maximized in a long run. To this end, an actor-critic-based DRL is adopted in ref. [403]. Specifically, it considers the state vector of user m of the system as:

$$\mathbf{s}_m(t) = [R_{m,s}(t), R_{m,c}(t), \omega_{m,s}, \omega_{m,c}, h_m(t)], \tag{16.9}$$

where $h_m(t)$ is the communication channel of user m at time t. It follows that the complete state vector of the system is given by $\mathbf{s} = [\mathbf{s}_1(t), ..., \mathbf{s}_M(t)]$. Then, since the optimization is on power and bandwidth allocation, the action at time t is given as:

$$\mathbf{a}(t) = [\mathbf{a}_1(t), ..., \mathbf{a}_M(t)], \tag{16.10}$$

where $\mathbf{a}_m(t) = [N_{m,s}(t), N_{m,c}(t), P_{m,s}(t), P_{m,c}(t)]$ with all possible candidates that satisfying the maximum power and bandwidth constraints. With all the above setup, the actor-critic-based DRL is directly used for obtaining the joint power and resource allocation strategy of the system.

In ref. [404], the DRL has been used for joint beamforming and resource management for integrated sensing, communication, and computing (ISCC) in uplink vehicular networks, where the computing task is introduced by the need of vehicles to send the acquired data back to the roadside unit (RSU) for conducting some inference task, e.g., the object recognition. The ISCC vehicular network model is illustrated in Figure 16.3, where M vehicles conduct the sensing and uplink communications to a RSU, and the over-the-air computing technique is used for the vehicles to enable computing at the RSU. Then, the joint beamforming and resource management problem is formulated in the way that the transmit communication beamforming of vehicles, the transmit sensing beamforming of vehicles, and receiving beamforming at the RSU are jointly optimized for providing effective long-term ISCC performance. We denote $\mathbf{W}_{k,c}(t)$ and $\mathbf{W}_{k,s}(t)$ as the communication and sensing beamforming matrices of vehicle k at time t, and denote $\mathbf{u}_k(t)$ and $\mathbf{s}_k(t)$ as the communication and sensing signals of vehicle k at time t. Then, as illustrated in Figure 16.4, the transmit signal is expressed as $\mathbf{x}(t) = \mathbf{W}_{k,c}(t)\mathbf{u}_k(t) + \mathbf{W}_{k,s}(t)\mathbf{s}_k(t)$. With this and denoting $\mathbf{P}(t)$ as the receiving beamforming matrix at RSU of time t, the joint beamforming and resource management problem is expressed as:

$$\begin{aligned} \max_{\mathbf{P}(t),\mathbf{W}_{k,c}(t),\mathbf{W}_{k,s}(t),\forall k,t} \quad & \sum_t \sum_k \sum_k R_{k,\text{com}}(t) \\ \text{s.t.} \quad & \text{MSE}(t) \leq \xi \\ & \text{sensing}_k(t) \geq \psi, \forall k, t \\ & p_k(t) \leq P_{\text{veh}}, \forall k, t, \end{aligned} \tag{16.11}$$

where $R_{k,\text{com}}(t)$ is the capacity of uplink communication of vehicle k at time t, $\text{MSE}(t)$ is the computational error at time t, $\text{sensing}_k(t)$ is the sensing information rate for vehicle k at time t, and $p_k(t) = \text{Tr}\left(\mathbf{W}_{k,c}(t)\mathbf{W}_{k,c}(t)^H + \mathbf{W}_{k,s}(t)\mathbf{W}_{k,s}(t)^H\right)$ is the total transmit power of vehicle k at time t.

To solve Eq. (16.11), a DRL learning approach has been proposed in ref. [404]. Specifically, it defines the state at time t as:

$$\mathbf{s}(t) = \{\mathbf{H}(t), \mathbf{P}(t), \mathbf{a}(t-1)\}. \tag{16.12}$$

where $\mathbf{H}(t)$ is the overall channel matrix that collects the uplink channel matrices of all vehicles and $\mathbf{a}(t-1)$ is the action vector at time $t-1$, which is defined as:

$$\mathbf{a}(t) = \{\mathbf{W}_{k,c}(t), \mathbf{W}_{k,s}(t), \forall k\}. \tag{16.13}$$

It should be noted that by the approach in ref. [404], the receiving beamforming matrix $\mathbf{P}(\mathbf{t})$ is determined directly by the channel matrix $\mathbf{H}(t)$. Thus,

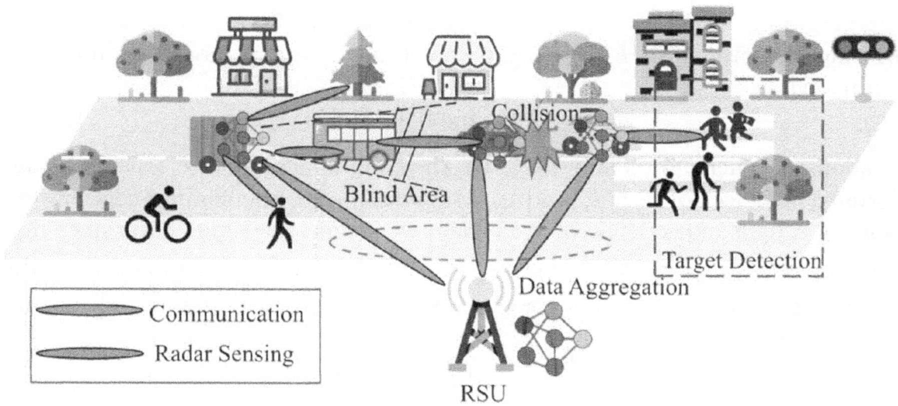

FIGURE 16.3
Illustration of the ISCC vehicular network model (copyright from ref. [404]).

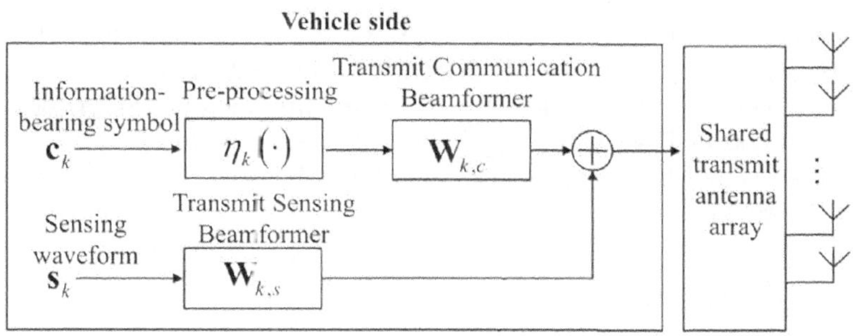

FIGURE 16.4
Transmission Model of the ISCC vehicular network (copyright from ref. [404]).

instead of being an action to determine by the DRL, it is treated as one of the parameters to define the system state at time t. Finally, the reward function at time t is defined as:

$$R_t = \lambda_1 \sum_k R_{k,\text{com}} + \lambda_2 \Phi_{\text{MSE}} + \lambda_3 \Phi_{\text{sen}} + \lambda_4 \Phi_{\text{pow}}, \tag{16.14}$$

where $\Phi_{\text{sen}} = \max(\text{MSE}(t) - \xi, 0)$, $\Phi_{\text{sen}} = \sum_k (\min(\text{sensing}_k(t) - \psi, 0))$ and $\Phi_{\text{pow}} = \sum_k \max(p_k(t) - P_{\text{veh}}, 0)$; and λ_1, λ_2, λ_3, λ_4 are some hyper-parameters to adjust for trade-off between the communication performance, computing MSE performance, sensing performance, and power. With the above definitions of state, action, and reward function, the policy-based deep deterministic policy gradient (DDPG) is directly adopted in ref. [404] to obtain the effective design of beamforming matrices $\mathbf{P}(t), \forall t$, $\mathbf{W}_{k,c}(t), \forall k, t$, and $\mathbf{W}_{k,s}(t), \forall k, t$.

Finally, by taking a different aspect of the DRL, ref. [405] designed the resource and interference management approach by using an unsupervised learning framework. Specifically, when considering a BS to serve K users while sensing the targets in L directions, the transmit signal model for user k is expressed as

$$\mathbf{x}_k(t) = \mathbf{W}_{k,c}\mathbf{u}_k(t) + \mathbf{W}_{k,s}\mathbf{s}_k(t) \tag{16.15}$$

is adopted, where $\mathbf{W}_{k,c}$ and $\mathbf{W}_{k,s}$ are the communication and sensing beamforming matrices of user k, and $\mathbf{u}_k(t)$ and $\mathbf{s}_k(t)$ are the communication and sensing signals of user k at time t. Then, we denote the signal covariance matrix matrix as $\mathbf{R} = \mathbb{E}\left[\sum_{k=1}^{K} \mathbf{x}_k(t)\mathbf{x}_k^T(t)\right]$ for the sensing design criterion, the beam pattern error and cross correlation pattern, denoted as $L_{r,1}$ and $L_{r,2}$ are used, where

$$\begin{aligned} L_{r,1} &= \frac{1}{L}\sum_{l=1}^{L} |d(\theta_l) - P(\theta_l, \mathbf{R})| \\ L_{r,2} &= \frac{1}{L^2 - L}\sum_{l=1}^{L-1}\sum_{r=l+1}^{L} |P'(\theta_l, \theta_r, \mathbf{R})|, \end{aligned} \tag{16.16}$$

where $d(\theta_l)$ is the ideal beam pattern power emission on angle θ_l, $P(\theta_l, \mathbf{R})| = \mathbf{a}^H(\theta_l)\mathbf{R}\mathbf{a}(\theta_l)$, $P'(\theta_l, \theta_r, \mathbf{R}) = \mathbf{a}^H(\theta_l)\mathbf{R}\mathbf{a}(\theta_r)$, and $\mathbf{a}(\theta_l)$ is the beam steering vector on angle θ_l. Subsequently, for the communication design criterion, the sum-rate is adopted, with $C = \sum_{k=1}^{K} \log_2(1 + \gamma_k)$, where γ_k is the SINR of the received signal for user k. With the above, the design problem can then be formulated as:

$$\begin{aligned} \max_{\mathbf{W}_{k,c}, \mathbf{W}_{k,s}, \forall k} \quad & \mathbb{E}\left[\sum_{k=1}^{K} \log_2(1 + \gamma_k)\right] \\ \text{s.t.} \quad & L_{r,1} + L_{r,2} \le \xi \\ & \sum_{k=1}^{K} \|\mathbf{W}_{k,c}\|_F^2 + \|\mathbf{W}_{k,s}\|_F^2 \le P_{\text{BS}}, \end{aligned} \tag{16.17}$$

where ξ is the sensing error performance threshold and P_{BS} is the total power of the BS. To solve Eq. (16.17), the unsupervised learning approach is adopted. Specifically, it first defines a NN $\chi(\mathbf{H}, \{\theta_l\}_l^L; \boldsymbol{\omega})$ parameterized by $\boldsymbol{\omega}$ which generates the precoders $\mathbf{W}_{k,c}$, $\mathbf{W}_{k,s}, \forall k$ as a function of the channel $\mathbf{H}$ and targeting angle $\{\theta_l\}_l^L$, and then obtains the Lagrangian problem given as:

$$\begin{aligned} \mathcal{L} = \mathbb{E}&\left[\sum_{k=1}^{K} \log_2(1 + \gamma_k)\right] + \lambda_1(L_{r,1} + L_{r,2} - \xi) \\ &+ \lambda_2\left(\sum_{k=1}^{K} \|\mathbf{W}_{k,c}\|_F^2 + \|\mathbf{W}_{k,s}\|_F^2 - P_{\text{BS}}\right). \end{aligned} \tag{16.18}$$

Subsequently, the primal-dual update of the NN is adopted, in which each iteration it conducted the updates of primal and dual variables. Since the design of the precoders is not directly generated by the NN $\chi(\mathbf{H}, \{\theta_l\}_l^L; \boldsymbol{\omega})$, the update of the primal variables is to update the parameters $\boldsymbol{\omega}$ of the NN. Thus, for the update of the primal variables at iteration t, the following is implemented:

$$\boldsymbol{\omega}^t = \boldsymbol{\omega}^{t-1} + \eta^t \nabla_{\boldsymbol{\omega}} \mathcal{L}(\boldsymbol{\omega}^{t-1}, \boldsymbol{\lambda}^{t-1}), \tag{16.19}$$

where η^t is the step-size at iteration t. Then, for the update of the dual variables, it implements the following:

$$\boldsymbol{\lambda}^t = \boldsymbol{\lambda}^{t-1} - \eta^t \nabla_{\boldsymbol{\lambda}} \mathcal{L}(\boldsymbol{\omega}^{t}, \boldsymbol{\lambda}^{t-1}). \tag{16.20}$$

The above updates continue until a NN that can generate sufficiently good precoders is obtained.

16.3 DL-Based Predictive Beamforming-Aided ISAC Systems

Finally, this chapter discusses a particular sensing-assisted communication system, namely, the predictive beamforming-aided ISAC system that is highly appropriate for the adoption of DL-based approach [406]. Specifically, in the conventional communication systems, the system needs to first estimation the channel of users before designing the beamforming for communications. However, such channel estimation creates overhead for the communication system. On the other hand, with the aid of sensing, the system can determine the locations and angles of users such that the beamforming of users can be predicted based on the sensing information. As a consequence, the channel estimation stage can be removed to reduce the overhead. Such an approach is known as the predictive beamforming with ISAC system, and since it requires the prediction of beamforming direction/weights, it is suitable for the use of DL-based approaches.

To further elaborate the predictive beamforming-aided ISAC system, the system and design approach proposed in ref. [406] is considered, where an ISAC-assisted vehicular-to-infrastructure (V2I) system illustrated in Figure 16.5 is considered. In the system, a RSU equipped with N_t transmit antennas and N_r receive antennas to serve K vehicles is considered. To conduct the predictive beamforming to improve the communication to vehicles with sensing, the dynamic optimization of beamforming weights for vehicles is considered. We denote $\mathbf{w}_{k,n}$ as the beamforming weight for vehicle k at time-slot n. Then, the transmitted signal is given as:

$$\tilde{\mathbf{s}}_n(t) = \mathbf{W}_n \mathbf{s}_n(t), \tag{16.21}$$

where $\mathbf{W}_n = [\mathbf{w}_{1,n}, ..., \mathbf{w}_{K,n}]$ and $\mathbf{s}_n(t) = [s_{1,n}(t), ..., s_{K,n}(t)]$ is the transmitted signal at time t of time-slot n with $s_{k,n}(t)$ being the corresponding transmitted symbol for vehicle k. Then, the protocol illustrated in Figure 16.6 is considered. In this protocol, we suppose that the beamforming matrix $\mathbf{W}$ for time-slot n is optimized at time-slot $n-1$ based on the

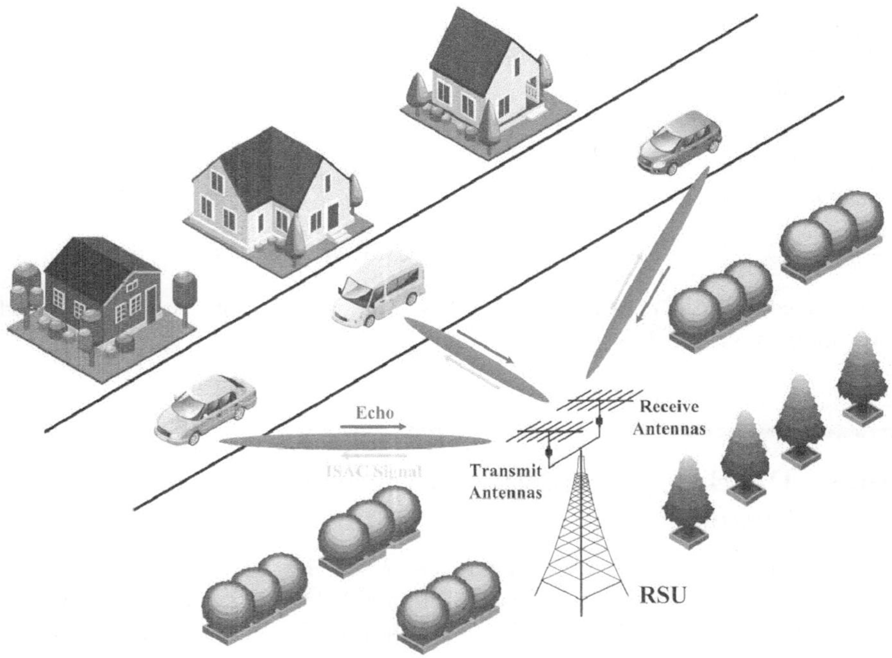

FIGURE 16.5
Illustration of the ISAC-assisted V2I system for predictive beamforming (copyright from ref. [406]).

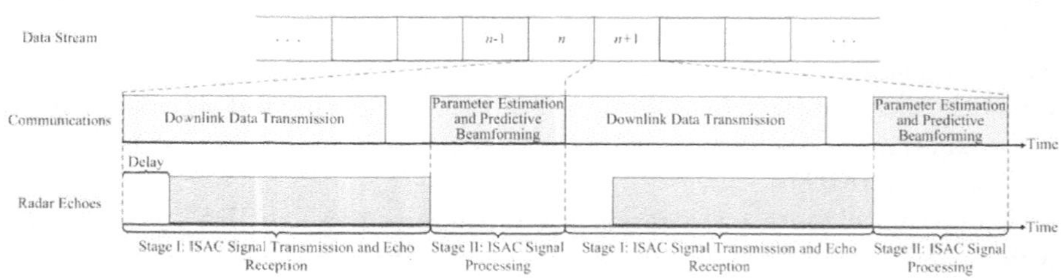

FIGURE 16.6
Predictive beamforming protocol for ISAC systems (copyright from ref. [406]).

sensing results so that the expected communication performance at time-slot n can be guaranteed. Specifically, the transmission within a time-slot is split into two stages. In the first stage, based on the beamforming matrix optimized at the previous time-slot, the ISAC signal is transmitted for communications to vehicles and simultaneously receives the echo signals from vehicles to conduct the estimations of their locations. Then, in stage II, the RSU estimates the motion parameters of vehicles at the current time-slot based on the echo signals, and then optimizes the beamforming matrix for the next time-slot, i.e., time-slot $n+1$. When compared with the conventional beamforming pro-

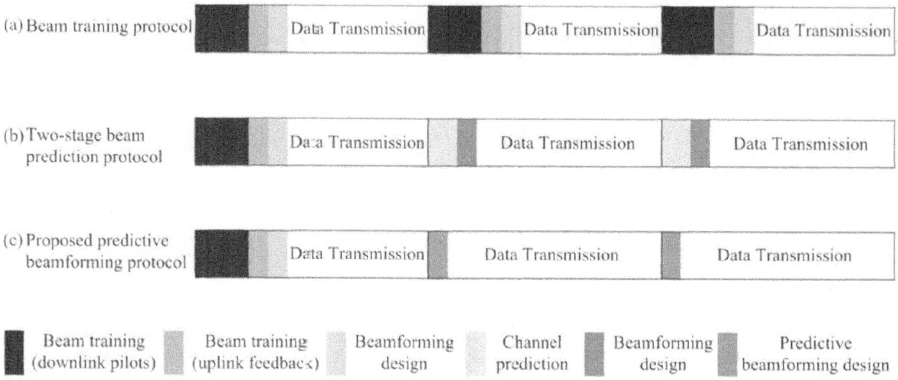

FIGURE 16.7
Comparison with conventional beamforming framework (copyright from ref. [406]).

tocol which requires the transmissions of pilots and beam training overhead as illustrated in Figure 16.7, the predictive beamforming-based approach can reduce the overhead as it does not need the periodic channel estimation and beam training process.

To design the beamforming matrix, the following vehicle mobility model is considered:

$$v_{k,n} = v_{k,n-1} + \Delta v_{k,n-1}, \tag{16.22}$$

where $v_{k,n}$ is the average velocity of vehicle k at time-slot n and $\Delta v_{k,n-1}$ is the variant between $v_{k,n}$ and $v_{k,n-1}$ which follows certain distribution. We assume the movements of vehicles follow the quasi-static assumption and consider the road scenario illustrated in Figure 16.8. As a consequence, the models for the change of angle $\Delta\theta_{k,n}$ and the change of distance $\Delta d_{k,n}$ for vehicle k between time-slots $n-1$ and n can be obtained. Then, by the derivations (see details in ref. [406]), the predictive beamforming design problem for $\mathbf{W}_m$ can be formulated as an expected communication performance optimization problem subject to the Cramer-Rao lower bound (CRLB) constraints for both angle and distance estimation of vehicles, which is expressed as:

$$\begin{aligned} \max_{\mathbf{W}_n} \quad & \mathbb{E}\left[\sum_{k=1}^{K} \log_2(1+\gamma_{k,n})\right] \\ \text{s.t.} \quad & \mathbb{E}\left[\frac{1}{K}\sum_{k=1}^{K} \mathrm{CRLB}(\vartheta_{k,n}, \mathbf{W}_n)\right] \le \gamma_\theta, \quad \mathbb{E}\left[\frac{1}{K}\sum_{k=1}^{K} \mathrm{CRLB}(d_{k,n}, \mathbf{W}_n)\right] \le \gamma_d, \\ & \|\mathbf{W}_n\|_F^2 \le P_{\max}, \end{aligned} \tag{16.23}$$

where $\gamma_{k,n}$ is the communication SINR for vehicle k at time-slot n; $\theta_{k,n}$ and $d_{k,n}$ are the angle and distance of vehicle k at time-slot n; and γ_θ, γ_d, and $P_{\max}$ are the CRLB thresholds for angle and distance, and the power budget,

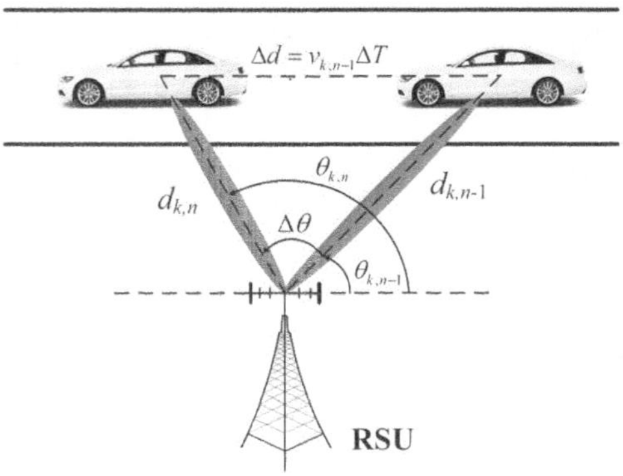

FIGURE 16.8
Illustration of vehicle movement in the system (copyright from ref. [406]).

respectively.

To solve Eq. (16.23), a learning-based approach has been proposed in ref. [406]. Specifically, it first converts Eq. (16.23) into a penalty-based problem given as:

$$\max_{\mathbf{W}_n} \quad F(\mathbf{W}_n), \tag{16.24}$$

where

$$\begin{aligned} F(\mathbf{W}_n) = & \ \mathbb{E}\left[\sum_{k=1}^{K} \log_2(1+\gamma_{k,n})\right] \\ & - \lambda_1 \left[\max\left(\mathbb{E}\left[\frac{1}{K}\sum_{k=1}^{K} \mathrm{CRLB}(\theta_{k,n},\mathbf{W}_n)\right] - \gamma_\theta, 0\right)\right]^2 \\ & - \lambda_2 \left[\max\left(\mathbb{E}\left[\frac{1}{K}\sum_{k=1}^{K} \mathrm{CRLB}(d_{k,n},\mathbf{W}_n)\right] - \gamma_d, 0\right)\right]^2 \\ & - \lambda_3 \left[\max\left(\|\mathbf{W}_n\|_F^2 - P_{\max}, 0\right)\right]^2, \end{aligned} \tag{16.25}$$

and λ_1, λ_2, and λ_3 are penalty parameters that should be much larger than 0. With Eqs. (16.24) and (16.25), the unsupervised learning is introduced to solve the problem by considering the following loss function of a sample $\omega_{n,i}$ and the corresponding output of the NN model $g_{\omega_{n,i}}\left(\tilde{\mathbf{H}}_{n,i}\right)$:

$$L = \sum_{i=1}^{N_\mathrm{S}} f\left(\omega_{n,i}, g_{\omega_{n,i}}\left(\tilde{\mathbf{H}}_{n,i}\right)\right), \tag{16.26}$$

where N_S is the number of learning samples; $\tilde{\mathbf{H}}_{n,i}$ is the estimated channel matrix with the estimated channel for vehicle k to be given as $\tilde{\mathbf{h}}_{k,n,i} = G\sqrt{\alpha_0 \tilde{d}_{k,n}^{-\xi}}\mathbf{a}(\tilde{\theta}_{k,n})$, where G, α_0, and ξ are pathloss modeling parameters, $\tilde{d}_{k,n,i}$

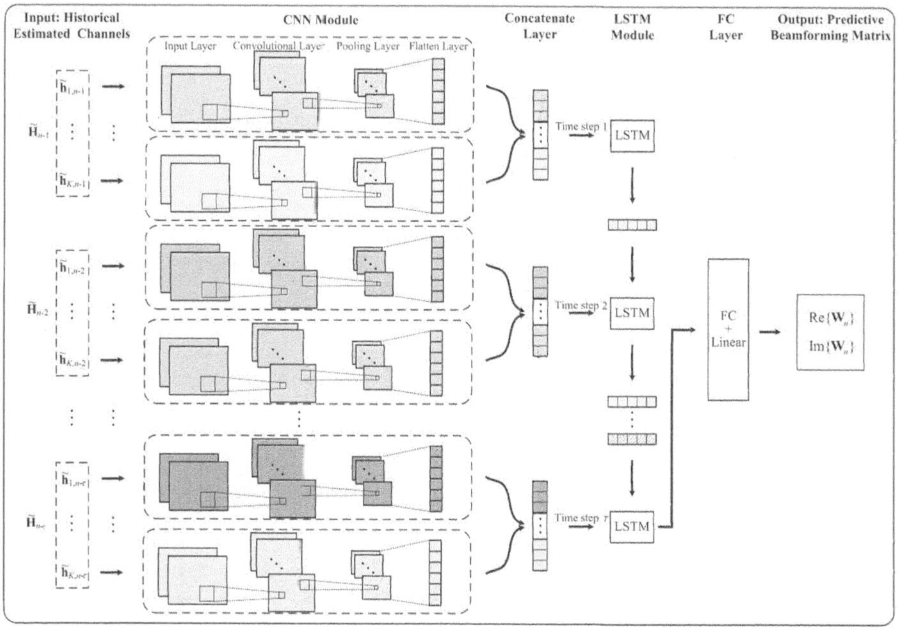

FIGURE 16.9
Illustration of the LSTM-based beamforming design learning model (copyright from ref. [406]).

and $\tilde{\theta}_{k,n,i}$ are the estimated distance and angle for vehicle k, and $\mathbf{a}(\tilde{\theta}_{k,n})$ is the steering vector with angle $\tilde{\theta}_{k,n}$; and

$$
\begin{aligned}
f\left(\omega_{n,i}, g_{\omega_{n,i}}\left(\tilde{\mathbf{H}}_{n,i}\right)\right) = & \sum_{k=1}^{K} \log_2\left(1 + \gamma_{k,n}\left(\omega_{n,i}, g_{\omega_{n,i}}\left(\tilde{\mathbf{H}}_{n,i}\right)\right)\right) \\
& - \lambda_1\left[\max\left(\frac{1}{K}\sum_{k=1}^{K}\mathrm{CRLB}\left(\omega_{n,i}, g_{\omega_{n,i}}\left(\tilde{\mathbf{H}}_{n,i}\right)\right) - \gamma_\theta, 0\right)\right]^2 \\
& - \lambda_2\left[\max\left(\frac{1}{K}\sum_{k=1}^{K}\mathrm{CRLB}\left(\omega_{n,i}, g_{\omega_{n,i}}\left(\tilde{\mathbf{H}}_{n,i}\right)\right) - \gamma_d, 0\right)\right]^2 \\
& - \lambda_3\left[\max\left(\left(\left\|g_{\omega_{n,i}}\left(\tilde{\mathbf{H}}_{n,i}\right)\right\|_F^2 - P_{\max}, 0\right)\right]^2 .
\end{aligned}
\tag{16.27}
$$

With the above loss function, the LSTM-based NN model illustrated in Figure 16.9 is adopted for the unsupervised learning and beamforming matrix inference after the training. We note that a similar approach is also discussed in ref. [407], where the more complex vehicle movements and road situations are considered.

Bibliography

[1] Claudine Badue, Rânik Guidolini, Raphael Vivacqua Carneiro, Pedro Azevedo, Vinicius B Cardoso, Avelino Forechi, Luan Jesus, Rodrigo Berriel, Thiago M Paixao, Filipe Mutz, et al. Self-driving cars: A survey. *Expert Systems with Applications*, 165:113816, 2021.

[2] Rong Liu, Jinling Wang, and Bingqi Zhang. High definition map for automated driving: Overview and analysis. *The Journal of Navigation*, 73(2):324–341, 2020.

[3] Yi Xiao, Felipe Codevilla, Akhil Gurram, Onay Urfalioglu, and Antonio M López. Multimodal end-to-end autonomous driving. *IEEE Transactions on Intelligent Transportation Systems*, 23(1):537–547, 2020.

[4] Agbotiname Lucky Imoize, Oluwadara Adedeji, Nistha Tandiya, and Sachin Shetty. 6G enabled smart infrastructure for sustainable society: Opportunities, challenges, and research roadmap. *Sensors*, 21(5):1709, 2021.

[5] Zixuan Zhang, Feng Wen, Zhongda Sun, Xinge Guo, Tianyiyi He, and Chengkuo Lee. Artificial intelligence-enabled sensing technologies in the 5G/internet of things era: from virtual reality/augmented reality to the digital twin. *Advanced Intelligent Systems*, 4(7):2100228, 2022.

[6] John Torous, Sandra Bucci, Imogen H Bell, Lars V Kessing, Maria Faurholt-Jepsen, Pauline Whelan, Andre F Carvalho, Matcheri Keshavan, Jake Linardon, and Joseph Firth. The growing field of digital psychiatry: current evidence and the future of apps, social media, chatbots, and virtual reality. *World Psychiatry*, 20(3):318–335, 2021.

[7] Inocencio Maramba, Arunangsu Chatterjee, and Craig Newman. Methods of usability testing in the development of eHealth applications: a scoping review. *International Journal of Medical Informatics*, 126:95–104, 2019.

[8] Lex van Velsen, Geke Ludden, Christiane Grünloh, et al. The limitations of user-and human-centered design in an ehealth context and how to move beyond them. *Journal of Medical Internet Research*, 24(10):e37341, 2022.

[9] Gongfan Chen, Min Liu, YuXiang Zhang, ZhiGao Wang, Simon M Hsiang, and Chuanni He. Using images to detect, plan, analyze, and coordinate a smart contract in construction. *Journal of Management in Engineering*, 39(2):04023002, 2023.

[10] Chuanni He, Min Liu, YuXiang Zhang, Zhigao Wang, Simon M Hsiang, Gongfan Chen, and Jiang Chen. Exploit social distancing in construction scheduling: Visualize and optimize space-time-workforce tradeoff. *Journal of Management in Engineering*, 38(4):04022027, 2022.

[11] Gongfan Chen, Min Liu, Huaming Li, Simon M Hsiang, and Ashtad Jarvamard. Motivating reliable collaboration for modular construction: Shapley value-based smart contract. *Journal of Management in Engineering*, 39(6):04023042, 2023.

[12] Chuanni He, Min Liu, Yuxiang Zhang, Zhigao Wang, Simon M Hsiang, Gongfan Chen, Weiqiang Li, and Gongfu Dai. Space–time–workforce visualization and conditional capacity synthesis in uncertainty. *Journal of Management in Engineering*, 39(2):04022071, 2023.

[13] Gongfan Chen, Chuanni He, S Hsiang, Min Liu, and Huaming Li. A mechanism for smart contracts to mediate production bottlenecks under constraints. In *Proc., 31st Annual Conf. of the Int. Group for Lean Construction (IGLC)*, pages 1232–1244, 2023.

[14] Gongfan Chen, Huaming Li, Min Liu, Simon M Hsiang, and Ashtad Jarvamardi. Knowing what is going on a smart contract for modular construction. *Canadian Journal of Civil Engineering*, 50(3):210–223, 2022.

[15] Chuanni He, Min Liu, Zhigao Wang, Gongfan Chen, Yuxiang Zhang, and Simon M Hsiang. Facilitating smart contract in project scheduling under uncertainty—A Choquet integral approach. In *Construction Research Congress 2022*, pages 930–939, 2022.

[16] Gongfan Chen, Abdullah Alsharef, and Edward Jaselskis. Construction jobsite image classification using an edge computing framework. *Sensors*, 24(20):6603, 2024.

[17] Khaled B Letaief, Wei Chen, Yuanming Shi, Jun Zhang, and Ying-Jun Angela Zhang. The roadmap to 6G: AI empowered wireless networks. *IEEE Communications Magazine*, 57(8):84–90, 2019.

[18] Zhengquan Zhang, Yue Xiao, Zheng Ma, Ming Xiao, Zhiguo Ding, Xianfu Lei, George K Karagiannidis, and Pingzhi Fan. 6G wireless networks: Vision, requirements, architecture, and key technologies. *IEEE Vehicular Technology Magazine*, 14(3):28–41, 2019.

[19] Mikko A Uusitalo, Patrik Rugeland, Mauro Renato Boldi, Emilio Calvanese Strinati, Panagiotis Demestichas, Mårten Ericson, Gerhard P Fettweis, Miltiadis C Filippou, Azeddine Gati, Marie-Helene Hamon, et al. 6G vision, value, use cases and technologies from European 6G flagship project Hexa-X. *IEEE Access*, 9:160004–160020, 2021.

[20] Khaled B Letaief, Yuanming Shi, Jianmin Lu, and Jianhua Lu. Edge artificial intelligence for 6G: Vision, enabling technologies, and applications. *IEEE Journal on Selected Areas in Communications*, 40(1):5–36, 2021.

[21] Guangyi Liu, Yuhong Huang, Na Li, Jing Dong, Jing Jin, Qixing Wang, and Nan Li. Vision, requirements and network architecture of 6G mobile network beyond 2030. *China Communications*, 17(9):92–104, 2020.

[22] David J Langley, Jenny van Doorn, Irene CL Ng, Stefan Stieglitz, Alexander Lazovik, and Albert Boonstra. The internet of everything: Smart things and their impact on business models. *Journal of Business Research*, 122:853–863, 2021.

[23] Hadeel Elayan, Osama Amin, Basem Shihada, Raed M Shubair, and Mohamed-Slim Alouini. Terahertz band: The last piece of rf spectrum puzzle for communication systems. *IEEE Open Journal of the Communications Society*, 1:1–32, 2019.

[24] Zhi Chen, Xinying Ma, Bo Zhang, Yaxin Zhang, Zhongqian Niu, Ningyuan Kuang, Wenjie Chen, Lingxiang Li, and Shaoqian Li. A survey on terahertz communications. *China Communications*, 16(2):1–35, 2019.

[25] Morio Toyoshima. Recent trends in space laser communications for small satellites and constellations. *Journal of Lightwave Technology*, 39(3):693–699, 2020.

[26] Arun K Majumdar and Jennifer C Ricklin. *Free-space laser communications: principles and advances*, volume 2. Springer Science & Business Media, 2010.

[27] Jing Xu, Aobo Lin, Xiangyu Yu, Yuhang Song, Meiwei Kong, Fengzhong Qu, Jun Han, Wei Jia, and Ning Deng. Underwater laser communication using an OFDM-modulated 520-nm laser diode. *IEEE Photonics Technology Letters*, 28(20):2133–2136, 2016.

[28] Ping Yang, Yue Xiao, Yong Liang Guan, KVS Hari, A Chockalingam, Shinya Sugiura, Harald Haas, Marco Di Renzo, Christos Masouros, Zilong Liu, et al. Single-carrier SM-MIMO: A promising design for broadband large-scale antenna systems. *IEEE Communications Surveys & Tutorials*, 18(3):1687–1716, 2016.

[29] Marco Di Renzo, Harald Haas, Ali Ghrayeb, Shinya Sugiura, and Lajos Hanzo. Spatial modulation for generalized MIMO: Challenges, opportunities, and implementation. *Proceedings of the IEEE*, 102(1):56–103, 2013.

[30] Rawan Alghamdi, Reem Alhadrami, Dalia Alhothali, Heba Almorad, Alice Faisal, Sara Helal, Rahaf Shalabi, Rawan Asfour, Noofa Hammad, Asmaa Shams, et al. Intelligent surfaces for 6G wireless networks: A survey of optimization and performance analysis techniques. *IEEE Access*, 2020.

[31] Yuanwei Liu, Xiao Liu, Xidong Mu, Tianwei Hou, Jiaqi Xu, Marco Di Renzo, and Naofal Al-Dhahir. Reconfigurable intelligent surfaces: Principles and opportunities. *IEEE Communications Surveys & Tutorials*, 23(3):1546–1577, 2021.

[32] Rui Chen, Hong Zhou, Marco Moretti, Xiaodong Wang, and Jiandong Li. Orbital angular momentum waves: generation, detection, and emerging applications. *IEEE Communications Surveys & Tutorials*, 22(2):840–868, 2019.

[33] Doohwan Lee, Hirofumi Sasaki, Hiroyuki Fukumoto, Ken Hiraga, and Tadao Nakagawa. Orbital angular momentum (OAM) multiplexing: An enabler of a new era of wireless communications. *IEICE Transactions on Communications*, 100(7):1044–1063, 2017.

[34] J Andrew Zhang, Fan Liu, Christos Masouros, Robert W Heath, Zhiyong Feng, Le Zheng, and Athina Petropulu. An overview of signal processing techniques for joint communication and radar sensing. *IEEE Journal of Selected Topics in Signal Processing*, 15(6):1295–1315, 2021.

[35] J Andrew Zhang, Md Lushanur Rahman, Kai Wu, Xiaojing Huang, Y Jay Guo, Shanzhi Chen, and Jinhong Yuan. Enabling joint communication and radar sensing in mobile networks: A survey. *IEEE Communications Surveys & Tutorials*, 24(1):306–345, 2021.

[36] Zhenyu Zhou, Xinyi Chen, Yan Zhang, and Shahid Mumtaz. Blockchain-empowered secure spectrum sharing for 5G heterogeneous networks. *IEEE Network*, 34(1):24–31, 2020.

[37] Praveen Gorla, Vinay Chamola, Vikas Hassija, and Nirwan Ansari. Blockchain based framework for modeling and evaluating 5G spectrum sharing. *IEEE Network*, 35(2):229–235, 2020.

[38] Syed Junaid Nawaz, Shree Krishna Sharma, Shurjeel Wyne, Mohammad N Patwary, and Md Asaduzzaman. Quantum machine learning for 6G communication networks: State-of-the-art and vision for the future. *IEEE Access*, 7:46317–46350, 2019.

[39] Chonggang Wang and Akbar Rahman. Quantum-enabled 6G wireless networks: Opportunities and challenges. *IEEE Wireless Communications*, 29(1):58–69, 2022.

[40] Ali Ahmed Mohammed, Kamarudin Ambak, Ahmed Mancy Mosa, and Deprizon Syamsunur. A review of traffic accidents and related practices worldwide. *The Open Transportation Journal*, 13(1), 2019.

[41] Alireza Razzaghi, Hamid Soori, Amir Kavousi, Alireza Abadi, Ardeshir Khosravi, and Abbas Alipour. Risk factors of deaths related to road traffic crashes in world health organization regions: A systematic review. *Archives of Trauma Research,*, 8(2):57–86, 2019.

[42] Muhammad Qasim Khan and Sukhan Lee. Gaze and eye tracking: Techniques and applications in ADAS. *Sensors*, 19(24):5540, 2019.

[43] Juan Borrego-Carazo, David Castells-Rufas, Ernesto Biempica, and Jordi Carrabina. Resource-constrained machine learning for ADAS: a systematic review. *IEEE Access*, 8:40573–40598, 2020.

[44] Matthew Veres and Medhat Moussa. Deep learning for intelligent transportation systems: A survey of emerging trends. *IEEE Transactions on Intelligent Transportation Systems*, 21(8):3152–3168, 2019.

[45] Ammar Haydari and Yasin Yılmaz. Deep reinforcement learning for intelligent transportation systems: A survey. *IEEE Transactions on Intelligent Transportation Systems*, 23(1):11–32, 2020.

[46] Yuchuan Fu, Changle Li, Fei Richard Yu, Tom H Luan, and Yao Zhang. A survey of driving safety with sensing, vehicular communications, and artificial intelligence-based collision avoidance. *IEEE Transactions on Intelligent Transportation Systems*, 2021.

[47] Kwonjong Lee, Joonki Kim, Yosub Park, Hanho Wang, and Daesik Hong. Latency of cellular-based V2X: Perspectives on TTI-proportional latency and TTI-independent latency. *IEEE Access*, 5:15800–15809, 2017.

[48] Fakhar Abbas, Gang Liu, Pingzhi Fan, Zahid Khan, and Muhammad Saleh Bute. A vehicle density based two-stage resource management scheme for 5G-V2X networks. In *2020 IEEE 91st Vehicular Technology Conference (VTC2020-Spring)*, pages 1–5. IEEE, 2020.

[49] Bo Lin, Chuanbin Zhao, Feifei Gao, Geoffrey Ye Li, Jing Qian, and Hao Wang. Environment reconstruction based on multi-user selection and multi-modal fusion in ISAC. *IEEE Transactions on Wireless Communications*, 2024.

[50] Wenjiang Ouyang, Qian Liu, Junsheng Mu, AI-Dulaimi Anwer, Xiaojun Jing, and Qilie Liu. Communication-efficient federated learning for large-scale multiagent systems in ISAC: Data augmentation with reinforcement learning. *IEEE Systems Journal*, 2024.

[51] Hammam Salem, MD Quamar, Adeb Mansoor, Mohammed Elrashidy, Nasir Saeed, and Mudassir Masood. Data-driven integrated sensing and communication: Recent advances, challenges, and future prospects. *arXiv preprint arXiv:2308.09090*, 2023.

[52] Chuanbin Zhao, Bo Lin, and Feifei Gao. Multi-sensor fusion based environment reconstruction in integrated sensing and communications. In *2024 IEEE/CIC International Conference on Communications in China (ICCC)*, pages 1116–1120. IEEE, 2024.

[53] Xiang Cheng, Haotian Zhang, Jianan Zhang, Shijian Gao, Sijiang Li, Ziwei Huang, Lu Bai, Zonghui Yang, Xinhu Zheng, and Liuqing Yang. Intelligent multi-modal sensing-communication integration: Synesthesia of machines. *IEEE Communications Surveys & Tutorials*, 2023.

[54] Kayhan Zrar Ghafoor, Mohsen Guizani, Linghe Kong, Halgurd S Maghdid, and Khalid Fadhil Jasim. Enabling efficient coexistence of DSRC and C-V2X in vehicular networks. *IEEE Wireless Communications*, 27(2):134–140, 2019.

[55] Zeeshan Hameed Mir, Jamal Toutouh, Fethi Filali, and Young-Bae Ko. Enabling DSRC and C-V2X integrated hybrid vehicular networks: Architecture and protocol. *IEEE Access*, 8:180909–180927, 2020.

[56] Zeeshan Hameed Mir, Jamal Toutouh, Fethi Filali, and Young-Bae Ko. Enabling DSRC and C-V2X integrated hybrid vehicular networks: Architecture and protocol. *IEEE Access*, 8:180909–180927, 2020.

[57] Christian Waldschmidt, Juergen Hasch, and Wolfgang Menzel. Automotive radar—from first efforts to future systems. *IEEE Journal of Microwaves*, 1(1):135–148, 2021.

[58] Mial E Warren. Automotive LIDAR technology. In *2019 Symposium on VLSI Circuits*, pages C254–C255. IEEE, 2019.

[59] Éloi Zablocki, Hédi Ben-Younes, Patrick Pérez, and Matthieu Cord. Explainability of deep vision-based autonomous driving systems: Review and challenges. *International Journal of Computer Vision*, pages 1–28, 2022.

[60] Xiaodan Liang, Tairui Wang, Luona Yang, and Eric Xing. Cirl: Controllable imitative reinforcement learning for vision-based self-driving. In *Proceedings of the European Conference on Computer Vision (ECCV)*, pages 584–599, 2018.

[61] Danny Kai Pin Tan, Jia He, Yanchun Li, Alireza Bayesteh, Yan Chen, Peiying Zhu, and Wen Tong. Integrated sensing and communication in 6G: Motivations, use cases, requirements, challenges and future directions. In *2021 1st IEEE International Online Symposium on Joint Communications & Sensing (JC&S)*, pages 1–6. IEEE, 2021.

[62] An Liu, Zhe Huang, Min Li, Yubo Wan, Wenrui Li, Tony Xiao Han, Chenchen Liu, Rui Du, Danny Kai Pin Tan, Jianmin Lu, et al. A survey on fundamental limits of integrated sensing and communication. *IEEE Communications Surveys & Tutorials*, 24(2):994–1034, 2022.

[63] Yuanhao Cui, Fan Liu, Xiaojun Jing, and Junsheng Mu. Integrating sensing and communications for ubiquitous IoT: Applications, trends, and challenges. *IEEE Network*, 35(5):158–167, 2021.

[64] Athanasios Voulodimos, Nikolaos Doulamis, Anastasios Doulamis, and Eftychios Protopapadakis. Deep learning for computer vision: A brief review. *Computational Intelligence and Neuroscience*, 2018.

[65] Niall O'Mahony, Sean Campbell, Anderson Carvalho, Suman Harapanahalli, Gustavo Velasco Hernandez, Lenka Krpalkova, Daniel Riordan, and Joseph Walsh. Deep learning vs. traditional computer vision. In *Science and Information Conference*, pages 128–144. Springer, 2020.

[66] Daniel W Otter, Julian R Medina, and Jugal K Kalita. A survey of the usages of deep learning for natural language processing. *IEEE Transactions on Neural Networks and Learning Systems*, 32(2):604–624, 2020.

[67] Thomas Wolf, Lysandre Debut, Victor Sanh, Julien Chaumond, Clement Delangue, Anthony Moi, Pierric Cistac, Tim Rault, Rémi Louf, Morgan Funtowicz, et al. Transformers: State-of-the-art natural language processing. In *Proceedings of the 2020 Conference on Empirical Methods in Natural Language Processing: System Demonstrations*, pages 38–45, 2020.

[68] Ian Goodfellow, Jean Pouget-Abadie, Mehdi Mirza, Bing Xu, David Warde-Farley, Sherjil Ozair, Aaron Courville, and Yoshua Bengio. Generative adversarial networks. *Communications of the ACM*, 63(11):139–144, 2020.

[69] Zhengwei Wang, Qi She, and Tomas E Ward. Generative adversarial networks in computer vision: A survey and taxonomy. *ACM Computing Surveys (CSUR)*, 54(2):1–38, 2021.

[70] Victor Uc-Cetina, Nicolás Navarro-Guerrero, Anabel Martin-Gonzalez, Cornelius Weber, and Stefan Wermter. Survey on reinforcement learning for language processing. *Artificial Intelligence Review*, pages 1–33, 2022.

[71] William Yang Wang, Jiwei Li, and Xiaodong He. Deep reinforcement learning for NLP. In *Proceedings of the 56th Annual Meeting of the Association for Computational Linguistics: Tutorial Abstracts*, pages 19–21, 2018.

[72] Guangxu Zhu, Dongzhu Liu, Yuqing Du, Changsheng You, Jun Zhang, and Kaibin Huang. Toward an intelligent edge: Wireless communication meets machine learning. *IEEE Communications Magazine*, 58(1):19–25, 2020.

[73] Tugba Erpek, Timothy J O'Shea, Yalin E Sagduyu, Yi Shi, and T Charles Clancy. Deep learning for wireless communications. In *Development and Analysis of Deep Learning Architectures*, pages 223–266. Springer, 2020.

[74] Tianqi Wang, Chao-Kai Wen, Hanqing Wang, Feifei Gao, Tao Jiang, and Shi Jin. Deep learning for wireless physical layer: Opportunities and challenges. *China Communications*, 14(11):92–111, 2017.

[75] Chia-Hung Lin, Yu-Chien Lin, Yen-Jung Wu, Wei-Ho Chung, and Ta-Sung Lee. A survey on deep learning-based vehicular communication applications. *Journal of Signal Processing Systems*, 93(4):369–388, 2021.

[76] Xingqin Lin. An overview of 5G advanced evolution in 3GPP release 18. *arXiv preprint arXiv:2201.01358*, 2022.

[77] Osvaldo Simeone. A very brief introduction to machine learning with applications to communication systems. *IEEE Transactions on Cognitive Communications and Networking*, 4(4):648–664, 2018.

[78] Masaharu Hata. Empirical formula for propagation loss in land mobile radio services. *IEEE Transactions on Vehicular Technology*, 29(3):317–325, 1980.

[79] Joram Walfisch and Henry L Bertoni. A theoretical model of UHF propagation in urban environments. *IEEE Transactions on Antennas and Propagation*, 36(12):1788–1796, 1988.

[80] Raymond Steele and Lajos Hanzo. *Mobile radio communications: Second and third generation cellular and WATM systems: 2nd.* IEEE Press-John Wiley, 1999.

[81] 3GPP TR 25.996 v16.0.0. Spatial channel model for multiple input multiple output (MIMO) simulations (Release 16). *3GPP Technical Report*, 2020.

[82] Seung Hee Han and Jae Hong Lee. An overview of peak-to-average power ratio reduction techniques for multicarrier transmission. *IEEE Wireless Communications*, 12(2):56–65, 2005.

[83] D Cox. Linear amplification with nonlinear components. *IEEE Transactions on Communications*, 22(12):1942–1945, 1974.

[84] Ye Geoffrey Li and Gordon L Stuber. *Orthogonal frequency division multiplexing for wireless communications.* Springer Science & Business Media, 2006.

[85] Richard van Nee and Ramjee Prasad. *OFDM for wireless multimedia communications.* Artech House, Inc., 2000.

[86] John Terry and Juha Heiskala. *OFDM wireless LANs: A theoretical and practical guide.* Sams publishing, 2002.

[87] David D Falconer and S Lek Ariyavisitakul. Broadband wireless using single carrier and frequency domain equalization. In *The 5th International Symposium on Wireless Personal Multimedia Communications*, volume 1, pages 27–36. IEEE, 2002.

[88] Harry L Van Trees. *Optimum array processing: Part IV of detection, estimation, and modulation theory.* John Wiley & Sons, 2002.

[89] David Tse and Pramod Viswanath. *Fundamentals of wireless communication.* Cambridge university press, 2005.

[90] Arogyaswami Paulraj, Rohit Nabar, and Dhananjay Gore. *Introduction to space-time wireless communications.* Cambridge university press, 2003.

[91] Gerard J Foschini. Layered space-time architecture for wireless communication in a fading environment when using multi-element antennas. *Bell labs technical journal*, 1(2):41–59, 1996.

[92] Andreas Burg, Moritz Borgmann, Markus Wenk, Martin Zellweger, Wolfgang Fichtner, and Helmut Bolcskei. VLSI implementation of MIMO detection using the sphere decoding algorithm. *IEEE Journal of solid-state circuits*, 40(7):1566–1577, 2005.

[93] Hemanth Sampath, Petre Stoica, and Arogyaswami Paulraj. Generalized linear precoder and decoder design for MIMO channels using the weighted MMSE criterion. *IEEE Transactions on Communications*, 49(12):2198–2206, 2001.

[94] David J Love, Robert W Heath, Vincent KN Lau, David Gesbert, Bhaskar D Rao, and Matthew Andrews. An overview of limited feedback in wireless communication systems. *IEEE Journal on selected areas in Communications*, 26(8):1341–1365, 2008.

[95] Amitav Mukherjee and A Lee Swindlehurst. Robust beamforming for security in MIMO wiretap channels with imperfect CSI. *IEEE Transactions on Signal Processing*, 59(1):351–361, 2010.

[96] Chao Shen, Tsung-Hui Chang, Kun-Yu Wang, Zhengding Qiu, and Chong-Yung Chi. Distributed robust multicell coordinated beamforming with imperfect CSI: An ADMM approach. *IEEE Transactions on Signal Processing*, 60(6):2988–3003, 2012.

[97] Kilian Roth, Hessam Pirzadeh, A Lee Swindlehurst, and Josef A Nossek. A comparison of hybrid beamforming and digital beamforming with low-resolution ADCs for multiple users and imperfect CSI. *IEEE Journal of Selected Topics in Signal Processing*, 12(3):484–498, 2018.

[98] Meng Gao, Jiaao Yang, Huafu Li, and Yang Wang. Robust beamforming optimization design for RIS-aided MIMO systems with practical phase shift model and imperfect CSI. *IEEE Internet of Things Journal*, 11(1):958–973, 2023.

[99] Alex B Gershman, Nicholas D Sidiropoulos, Shahram Shahbazpanahi, Mats Bengtsson, and Bjorn Ottersten. Convex optimization-based beamforming. *IEEE Signal Processing Magazine*, 27(3):62–75, 2010.

[100] Christian B Peel. On "dirty-paper coding". *IEEE Signal Processing Magazine*, 20(3):112–113, 2003.

[101] AB Gershman and ND Sidiropoulos. Space-time processing for MIMO communications. 2005.

[102] Goran Dimic and Nicholas D Sidiropoulos. On downlink beamforming with greedy user selection: performance analysis and a simple new algorithm. *IEEE Transactions on Signal Processing*, 53(10):3857–3868, 2005.

[103] Jianqi Wang, David J Love, and Michael D Zoltowski. User selection with zero-forcing beamforming achieves the asymptotically optimal sum rate. *IEEE Transactions on Signal Processing*, 56(8):3713–3726, 2008.

[104] Boon Chin Lim, Witold A Krzymien, and Christian Schlegel. Efficient sum rate maximization and resource allocation in block-diagonalized space-division multiplexing. *IEEE Transactions on Vehicular Technology*, 58(1):478–484, 2008.

[105] Zhenyu Tu and Rick S Blum. Multiuser diversity for a dirty paper approach. *IEEE Communications letters*, 7(8):370–372, 2003.

[106] Giuseppe Caire and Shlomo Shamai. On the achievable throughput of a multiantenna gaussian broadcast channel. *IEEE Transactions on Information Theory*, 49(7):1691–1706, 2003.

[107] Xiaojie Zhang and Jungwoo Lee. Low complexity MIMO scheduling with channel decomposition using capacity upperbound. *IEEE Transactions on Communications*, 56(6):871–876, 2008.

[108] Wessam Ajib and David Haccoun. An overview of scheduling algorithms in MIMO-based fourth-generation wireless systems. *IEEE network*, 19(5):43–48, 2005.

[109] Hooman Shirani-Mehr, Giuseppe Caire, and Michael J Neely. MIMO downlink scheduling with non-perfect channel state knowledge. *IEEE Transactions on Communications*, 58(7):2055–2066, 2010.

[110] Ashok Bandi, Symeon Chatzinotas, Björn Ottersten, et al. A joint solution for scheduling and precoding in multiuser MISO downlink channels. *IEEE Transactions on Wireless Communications*, 19(1):475–490, 2019.

[111] Jiwook Choi, Namyoon Lee, Song-Nam Hong, and Giuseppe Caire. Joint user scheduling, power allocation, and precoding design for massive MIMO systems: A principal component analysis approach. In *2018 IEEE International Symposium on Information Theory (ISIT)*, pages 396–400. IEEE, 2018.

[112] Yuan-Xin Zhu, Do-Yup Kim, and Jang-Won Lee. Joint antenna and user scheduling in the massive MIMO system over time-varying fading channels. *IEEE Access*, 9:92431–92445, 2021.

[113] Omar Maraqa, Aditya S Rajasekaran, Saad Al-Ahmadi, Halim Yanikomeroglu, and Sadiq M Sait. A survey of rate-optimal power domain NOMA with enabling technologies of future wireless networks. *IEEE Communications Surveys & Tutorials*, 22(4):2192–2235, 2020.

[114] Shipon Ali, Ekram Hossain, and Dong In Kim. Non-orthogonal multiple access (NOMA) for downlink multiuser MIMO systems: User clustering, beamforming, and power allocation. *IEEE Access*, 5:565–577, 2016.

[115] Behrooz Makki, Krishna Chitti, Ali Behravan, and Mohamed-Slim Alouini. A survey of NOMA: Current status and open research challenges. *IEEE Open Journal of the Communications Society*, 1:179–189, 2020.

[116] Aamina Akbar, Sobia Jangsher, and Farrukh A Bhatti. NOMA and 5G emerging technologies: A survey on issues and solution techniques. *Computer Networks*, 190:107950, 2021.

[117] Mojtaba Vaezi, Gayan Amarasuriya Aruma Baduge, Yuanwei Liu, Ahmed Arafa, Fang Fang, and Zhiguo Ding. Interplay between NOMA and other emerging technologies: A survey. *IEEE Transactions on Cognitive Communications and Networking*, 5(4):900–919, 2019.

[118] Nagendra Kumar Vishwakarma and Rakesh Kumar Singh. Design and implementation of FHSS (frequency hopping spread spectrum) synthesizer. In *2021 7th International Conference on Signal Processing and Communication (ICSC)*, pages 151–155. IEEE, 2021.

[119] Xin Liu, Yuhua Xu, Luliang Jia, Qihui Wu, and Alagan Anpalagan. Anti-jamming communications using spectrum waterfall: A deep reinforcement learning approach. *IEEE Communications Letters*, 22(5):998–1001, 2018.

[120] Khalid El-Darymli, Peter McGuire, Desmond Power, and Cecilia Moloney. Target detection in synthetic aperture radar imagery: A state-of-the-art survey. *Journal of Applied Remote Sensing*, 7(1):071598–071598, 2013.

[121] Christian R Berger, Bruno Demissie, Jörg Heckenbach, Peter Willett, and Shengli Zhou. Signal processing for passive radar using OFDM waveforms. *IEEE Journal of Selected Topics in Signal Processing*, 4(1):226–238, 2010.

[122] Werner Wiesbeck and Leen Sit. Radar 2020: The future of radar systems. In *2014 International Radar Conference*, pages 1–6. IEEE, 2014.

[123] YT Chan, AGC Hu, and JB Plant. A Kalman filter based tracking scheme with input estimation. *IEEE transactions on Aerospace and Electronic Systems*, (2):237–244, 1979.

[124] Eli Brookner. Tracking and Kalman filtering made easy. 1998.

[125] Gennady Yu Kulikov and Maria V Kulikova. The accurate continuous-discrete extended Kalman filter for radar tracking. *IEEE Transactions on Signal Processing*, 64(4):948–958, 2015.

[126] Shishan Yang and Marcus Baum. Extended Kalman filter for extended object tracking. In *2017 IEEE International Conference on Acoustics, Speech and Signal Processing (ICASSP)*, pages 4386–4390. IEEE, 2017.

[127] Masoud Khodarahmi and Vafa Maihami. A review on Kalman filter models. *Archives of Computational Methods in Engineering*, 30(1):727–747, 2023.

[128] Bo Tang, Jun Tang, and Yingning Peng. MIMO radar waveform design in colored noise based on information theory. *IEEE Transactions on Signal Processing*, 58(9):4684–4697, 2010.

[129] YANG Yang and Rick S Blum. Minimax robust MIMO radar waveform design. *IEEE Journal of Selected Topics in Signal Processing*, 1(1):147–155, 2007.

[130] Xianxiang Yu, Khaled Alhujaili, Guolong Cui, and Vishal Monga. MIMO radar waveform design in the presence of multiple targets and practical constraints. *IEEE Transactions on Signal Processing*, 68:1974–1989, 2020.

[131] Yang Yang and Rick S Blum. MIMO radar waveform design based on mutual information and minimum mean-square error estimation. *IEEE Transactions on Aerospace and Electronic Systems*, 43(1):330–343, 2007.

[132] CUI Guolong, YU Xianxiang, WEI Wenqiang, XIONG Kui, KONG Yukai, and KONG Lingjiang. An overview of antijamming methods and future works on cognitive intelligent radar. , 11(6):974–1002, 2022.

[133] Sevgi Zubeyde Gurbuz, Hugh D Griffiths, Alexander Charlish, Muralidhar Rangaswamy, Maria Sabrina Greco, and Kristine Bell. An overview of cognitive radar: Past, present, and future. *IEEE Aerospace and Electronic Systems Magazine*, 34(12):6–18, 2019.

[134] Mingliang Tao, Jia Su, Yan Huang, and Ling Wang. Mitigation of radio frequency interference in synthetic aperture radar data: Current status and future trends. *Remote Sensing*, 11(20):2438, 2019.

[135] Anum Ahmed Pirkani. *Characterisation and Mitigation of Radar Interference in Automotive Applications.* PhD thesis, University of Birmingham, 2024.

[136] Sian Jin and Sumit Roy. FMCW radar network: Multiple access and interference mitigation. *IEEE Journal of Selected Topics in Signal Processing*, 15(4):968–979, 2021.

[137] Scott D Berger. The spectrum of a digital radio frequency memory linear range gate stealer electronic attack signal. In *Proceedings of the 2001 IEEE Radar Conference (Cat. No. 01CH37200)*, pages 27–30. IEEE, 2001.

[138] Scott D Berger. Digital radio frequency memory linear range gate stealer spectrum. *IEEE Transactions on Aerospace and Electronic Systems*, 39(2):725–735, 2003.

[139] Kunteng Wang, Zeyu Dong, Tao Wan, Kaili Jiang, Wanan Xiong, and Xueli Fang. Research on radar active deception jamming identification method based on RESNET and bispectrum features. In *2021 International Conference on Computer Engineering and Application (ICCEA)*, pages 491–495. IEEE, 2021.

[140] Michael Barjenbruch, Dominik Kellner, Klaus Dietmayer, Jens Klappstein, and Juergen Dickmann. A method for interference cancellation in automotive radar. In *2015 IEEE MTT-S International Conference on Microwaves for Intelligent Mobility (ICMIM)*, pages 1–4. IEEE, 2015.

[141] Muyao Yu, Shengbo Dong, Xiangyu Duan, and Shangchao Liu. A novel interference suppression method for interrupted sampling repeater jamming based on singular spectrum entropy function. *Sensors*, 19(1):136, 2019.

[142] Dezhi Tian, Wei Ren, Lu Zhang, Zhennan Liang, and Quanhua Liu. An adaptive scheme for chopping and interleaving jamming suppression based on the mismatched filter. In *2021 International Conference on Control, Automation and Information Sciences (ICCAIS)*, pages 78–83. IEEE, 2021.

[143] Joshua GW Bernstein and Douglas S Brungart. Effects of spectral smearing and temporal fine-structure distortion on the fluctuating-masker benefit for speech at a fixed signal-to-noise ratio. *The Journal of the Acoustical Society of America*, 130(1):473–488, 2011.

[144] Jie Xiao, Xizhang Wei, and Jia Sun. Research on interrupted sampling repeater jamming performance based on joint frequency shift/phase modulation. *Sensors*, 23(5):2812, 2023.

[145] Yang Zhang, Yinsheng Wei, and Lei Yu. Interrupted sampling repeater jamming recognition and suppression based on phase-coded signal processing. *Signal Processing*, 198:108596, 2022.

[146] Hui Yuan, Chun-yang Wang, Xin Li, and Lei An. A method against interrupted-sampling repeater jamming based on energy function detection and band-pass filtering. *International Journal of Antennas and Propagation*, 2017(1):6759169, 2017.

[147] Chao Zhou, Quanhua Liu, and Xinliang Chen. Parameter estimation and suppression for DRFM-based interrupted sampling repeater jammer. *IET Radar, Sonar & Navigation*, 12(1):56–63, 2018.

[148] Jing Chen, Xinliang Chen, Honggang Zhang, Kaixiang Zhang, and Quanhua Liu. Suppression method for main-lobe interrupted sampling repeater jamming in distributed radar. *IEEE Access*, 8:139255–139265, 2020.

[149] Simon Haykin, Yanbo Xue, and Timothy N Davidson. Optimal waveform design for cognitive radar. In *2008 42nd Asilomar Conference on Signals, Systems and Computers*, pages 3–7. IEEE, 2008.

[150] Zora Slavik and Kumar Vijay Mishra. Cognitive interference mitigation in automotive radars. In *2019 IEEE Radar Conference (RadarConf)*, pages 1–6. IEEE, 2019.

[151] Mojtaba Soltanalian, Bo Tang, Jian Li, and Petre Stoica. Joint design of the receive filter and transmit sequence for active sensing. *IEEE Signal Processing Letters*, 20(5):423–426, 2013.

[152] Bo Tang, Jonathan Tuck, and Petre Stoica. Polyphase waveform design for MIMO radar space time adaptive processing. *IEEE Transactions on Signal Processing*, 68:2143–2154, 2020.

[153] Lan Lan, Jingwei Xu, Guisheng Liao, Yuhong Zhang, Francesco Fioranelli, and Hing Cheung So. Suppression of mainbeam deceptive jammer with FDA-MIMO radar. *IEEE Transactions on Vehicular Technology*, 69(10):11584–11598, 2020.

[154] Xiang Feng, Yang-chun Song, Zhi-quan Zhou, and Yi-nan Zhao. Designing unimodular waveform with low range sidelobes and stopband for cognitive radar via relaxed alternating projection. *International Journal of Antennas and Propagation*, 2016(1):6280508, 2016.

[155] Arindam Bose, Bo Tang, Wenjie Huang, Mojtaba Soltanalian, and Jian Li. Waveform design for mutual interference mitigation in automotive radar. *arXiv preprint arXiv:2208.04398*, 2022.

[156] Yi Bu, Xianxiang Yu, Jing Yang, Tao Fan, and Guolong Cui. A new approach for design of constant modulus discrete phase radar waveform with low WISL. *Signal Processing*, 187:108145, 2021.

[157] Steven I Krich, Monica Montanari, Vincent Amendolare, and Paul Berestesky. Wind turbine interference mitigation using a waveform diversity radar. *IEEE Transactions on Aerospace and Electronic Systems*, 53(2):805–815, 2017.

[158] Ashish Basireddy and Hassan Moradi. OFDM waveform design for interference resistant automotive radars. *IEEE Sensors Journal*, 21(14):15670–15678, 2021.

[159] Mohsen Eslami Nazari, Weimin Huang, and Chen Zhao. Radio frequency interference suppression for hf surface wave radar using CEMD and temporal windowing methods. *IEEE Geoscience and Remote Sensing Letters*, 17(2):212–216, 2019.

[160] Christoph Fischer, Hans Ludwig Blöcher, Jürgen Dickmann, and Wolfgang Menzel. Robust detection and mitigation of mutual interference in automotive radar. In *2015 16th International Radar Symposium (IRS)*, pages 143–148. IEEE, 2015.

[161] Muhammad Rameez, Mattias Dahl, and Mats I Pettersson. Adaptive digital beamforming for interference suppression in automotive FMCW radars. In *2018 IEEE Radar Conference (RadarConf18)*, pages 0252–0256. IEEE, 2018.

[162] Yuan Feng, Tao Shan, Shengheng Liu, and Ran Tao. Interference suppression using joint spatio-temporal domain filtering in passive radar. In *2015 IEEE Radar Conference (RadarCon)*, pages 1156–1160. IEEE, 2015.

[163] Francisco Paisana, Nicholas J Kaminski, Nicola Marchetti, and Luiz A DaSilva. Signal processing for temporal spectrum sharing in a multi-radar environment. *IEEE Transactions on Cognitive Communications and Networking*, 3(2):123–137, 2017.

[164] Robert Muja, Andrei Anghel, Remus Cacoveanu, and Silviu Ciochina. Real-time interference mitigation in automotive radars using the short-time Fourier transform and L-Statistics. *IEEE Transactions on Vehicular Technology*, 2024.

[165] Xing-Peng Mao, Yun-Long Yang, Hong Hong, and Wei-Bo Deng. Multi-domain collaborative filter for interference suppressing. *IET Signal Processing*, 10(9):1157–1168, 2016.

[166] Suba R Subramaniam, Bingo Wing-Kuen Ling, and Apostolos Georgakis. Filtering in rotated time-frequency domains with unknown noise statistics. *IEEE transactions on signal processing*, 60(1):489–493, 2011.

[167] Irena Orović and Srdjan Stanković. L-statistics based space/spatial-frequency filtering of 2D signals in heavy tailed noise. *Signal processing*, 96:190–202, 2014.

[168] Xirui Zhang, Zhiwen Liu, Yougen Xu, and Wei Liu. Adaptive tensorial beamformer based on electromagnetic vector-sensor arrays with coherent interferences. *Multidimensional Systems and Signal Processing*, 26:803–821, 2015.

[169] Jian-Wu Tao and Wen-Xiu Chang. A novel combined beamformer based on hypercomplex processes. *IEEE Transactions on Aerospace and Electronic Systems*, 49(2):1276–1289, 2013.

[170] Hong Hong, Xing-Peng Mao, and Cui Hu. A multi-domain collaborative filter for HFSWR based on oblique projection. In *2012 IEEE Radar Conference*, pages 0907–0912. IEEE, 2012.

[171] Xiufeng Song, Peter Willett, Shengli Zhou, and Peter B Luh. The MIMO radar and jammer games. *IEEE Transactions on Signal Processing*, 60(2):687–699, 2011.

[172] Marco Piezzo, Augusto Aubry, Stefano Buzzi, Antonio De Maio, and Alfonso Farina. Non-cooperative code design in radar networks: A game-theoretic approach. *EURASIP Journal on Advances in Signal Processing*, 2013:1–17, 2013.

[173] Xiaofeng Jiang, Feng Zhou, Shuangwu Chen, Huasen He, and Jian Yang. Jamming resilient tracking using POMDP-based detection of hidden targets. *IEEE Transactions on Information Forensics and Security*, 16:983–998, 2020.

[174] Jingxuan Huang, Zesong Fei, Tianxiong Wang, Xinyi Wang, Fan Liu, Haijun Zhou, J Andrew Zhang, and Guohua Wei. V2X-communication assisted interference minimization for automotive radars. *China Communications*, 16(10):100–111, 2019.

[175] Sruthy Skaria, Akram Al-Hourani, Robin J. Evans, Kandeepan Sithamparanathan, and Udaya Parampalli. Interference mitigation in automotive radars using pseudo-random cyclic orthogonal sequences. *Sensors*, 19(20):4459, 2019.

[176] Yi Wang, Qixun Zhang, Zhiqing Wei, Yuewei Lin, and Zhiyong Feng. Performance analysis of coordinated interference mitigation approach for automotive radar. *IEEE Internet of Things Journal*, 10(13):11683–11695, 2023.

[177] Canan Aydogdu, Musa Furkan Keskin, Nil Garcia, Henk Wymeersch, and Daniel W Bliss. RadChat: Spectrum sharing for automotive radar interference mitigation. *IEEE Transactions on Intelligent Transportation Systems*, 22(1):416–429, 2019.

[178] Yunxuan Wang, Yan Huang, Jiang Liu, Ruizhe Zhang, Hui Zhang, and Wei Hong. Interference mitigation for automotive FMCW radar with tensor decomposition. *IEEE Transactions on Intelligent Transportation Systems*, 2024.

[179] M Kay Steven. Fundamentals of statistical signal processing. *PTR Prentice-Hall, Englewood Cliffs, NJ*, 10(151045):148, 1993.

[180] Hermann Rohling. Radar CFAR thresholding in clutter and multiple target situations. *IEEE transactions on aerospace and electronic systems*, (4):608–621, 1983.

[181] Fan Liu, Ya-Feng Liu, Ang Li, Christos Masouros, and Yonina C Eldar. Cramér-Rao bound optimization for joint radar-communication beamforming. *IEEE Transactions on Signal Processing*, 70:240–253, 2021.

[182] Mahvish Nazir. *Automotive radar target detection using ambiguity function.* PhD thesis, University of Birmingham, 2016.

[183] Nadav Levanon and Eli Mozeson. *Radar signals.* John Wiley & Sons, 2004.

[184] Chenguang Shi, Fei Wang, Mathini Sellathurai, Jianjiang Zhou, and Sana Salous. Power minimization-based robust OFDM radar waveform design for radar and communication systems in coexistence. *IEEE Transactions on Signal Processing*, 66(5):1316–1330, 2017.

[185] Chenguang Shi, Yijie Wang, Fei Wang, Sana Salous, and Jianjiang Zhou. Joint optimization scheme for subcarrier selection and power allocation

in multicarrier dual-function radar-communication system. *IEEE Systems Journal*, 15(1):947–958, 2020.

[186] Renhui Xu, Laixian Peng, Wendong Zhao, and Zhichao Mi. Radar mutual information and communication channel capacity of integrated radar-communication system using MIMO. *ICT Express*, 1(3):102–105, 2015.

[187] Andrea Goldsmith. *Wireless communications.* Cambridge university press, 2005.

[188] Batu K Chalise, Moeness G Amin, and Braham Himed. Performance tradeoff in a unified passive radar and communications system. *IEEE Signal Processing Letters*, 24(9):1275–1279, 2017.

[189] Steven M Kay. *Fundamentals of statistical signal processing: estimation theory.* Prentice-Hall, Inc., 1993.

[190] Zhen Du, Fan Liu, Yifeng Xiong, Tony Xiao Han, Yonina C Eldar, and Shi Jin. Reshaping the ISAC tradeoff under OFDM signaling: A probabilistic constellation shaping approach. *IEEE Transactions on Signal Processing*, 2024.

[191] Zixiang Ren, Yunfei Peng, Xianxin Song, Yuan Fang, Ling Qiu, Liang Liu, Derrick Wing Kwan Ng, and Jie Xu. Fundamental CRB-rate tradeoff in multi-antenna ISAC systems with information multicasting and multi-target sensing. *IEEE Transactions on Wireless Communications*, 2023.

[192] Yifeng Xiong, Fan Liu, Yuanhao Cui, Weijie Yuan, and Tony Xiao Han. Flowing the information from Shannon to Fisher: Towards the fundamental tradeoff in ISAC. In *GLOBECOM 2022-2022 IEEE Global Communications Conference*, pages 5601–5606. IEEE, 2022.

[193] Canan Aydogdu, Fan Liu, Christos Masouros, Henk Wymeersch, and Mats Rydström. Distributed radar-aided vehicle-to-vehicle communication. In *2020 IEEE Radar Conference (RadarConf20)*, pages 1–6. IEEE, 2020.

[194] Nanchi Su, Fan Liu, and Christos Masouros. Secure radar-communication systems with malicious targets: Integrating radar, communications and jamming functionalities. *IEEE Transactions on Wireless Communications*, 20(1):83–95, 2020.

[195] Fan Liu, Weijie Yuan, Christos Masouros, and Jinhong Yuan. Radar-assisted predictive beamforming for vehicular links: Communication served by sensing. *IEEE Transactions on Wireless Communications*, 19(11):7704–7719, 2020.

[196] Nabil Mohamed Drawil and Otman Basir. Intervehicle-communication-assisted localization. *IEEE Transactions on Intelligent Transportation Systems*, 11(3):678–691, 2010.

[197] Xiaoye Jing, Fan Liu, Christos Masouros, and Yong Zeng. ISAC from the sky: UAV trajectory design for joint communication and target localization. *arXiv preprint arXiv:2207.02904*, 2022.

[198] Ahmad Bazzi and Marwa Chafii. On outage-based beamforming design for dual-functional radar-communication 6G systems. *IEEE Transactions on Wireless Communications*, 2023.

[199] Rafael Cerna Loli, Onur Dizdar, and Bruno Clerckx. Rate-splitting multiple access for multi-antenna joint radar and communications with partial CSIT: Precoder optimization and link-level simulations. *arXiv preprint arXiv:2201.10621*, 2022.

[200] Fan Liu, Longfei Zhou, Christos Masouros, Ang Li, Wu Luo, and Athina Petropulu. Toward dual-functional radar-communication systems: Optimal waveform design. *IEEE Transactions on Signal Processing*, 66(16):4264–4279, 2018.

[201] Jingyu He, Li Chen, and Guo Wei. Waveform design for dual-functional radar and communication design based on conditional MI. In *2021 7th International Conference on Computer and Communications (ICCC)*, pages 1026–1030. IEEE, 2021.

[202] Xin Yuan, Zhiyong Feng, J Andrew Zhang, Wei Ni, Ren Ping Liu, Zhiqing Wei, and Changqiao Xu. Spatio-temporal power optimization for MIMO joint communication and radio sensing systems with training overhead. *IEEE Transactions on Vehicular Technology*, 70(1):514–528, 2020.

[203] Chenguang Shi, Yijie Wang, Fei Wang, Sana Salous, and Jianjiang Zhou. Power resource allocation scheme for distributed MIMO dual-function radar-communication system based on low probability of intercept. *Digital Signal Processing*, 106:102850, 2020.

[204] Christian Sturm, Elena Pancera, Thomas Zwick, and Werner Wiesbeck. A novel approach to OFDM radar processing. In *2009 IEEE Radar Conference*, pages 1–4. IEEE, 2009.

[205] Tuanwei Tian, Tianxian Zhang, Lingjiang Kong, and Yanhong Deng. Transmit/receive beamforming for MIMO-OFDM based dual-function radar and communication. *IEEE Transactions on Vehicular Technology*, 70(5):4693–4708, 2021.

[206] Xiaoyan Hu, Christos Masouros, Fan Liu, and Ronald Nissel. Low-PAPR DFRC MIMO-OFDM waveform design for integrated sensing

and communications. In *ICC 2022-IEEE International Conference on Communications*, pages 1599–1604. IEEE, 2022.

[207] Jeremy Johnston, Luca Venturino, Emanuele Grossi, Marco Lops, and Xiaodong Wang. MIMO OFDM dual-function radar-communication under error rate and beampattern constraints. *IEEE Journal on Selected Areas in Communications*, 40(6):1951–1964, 2022.

[208] Tianyao Huang, Nir Shlezinger, Xingyu Xu, Yimin Liu, and Yonina C Eldar. MAJoRCom: A dual-function radar communication system using index modulation. *IEEE Transactions on Signal Processing*, 68:3423–3438, 2020.

[209] Aboulnasr Hassanien, Braham Himed, and Brian D Rigling. A dual-function MIMO radar-communications system using frequency-hopping waveforms. In *2017 IEEE Radar Conference (RadarConf)*, pages 1721–1725. IEEE, 2017.

[210] Dingyou Ma, Nir Shlezinger, Tianyao Huang, Yimin Liu, and Yonina C Eldar. FRaC: FMCW-based joint radar-communications system via index modulation. *IEEE Journal of Selected Topics in Signal Processing*, 15(6):1348–1364, 2021.

[211] Haonan He, Haojian Zhang, and Tingting Zhang. Power allocation for OFDM-based radcom systems. In *2021 IEEE 22nd International Workshop on Signal Processing Advances in Wireless Communications (SPAWC)*, pages 461–465. IEEE, 2021.

[212] Chenguang Shi, Fei Wang, Sana Salous, and Jianjiang Zhou. Joint subcarrier assignment and power allocation strategy for integrated radar and communications system based on power minimization. *IEEE Sensors Journal*, 19(23):11167–11179, 2019.

[213] Hao-Wei Hsu, Ming-Chun Lee, Meng-Xun Gu, Yu-Chein Lin, and Ta-Sung Lee. Analysis and design for pilot power allocation and placement in OFDM based integrated radar and communication in automobile systems. *IEEE Transactions on Vehicular Technology*, 71(2):1519–1535, 2021.

[214] Irfan Ahmed, Hedi Khammari, Adnan Shahid, Ahmed Musa, Kwang Soon Kim, Eli De Poorter, and Ingrid Moerman. A survey on hybrid beamforming techniques in 5G: Architecture and system model perspectives. *IEEE Communications Surveys & Tutorials*, 20(4):3060–3097, 2018.

[215] Ziyang Cheng, Linlong Wu, Bowen Wang, MR Bhavani Shankar, and Björn Ottersten. Double-phase-shifter based hybrid beamforming for mmWave DFRC in the presence of extended target and clutters. *IEEE Transactions on Wireless Communications*, 2022.

[216] Ziyang Cheng, Zishu He, and Bin Liao. Hybrid beamforming design for OFDM dual-function radar-communication system. *IEEE Journal of Selected Topics in Signal Processing*, 15(6):1455–1467, 2021.

[217] Ahmet M Elbir, Kumar Vijay Mishra, and Symeon Chatzinotas. Terahertz-band joint ultra-massive MIMO radar-communications: Model-based and model-free hybrid beamforming. *IEEE Journal of Selected Topics in Signal Processing*, 15(6):1468–1483, 2021.

[218] Rang Liu, Ming Li, Qian Liu, and A Lee Swindlehurst. Dual-functional radar-communication waveform design: A symbol-level precoding approach. *IEEE Journal of Selected Topics in Signal Processing*, 15(6):1316–1331, 2021.

[219] Yunfei Chen and Xueyun Gu. Time allocation for integrated bi-static radar and communication systems. *IEEE Communications Letters*, 25(3):1033–1036, 2020.

[220] Emanuele Grossi, Marco Lops, Luca Venturino, and Alessio Zappone. Opportunistic radar in IEEE 802.11 ad networks. *IEEE Transactions on Signal Processing*, 66(9):2441–2454, 2018.

[221] Qixun Zhang, Hongzhuo Sun, Xinye Gao, Xinna Wang, and Zhiyong Feng. Time-division ISAC enabled connected automated vehicles cooperation algorithm design and performance evaluation. *IEEE Journal on Selected Areas in Communications*, 40(7):2206–2218, 2022.

[222] Fan Liu, Christos Masouros, Ang Li, Huafei Sun, and Lajos Hanzo. MU-MIMO communications with MIMO radar: From co-existence to joint transmission. *IEEE Transactions on Wireless Communications*, 17(4):2755–2770, 2018.

[223] Xu Chen, Zhiyong Feng, Zhiqing Wei, Ping Zhang, and Xin Yuan. Code-division OFDM joint communication and sensing system for 6g machine-type communication. *IEEE Internet of Things Journal*, 8(15):12093–12105, 2021.

[224] Fangzhou Wang and Hongbin Li. Joint power allocation for radar and communication co-existence. *IEEE Signal Processing Letters*, 26(11):1608–1612, 2019.

[225] Fangzhou Wang, Hongbin Li, and Mark A Govoni. Power allocation and co-design of multicarrier communication and radar systems for spectral coexistence. *IEEE Transactions on Signal Processing*, 67(14):3818–3831, 2019.

[226] Fangzhou Wang and Hongbin Li. Power allocation for coexisting multicarrier radar and communication systems in cluttered environments. *IEEE Transactions on Signal Processing*, 69:1603–1613, 2021.

[227] Chenguang Shi, Fei Wang, Mathini Sellathurai, and Jianjiang Zhou. Non-cooperative game theoretic power allocation strategy for distributed multiple-radar architecture in a spectrum sharing environment. *IEEE Access*, 6:17787–17800, 2018.

[228] Chenguang Shi, Fei Wang, Sana Salous, and Jianjiang Zhou. A robust stackelberg game-based power allocation scheme for spectral coexisting multistatic radar and communication systems. In *2019 IEEE Radar Conference (RadarConf)*, pages 1–5. IEEE, 2019.

[229] Emanuele Grossi, Marco Lops, and Luca Venturino. Energy efficiency optimization in radar-communication spectrum sharing. *IEEE Transactions on Signal Processing*, 69:3541–3554, 2021.

[230] Warren S McCulloch and Walter Pitts. A logical calculus of the ideas immanent in nervous activity. *The bulletin of mathematical biophysics*, 5:115–133, 1943.

[231] Frank Rosenblatt. *The perceptron, a perceiving and recognizing automaton Project Para.* Cornell Aeronautical Laboratory, 1957.

[232] Marvin Minsky and Seymour A Papert. *Perceptrons, Reissue of the 1988 Expanded Edition with a new foreword by Léon Bottou: An Introduction to Computational Geometry.* MIT press, 2017.

[233] David E Rumelhart, Geoffrey E Hinton, and Ronald J Williams. Learning representations by back-propagating errors. *Nature*, 323(6088):533–536, 1986.

[234] George Cybenko. Approximation by superpositions of a sigmoidal function. *Mathematics of control, signals and systems*, 2(4):303–314, 1989.

[235] Sepp Hochreiter. The vanishing gradient problem during learning recurrent neural nets and problem solutions. *International Journal of Uncertainty, Fuzziness and Knowledge-Based Systems*, 6(02):107–116, 1998.

[236] Marti A. Hearst, Susan T Dumais, Edgar Osuna, John Platt, and Bernhard Scholkopf. Support vector machines. *IEEE Intelligent Systems and their Applications*, 13(4):18–28, 1998.

[237] Geoffrey E Hinton, Simon Osindero, and Yee-Whye Teh. A fast learning algorithm for deep belief nets. *Neural Computation*, 18(7):1527–1554, 2006.

[238] Rajat Raina, Anand Madhavan, and Andrew Y Ng. Large-scale deep unsupervised learning using graphics processors. In *Proceedings of the 26th Annual International Conference on Machine Learning*, pages 873–880, 2009.

[239] Jia Deng, Wei Dong, Richard Socher, Li-Jia Li, Kai Li, and Li Fei-Fei. Imagenet: A large-scale hierarchical image database. In *2009 IEEE Conference on Computer Vision and Pattern Recognition*, pages 248–255. Ieee, 2009.

[240] Xavier Glorot, Antoine Bordes, and Yoshua Bengio. Deep sparse rectifier neural networks. In *Proceedings of the Fourteenth International Conference on Artificial Intelligence and Statistics*, pages 315–323. JMLR Workshop and Conference Proceedings, 2011.

[241] Alex Krizhevsky, Ilya Sutskever, and Geoffrey E Hinton. Imagenet classification with deep convolutional neural networks. *Communications of the ACM*, 60(6):84–90, 2017.

[242] David Silver, Aja Huang, Chris J Maddison, Arthur Guez, Laurent Sifre, George Van Den Driessche, Julian Schrittwieser, Ioannis Antonoglou, Veda Panneershelvam, Marc Lanctot, et al. Mastering the game of Go with deep neural networks and tree search. *Nature*, 529(7587):484–489, 2016.

[243] Jürgen Schmidhuber. Deep learning in neural networks: An overview. *Neural Networks*, 61:85–117, 2015.

[244] Yann LeCun and M Ranzato. Deep learning tutorial. In *Tutorials in International Conference on Machine Learning (ICML'13)*, pages 1–29. Citeseer, 2013.

[245] Ori Shental and Jakob Hoydis. "machine LLRning": Learning to softly demodulate. In *2019 IEEE Globecom Workshops (GC Wkshps)*, pages 1–7. IEEE, 2019.

[246] Hazem Barka, Md Sahabul Alam, Georges Kaddoum, Fabien Sacuto, and Basile L Agba. BNET: A neural network approach for LLR-based detection in the presence of bursty impulsive noise. *IEEE Wireless Communications Letters*, 12(1):80–84, 2022.

[247] Hao Huang, Yang Peng, Jie Yang, Wenchao Xia, and Guan Gui. Fast beamforming design via deep learning. *IEEE Transactions on Vehicular Technology*, 69(1):1065–1069, 2019.

[248] Kuan-Fu Chen, Ming-Chun Lee, Chia-Hung Lin, Wan-Chi Yeh, and Ta-Sung Lee. Multi-fault and severity diagnosis for self-organizing networks using deep supervised learning and unsupervised transfer learning. *IEEE Transactions on Wireless Communications*, 2023.

[249] YCAP Reddy, P Viswanath, and B Eswara Reddy. Semi-supervised learning: A brief review. *Int. J. Eng. Technol*, 7(1.8):81, 2018.

[250] Yanbei Chen, Massimiliano Mancini, Xiatian Zhu, and Zeynep Akata. Semi-supervised and unsupervised deep visual learning: A survey. *IEEE Transactions on Pattern Analysis and Machine Intelligence*, 46(3):1327–1347, 2022.

[251] Andreas Lugmayr, Martin Danelljan, and Radu Timofte. Unsupervised learning for real-world super-resolution. In *2019 IEEE/CVF International Conference on Computer Vision Workshop (ICCVW)*, pages 3408–3416. IEEE, 2019.

[252] Chia-Hung Lin, Yen-Ting Lee, Wei-Ho Chung, Shih-Chun Lin, and Ta-Sung Lee. Unsupervised ResNet-inspired beamforming design using deep unfolding technique. In *GLOBECOM 2020-2020 IEEE Global Communications Conference*, pages 1–7. IEEE, 2020.

[253] Xu Wang, Sen Wang, Xingxing Liang, Dawei Zhao, Jincai Huang, Xin Xu, Bin Dai, and Qiguang Miao. Deep reinforcement learning: A survey. *IEEE Transactions on Neural Networks and Learning Systems*, 35(4):5064–5078, 2022.

[254] Pawel Ladosz, Lilian Weng, Minwoo Kim, and Hyondong Oh. Exploration in deep reinforcement learning: A survey. *Information Fusion*, 85:1–22, 2022.

[255] Sven Gronauer and Klaus Diepold. Multi-agent deep reinforcement learning: a survey. *Artificial Intelligence Review*, 55(2):895–943, 2022.

[256] Wayne Xin Zhao, Kun Zhou, Junyi Li, Tianyi Tang, Xiaolei Wang, Yupeng Hou, Yingqian Min, Beichen Zhang, Junjie Zhang, Zican Dong, et al. A survey of large language models. *arXiv preprint arXiv:2303.18223*, 2023.

[257] Humza Naveed, Asad Ullah Khan, Shi Qiu, Muhammad Saqib, Saeed Anwar, Muhammad Usman, Naveed Akhtar, Nick Barnes, and Ajmal Mian. A comprehensive overview of large language models. *arXiv preprint arXiv:2307.06435*, 2023.

[258] Lei Wang, Chen Ma, Xueyang Feng, Zeyu Zhang, Hao Yang, Jingsen Zhang, Zhiyuan Chen, Jiakai Tang, Xu Chen, Yankai Lin, et al. A survey on large language model based autonomous agents. *Frontiers of Computer Science*, 18(6):186345, 2024.

[259] Weihao Xia, Yulun Zhang, Yujiu Yang, Jing-Hao Xue, Bolei Zhou, and Ming-Hsuan Yang. GAN inversion: A survey. *IEEE Transactions on Pattern Analysis and Machine Intelligence*, 45(3):3121–3138, 2022.

[260] Ali Borji. Pros and cons of GAN evaluation measures: New developments. *Computer Vision and Image Understanding*, 215:103329, 2022.

[261] Yihan Cao, Siyu Li, Yixin Liu, Zhiling Yan, Yutong Dai, Philip S Yu, and Lichao Sun. A comprehensive survey of AI-generated content (AIGC): A history of generative AI from GAN to chatGPT. *arXiv preprint arXiv:2303.04226*, 2023.

[262] Mark R Baker and Rajendra B Patil. Universal approximation theorem for interval neural networks. *Reliable Computing*, 4:235–239, 1998.

[263] Nan Wu, Xudong Wang, Bin Lin, and Kaiyao Zhang. A CNN-based end-to-end learning framework toward intelligent communication systems. *IEEE Access*, 7:110197–110204, 2019.

[264] Ade Pitra Hermawan, Rizki Rivai Ginanjar, Dong-Seong Kim, and Jae-Min Lee. CNN-based automatic modulation classification for beyond 5G communications. *IEEE Communications Letters*, 24(5):1038–1041, 2020.

[265] Abdul Karim Gizzini, Marwa Chafii, Ahmad Nimr, Raed M Shubair, and Gerhard Fettweis. CNN aided weighted interpolation for channel estimation in vehicular communications. *IEEE Transactions on Vehicular Technology*, 70(12):12796–12811, 2021.

[266] Yong Yu, Xiaosheng Si, Changhua Hu, and Jianxun Zhang. A review of recurrent neural networks: LSTM cells and network architectures. *Neural Computation*, 31(7):1235–1270, 2019.

[267] Rahul Dey and Fathi M Salem. Gate-variants of gated recurrent unit (GRU) neural networks. In *2017 IEEE 60th International Midwest Symposium on Circuits and Systems (MWSCAS)*, pages 1597–1600. IEEE, 2017.

[268] Koyel Datta Gupta, Ritu Nigam, Deepak Kumar Sharma, and Sanjay K Dhurandher. LSTM-based energy-efficient wireless communication with reconfigurable intelligent surfaces. *IEEE Transactions on Green Communications and Networking*, 6(2):704–712, 2021.

[269] Sachi Nandan Mohanty, E Laxmi Lydia, Mohamed Elhoseny, Majid M Gethami Al Otaibi, and K Shankar. Deep learning with LSTM based distributed data mining model for energy efficient wireless sensor networks. *Physical Communication*, 40:101097, 2020.

[270] Seungnyun Kim, Junwon Son, and Byonghyo Shim. Energy-efficient ultra-dense network using LSTM-based deep neural networks. *IEEE Transactions on Wireless Communications*, 20(7):4702–4715, 2021.

[271] Yifei Shen, Jun Zhang, SH Song, and Khaled B Letaief. Graph neural networks for wireless communications: From theory to practice. *IEEE Transactions on Wireless Communications*, 22(5):3554–3569, 2022.

[272] Mengyuan Lee, Guanding Yu, and Huaiyu Dai. Decentralized inference with graph neural networks in wireless communication systems. *IEEE Transactions on Mobile Computing*, 22(5):2582–2598, 2021.

[273] Tianrui Chen, Xinruo Zhang, Minglei You, Gan Zheng, and Sangarapillai Lambotharan. A GNN-based supervised learning framework for resource allocation in wireless IoT networks. *IEEE Internet of Things Journal*, 9(3):1712–1724, 2021.

[274] Soha Mohamed, Jian Dong, Allah Rakhio Junejo, et al. Model-based: End-to-end molecular communication system through deep reinforcement learning auto encoder. *IEEE Access*, 7:70279–70286, 2019.

[275] Max Bartunik, Oliver Keszocze, Benjamin Schiller, and Jens Kirchner. Using deep learning to demodulate transmissions in molecular communication. In *2022 IEEE 16th International Symposium on Medical Information and Communication Technology (ISMICT)*, pages 1–6. IEEE, 2022.

[276] Bon-Hong Koo, Ho Joong Kim, Jang-Yeon Kwon, and Chan-Byoung Chae. Deep learning-based human implantable nano molecular communications. In *ICC 2020-2020 IEEE International Conference on Communications (ICC)*, pages 1–7. IEEE, 2020.

[277] Chao-Kai Wen, Wan-Ting Shih, and Shi Jin. Deep learning for massive MIMO CSI feedback. *IEEE Wireless Communications Letters*, 7(5):748–751, 2018.

[278] Muhan Chen, Jiajia Guo, Chao-Kai Wen, Shi Jin, Geoffrey Ye Li, and Ang Yang. Deep learning-based implicit CSI feedback in massive MIMO. *IEEE Transactions on Communications*, 70(2):935–950, 2021.

[279] Yu-Chien Lin, Zhenyu Liu, Ta-Sung Lee, and Zhi Ding. Deep learning phase compression for MIMO CSI feedback by exploiting FDD channel reciprocity. *IEEE Wireless Communications Letters*, 10(10):2200–2204, 2021.

[280] Yu-Chien Lin, Ta-Sung Lee, and Zhi Ding. A scalable deep learning framework for dynamic CSI feedback with variable antenna port numbers. *IEEE Transactions on Wireless Communications*, 2023.

[281] Mehran Soltani, Vahid Pourahmadi, Ali Mirzaei, and Hamid Sheikhzadeh. Deep learning-based channel estimation. *IEEE Communications Letters*, 23(4):652–655, 2019.

[282] Yong Liao, Yuanxiao Hua, and Yunlong Cai. Deep learning based channel estimation algorithm for fast time-varying MIMO-OFDM systems. *IEEE Communications Letters*, 24(3):572–576, 2019.

[283] Peihao Dong, Hua Zhang, Geoffrey Ye Li, Ivan Simoes Gaspar, and Navid NaderiAlizadeh. Deep CNN-based channel estimation for mmWave massive MIMO systems. *IEEE Journal of Selected Topics in Signal Processing*, 13(5):989–1000, 2019.

[284] Siqiang Zhao, Yuan Fang, and Ling Qiu. Deep learning-based channel estimation with SRGAN in OFDM systems. In *2021 IEEE Wireless Communications and Networking Conference (WCNC)*, pages 1–6. IEEE, 2021.

[285] Akash S Doshi, Manan Gupta, and Jeffrey G Andrews. Over-the-Air design of GAN training for mmWave MIMO channel estimation. *IEEE Journal on Selected Areas in Information Theory*, 3(3):557–573, 2022.

[286] Javad Mirzaei, Shahram Shahbaz Panahi, Raviraj S Adve, and Navaneetha Krishna Madan Gopal. Deep generative models for downlink channel estimation in FDD massive MIMO systems. *IEEE Transactions on Signal Processing*, 70:2000–2014, 2022.

[287] Caoqi Gong and Die Hu. A GAN-based channel estimation method for MIMO OFDM systems. In *2023 8th International Conference on Communication, Image and Signal Processing (CCISP)*, pages 410–415. IEEE, 2023.

[288] Chia-Hung Lin, Wei-Cheng Kao, Shi-Qing Zhan, and Ta-Sung Lee. BsNet: A deep learning-based beam selection method for mmWave communications. In *2019 IEEE 90th Vehicular Technology Conference (VTC2019-Fall)*, pages 1–6. IEEE, 2019.

[289] Chang Min Hyun, Hwa Pyung Kim, Sung Min Lee, Sungchul Lee, and Jin Keun Seo. Deep learning for undersampled MRI reconstruction. *Physics in Medicine & Biology*, 63(13):135007, 2018.

[290] Cagla D Bahadir, Alan Q Wang, Adrian V Dalca, and Mert R Sabuncu. Deep-learning-based optimization of the under-sampling pattern in MRI. *IEEE Transactions on Computational Imaging*, 6:1139–1152, 2020.

[291] Ferdia Sherry, Martin Benning, Juan Carlos De los Reyes, Martin J Graves, Georg Maierhofer, Guy Williams, Carola-Bibiane Schönlieb, and Matthias J Ehrhardt. Learning the sampling pattern for MRI. *IEEE Transactions on Medical Imaging*, 39(12):4310–4321, 2020.

[292] Kashvi Taunk, Sanjukta De, Srishti Verma, and Aleena Swetapadma. A brief review of nearest neighbor algorithm for learning and classification. In *2019 International Conference on Intelligent Computing and Control Systems (ICCS)*, pages 1255–1260. IEEE, 2019.

[293] Kaiming He, Xiangyu Zhang, Shaoqing Ren, and Jian Sun. Deep residual learning for image recognition. In *Proceedings of the IEEE Conference on Computer Vision and Pattern Recognition*, pages 770–778, 2016.

[294] Kaiming He, Xiangyu Zhang, Shaoqing Ren, and Jian Sun. Delving deep into rectifiers: Surpassing human-level performance on imagenet classification. In *Proceedings of the IEEE International Conference on Computer Vision*, pages 1026–1034, 2015.

[295] Jingru Tan, Tom H Luan, Wenbo Guan, Yuntao Wang, Haixia Peng, Yao Zhang, Dongmei Zhao, and Ning Lu. Beam alignment in mmWave V2X communications: A survey. *IEEE Communications Surveys & Tutorials*, 2024.

[296] Qiaoyu Li, Philip Sisk, Arumugam Kannan, Taesang Yoo, Tao Luo, Gaurav Shah, Badri Manjunath, Chanaka Samarathungage, Mahmoud Taherzadeh Boroujeni, Hamed Pezeshki, et al. Machine learning based time domain millimeter-wave beam prediction for 5G-advanced and beyond: design, analysis, and over-the-air experiments. *IEEE Journal on Selected Areas in Communications*, 41(6):1787–1809, 2023.

[297] Sajad Rezaie, Abolfazl Amiri, Elisabeth De Carvalho, and Carles Navarro Manchón. Deep transfer learning for location-aware millimeter wave beam selection. *IEEE Communications Letters*, 25(9):2963–2967, 2021.

[298] Ziyan Lin, Jiamin Li, Pengcheng Zhu, and Dongming Wang. Fast adaptive intelligent beam training method for mmWave networks. *IEEE Communications Letters*, 2024.

[299] Shih-Chun Lin and Kwang-Cheng Chen. Cognitive and opportunistic relay for QoS guarantees in machine-to-machine communications. *IEEE Transactions on Mobile Computing*, 15(3):599–609, 2015.

[300] Shih-Chun Lin and Kwang-Cheng Chen. Improving spectrum efficiency via in-network computations in cognitive radio sensor networks. *IEEE Transactions on Wireless Communications*, 13(3):1222–1234, 2014.

[301] Yuan Fang, Lixiang Li, Yixiao Li, Haipeng Peng, and Yixian Yang. Low energy consumption compressed spectrum sensing based on channel energy reconstruction in cognitive radio network. *Sensors*, 20(5):1264, 2020.

[302] Jingyang Lu, Lun Li, Dan Shen, Genshe Chen, Bin Jia, Erik Blasch, and Khanh Pham. Dynamic multi-arm bandit game based multi-agents spectrum sharing strategy design. In *IEEE/AIAA 36th Digital Avionics Systems Conference (DASC)*, pages 1–6. IEEE, 2017.

[303] Md Ferdous Pervej and Shih-Chun Lin. Eco-vehicular edge networks for connected transportation: A distributed multi-agent reinforcement learning approach. In *Proc. IEEE 92nd Vehicular Technology Conference (VTC-Fall)*, 2020.

[304] Chia-Hung Lin, Shih-Chun Lin, and Erik Blasch. TULVCAN: Terahertz ultrabroadband learning vehicular channel-aware networking. In *IEEE INFOCOM 2021 - IEEE Conference on Computer Communications Workshops (INFOCOM WKSHPS)*, pages 1–6, 2021.

[305] Xiangyue Meng, Hazer Inaltekin, and Brian Krongold. End-to-End deep learning-based compressive spectrum sensing in cognitive radio networks. In *IEEE International Conference on Communications (ICC)*, pages 1–6. IEEE, 2020.

[306] Jitong Ma, Shih-Chun Lin, Hongjie Gao, and Tianshuang Qiu. Automatic modulation classification under non-Gaussian noise: A deep residual learning approach. In *IEEE International Conference on Communications (ICC)*, pages 1–6. IEEE, 2019.

[307] Byeoungdo Kim, Jaekyum Kim, Hyunmin Chae, Dongweon Yoon, and Jun Won Choi. Deep neural network-based automatic modulation classification technique. In *2016 International Conference on Information and Communication Technology Convergence (ICTC)*, pages 579–582. IEEE, 2016.

[308] JungHwan Lee, Byeoungdo Kim, Jaekyum Kim, Dongweon Yoon, and Jun Won Choi. Deep neural network-based blind modulation classification for fading channels. In *2017 International Conference on Information and Communication Technology Convergence (ICTC)*, pages 551–554. IEEE, 2017.

[309] Sang Hoon Lee, Kwang-Yul Kim, and Yoan Shin. Effective feature selection method for deep learning-based automatic modulation classification scheme using higher-order statistics. *Applied Sciences*, 10(2):588, 2020.

[310] Seung-Hwan Kim, Jae-Woo Kim, Williams-Paul Nwadiugwu, and Dong-Seong Kim. Deep learning-based robust automatic modulation classification for cognitive radio networks. *IEEE Access*, 9:92386–92393, 2021.

[311] Sreeraj Rajendran, Wannes Meert, Domenico Giustiniano, Vincent Lenders, and Sofie Pollin. Deep learning models for wireless signal classification with distributed low-cost spectrum sensors. *IEEE Transactions on Cognitive Communications and Networking*, 4(3):433–445, 2018.

[312] Shivam Chandhok, Himani Joshi, Sumit Jagdish Darak, and A Venkata Subramanyam. LSTM guided modulation classification and experimental validation for sub-nyquist rate wideband spectrum sensing. In *2019*

11th International Conference on Communication Systems & Networks (COMSNETS), pages 458–460. IEEE, 2019.

[313] Shisheng Hu, Yiyang Pei, Paul Pu Liang, and Ying-Chang Liang. Robust modulation classification under uncertain noise condition using recurrent neural network. In *2018 IEEE global communications conference (GLOBECOM)*, pages 1–7. IEEE, 2018.

[314] Timothy James O'Shea, Tamoghna Roy, and T Charles Clancy. Over-the-air deep learning based radio signal classification. *IEEE Journal of Selected Topics in Signal Processing*, 12(1):168–179, 2018.

[315] Fan Meng, Peng Chen, Lenan Wu, and Xianbin Wang. Automatic modulation classification: A deep learning enabled approach. *IEEE Transactions on Vehicular Technology*, 67(11):10760–10772, 2018.

[316] Xiaolei Shang, Honglin Hu, Xiaoqiang Li, Tianheng Xu, and Ting Zhou. Dive into deep learning based automatic modulation classification: A disentangled approach. *IEEE Access*, 8:113271–113284, 2020.

[317] Weisong Liu, Zhitao Huang, Xueqiong Li, Xiang Wang, and Baoguo Li. Automatic modulation classification with deep learning-based frequency selection filters. *Electronics Letters*, 56(21):1144–1145, 2020.

[318] Wei-Tao Zhang, Dan Cui, and Shun-Tian Lou. Training images generation for CNN based automatic modulation classification. *IEEE Access*, 9:62916–62925, 2021.

[319] Shengliang Peng, Hanyu Jiang, Huaxia Wang, Hathal Alwageed, Yu Zhou, Marjan Mazrouei Sebdani, and Yu-Dong Yao. Modulation classification based on signal constellation diagrams and deep learning. *IEEE Transactions on Neural Networks and Learning Systems*, 30(3):718–727, 2018.

[320] Alex Krizhevsky, Ilya Sutskever, and Geoffrey E Hinton. Imagenet classification with deep convolutional neural networks. *Advances in Neural Information Processing Systems*, 25, 2012.

[321] Christian Szegedy, Wei Liu, Yangqing Jia, Pierre Sermanet, Scott Reed, Dragomir Anguelov, Dumitru Erhan, Vincent Vanhoucke, and Andrew Rabinovich. Going deeper with convolutions. In *Proceedings of the IEEE Conference on Computer Vision and Pattern Recognition*, pages 1–9, 2015.

[322] Sai Huang, Lu Chai, Zening Li, Di Zhang, Yuanyuan Yao, Yifan Zhang, and Zhiyong Feng. Automatic modulation classification using compressive convolutional neural network. *IEEE Access*, 7:79636–79643, 2019.

[323] Yu Wang, Miao Liu, Jie Yang, and Guan Gui. Data-driven deep learning for automatic modulation recognition in cognitive radios. *IEEE Transactions on Vehicular Technology*, 68(4):4074–4077, 2019.

[324] Hao Huang, Wenchao Xia, Jian Xiong, Jie Yang, Gan Zheng, and Xiaomei Zhu. Unsupervised learning-based fast beamforming design for downlink MIMO. *IEEE Access*, 7:7599–7605, 2018.

[325] Qing An, Santiago Segarra, Chris Dick, Ashutosh Sabharwal, and Rahman Doost- Mohammady. A deep reinforcement learning-based resource scheduler for massive MIMO networks. *IEEE Transactions on Machine Learning in Communications and Networking*, 2023.

[326] Weskley VF Mauricio, Tarcisio F Maciel, Anja Klein, and F Rafael M Lima. Learning-based scheduling: Contextual bandits for massive MIMO systems. In *2020 IEEE International Conference on Communications Workshops (ICC Workshops)*, pages 1–6. IEEE, 2020.

[327] Hongji Huang, Yuchun Yang, Zhiguo Ding, Hong Wang, Hikmet Sari, and Fumiyuki Adachi. Deep learning-based sum data rate and energy efficiency optimization for MIMO-NOMA systems. *IEEE Transactions on Wireless Communications*, 19(8):5373–5388, 2020.

[328] Syed Agha Hassnain Mohsan, Yanlong Li, Alexey V Shvetsov, José Varela-Aldás, Samih M Mostafa, and Abdelrahman Elfikky. A survey of deep learning based NOMA: State of the art, key aspects, open challenges and future trends. *Sensors*, 23(6):2946, 2023.

[329] Ranjan Kumar Senapati and Paresh J Tanna. Deep learning-based NOMA system for enhancement of 5G networks: A review. *IEEE Transactions on Neural Networks and Learning Systems*, 2022.

[330] Vasuki Andiappan and Vijayakumar Ponnusamy. Deep learning enhanced NOMA system: A survey on future scope and challenges. *Wireless Personal Communications*, 123(1):839–877, 2022.

[331] Songyi Liu, Yifan Xu, Xueqiang Chen, Ximing Wang, Meng Wang, Wen Li, Yangyang Li, and Yuhua Xu. Pattern-aware intelligent anti-jamming communication: A sequential deep reinforcement learning approach. *IEEE Access*, 7:169204–169216, 2019.

[332] Luliang Jia, Nan Qi, Zhe Su, Feihuang Chu, Shengliang Fang, Kai-Kit Wong, and Chan-Byoung Chae. Game theory and reinforcement learning for anti-jamming defense in wireless communications: Current research, challenges, and solutions. *IEEE Communications Surveys & Tutorials*, 2024.

[333] Ning Rao, Hua Xu, Dan Wang, Zisen Qi, Yue Zhang, Wanyi Gu, and Xiang Peng. Efficient jamming resource allocation against frequency-hopping spread spectrum in WSNs with asynchronous deep reinforcement learning. *IEEE Sensors Journal*, 2024.

[334] Jie Qi, Hongming Zhang, Xiaolei Qi, and Mugen Peng. Deep reinforcement learning based hopping strategy for wideband anti-jamming wireless communications. *IEEE Transactions on Vehicular Technology*, 2023.

[335] Wen Li, Jin Chen, Xin Liu, Ximing Wang, Yangyang Li, Dianxiong Liu, and Yuhua Xu. Intelligent dynamic spectrum anti-jamming communications: A deep reinforcement learning perspective. *IEEE Wireless Communications*, 29(5):60–67, 2022.

[336] Anni Lin, Zhiyuan Ma, Zhi Huang, Yan Xia, and Wenting Yu. Unknown radar waveform recognition based on transferred deep learning. *IEEE Access*, 8:184793–184807, 2020.

[337] Qing Wang, Panfei Du, Jingyu Yang, Guohua Wang, Jianjun Lei, and Chunping Hou. Transferred deep learning based waveform recognition for cognitive passive radar. *Signal processing*, 155:259–267, 2019.

[338] Chao Wang, Jian Wang, and Xudong Zhang. Automatic radar waveform recognition based on time-frequency analysis and convolutional neural network. In *2017 IEEE International Conference on Acoustics, Speech and Signal Processing (ICASSP)*, pages 2437–2441. IEEE, 2017.

[339] Qiang Guo, Xin Yu, and Guoqing Ruan. Lpi radar waveform recognition based on deep convolutional neural network transfer learning. *Symmetry*, 11(4):540, 2019.

[340] Charles E Thornton and R Michael Buehrer. On the value of online learning for radar waveform selection. *IEEE Transactions on Radar Systems*, 1:505–519, 2023.

[341] Charles E Thornton, R Michael Buehrer, and Anthony F Martone. Constrained contextual bandit learning for adaptive radar waveform selection. *IEEE Transactions on Aerospace and Electronic Systems*, 58(2):1133–1148, 2021.

[342] Xin Cao, Zhe Zheng, and Di An. Adaptive waveform selection algorithm based on reinforcement learning for cognitive radar. In *2019 IEEE 2nd International Conference on Automation, Electronics and Electrical Engineering (AUTEEE)*, pages 208–213. IEEE, 2019.

[343] Charles E Thornton, R Michael Buehrer, Harpreet S Dhillon, and Anthony F Martone. Universal learning waveform selection strategies for adaptive target tracking. *IEEE Transactions on Aerospace and Electronic Systems*, 58(6):5798–5814, 2022.

[344] Matthew R Ziemann and Christopher A Metzler. Adaptive LPD radar waveform design with generative deep learning. *IEEE Transactions on Radar Systems*, 3:417-429, 2025.

[345] Wei Jiang, Alexander M Haimovich, and Osvaldo Simeone. Joint design of radar waveform and detector via end-to-end learning with waveform constraints. *IEEE Transactions on Aerospace and Electronic Systems*, 58(1):552–567, 2021.

[346] Wei Jiang, Alexander M Haimovich, and Osvaldo Simeone. End-to-end learning of waveform generation and detection for radar systems. In *2019 53rd Asilomar Conference on Signals, Systems, and Computers*, pages 1672–1676. IEEE, 2019.

[347] Carl R. Barrett. *Adaptive thresholding and automatic detection*, pages 368–393. Springer US, Boston, MA, 1987.

[348] V. G. Hansen and J. H. Sawyers. Detectability loss due to "greatest of" selection in a cell-averaging CFAR. *IEEE Trans. Aerosp. Electron. Syst.*, AES-16(1):115–118, Jan. 1980.

[349] G. V. Trunk. Range resolution of targets using automatic detectors. *IEEE Trans. Aeros. Electron. Syst.*, AES-14(5):750–755, Sept. 1978.

[350] J. T. Rickard and G. M. Dillard. Adaptive detection algorithms for multiple-target situations. *IEEE Trans. Aeros. Electron. Syst.*, AES-13(4):338–343, July 1977.

[351] J. Akhtar and K. E. Olsen. A neural network target detector with partial CA-CFAR supervised training. In *Int. Conf. on Radar (RADAR)*, pages 1–6, Aug. 2018.

[352] L. Wang, D. Wang, and C. Hao. Intelligent CFAR detector based on support vector machine. *IEEE Access*, 5:26965–26972, 2017.

[353] K. Zhang, W. Zuo, Y. Chen, D. Meng, and L. Zhang. Beyond a Gaussian denoiser: residual learning of deep CNN for image denoising. *IEEE Trans. Image Process.*, 26(7):3142–3155, July 2017.

[354] K. Zhang, W. Zuo, and L. Zhang. FFDNet: toward a fast and flexible solution for CNN-based image denoising. *IEEE Trans. Image Process.*, 27(9):4608–4622, Sept. 2018.

[355] Chia-Hung Lin, Yu-Chien Lin, Yue Bai,Wei-Ho Chung, Ta-Sung Lee, and Heikki Huttunen. DL-CFAR: A novel CFAR target detection method based on deep learning. In *2019 IEEE 90th Vehicular Technology Conference (VTC2019-Fall)*, pages 1–6. IEEE, 2019.

[356] S. Ioffe, and C. Szegedy,. Batch normalization: accelerating deep network training by reducing internal covariate shift. *arXiv:1502.03167*, 2015.

[357] K. He, X. Zhang, S. Ren, and J. Sun. Delving deep into rectifiers: surpassing human-level performance on ImageNet classification. In *IEEE Int. Conf. Comput. Vis (ICCV)*, pages 1026–1034, 2015.

[358] K. He, X. Zhang, S. Ren, and J. Sun. Deep residual learning for image recognition. In *IEEE Conf. on Comput. Vision and Pattern Recognition (CVPR)*, pages 770–778, June 2016.

[359] C. Wen, W. Shih, and S. Jin. Deep learning for massive MIMO CSI feedback. *IEEE Wireless Commun. Lett.*, 7(5):748–751, Oct. 2018.

[360] David Silver, Julian Schrittwieser, Karen Simonyan, Ioannis Antonoglou, Aja Huang, Arthur Guez, Thomas Hubert, Lucas Baker, Matthew Lai, Adrian Bolton, Yutian Chen, Timothy Lillicrap, Fan Hui, Laurent Sifre, George van den Driessche, Thore Graepel, and Demis Hassabis. Mastering the game of Go without human knowledge. *Nature*, 550:354–, October 2017.

[361] Diederik P. Kingma and Jimmy Ba. Adam: A method for stochastic optimization. In *Int. Conf. Learning Representations (ICLR)*, 2014.

[362] Zhihui Cao, Wenwei Fang, Yuying Song, Lai He, Chunyi Song, and Zhiwei Xu. DNN-based peak sequence classification CFAR detection algorithm for high-resolution FMCW radar. *IEEE Transactions on Geoscience and Remote Sensing*, 60:1–15, 2021.

[363] Stefan Feintuch, Haim H Permuter, Igal Bilik, and Joseph Tabrikian. Neural network-based multitarget detection within correlated heavy-tailed clutter. *IEEE Transactions on Aerospace and Electronic Systems*, 59(5):5684–5698, 2023.

[364] Jacopo Pegoraro, Domenico Solimini, Federico Matteo, Enver Bashirov, Francesca Meneghello, and Michele Rossi. Deep learning for accurate indoor human tracking with a mm-Wave radar. In *2020 IEEE Radar Conference (RadarConf20)*, pages 1–6, 2020.

[365] Ji-He Kim, Ming-Chun Lee, and Ta-Sung Lee. Deep-learning based multi-object detection and tracking using range-angle map in automotive radar systems. In *2022 IEEE 95th Vehicular Technology Conference: (VTC2022-Spring)*, pages 1–6, 2022.

[366] Xinyu Hou, Yi Wang, and Lap-Pui Chau. Vehicle tracking using deep SORT with low confidence track filtering. In *2019 16th IEEE International Conference on Advanced Video and Signal Based Surveillance (AVSS)*, pages 1–6, 2019.

[367] Tai-Yuan Huang, Ming-Chun Lee, Chia-Hsing Yang, and Ta-Sung Lee. YOLO-ORE: A deep learning-aided object recognition approach for radar systems. IEEE Transactions on Vehicular Technology, 72(5):5715–5731, 2023.

[368] Jianan Liu, Qiuchi Zhao, Weiyi Xiong, Tao Huang, Qing-Long Han, and Bing Zhu. SMURF: Spatial multi-representation fusion for 3D object detection with 4D imaging radar. *IEEE Transactions on Intelligent Vehicles*, 9(1):799–812, 2024.

[369] Ali Hanif, Muhammad Muaz, Azhar Hasan, and Muhammad Adeel. Micro-doppler based target recognition with radars: A review. *IEEE Sensors Journal*, 22(4):2948–2961, 2022.

[370] Hsin-Yuan Chang, Chia-Hung Lin, Yu-Chien Lin, Wei-Ho Chung, and Ta-Sung Lee. DL-aided NOMP: A deep learning-based vital sign estimating scheme using FMCW radar. In *2020 IEEE 91st vehicular technology conference (VTC2020-Spring)*, pages 1–7. IEEE, 2020.

[371] Jiwoo Mun, Seokhyeon Ha, and Jungwoo Lee. Automotive radar signal interference mitigation using RNN with self attention. In *ICASSP 2020-2020 IEEE International Conference on Acoustics, Speech and Signal Processing (ICASSP)*, pages 3802–3806. IEEE, 2020.

[372] Jianping Wang, Runlong Li, Yuan He, and Yang Yang. Prior-guided deep interference mitigation for FMCW radars. *IEEE Transactions on Geoscience and Remote Sensing*, 60:1–16, 2022.

[373] Jiwoo Mun, Heasung Kim, and Jungwoo Lee. A deep learning approach for automotive radar interference mitigation. In *2018 IEEE 88th Vehicular Technology Conference (VTC-Fall)*, pages 1–5. IEEE, 2018.

[374] Johanna Rock, Wolfgang Roth, Mate Toth, Paul Meissner, and Franz Pernkopf. Resource-efficient deep neural networks for automotive radar interference mitigation. *IEEE Journal of Selected Topics in Signal Processing*, 15(4):927–940, 2021.

[375] Jianping Wang, Runlong Li, Xinqi Zhang, and Yuan He. Interference mitigation for automotive FMCW radar based on contrastive learning with dilated convolution. *IEEE Transactions on Intelligent Transportation Systems*, 2023.

[376] Shengyi Chen, Wangyi Shangguan, Jalal Taghia, Uwe Kühnau, and Rainer Martin. Automotive radar interference mitigation based on a generative adversarial network. In *2020 IEEE Asia-Pacific Microwave Conference (APMC)*, pages 728–730. IEEE, 2020.

[377] Qinzhe Lv, Yinghui Quan, Wei Feng, Minghui Sha, Shuxian Dong, and Mengdao Xing. Radar deception jamming recognition based on weighted ensemble cnn with transfer learning. *IEEE Transactions on Geoscience and Remote Sensing*, 60:1–11, 2021.

[378] Guangqing Shao, Yushi Chen, and Yinsheng Wei. Convolutional neural network-based radar jamming signal classification with sufficient and limited samples. *IEEE Access*, 8:80588–80598, 2020.

[379] Yu Zhang, Bo Jiu, Penghui Wang, Hongwei Liu, and Siyuan Liang. An end-to-end anti-jamming target detection method based on CNN. *IEEE Sensors Journal*, 21(19):21817–21828, 2021.

[380] Peiyuan Jiang, Daji Ergu, Fangyao Liu, Ying Cai, and Bo Ma. A review of Yolo algorithm developments. *Procedia Computer Science*, 199:1066–1073, 2022.

[381] Jiaxiang Zhang, Zhennan Liang, Chao Zhou, Quanhua Liu, and Teng Long. Radar compound jamming cognition based on a deep object detection network. *IEEE Transactions on Aerospace and Electronic Systems*, 59(3):3251–3263, 2022.

[382] Hamda Bouzabia, Tri Nhu Do, and Georges Kaddoum. Deep learning-enabled deceptive jammer detection for low probability of intercept communications. *IEEE Systems Journal*, 17(2):2166–2177, 2022.

[383] Guangqing Shao, Yushi Chen, and Yinsheng Wei. Deep fusion for radar jamming signal classification based on CNN. *IEEE Access*, 8:117236–117244, 2020.

[384] Shengyi Chen, Jalal Taghia, Tai Fei, Uwe Kühnau, Nils Pohl, and Rainer Martin. A DNN autoencoder for automotive radar interference mitigation. In *ICASSP 2021-2021 IEEE International Conference on Acoustics, Speech and Signal Processing (ICASSP)*, pages 4065–4069. IEEE, 2021.

[385] Cheng Qi, Junwei Xie, Haowei Zhang, Weijian Liu, and Ruijun Wang. Cooperative game theoretic power allocation method to distributed MIMO radar sensor network for multi-target detection in low-altitude. *IEEE Sensors Journal*, 2024.

[386] Pengfei Liu, Yimin Liu, Tianyao Huang, Yuxiang Lu, and Xiqin Wang. Decentralized automotive radar spectrum allocation to avoid mutual interference using reinforcement learning. *IEEE Transactions on Aerospace and Electronic Systems*, 57(1):190–205, 2020.

[387] Marius Schwarz, Axel Acosta Aponte, and Gor Hakobyan. Deep-learning based spectrum prediction for cognitive automotive radar interference mitigation. In *2023 IEEE Radar Conference (RadarConf23)*, pages 1–6. IEEE, 2023.

[388] Gor Hakobyan, Karim Armanious, and Bin Yang. Interference-aware cognitive radar: A remedy to the automotive interference problem. *IEEE Transactions on Aerospace and Electronic Systems*, 56(3):2326–2339, 2019.

[389] Gor Hakobyan, Martin Fink, Ayinhu Soyolyn, Nour Mansour, and Dirk Dahlhaus. Sweep-based spectrum sensing method for interference-aware

cognitive automotive radar. In *2020 IEEE Radar Conference (RadarConf20)*, pages 1–6. IEEE, 2020.

[390] Wei Yi, Pramod K Varshney, et al. Adaptation of frequency hopping interval for radar anti-jamming based on reinforcement learning. *IEEE Transactions on Vehicular Technology*, 71(12):12434–12449, 2022.

[391] Stephen Machuzak and Sudharman K Jayaweera. Reinforcement learning based anti-jamming with wideband autonomous cognitive radios. In *2016 IEEE/CIC International Conference on Communications in China (ICCC)*, pages 1–5. IEEE, 2016.

[392] Xiang Zhang, Lan Lan, Shengqi Zhu, Ximin Li, Guisheng Liao, and Jingwei Xu. Intelligent suppression of interferences based on reinforcement learning. *IEEE Transactions on Aerospace and Electronic Systems*, 2023.

[393] Shane A Flandermeyer, Rylee G Mattingly, and Justin G Metcalf. Deep reinforcement learning for cognitive radar spectrum sharing: A continuous control approach. *IEEE Transactions on Radar Systems*, 2024.

[394] Vesa Saarinen and Visa Koivunen. Radar waveform synthesis using generative adversarial networks. In *2020 IEEE Radar Conference (RadarConf20)*, pages 1–6. IEEE, 2020.

[395] Matthew R Ziemann and Christopher A Metzler. Adaptive LPD radar waveform design with generative deep learning. *arXiv preprint arXiv:2403.12254*, 2024.

[396] Matthew R Ziemann and Christopher A Metzler. Adaptive LPD radar waveform design with generative adversarial neural networks. In *2023 57th Asilomar Conference on Signals, Systems, and Computers*, pages 1039–1043. IEEE, 2023.

[397] Qiao Qi, Xiaoming Chen, Caijun Zhong, Chau Yuen, and Zhaoyang Zhang. Deep learning-based design of uplink integrated sensing and communication. *IEEE Transactions on Wireless Communications*, 23(9):10639–10652, 2024.

[398] Peng Jiang, Rang Liu, Ming Li, Zichao Xiao, and Qian Liu. Deep learning for SLPbased ISAC waveform design. In *ICC 2024 - IEEE International Conference on Communications*, pages 2270–2275, 2024.

[399] Rang Liu, Ming Li, Qian Liu, and A. Lee Swindlehurst. Dual-functional radar-communication waveform design: A symbol-level precoding approach. *IEEE Journal of Selected Topics in Signal Processing*, 15(6):1316–1331, 2021.

[400] Jifa Zhang, Mingqian Liu, Jie Tang, Nan Zhao, Dusit Niyato, and Xianbin Wang. Joint design for RIS-aided ISAC via deep unfolding learning. *IEEE Transactions on Cognitive Communications and Networking*, 11(1):349–361, 2025.

[401] Lifan Xu, Ruxin Zheng, and Shunqiao Sun. A deep reinforcement learning approach for integrated automotive radar sensing and communication. In *2022 IEEE 12th Sensor Array and Multichannel Signal Processing Workshop (SAM)*, pages 316–320, 2022.

[402] Ziyang Lu, M. Cenk Gursoy, Chilukuri K. Mohan, and Pramod K. Varshney. Learning-based resource management in integrated sensing and communication systems. In *IEEE INFOCOM 2024 - IEEE Conference on Computer Communications Workshops (INFOCOM WKSHPS)*, pages 1–6, 2024.

[403] Biwei Li, Xianbin Wang, Sungjun Ahn, Sung-Ik Park, and Yiyan Wu. Successive resource allocation in multi-user ISAC system through deep reinforcement learning. In *ICC 2024 - IEEE International Conference on Communications*, pages 5577–5583, 2024.

[404] Liu Yang, Yifei Wei, Zhiyong Feng, Qixun Zhang, and Zhu Han. Deep reinforcement learning-based resource allocation for integrated sensing, communication, and computation in vehicular network. *IEEE Transactions on Wireless Communications*, 23(12):18608–18622, 2024.

[405] Xiangnan Liu, Haijun Zhang, Keping Long, Arumugam Nallanathan, and Victor C. M. Leung. Distributed unsupervised learning for interference management in integrated sensing and communication systems. *IEEE Transactions on Wireless Communications*, 22(12):9301–9312, 2023.

[406] Chang Liu, Weijie Yuan, Shuangyang Li, Xuemeng Liu, Husheng Li, Derrick Wing Kwan Ng, and Yonghui Li. Learning-based predictive beamforming for integrated sensing and communication in vehicular networks. *IEEE Journal on Selected Areas in Communications*, 40(8):2317–2334, 2022.

[407] Xiaoqi Zhang, Weijie Yuan, Chang Liu, Jun Wu, and Derrick Wing Kwan Ng. Predictive beamforming for vehicles with complex behaviors in ISAC systems: A deep learning approach. *IEEE Journal of Selected Topics in Signal Processing*, 18(5):828–841, 2024.

Index

Note: *Italic* and **Bold** page numbers refer to *figures* and **tables**. Page numbers followed by 'n' refer to notes.

D

J

K

L

M

P